AF388663

Life Cycle Assessment (LCA) of Environmental and Energy Systems

Life Cycle Assessment (LCA) of Environmental and Energy Systems

Editors

Fabrizio Passarini
Luca Ciacci

MDPI • Basel • Beijing • Wuhan • Barcelona • Belgrade • Manchester • Tokyo • Cluj • Tianjin

Editors
Fabrizio Passarini
University of Bologna
Italy

Luca Ciacci
University of Bologna
Italy

Editorial Office
MDPI
St. Alban-Anlage 66
4052 Basel, Switzerland

This is a reprint of articles from the Special Issue published online in the open access journal *Energies* (ISSN 1996-1073) (available at: https://www.mdpi.com/journal/energies/special_issues/LCA_environmental_energy).

For citation purposes, cite each article independently as indicated on the article page online and as indicated below:

LastName, A.A.; LastName, B.B.; LastName, C.C. Article Title. *Journal Name* **Year**, *Volume Number*, Page Range.

ISBN 978-3-0365-0080-5 (Hbk)
ISBN 978-3-0365-0081-2 (PDF)

Contents

About the Editors

Fabrizio Passarini is Associate Professor in the Department of Industrial Chemistry "Toso Montanari", University of Bologna, in the of Environmental and Cultural Heritage Chemistry. He is presently Director of Interdepartmental Centre for Industrial Research "Renewable Sources, Environment, Sea, and Energy" (University of Bologna) and Vice-President of the Environmental and Cultural Heritage Chemistry Division of the Italian Chemical Society. His research activity mainly concerns the environmental impacts of waste management; environmental monitoring; the application of LCA; and material flow analysis.

Luca Ciacci is researcher at the Department of Industrial Chemistry "Toso Montanari" at the University of Bologna. His research interests include the anthropogenic metabolism of materials and energy and their impact on the environment from an industrial ecology perspective. To these goals, material flow analysis (MFA) and life cycle assessment (LCA) are commonly applied. His experience with LCA and MFA modeling comprises the assessment of products, chemical processes, end-of-life products, and waste treatment. He is also involved in investigating the present and future criticality of metals as well as the sustainability implications of their demand and recycling potential.

 *energies*

Editorial

Life Cycle Assessment (LCA) of Environmental and Energy Systems

Luca Ciacci [1,2,*] and Fabrizio Passarini [1,2,*]

[1] Department of Industrial Chemistry "Toso Montanari", Alma Mater Studiorum—University of Bologna, Viale del Risorgimento 4, 40136 Bologna, Italy

[2] Interdepartmental Centre for Industrial Research "Renewable Sources, Environment, Sea, Energy", Alma Mater Studiorum—University of Bologna, Via Dario Campana 71, 47922 Rimini, Italy

* Correspondence: luca.ciacci5@unibo.it (L.C.); fabrizio.passarini@unibo.it (F.P.)

Received: 2 November 2020; Accepted: 6 November 2020; Published: 12 November 2020

1. Introduction

The transition towards renewable energy sources and "green" technologies for energy generation and storage is expected to mitigate the climate emergency in the coming years. However, in many cases, this progress has been hampered by our dependency on critical materials or other resources that are often processed with high environmental burdens. Yet, beyond global warming, several global challenges have to be promptly addressed, including the loss of biodiversity, environmental pollution, water scarcity, and energy security.

Environmental and energy issues are strictly interconnected and require a comprehensive understanding of resource management strategies and their implications. For instance, the depletion and contamination of a vital resource such as water has been related to possible shortages in heat and power generation, distribution and use; on the other hand, water supply requires energy inputs, particularly if the most common sources of natural provision (e.g., groundwater) are not easily accessible. Actions undertaken in separately considered systems may hinder the achievement of optimized benefits and reduction in adverse consequences.

A system perspective is therefore needed to identify and quantify the impact of human activity on the environment. Life cycle assessment (LCA) is among the most inclusive analytical techniques to analyze sustainability benefits and trade-offs resulting from complex systems. This Special Issue presents a collection of original articles, reviews, and case studies focusing on mutual influences of environmental and energy systems. A brief description and discussion of the contributions to this Special Issue is reported hereafter. It is worth noting that the order in which the contributions are presented does not imply any judgment of merit and it is dictated only by narrative purposes.

2. Brief Overview of Contributions to This Special Issue

The selection of macro-economic sectors covered by the articles in this Special Issue is well representative of the main driving applications for energy requirements and greenhouse gas (GHG) emissions, including power generation, bioenergy and biorefinery, building, and transportation.

Renewable energy sources as carriers for electricity generation have attracted quite a lot of attention, particularly photovoltaics. Maranghi and colleagues [1] harmonized the LCA results of several studies in the literature, focusing on 12 different configurations of perovskite solar cell (PSC) technology and identified the main environmental hotspots in PSC technology manufacturing. The development of PSC technology is particularly attractive for photovoltaic energy production thanks to the high photoconversion efficiency. However, the related environmental impacts may vary remarkably depending on technological configurations and materials employed in PSC manufacturing. In particular, the cradle-to-gate results of this harmonization effort highlight (i) gold used as the back contact, (ii) the conductive solar glass, and (iii) the consumption of organic solvents utilized in the synthesis of the electron transport layer as the materials embedding the highest environmental impacts of PSC manufacturing. The harmonized results provided by these authors constitute a benchmark for PSC technological advancements and they may also orient future efforts towards integrative research between LCA and toxicological assessment tools to better understand and model toxicity implications of perovskite compounds.

Piasecka and colleagues [2] explored the environmental impacts of a 1 MW photovoltaic (PV) power plant by cradle-to-grave LCA with the dual goal of analyzing the state-of-the-art and proposing potential improvements of its overall environmental performance. While Poland was set as the geographical level in which the power plant is located, several life cycle stages (e.g., production and waste management) usually occur in other European countries so that the main results can be considered to be representative of the average situation in this region. Among the analyzed variables, the greatest environmental burdens resulted from the intensive material and energy requirements employed in the manufacture of PV panels and loss through final disposal in landfill. In particular, potential harmful effects on the health and the ecosystem were associated with the presence of metals (e.g., silver, nickel, copper, lead and cadmium) and polymers (e.g., polyamides 6) in PV panels. These effects can be significantly reduced if pro-environmental management strategies, such as ecodesign practices and efficient recovery and recycling of obsolete PV panels, are successfully implemented.

PV waste management will gain relevance proportionally to the amounts of waste that are expected to arise with the phasing-out of old installations in the upcoming years and decades. Herceg and colleagues [3] applied LCA to compare the environmental performance of different recycling processes for PV systems and assess their contribution to the overall environmental footprint of electricity produced by a standard PV system in Germany. The waste management approaches considered in this study include state-of-the-art recycling technology, further improved by supplemental material recovery, and advanced recycling processes discussed in the literature. The results demonstrate that recycling has a significant potential to improve the environmental profile of PV electricity, particularly by means of climate change mitigation achievements. Beyond benefiting the environment and resource conservation, the establishment of an appropriate recycling scheme will also positively impact the economic and financial balances of the logistic network.

A general agreement on the progressive reduction of thermal power plants in electricity grid mixes does exist. However, in transition times or when the greening of electricity is lagging behind, carbon capture and sequestration (CCS) technology can be suitable solutions to contrast the climate emergency. In this context, Zakuchova and colleagues [4] compared the environmental burdens of a 250 MW coal power plant in Czech Republic, set as the reference scenario, with a hypothetical improvement of the same plant upon implementation of a CCS technology based on activated carbons (AC). An economic feasibility evaluation of the AC-based CCS system was also included in the analysis. The results show that the implementation of an AC-based CCS system in the targeted coal power plant would determine a sensible reduction in environmental impacts, mainly in terms of GHG emissions, compared to the baseline scenario. The payback period of six years estimated in this study is particularly attractive as well as the envisaged economic implications of using CO_2 as a building block in agriculture or industry sectors. Although CCS appears to be a promising technology for the environment and for the economy of the whole process, context-specific parameters such as quality standards for material inputs, combustion efficiency, and CO_2 purification stages may affect the environmental gains achievable through CCS. Ultimately, the highest impact to the ecosystem is attributable to the raw material (i.e., hard coal) extraction that constitutes the input fuel in both scenarios. According to the authors, CCS may provide a viable option in countries where the transition to renewable energy sources is lagging. Further investigation to avert unpredictable and potentially harmful effects associated with the management of CCS systems is also required.

Raugei and colleagues [5] combined net energy analysis and LCA to explore energy and environmental implications of an aggressive decarbonization scenario of the electricity mix in the United

Kingdom to 2050. Energy return on investment (EROI), net-to-gross primary energy ratio, and life cycle impact assessment results are computed for fossil and renewable energy sources, carbon storage and sequestration technologies, energy storage systems, and transmission to the grid. The results show that the aggressive decarbonization scenario can be a very promising pathway to mitigate the climate emergency mainly thanks to a larger share of renewable energy sources in the national grid mix. The marginal contribution of energy storage systems for the EROI and LCA results is particularly comforting under a prospective transition to a central presence of variable renewable energy sources (e.g., wind, tidal, and solar) in the future electricity grid mix. However, some trade-offs may result due to detrimental effects of resource depletion potential and human toxicity potential due to the intensive demand for specialty metals essential to, for instance, PV panels and wind turbines and the spatial distribution of these installations, which requires more (copper) transmission lines. Biogas and biomass feedstocks result in low EROI values and they are also responsible for acid emissions, biogas upgrade through scrubbing and membrane filtration may lead to significant improvements in this sense. The planned reduction reliance on biomass as an energy source seems to be supported by these outcomes; however, according to Raugei et and colleagues [5], the related debate should include a comprehensive discussion on viable alternatives for organic waste streams.

Four papers addressed, instead, the use of biomass as a raw material for bioenergy and biorefinery purposes. Pergola and colleagues [6] focused on a case study of spatial LCA to identify the most promising locations for the construction of bioenergy power plants in the south of Italy. To this aim, a geographical information system (GIS) was applied to characterize the potential biomass availability in this region as well as to map the proximity to main roads, the existing connection to the electricity grid, the presence of residential areas and of protected areas. (Spatial) LCA was then combined to assess the environmental impacts associated with the loading and transport of harvesting biomass residues from local forest management plans. The use of fossil fuels and the release of pollutant emissions from tire abrasion were detected as the main causes of impacts to the ecosystem and the human health resulting from the movement of logging residues. Based on the spatial LCA results, three sites were selected as the most promising areas, providing essential information for the construction of environmentally preferable options for cogeneration or trigeneration bioenergy plants.

Sharara and colleagues [7] applied LCA to assess the environmental impacts of swine manure management within a thermal gasification scenario that includes drying, syngas production, and biochar field application. Hot-gas efficiency and boiler efficiency were also varied according to alternative models of potential improvements to better understand the implications of thermochemical conversion parameters on the overall environmental performance of the proposed management scenario. The results demonstrate that storage of swine manure liquids contributes for about 60% of the total carbon profile of the investigated system. Manure drying demands energy inputs higher than energy outputs generated from the gasification stage so that innovative drying technologies and the utilization of renewable energy sources may considerably reduce energy requirements and the associate greenhouse gas emissions. Further environmental benefits are achieved through land application of biochar in replacement of traditional fertilizers. In areas where swine manure land application is particularly intensive, thermochemical processing represents a suitable alternative to reduce pressure on the environment and improve the energy performance of common manure management scenarios.

Livestock production and the related manure management systems are highlighted as main contributors to global warming and ecosystem degradation. The development of renewable and low carbon fuel standards has shaped the way in which LCA tools are used to assess a fuel's "greenness," specifically for addressing GHG mitigation. Based on LCA, the United States (US) have defined advanced designation under the Renewable Fuel Standard (RFS2) for biofuels and set a 50% GHG emissions reduction target compared to gasoline. Spatari and colleagues [8] investigated the role of biorefinery co-products in the context of US bioenergy policy goals. More specifically, attributional and consequential LCAs were applied to assess four alternative winter barley-to-ethanol scenarios and commercial dry-grind technology, in order to meet the advanced designation under the RFS2 for biofuels produced in the proximity of highly populated areas in the mid-Atlantic US. Monte Carlo analysis was applied to estimate confidential levels for stochastic soil and foreground GHG emissions. The results show that co-products are essential to biorefinery economics as well as for meeting the RFS2 designations and the national energy policy goals. Co-products' credits mainly result from avoided fossil-based energy generation employed in traditional scenarios and, even under conservative assumptions, ethanol derived from winter barley would exceed the 50% GHG emissions reduction target set for RFS2 advanced fuels compared to gasoline. In highly populated areas with climate and agronomic conditions comparable to those investigated in this study, biorefinery may constitute a profitable and environmentally preferable solution for advanced biofuels.

Alternative exploitation of agricultural sludges can be aimed at chemical products such as biopolymers. This option was explored by Vogli and colleagues [9], who analyzed the environmental impacts associated with polyhydroxyalkanoate (PHA) production through hybrid thermochemical–biological processes using anaerobically digested sewage sludge as the material input. Five scenarios were developed and compared, of which three scenarios modelled alternative uses of syngas from pyrolysis and biochar gasification, and two scenarios modelled different energy sources for the system. From the energy perspective, the amount of sewage sludge used for onsite energy production, the energy recovery from PHAs at end-of-life, and the fraction of renewables in energy production are the main explanatory variables of the overall energy balance. The outcomes indicate that the selection of the most environmentally sustainable scenario ultimately depends on the priority of regional versus global scale impact effects. The scenario modelling the use of syngas to energy production for the satisfaction of internal electrical and thermal energy requirements is preferable for global scale impact categories such as global warming potential and resource depletion. In contrast, the supply of the total syngas produced for PHA production is preferable for acidification, eutrophication, and similar regional scale effects. In any case, a higher fraction of renewable energy sources can significantly reduce environmental impacts at both the regional and global scale. Further improvements are achievable through the expected technological advancements and the process scale up.

Building and transportation are also major sectors where energy-related advancements (e.g., insulation and battery systems) may play a pivotal role to move modern society towards sustainable production and consumption patterns. Space conditioning is responsible for the majority of carbon dioxide emissions and fossil fuel consumption during a building's life cycle. The environmental impacts resulting from innovative ground-source heat pumps for air conditioning in buildings were explored by Bonamente and Aquino [10]. Based on past work of the same authors, this study compared three scenarios including (i) a conventional ground-source heat pump system (set as the baseline scenario), further upgraded, respectively, with (ii) upstream sensible-heat thermal storage unit (SH-TES), and (iii) upstream latent-heat thermal storage unit containing phase-change materials (PCMs-TES). For all scenarios, two hypotheses were modeled for electrical energy input, namely either it being supplied from the national grid or from PV panels. Enabling the switch from sensible heat storage to latent heat storage, the PCMs-TES system resulted in being the most promising solution for space conditioning compared to other scenarios thanks to the reduction of the volume storage and of electrical energy inputs. These benefits particularly amplify when the enhancement of the overall efficiency performance is pursued along with the exploitation of renewable energy sources for electricity generation.

Materials with thermal storage potential such as, for instance, PCMs, are main means for reducing energy demand in the building sector. To this aim, Di Bari and colleagues [11] coupled LCA and a building simulation to assess the environmental impacts of PCM systems. A new developed software named "Storage LCA Tool" was applied to simulate energy implications at the storage material, component and building levels for a wide set of case studies. The system boundaries are set from cradle-to-grave and different climate zones are investigated to provide representative results of the European situation for heating, cooling, and ventilation needs. The outcomes demonstrate that PCMs systems are often, but not always, preferable to traditional systems. Environmental impacts are context-specific so the choice of materials, building typologies, insulation levels, and location dictate the ultimate preference for the best solutions. To this aim, the storage LCA tool is highly versatile and informative, and it can be successfully used to conceptualize environmentally preferable solutions for energy efficient buildings.

Hu [12] proposed instead a new metric to describe the embodied environmental performance of buildings. This metric is named life cycle embodied performance (LCEP) and it is defined as the ratio between embodied energy and embodied carbon. LCEP was applied to a training set of buildings in the US. In addition, the results for four environmental impact categories (i.e., acidification potential, eutrophication potential, smog formation potential, and ozone depletion potential) as well as their correlation with LCEP scores were quantified. The results highlight that LCEP is a better indicator for the life-cycle embodied performance of building assemblies and materials than individual embodied energy and embodied carbon metrics. In addition, the environmental impact categories considered in this study are proportional to LCEP and, particularly, ozone depletion potential shows correlation with LCEP results. Exterior building walls and assemblies are the main factor in reducing embodied energy and embodied carbon. LCEP demonstrated to be a suitable indicator of the overall embodied environmental performance of buildings and it can be considered as a novel criterion for designing new buildings oriented to environmental sustainability.

Sustainable and smart mobility as well as associated energy systems are also vital to decarbonize economies and develop a clean, resource efficient, circular, and carbon-neutral future. Bobba and colleagues [13] combined LCA with material flow analysis (MFA) and the EU criticality assessment to provide a systemic overview of strategies to secure the supply of materials and environmental protection associated with mobility batteries in this region. In more detail, the environmental performance of lithium-ions batteries (LIBs) in the EU fleet was explored through a prospective assessment of anthropogenic flows and stock of selected raw materials (i.e., lithium and nickel) in different LIBs chemistries and electric vehicle types (i.e., battery electric vehicles and plug-in electric vehicles) according to different scenarios. The results highlight that the future life-cycle global warming potential of traction LIBs may considerably lessen from an increased share of renewables in the electricity production mix, the adoption of resource efficiency strategies in the manufacturing phase, and improved end-of-life recycling performance. Furthermore, the extension of LIBs' lifetime extension though reuse practices may further enhance the energy storage capacity in the EU, but it might also constrain considerable amounts of lithium and nickel in in-use stock, limiting their availability for approaching material circularity and securing stable material supply in this region.

Kosai and colleagues [14] investigated the transport energy intensity (TEI) for different transportation means (i.e., walking, bicycle, conventional automobile, electrical vehicle, hybrid vehicle, fuel-cell vehicle, bus, and electric train) in 38 Japanese cities between 1987–2015. Depending on the annual modal distribution, the relationship between the intracity transport energy intensity and population density has been analyzed, and the diachronic transition in each city visually interpreted. The results show that TEI decreases according to the following order: automobile, bus, train, bicycle, and walking. Material breakdown and manufacturing contribute the most for small-scale transportation options such as bicycles and automobiles, while it is less relevant for buses and trains. The detected negative correlation between population density and intracity TEI has motivated the authors' hypothesis that cities with low population density have mainly relied on automobiles with a consequential increase in TEI. In particular, this aspect has intensified due to a greater share of fossil fuels in the national electricity grid mix after the nuclear accident in Fukushima. In contrast, medium to highly density populated cities have benefitted from the development and consolidation of public transportation systems such as buses and trains. The strategic implications of these results in regional areas have been also discussed by the authors for the improvement of the intracity lifecycle transport energy efficiency.

Lastly, Nwodo and Anumba [15] provided a review of existing exergetic LCA studies and discussed potential improvements for integration between exergy analysis and traditional LCA.

Furthermore, 25 peer-reviewed journal articles that focused on the use of exergy and its methodological advancements in LCA were selected over a time frame from 1990 to 2018. The results indicate that exergy analysis is a multi-disciplinary and emerging field, finding its main application on assessing the use of resources. Particularly, both for energetic and non-energetic resources, exergy analysis drives sustainability assessment on thermodynamic properties and parameters, which usually require less subjective choices compared to fate, exposure, and effects models underlying most LCA methods. Nwodo and Anumba also commented on the opportunity of extending traditional boundaries of exergetic LCA to emissions as well, as they ultimately describe an exergy loss to the environment along with the substance release. This measurement would make characterization independent from a reference substance (e.g., carbon dioxide for global warming potential) and it would also enable the combination of the results of different impact categories into a cumulative value thanks to the use of the same unit of exergy. On the other hand, extensive application of exergy analysis to conventional LCA comes through a systematic and comprehensive determination of exergies covering standard thermodynamic conditions, pure state of resources and emissions, and individual emission amounts. These authors expect that an increasing diffusion of systematic integration between exergy analysis and LCA will enable the overcoming of these limitations, and they propose the term exergy-based life cycle assessment (Exe-LCA) as a new field of research to characterize resource depletion and life cycle emissions.

3. Discussion

This Special Issue comprehends 14 research articles and one review, with the scope investigated being mainly representative of the EU or EU countries (10 papers). Other geographical boundaries discussed include the US (three papers), Japan (one paper), and the world (one paper). Overall, the resulting distribution of geographical scopes provides a selective characterization of energy system-related issues in developed countries and, in one case, of the world, but it lacks inclusion of other major economies such as China or developing countries. Ultimately, a joint and global effort to tackle the energy challenge successfully is needed.

All the papers assessed the environmental burdens related to a selected set of impact categories, among which global warming potential is the most common one. Acidification potential, eutrophication potential, toxicity effects, and resource consumption are also frequently considered. Impact assessment methods reflect the articles' geographical scope distribution, with European methods being applied most (e.g., ILCD, CML, ReCiPe, Ecoindicator99). Impact World method, Cumulative Energy Demand, and exergy indicators are also considered. The prevalence of European methods and case studies also results in the reference to Product Environmental Footprint (PEF) guidelines. This aspect attests some effort to increase comparability among the plethora of impact assessment methods through more coherent, exhaustive, and reproducible LCA applications.

Apart two studies that explicitly focus on end-of-life management, the published articles distribute almost evenly between cradle-to-grave and cradle-to-gate approaches, with the latter extending to the operational phase in a couple of works. This aspect demonstrates the preference for adopting overarching and holistic viewpoints to address environmental assessments and problems. A general agreement is also detectable for combining or integrating LCA with complementary tools and methodologies to tackle the sustainability challenges from multiple angles. More in detail, integrative methodologies include GIS [6], MFA and criticality assessment [13], stochastic modelling for spatial emissions [8], material engineering tools [11], and exergy analysis [15]. While attributional LCA is often the preferred methodological baseline approach, several studies have adopted consequential LCA settings or applied scenario analysis to analyze possible future and context-specific implications in the systems investigated.

Overall, a general agreement clearly emerges on the positive effect of (i) decarbonization of energy sources, and (ii) improvement of process efficiencies through technological progress as key strategies to future environmental sustainability. However, while these strategies are certainly embraceable and overarching at the global level, context-specific analyses are fundamental to capture the inherent variability of local and regional parameters, which ultimately influence the choice of the most preferable alternative to select. The set of case studies addressed in this Special Issue covers a wide range of topics and technologies and provides an insightful perspective on the current research needs. However, it is not comprehensive nor exhaustive and certainly calls for complementary research studies to approach achieving environmental sustainability in energy systems. To this end, alone or in combination with integrative methodologies, LCA can be of pivotal importance and constitute the scientific foundation on which a full system understanding can be reached.

We wish to thank Mr. Howland Wu, the authors, and the reviewers for their outstanding work and support in fulfilling the proposed goal of this Special Issue.

Funding: This research received no external funding.

Conflicts of Interest: The authors declare no conflict of interest.

References

1. Maranghi, S.; Parisi, M.L.; Basosi, R.; Sinicropi, A. Environmental profile of the manufacturing process of perovskite photovoltaics: Harmonization of life cycle assessment studies. *Energies* **2019**, *12*, 3746. [CrossRef]
2. Piasecka, I.; Bałdowska-Witos, P.; Piotrowska, K.; Tomporowski, A. Eco-energetical life cycle assessment of materials and components of photovoltaic power plant. *Energies* **2020**, *13*, 1385. [CrossRef]
3. Herceg, S.; Pinto Bautista, S.; Weiß, K.-A. Influence of waste management on the environmental footprint of electricity produced by photovoltaic systems. *Energies* **2020**, *13*, 2146. [CrossRef]
4. Zakuciová, K.; Štefanica, J.; Carvalho, A.; Kočí, V. Environmental assessment of a coal power plant with carbon dioxide capture system based on the activated carbon adsorption process: A case study of the Czech Republic. *Energies* **2020**, *13*, 2251. [CrossRef]
5. Raugei, M.; Kamran, M.; Hutchinson, A. A prospective net energy and environmental life-cycle assessment of the uk electricity grid. *Energies* **2020**, *13*, 2207. [CrossRef]
6. Pergola, M.; Rita, A.; Tortora, A.; Castellaneta, M.; Borghetti, M.; De Franchi, A.S.; Lapolla, A.; Moretti, N.; Pecora, G.; Pierangeli, D.; et al. Identification of suitable areas for biomass power plant construction through environmental impact assessment of forest harvesting residues transportation. *Energies* **2020**, *13*, 2699. [CrossRef]
7. Sharara, M.; Kim, D.; Sadaka, S.; Thoma, G. Consequential life cycle assessment of swine manure management within a thermal gasification scenario. *Energies* **2019**, *12*, 4081. [CrossRef]
8. Spatari, S.; Stadel, A.; Adler, P.R.; Kar, S.; Parton, W.J.; Hicks, K.B.; McAloon, A.J.; Gurian, P.L. The role of biorefinery co-products, market proximity and feedstock environmental footprint in meeting biofuel policy goals for winter barley-to-ethanol. *Energies* **2020**, *13*, 2236. [CrossRef]
9. Vogli, L.; Macrelli, S.; Marazza, D.; Galletti, P.; Torri, C.; Samorì, C.; Righi, S. Life cycle assessment and energy balance of a novel polyhydroxyalkanoates production process with mixed microbial cultures fed on pyrolytic products of wastewater treatment sludge. *Energies* **2020**, *13*, 2706. [CrossRef]
10. Bonamente, E.; Aquino, A. Environmental performance of innovative ground-source heat pumps with pcm energy storage. *Energies* **2020**, *13*, 117. [CrossRef]
11. Di Bari, R.; Horn, R.; Nienborg, B.; Klinker, F.; Kieseritzky, E.; Pawelz, F. The environmental potential of phase change materials in building applications. A multiple case investigation based on life cycle assessment and building simulation. *Energies* **2020**, *13*, 3045. [CrossRef]
12. Hu, M. A Building life-cycle embodied performance index—The relationship between embodied energy, embodied carbon and environmental impact. *Energies* **2020**, *13*, 1905. [CrossRef]
13. Bobba, S.; Bianco, I.; Eynard, U.; Carrara, S.; Mathieux, F.; Blengini, G.A. Bridging tools to better understand environmental performances and raw materials supply of traction batteries in the future EU fleet. *Energies* **2020**, *13*, 2513. [CrossRef]
14. Kosai, S.; Yuasa, M.; Yamasue, E. Chronological transition of relationship between intracity lifecycle transport energy efficiency and population density. *Energies* **2020**, *13*, 2094. [CrossRef]
15. Nwodo, M.N.; Anumba, C.J. Exergetic Life Cycle Assessment: A Review. *Energies* **2020**, *13*, 2684. [CrossRef]

Publisher's Note: MDPI stays neutral with regard to jurisdictional claims in published maps and institutional affiliations.

Article

Environmental Profile of the Manufacturing Process of Perovskite Photovoltaics: Harmonization of Life Cycle Assessment Studies

Simone Maranghi [1,2], **Maria Laura Parisi** [1,2,3,*], **Riccardo Basosi** [1,2,3] **and Adalgisa Sinicropi** [1,2,3,*]

[1] R²ES Lab, Department of Biotechnology, Chemistry and Pharmacy, University of Siena, 53100 Siena, Italy; simone.maranghi@unisi.it (S.M.); riccardo.basosi@unisi.it (R.B.)

[2] Center for Colloid and Surface Science (CSGI), 50019 Firenze, Italy

[3] Italian National Council for Research, Institute for the Chemistry of OrganoMetallic Compounds (CNR-ICCOM), 50019 Firenze, Italy

[*] Correspondence: marialaura.parisi@unisi.it (M.L.P.); adalgisa.sinicropi@unisi.it (A.S.)

Received: 5 September 2019; Accepted: 27 September 2019; Published: 30 September 2019

Abstract: The development of perovskite solar cell technology is steadily increasing. The extremely high photoconversion efficiency drives factor that makes these devices so attractive for photovoltaic energy production. However, the environmental impact of this technology could represent a crucial matter for industrial development, and the sustainability of perovskite solar cell is at the center of the scientific debate. The life cycle assessment studies available in the literature evaluate the environmental profile of this technology, but the outcomes vary consistently depending on the methodological choices and assumptions made by authors. In this work, we performed the harmonization of these life cycle assessment results to understand which are effectively the environmental hotspots of the perovskite solar cell fabrication. The outcomes of this analysis allowed us to outline an environmental ranking of the profiles of the several cell configurations investigated and, most importantly, to identify the material and energy flows that mostly contribute to the technology in terms of environmental impact.

Keywords: life cycle assessment; harmonization; photovoltaic; perovskite solar cell; manufacturing process

---

1. Introduction

Since they were first reported as a promising novel photovoltaic (PV) technology [1], the perovskite solar cells (PSCs) showed impressive technological growth, especially concerning the improvement of photoconversion efficiency. Thanks to their remarkable versatility, the possibility to tune the energy band gap, and the opportunity to make tandem solar devices with other photovoltaic technologies (e.g., silicon-based solar cells, copper indium gallium selenide -CIGS- solar cells), PSCs gained a leading role among the last generation photovoltaics. Recently, they broke a new laboratory photoconversion efficiency record by reaching 25.2% photoconversion efficiency [2]. This value overcomes those recorded for thin-film PV, as well as the record obtained by multi-crystalline silicon solar cells [3].

The technology development of PSC has already led to the fabrication of some perovskite solar module (PSM) prototypes [4–8]. Nowadays, the industrialization and commercialization phase of these devices is at the center of the scientific debate [9–11].

However, for a final mass production of this innovative technology, several limitations must be overcome in the next future [12]. Researchers must increase the stability of some chemical compound used in the cell [13,14], improve the scalability of manufacturing processes and techniques [15,16], and reach adequate environmental performances [17–20]. Concerning the last issue and in order to assess the sustainability of the devices based on PSCs, it is pivotal to investigate the environmental hotspots of their potential industrial production, which is already at an early-stage development [21,22].

One of the most powerful methods available to perform environmental sustainability assessment is provided by the life cycle assessment (LCA), as recommended by the European Commission [23,24]. The LCA analytical approach allows an outlining of the environmental profile of products, processes, or services across all life cycle stages, modeling its interaction with the environment and considering all steps from raw material extraction to the end-of-life (EoL) phase [25]. PSCs should undergo such environmental evaluation to be classified as a sustainable technology. A fair number of LCA studies on PSCs have been published in the scientific literature to date [26–33].

The reported results vary consistently depending on the methodological choices and assumptions made by authors, thus generating confusion about the environmental consequences connected with the implementation of PSC technology. In this context, the usefulness of LCA in informing the stakeholders and/or in supporting the scientific community in the development of PCS technology toward environmental sustainability is limited.

The primary goal of this study was to perform a critical review and an in-depth harmonization of LCA studies on PSCs available in the literature. The objective was to align the methodological approaches, the LCI datasets, and the LCIA methods, as well as to highlight the environmental hotspots of PSC production processes.

2. Methodological Approach of This Study

The LCA is a standardized methodology suitable for assessing the environmental performance of a product, a process, or an activity. Furthermore, it can be managed as a systematic analysis performed to determine and quantify all the resources used (input flows of raw materials, energy, water, etc.) during the whole life cycle of a product, process, service, or supply chain useful to evaluate the potential environmental impacts caused by consumption, emissions, and waste generation (output flows). The LCA approach is important to identify opportunities for improvement of the eco-profile of the investigated system. This representation, usually defined as the *cradle-to-grave* approach, represents the more extended and beneficial approach that can be implemented for a complete LCA. Other options are the *cradle-to-gate* or the *gate-to-gate* approaches that can be set up depending on the focus of the.

According to the guidelines provided by the ISO 14040 family standards [34,35], the LCA analysis is divided into four phases: Goal and scope definition, life cycle inventory (LCI) analysis, life cycle impact assessment (LCIA), and interpretation of results. Throughout the first phase the system model is defined, specifying the methodological elements (system boundaries, quality of data, functional unit, cut-off criteria, etc.) which refer all input and output flows. The second phase outlines each product system by quantifying all the input and output flows for each stage in the life cycle. The impact assessment phase quantifies the relative magnitude of the system' impact from an environmental point of view. Finally, in the interpretation phase, recommendations and conclusion are outlined based on the scientific findings to support the overall improvement of the system.

As previously described, we performed the harmonization of LCI datasets and LCIA methods reported in the literature. In fact, the massive amount of different results and findings already published in the literature do not allow for an exhaustive assessment of the environmental footprint associated with the numerous PSC configurations. In this context, the present paper focused on the harmonization of previous LCA studies, aiming at establishing a homogenized methodological framework. This allowed for a consistent comparison of results obtained by different studies and the definition of baseline information useful for the development of new LCA analyses on PSCs.

Among all the LCA studies of PSC published in literature so far, we limited ourselves to studies on the single junction PSC and did not consider papers dealing with PSC/silicon solar cells. As a further criterion to collect high-quality information, we decided to select only the studies reporting detailed model systems and complete life cycle inventories (i.e., raw materials and energy input and output flows) of the manufacturing process of the cell [26–31]. The selected studies and the related PSC configurations investigated therein are reported in Table 1.

Table 1. Selected life cycle assessment (LCA) studies for the harmonization of life cycle impact (LCI) and results.

PAPER	REF	PSC CONFIGURATION (1 cm^2)	Label
Gong 2015	[26]	FTO glass/TiO$_2$/MAPbI$_3$/Spiro-OMeTAD/Au	G1
Gong 2015	[26]	ITO glass/ZnO/MAPbI$_3$/Spiro-OMeTAD/Ag	G2
Espinosa 2015	[27]	FTO glass/TiO$_2$/MAPbI$_3$/Spiro-OMeTAD/Ag	E
Serrano-Lujan 2015	[28]	FTO glass/TiO$_2$/MASnI$_3$/Spiro-OMeTAD/Au	S
Zhang 2015	[29]	FTO glass/TiO$_2$ nanotube (TNT)/MAPbI$_3$/Iodine liq el./Pt glass	Z
Celik 2016	[30]	FTO glass/SnO$_2$/MAPbI$_3$/CuSCN/MoOx-Al (*solution-based dep.*)	C1
Celik 2016	[30]	FTO glass/SnO$_2$/MAPbI$_3$/CuSCN/MoOx-Al (*vacuum-based dep.*)	C2
Celik 2016	[30]	FTO glass/SnO$_2$/MAPbI$_3$/C-Paste (*HTL free*)	C3
Alberola-Borràs 2018	[31]	FTO glass/TiO$_2$/MAPbI$_3$ (Solv1)/Spiro-OMeTAD/Au	AB1
Alberola-Borràs 2018	[31]	FTO glass/TiO$_2$/MAPbI$_3$ (Solv2)/Spiro-OMeTAD/Au	AB2
Alberola-Borràs 2018	[31]	FTO glass/TiO$_2$/MAPbI$_3$ (Solv3)/Spiro-OMeTAD/Au	AB3
Alberola-Borràs 2018	[31]	FTO glass/TiO$_2$/MAPbI$_3$ (Solv2 + TiO$_2$ scaffold)/Spiro-OMeTAD/Au	AB4

A careful inspection of the six papers reported in Table 1 allowed to us point out the high degree of diversity of the LCI datasets. This is mainly due to the variety of PSC configurations and fabrication processes which have been proposed during recent years. Indeed, the early development stage of PSC technology has led to the development of a multitude of cell components, structures, and manufacturing procedures that, in turn, determine a large number of combinations. Therefore, as LCI datasets available in the literature are referred to the lab-scale fabrication of different cell configurations, they show a wide variability in featuring energy and material input and output flows, both in terms of type and quantity of employed materials.

The harmonization procedure focused on the rebuilding of the LCIs reported in the studies listed in Table 1 and on the alignment and update of the environmental footprint evaluation. Each input and output flow was recreated with an updated version of the LCA database and all the environmental profiles were recalculated with the most updated version of the LCIA method defined by the Joint Research Centre—European Commission, namely the ILCD 2011 method [36]. The latter has been developed based on a detailed examination of several LCIA methods, models, and indicators and is now widely recommended for the environmental footprint calculation of systems located in the European area.

All calculations were performed with the SimaPro software version 8.5.2 [37], and the main database used for gathering secondary data and average information was the Ecoinvent version 3.4 [38].

In order to harmonize all the datasets, the different assumptions concerning the LCI building phase implemented in the selected LCA studies were modified and aligned as follows:

- All the manufacturing processes were considered to take place in the same geographical context. Thus, all the input and output flows were deemed to be located in the European area.
- All secondary data were taken from the Ecoinvent database, employing average European production flows as the primary selection criterion. When such averages were missing and

the Ecoinvent production flows referred only to specific European manufacturing procedures, global average production flows were chosen.

- Product system models built for raw materials that are involved in all the PSCs' LCI (such as fluorine-doped tin oxide (FTO) glass) are the same for all the configurations.
- When some specific data and information concerning a product were missing (such as for the Spiro-MeOTAD compound in the study by Alberola-Borràs et al.), we modelled meta-datasets which were calculated as the average of the same product system models employed in the LCIs of the other studies.

3. Harmonization of the Selected Studies

3.1. System Boundaries

In LCA, a product system is defined as a set of various subcomponents, namely processes and flows. The system boundaries describe which subset of the overall collection of processes and flows of the product system life cycle is part of the study according to the scope and goal stated in the first phase of an LCA study.

In the papers reported in Table 1, the selected system boundaries are very different from one study to another (Figure 1). Indeed, some of the studies investigated the environmental impacts connected to the manufacturing process without considering the operative and End-of-Life phases of the system (i.e., as in the cradle-to-gate approach). Other studies drew the system boundaries to include the manufacturing and operative stages, thus setting several technical parameters for the PSCs (such as photoconversion efficiency, lifetime, active area, degradation rate) that are required to assess the potential electricity production of photovoltaic devices (i.e., cradle-to-gate approach). Finally, other studies also considered the EoL phase by making some *a priori* assumptions, thus modelling processes concerning the potential disposal and recovery modalities for the PSC technology (i.e., cradle-to-grave approach).

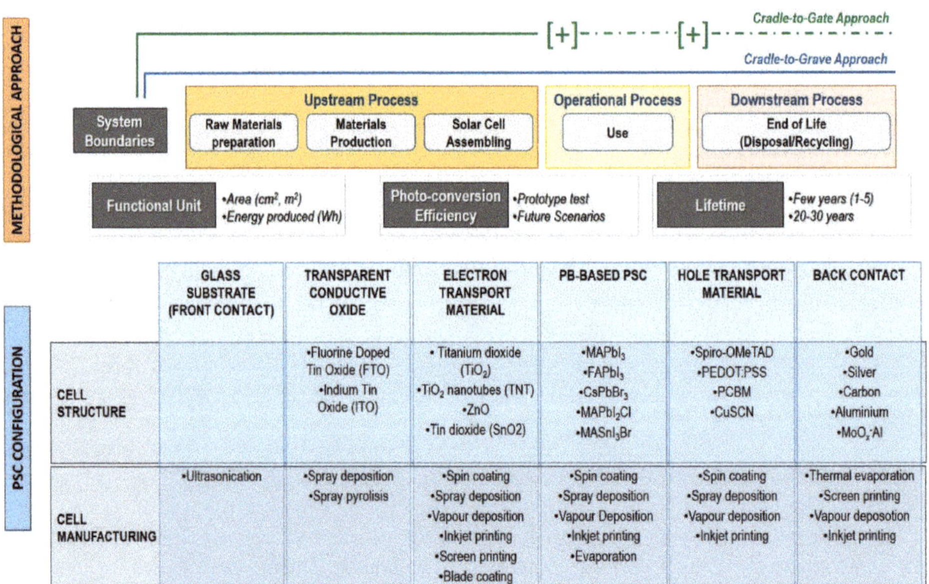

	GLASS SUBSTRATE (FRONT CONTACT)	TRANSPARENT CONDUCTIVE OXIDE	ELECTRON TRANSPORT MATERIAL	PB-BASED PSC	HOLE TRANSPORT MATERIAL	BACK CONTACT
CELL STRUCTURE		•Fluorine Doped Tin Oxide (FTO) •Indium Tin Oxide (ITO)	• Titanium dioxide (TiO$_2$) •TiO$_2$ nanotubes (TNT) •ZnO •Tin dioxide (SnO2)	•MAPbI$_3$ •FAPbI$_3$ •CsPbBr$_3$ •MAPbI$_2$Cl •MASnI$_3$Br	•Spiro-OMeTAD •PEDOT:PSS •PCBM •CuSCN	•Gold •Silver •Carbon •Aluminium •MoO$_x$;Al
CELL MANUFACTURING	•Ultrasonication	•Spray deposition •Spray pyrolisis	•Spin coating •Spray deposition •Vapour deposition •Inkjet printing •Screen printing •Blade coating	•Spin coating •Spray deposition •Vapour Deposition •Inkjet printing •Evaporation	•Spin coating •Spray deposition •Vapour deposition •Inkjet printing	•Thermal evaporation •Screen printing •Vapour deposotion •Inkjet printing

Figure 1. Overview of the methodological assumptions and perovskite solar cell (PSC) configurations reported in the selected LCA studies.

In the harmonization procedure employed in this work, only the manufacturing phase of the PSCs was considered, thus applying a cradle-to-gate approach. Given the very early stage of development of this PV technology, this approach was taken in order to reduce the uncertainty of the results. In fact, although a complete life cycle analysis should be evaluated to obtain a fair comparison of different options, the modelling of prospective scenarios for operational and End-of-Life phases at this level would not add reliable information for the purpose of the comparison developed in this study. All the raw materials employed in the production process of a PSC were modelled and included in this study.

3.2. Functional Unit

The functional unit is the reference flow to which all the product systems modelled during the LCI phase are referred. It must be representative of a quantifiable function of the product, process, or service under study to allow for the comparison among different LCA studies. Its choice is directly connected with the outlined system boundaries.

There were two functional units reported in the LCA studies published in the considered literature, as shown in Figure 1. They are the area of the cell or module (for the manufacturing phase) and the amount of electricity produced (for the operative and disposal phases). By setting the system boundaries of our analysis with a focus on the manufacturing phase, we consequently selected the area of PSC (1 cm^2) as the functional unit. The studies employing this dimensional functional unit (1 m^2) had to apply a linear scale-up starting from primary or secondary data referring to the production of a PSC (1 cm^2). Thus, the choice of 1 cm^2 as the functional unit forced us to scale down some of the functional units reported in LCA studies that modelled the LCA for the production of 1 m^2 of perovskite solar module [26,30]. Moreover, in all those cases, a linear scale-down of all the input and output flows was performed by changing the functional unit.

3.3. LCIA Methods

In the LCIA phase, the environmental burdens connected with all the LCI data are identified and evaluated. The aim is to assign a value judgment to the energy and material input and output flows data collected in the inventory in order to assess the magnitude of their critical effects on impact categories. This task is generally addressed by summing up the whole environmental load in a few clear LCIA indicators. each one focusing on a different environmental issue.

LCA papers reported in Table 1 applied several LCIA methods (Figure 1) such as the ILCD 2011 Midpoint [36], the CML-IA [39], the Eco-indicator 99 [40], TRACI [41], and the IPCC 2013 [42]. Here, the chosen LCIA method for the harmonization procedure was the ILCD 2011 Midpoint, in accordance to the European Commission Recommendation [43] and Communication [44] on the use of the Product Environmental Footprint (PEF) to measure and communicate the life cycle environmental performance. The 16 impact categories included in the ILCD 2011 method are:

- climate change
- ozone depletion
- human toxicity non-cancer effect
- human toxicity cancer effect
- particulate matter
- ionizing radiation Human Health
- ionizing radiation Ecosystem (interim)
- photochemical ozone formation
- acidification
- terrestrial eutrophication
- freshwater eutrophication
- marine eutrophication
- freshwater ecotoxicity

- land use
- water resource depletion
- mineral fossil and renewable resource depletion

Among these impact categories, the ionizing radiation ecosystem (interim) and the water resource depletion were not taken into account for the harmonization procedure, following the suggestion reported in the PEF category rules (PEFCR) document [43]. Furthermore, in order to evaluate all the energy consumption associated with PSC manufacturing, the single-issue method cumulative energy demand (CED) was employed [45] as well. The CED indicator quantifies the whole energy requirement during the system life cycle. Its measuring unit is theMJ equivalent of primary energy and it accounts for both direct energy (like electricity and thermal energy) and indirect energy contributions (embodied energy of materials).

4. Results

The comparison among the environmental profiles of the 12 different PSC configurations investigated in the selected LCA studies and calculated with ILCD 2011 method is shown in Figure 2.

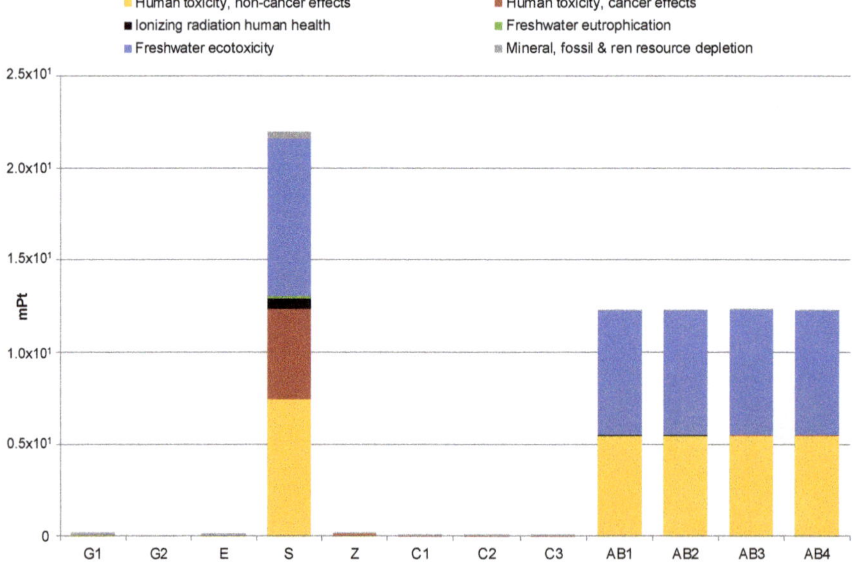

Figure 2. Single Score results comparing the environmental profiles of PSC configurations. Calculation was performed with ILCD 2011 method/Normalization. Only categories with a percentage impact higher than 1% are shown (mPt = milliPoints).

The graph in Figure 2 clearly shows that among the PSC configurations analyzed, five of them have an extremely high environmental impact [28,31]. This is due to the PSC device structures, which are characterized by various raw materials usage and manufacturing procedures. In detail, the S configuration [28] reports an amount of gold in the back contact that is two orders of magnitude higher than in the other studies. Furthermore, the AB1-4 configurations [31] have a direct metallic zinc (Zn) emission into the water that leads to a remarkable environmental load on categories dealing with toxicity.

The graph in Figure 3 shows the differences among the environmental impacts of the seven PSC configurations whose profiles are undetectable in Figure 2.

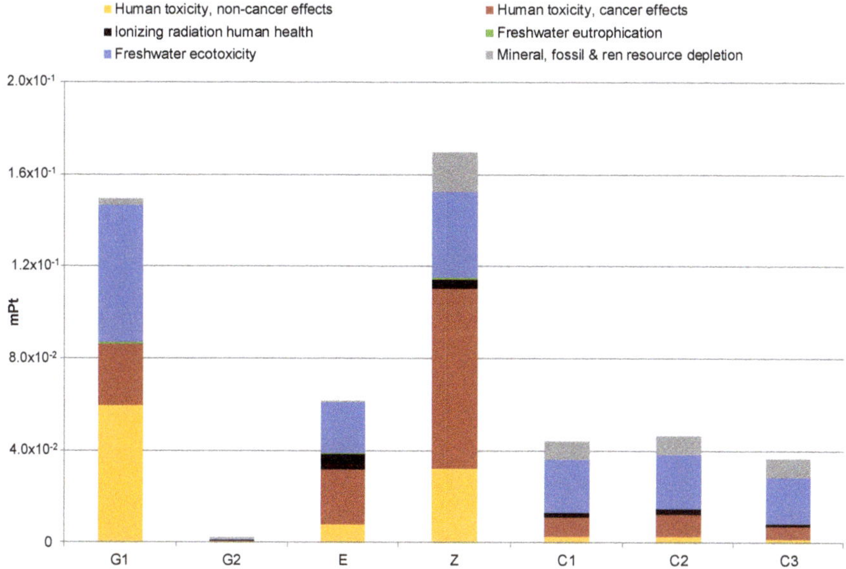

Figure 3. Single Score results comparing the environmental profiles of PSC configurations [26,27,29,30]. Calculation performed with ILCD 2011 method/Normalization. Only categories with a percentage impact higher than 1% are shown (mPt = milliPoints).

The G1 and Z configurations reported in [26,29] have a high impact on the environmental categories that deal with toxicity issues. Concerning G1, the gold used as back contact has the highest load on the environmental profile of the cell, followed by direct electricity consumption and the solar glass (see Figure S1 in SI). Regarding the Z configuration, the consumption of organic solvents and eluents (mainly diethyl ether) used in the synthesis of the TiO_2 nanotubes (TNT) electron transport layer (ETL) shows a remarkable environmental impact, followed by disposal (incineration) of average residues, the metal use (Pt), the iodine employment, and the electricity and water consumption [29] (see Figure S2 in SI). The G2 configuration [26] shows the lowest environmental profile due to the replacement of gold with silver and TiO_2 with ZnO, respectively. These technological implementations correspond to an environmental benefit that overcomes the increase of the impact due to the employ of indium tin oxide (ITO) instead of the FTO glass.

By analyzing the PSC configurations (Table 1) and the environmental profiles shown in Figure 3, the highest impact of the G1 and Z configurations appear to be mainly related with the manufacturing procedure and the materials used for the ETL, as well as with the metal used for the back contact. In order to investigate to what extent the unit processes and input flows contribute to the environmental profile of each PSC, a detailed inspection of the impacts connected with the manufacturing phase was performed with the characterization step of the LCIA method. In Figure 4, the results for the G1 configuration [26] are shown in detail as it is the most common configuration reported in the literature and can therefore be taken as a reference.

It is evident from Figure 4 that the large impact is given by gold that is used as raw material for the back contact. This is mainly due to the environmental burden of the upstream processes employed for the production of gold.

The comparison of the environmental profiles of PSC configurations, calculated with the CED—Single Score impact assessment method, is shown in Figure 5.

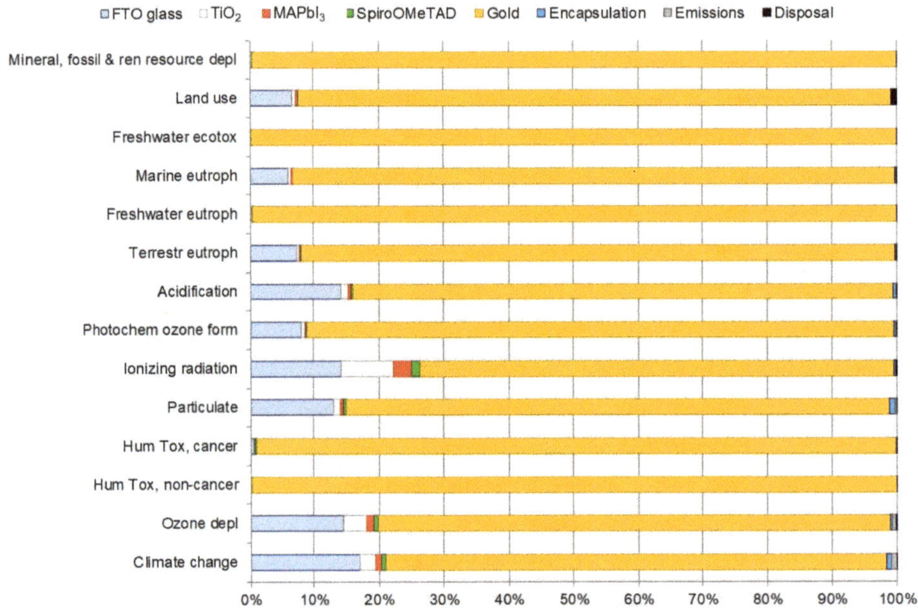

Figure 4. Characterization of the environmental profile of the G1 PSC configuration [26] calculated with the ILCD 2011 method.

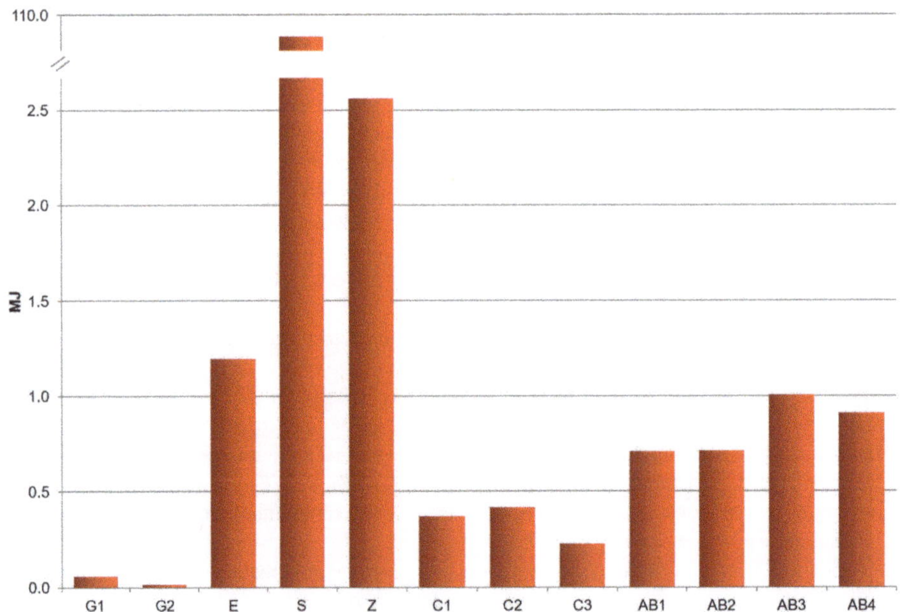

Figure 5. Comparison of energy requirements for the production process of PSC configurations, calculated with the CED—Single Score impact assessment method.

The calculated CED indicator value for the S configuration [28] is clearly out of scale compared to the results obtained for the other solar cells. This is due to direct LCI energy input flows that are about two or three orders of magnitude higher than for the other PSC configurations. The sum of all the energy input flows for the S configuration gives a total value of 104.86 MJ. Truthfully, in the paper that reports S configuration [28], there are no clear indication of the reason why the manufacturing of a PSC is so energy intensive.

Analyzing the contributions to the CED of each unit processes and input flow, the most relevant impact is caused by the direct electricity consumption during the manufacturing of the PSCs (see Table S1 in SI). Despite that, the G1 and Z configurations [26,29] show a different behavior: The employment of gold as the back contact in the G1 cell and the consumption of organic solvents and eluents in the Z cell are the most energy-intensive processes of the respective PSC configurations.

In Figure 6, the graph reports the CED profile calculated for the G1 configuration [26], which is potentially better for future industrial development than the Z configuration.

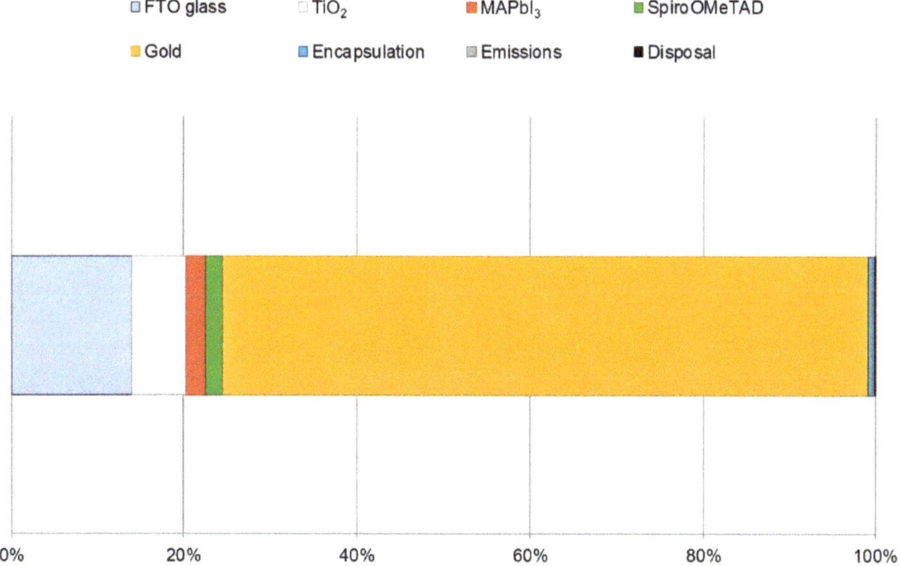

Figure 6. Environmental profiles of the G1 PSC configuration 25 calculated with the CED—Single Score impact assessment method.

The impact of gold deposition is remarkable due to the high energy embedded in the material. The gold production process in the Ecoinvent database considers all the upstream input and output flows related to all the steps characterizing its life cycle, such as ore extraction, metal purification, and ingot manufacturing. Other materials seem to have limited loads on the PSC environmental profile, except solar glass and ETL, and this is valid both for the CED and ILCD 2011 methods.

Calculating the burden of the direct electricity consumption during the manufacturing processes of the G1 configuration with the CED method, the most energy-intensive processes are the FTO glass substrate preparation, the TiO$_2$ electron layer deposition, and the gold back contact deposition (Figure 7).

Besides the calculation of the environmental profile of the PSC configurations and their comparison, the potential environmental impact connected with the use of materials the manufacturing procedures employed in the cells fabrication should also be investigated. Such assessment, performed in the framework of the harmonization procedure, is useful for the identification of the environmental hotspots in the first life cycle stage of the PSC technology.

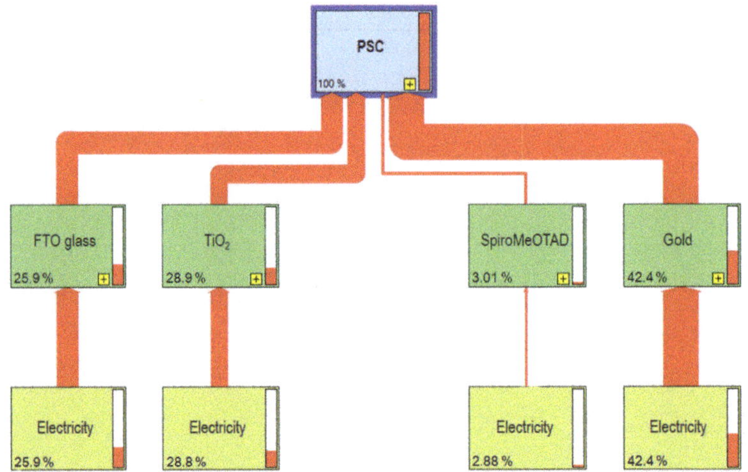

Figure 7. The G1 PSC configuration calculated with the CED—Single Score impact assessment method. Single product flow: Direct energy consumption.

4.1. FTO Glass Substrate

Two LCA studies reported specific data and information concerning the production process of the FTO glass [26,27]. Other studies assumed that the Ecoinvent process called *Solar glass, low-iron {RER}|production* can be taken as a reliable proxy for the FTO glass. However, this process is equal to *Flat glass, uncoated {RER}|production*, so the assumption made by several LCA studies excludes *a priori* the environmental burdens caused by the materials and the energy consumption related to the glass coating process with FTO. In Figure 8, the comparison between the FTO glass modeled by the LCA studies [26,27] and the uncoated glass process taken from the Ecoinvent database is shown.

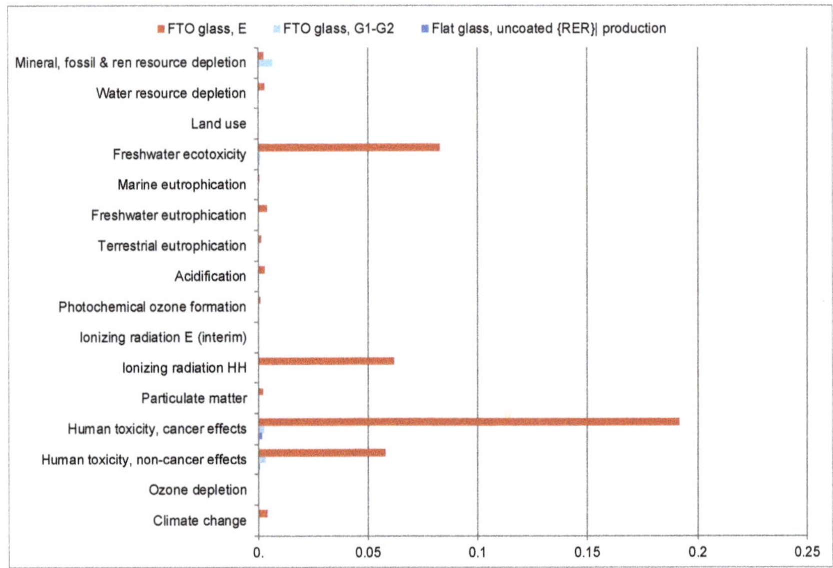

Figure 8. Comparison among glass production processes, calculated with the ILCD 2011/Normalization method.

A high consumption of electricity characterizes the production process of FTO glass modelled by Espinosa et al. [27] for the E configuration due to the oxygen plasma treatment process used for FTO deposition. The high energy requirement leads to a remarkable environmental impact, higher than the other two processes for all the environmental impact categories. The environmental load of the FTO glass manufacturing procedure employed in the G1 configuration [26] is correlated with the usage of tin in the FTO compound. This results in the calculation of a non-negligible impact, higher than the glass production taken from the Ecoinvent.

4.2. ETL Deposition

The calculation of the environmental profiles of the different ETL deposition processes reported in the selected LCA studies is shown in Figure 9.

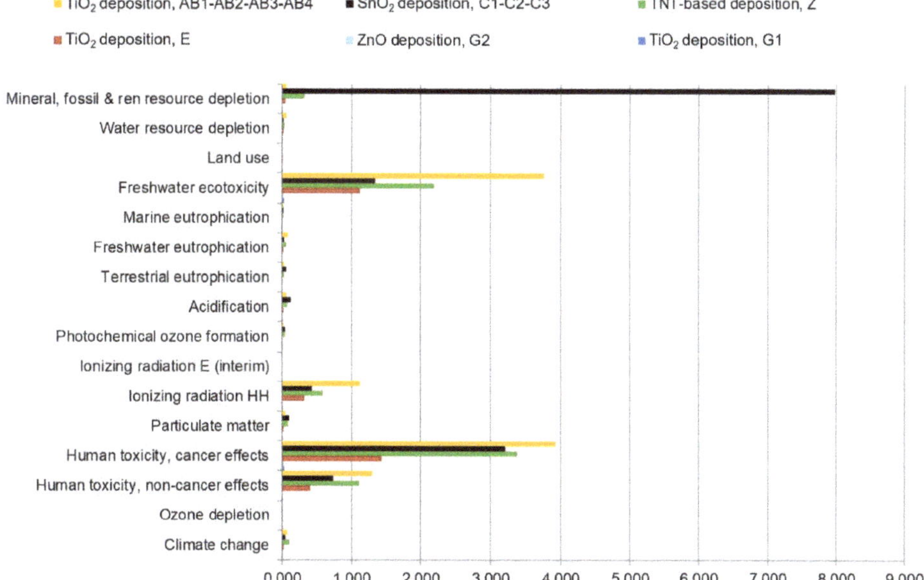

Figure 9. Comparison between ETL deposition techniques, calculated with the ILCD 2011/Normalization method. In order to magnify the differences among the PSC configurations, the environmental profile of ETL deposition employed in the S configuration is not shown here (see Table S2 in SI for the comparison among all ETL deposition techniques).

The environmental profile of the ETL deposition of the S configuration reported by Serrano-Lujan et al. [28] shows a remarkable impact due to a high energy consumption (see Table S2 in SI). In order to appreciate the differences among the environmental profiles of the PSCs' configurations, the results obtained for the system model described by Serrano-Lujan et al. [28] for the S configuration are not discussed here.

The impact of ETL deposition described by Celik et al. [30] is mainly caused by the presence of Sn in the SnO$_2$. This affects the whole environmental profile, especially the category mineral, fossil & renewable resource depletion. Alberola-Borràs et al. [31] reported an ETL deposition featuring a high energy consumption due to the spin-coating, heating, and annealing of TiO$_2$ for the AB1-4 configurations. The environmental burdens associated to the deposition process of the Z configuration [29] are mainly related to the energy consumption during TNT deposition (19.6%) and production (24.5%). The synthesis of TNT-film accounts for 79.6% of the whole environmental impact of the TNT-based deposition procedure. The impact of titanium as raw material in the TNT-film synthesis is the highest after that connected with the energy consumption. Concerning the E configuration [27], the highest environmental impact is related to the use of electricity during the spin-coating process. The authors reported a synthetic procedure for TiO_2 production, but this was not considered in the harmonization procedure because the TiO_2 present in the updated version of the Ecoinvent database is taken as an input flow for all the PSC configurations.

4.3. ETL Production

In order to investigate to which extent the raw materials affect the environmental profile of the ETL deposition techniques reported in Figure 9, a comparison among the production processes of ETL described in the LCA studies is shown in Figure 10. The functional unit is the production of 1 g of ETL.

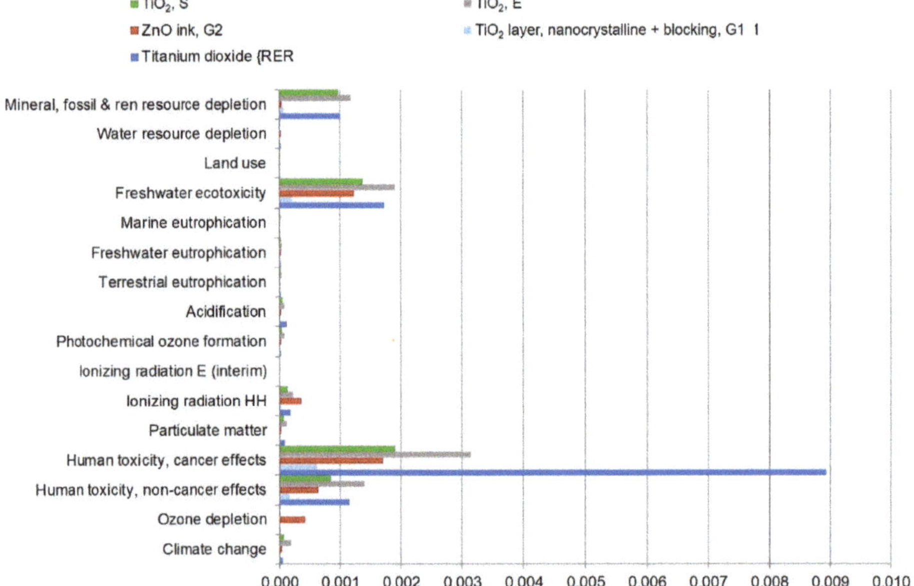

Figure 10. Comparison among ETL production process, calculated with the ILCD 2011/Normalization method. The functional unit is 1 g of ETL produced. In order to magnify the differences among the PSC configurations, the environmental profile of ETL employed in the Z configuration is not shown here (see Table S3 in SI for the comparison among all ETL production processes).

The SnO_2 employed in the C1-C2-C3 configuration [30] as a raw material is not reported in Figure 10 because the LCI data provided by the authors are specific for the production of 1 m² of perovskite solar module. Thus, it was not possible to make robust assumptions to support the modeling of SnO_2 required for a 1 cm² of PSC.

The impact of TNT production [29] overwhelms the environmental profile of other ETM manufacturing procedures, including the TiO_2 process taken from Ecoinvent database (see Table S3 in SI). Besides TNT, other ETM production processes exhibit environmental advantages compared with the TiO_2 process present in the database (Figure 10). In particular, TiO_2 manufacturing process employed in the G1-G2 configurations [26] shows a very low environmental impact, even if it is the only process that takes TiO_2 as a direct input flow from the database. Concerning other ETM, the use of $TiCl_4$ as a precursor for the synthesis of TiO_2 is the main source of the impact of the process employed in the E [27] and S [28] configurations. The electricity consumption is the principal source of impact in the ZnO production process reported for the G2 configuration [26], and the use of TiO_2 of the Ecoinvent database shows the higher load on the environmental profile of the TiO_2-based ETM employed in G1 configuration [26].

4.4. Perovskite Deposition

In Figure 11, the comparison among the lead-based perovskite deposition processes is provided. In order to magnify the differences resulting from the analysis, the environmental profile of perovskite deposition employed in the S configuration [28] is not reported here (see Table S4 in SI). Indeed, the results of the analysis show that the Sn-based perovskite deposition technique reported by Serrano-Lujan et al. [28] has a remarkable environmental impact due to direct energy consumption.

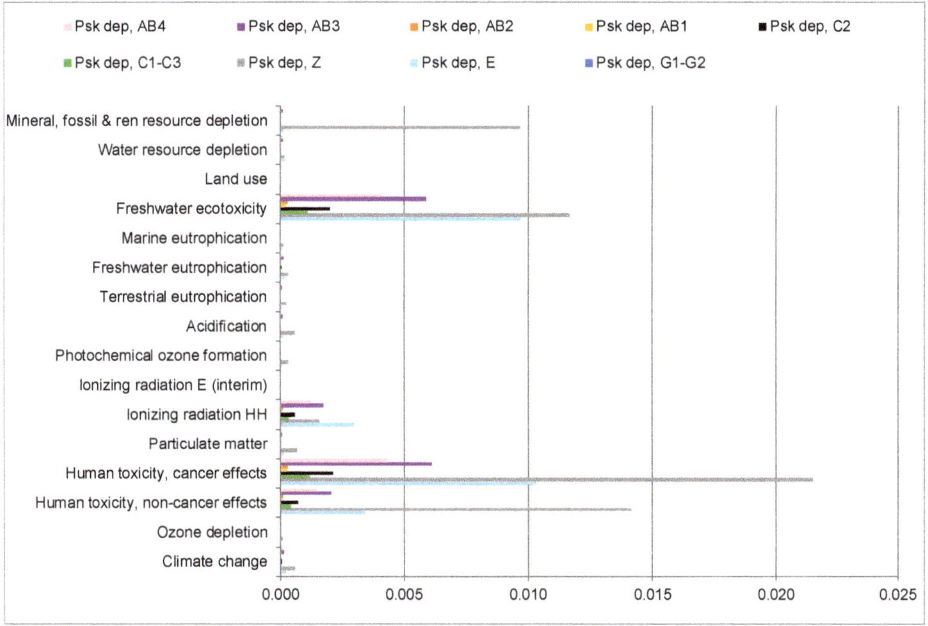

Figure 11. Comparison between Pb-based perovskite (psk) deposition techniques calculated with the ILCD 2011/Normalization method. In order to magnify the differences among the PSC configurations, the environmental profile of perovskite deposition employed in the S configuration is not shown here (see Table S4 in SI for the comparison among all perovskite deposition processes).

The environmental impact calculated for the G1-G2 [26] and AB1-AB2 [31] PSC configurations is very limited. The main reason is the low energy and materials consumption during the perovskite deposition process. Regarding G1-G2 configuration [26], a PbI_2 spin-coating deposition followed by sintering and dipping in methylammonium iodide CH_3NH_3I in isopropanol is reported.

The deposition process employed by Alberola-Borràs et al. [31] is a spin-coating technique for AB1-4 PSC configurations. Thus, there are some little changes in the procedure steps and the materials used. However, the direct consumption of energy for the spin-coating is the major cause of environmental impact for all configurations. The amount of organic solvent (i.e., diethyl ether) consumed during the synthesis of perovskite compound has a remarkable load on the environmental profile of the Z PSC configuration and leads to a high impact on the toxicity-related categories.

The energy used during the perovskite vapor deposition step reported by Espinosa et al. [27] represents the process that produces a high burden on the PSC environmental profile. It is the same for the C1-C3 and C2 configurations described by Celik et al. [30], and the energy requirement causes the highest impact for both deposition methods, solution-based spray for C1-C3 and under vacuum vapor-based for C2.

4.5. HTL Deposition

The results reported in Figure 12 allow the assessment of the major contributions to the environmental profiles of the hole transport layer (HTL) deposition methods. The comparison among HTL depositions described in the selected LCA studies again shows that the impact obtained for the S configuration of [28] overwhelms the other processes due to the high energy consumption (see Table S5 in SI). Thus, their environmental profiles are not included in Figure 12.

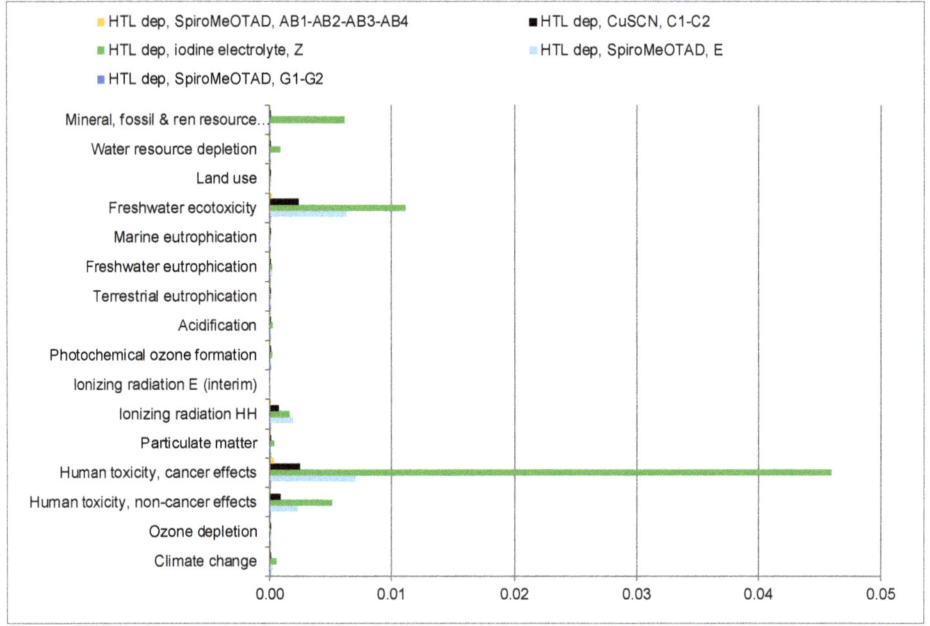

Figure 12. Hole transport layer (HTL) deposition techniques, calculated with the ILCD 2011/Normalization method. In order to magnify the differences among the PSC configurations, the environmental profile of HTL deposition employed in the S configuration is not shown here (see Table S5 in SI for the comparison among all HTL deposition processes).

The use of the iodine electrolyte shows a significant impact due to the synthetic procedure of LiI employed in the Z configuration [29]. In particular, the waste treatment is responsible for the substantial environmental burden on the process. Furthermore, the employment of iodine as a raw material has a considerable impact compared to raw materials used in other HTL deposition procedures.

The consumption of energy during the processes is the most critical contribution on the environmental profile of other HTL depositions, despite the different techniques described in the selected studies (e.g., Gong et al. for G1-G2, Espinosa et al. for E, and Alberola-Borràs et al. for AB1-4 modelled a spin-coating deposition process, while Celik et al. C1-3 reported a screen-printing procedure).

4.6. Back Contact Layer Deposition

As stated in Section 4, employing gold as the metal for back contact layer production leads to a massive load on the environmental profile of the whole PSC, and this is essentially due to the upstream production processes of gold that are present in the database. The large amount of gold reported by

Serrano-Lujan et al. [28] leads to the calculation of an impact value that is basically out of scale, so it is not included in following Figure 13 (see Table S6 in SI for further details).

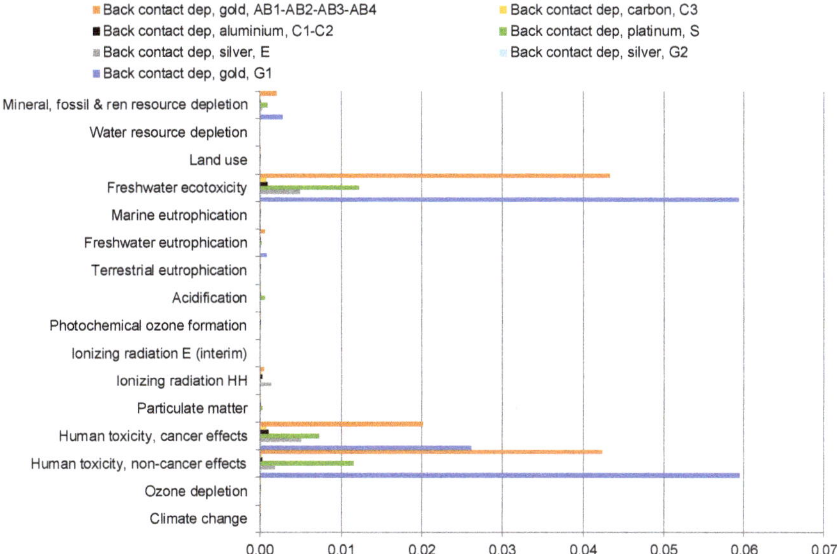

Figure 13. Back contact layer deposition techniques calculated with the ILCD 2011/Normalization method. In order to magnify the differences among the PSC configurations, the environmental profile of back contact deposition employed in the S configuration is not shown here (see Table S6 in SI for the comparison among all back contact deposition processes).

Despite the PSC configuration containing gold, the employment of Pt in the back contact of the Z configuration reported by Zhang et al. [29], the use of silver in the G2 configuration described by Gong et al., and the energy consumption of the other back contact deposition processes exhibits a considerable contribution on the environmental profiles.

Concerning the employed techniques, Gong et al. [26] for G1 and G2, Espinosa et al. for E [27], and Alberola-Borràs et al. for AB1-4 [31] considered the deposition if gold and silver layer by thermal evaporation, whereas Celik et al. [30] reported an under vacuum evaporation for C1 and C2 PSC configurations.

5. Discussion

The results of the harmonization of the selected six LCA studies investigated in this work allowed us to identify the main environmental hotspots of the manufacturing phase of the PSC technology. The most critical raw materials used in the cells resulted to be the metals employed to manufacture the back contact (in particular gold), the conductive solar glass, and the ETL (mainly TiO_2 and TNT, due to the high consumption of organic compounds during their synthesis and preparation). Other materials, like lead-based perovskite and hole transport layer, seem to have limited load on the environmental footprint of PSC due to the minimal quantities of this metal present in the cell.

The PSC manufacturing processes that show the highest environmental impact are the back contact deposition, the ETL deposition, and glass substrate preparation.

The energy demand calculated with the CED impact assessment method is directly related with the use and consumption of the raw materials. However, looking at the direct energy requirement of the PSC manufacturing procedure reported in the analyzed LCA studies, several improvements could be obtained in the FTO glass substrate preparation, in the ETL production and deposition, and all the various deposition procedures (i.e., perovskite, back contact, and hole transport material).

One of the most relevant issues discussed in LCA studies on PSC is their toxicity. As also detected by the harmonization performed in this work, despite the relatively low calculated impact of lead, the presence of such metal highly contributes to the toxicity-related environmental categories of the ILCD 2011 method. Indeed, the human toxicity non-cancer effect, the human toxicity cancer effect, and the freshwater ecotoxicity categories show the highest contribution to the environmental profiles of PSCs. In this context, it should be underlined that the high impact shown by these toxicity-related environmental categories is mainly due to the inherent uncertainty of the USEtox model [46–48] (explicitly declared by developers) employed by the ILCD 2011 method. In detail, the characterization of the inorganic compounds' toxicity (in particular, heavy metals) is affected by a high uncertainty in the definition of their environmental mechanism models. In fact, in the USEtox method, they are defined "as interim" and their characterization factors should be employed cautiously. Basically, the USEtox model is useful to identify the 10 or 20 main compounds or processes that have a non-negligible load on the toxicity-related environmental categories, but a further precise analysis should be carried out in order to investigate their fate, exposure, and effects (i.e., ecotoxicological analysis).

Nevertheless, the results provided by the papers published in the literature, should be taken with care in order not to underestimate the problem related to the toxicity and the potential environmental impact of lead. The most common statement combined with the results of the LCA analysis is that the presence of lead in the PSCs is not a relevant environmental hotspot due to the very low amount of lead contained in the perovskite compounds [26–31]. However, some studies reported that the emission of PbI_2 in water could lead to a high level of pollution in the surrounding areas of the site of production or use [17–20]. The absence of the specific characterization factor for PbI_2 and the high uncertainty in the toxicity modelling of inorganic compound in the USEtox model, together with the lack of primary data concerning the direct emission of lead, PbI_2, and other lead-based compounds that may occur during the whole life cycle of PSC, do not allow a proper evaluation of the real toxicological risk of production and use of PSCs technology.

6. Conclusions

The harmonization of LCA studies on PSCs performed in this study allowed us to outline an environmental ranking of the profiles of PSCs configurations in their manufacturing phase, and, most importantly, to identify the material and energy flows that more contribute to the technology profiles in terms of environmental impact. Results obtained from the harmonization highlight that, especially from the electricity consumption point of view, there is much room for improvement of production processes.

It is noteworthy that the analysis of environmental risk issues associated with the use of large-scale PSCs is one of the main aspects that should need to be addressed. In this context, it is essential that LCA analysts continue to work supporting the technological development of PSC from the earliest stages and that comprehensive life cycle data inventories of primary data are made public and accessible in order to maximize the scientific efforts to minimizing the environmental impact of novel PV technologies.

More importantly, LCA should be used in synergistic way with other methodological and toxicological tools that can properly contribute to assess all environmental impacts linked with the introduction in the market of this promising photovoltaic technology.

Supplementary Materials: The following are available online at http://www.mdpi.com/1996-1073/12/19/3746/s1.

Author Contributions: Conceptualization, S.M., M.L.P., R.B. and A.S.; investigation, S.M., M.L.P.; writing—original draft preparation, S.M., M.L.P.; writing—review and editing, M.L.P, R.B. and A.S.; funding acquisition, R.B. and A.S.

Funding: This research was funded by European Union's Horizon 2020 Framework Program grant number 764047

Acknowledgments: Authors acknowledge MIUR Grant—Department of Excellence 2018–2022 and the European Union's Horizon 2020 Framework Program for funding Research and Innovation under Grant agreement no. 764047 (ESPResSO) for funding.

Conflicts of Interest: The authors declare no conflict of interest.

References

1. Kojima, A.; Teshima, K.; Shirai, Y.; Miyasaka, T. Organometal Halide Perovskites as Visible-Light Sensitizers for Photovoltaic Cells. *J. Am. Chem Soc.* **2009**, *131*, 6050–6051. [CrossRef] [PubMed]
2. NREL. Best Research-Cell Efficiencies. Available online: https://www.nrel.gov/pv/assets/pdfs/best-reserch-cell-efficiencies.20190411.pdf (accessed on 23 September 2019).
3. Green, M.A.; Hishikawa, Y.; Dunlop, E.D.; Levi, D.H.; Hohl-Ebinger, J.; Yoshita, M.; Ho-Baillie, A.W.Y. Solar Cell Efficiency Tables (Version 53). *Prog. Photovol. Res. Appl.* **2019**, *27*, 3–12. [CrossRef]
4. Razza, S.; Di Giacomo, F.; Matteocci, F.; Cinà, L.; Palma, A.L.; Casaluci, S.; Cameron, P.; D'Epifanio, A.; Licoccia, S.; Reale, A.; et al. Perovskite Solar Cells and Large Area Modules (100 cm^2) Based on an Air Flow-Assisted PbI2 Blade Coating Deposition Process. *J. Power Sour.* **2015**, *277*, 286–291. [CrossRef]
5. Matteocci, F.; Cinà, L.; Di Giacomo, F.; Razza, S.; Palma, A.L.; Guidobaldi, A.; D'Epifanio, A.; Licoccia, S.; Brown, T.M.; Reale, A.; et al. High Efficiency Photovoltaic Module Based on Mesoscopic Organometal Halide Perovskite. *Prog. Photovolt. Res. Appl.* **2016**, *24*, 436–445. [CrossRef]
6. Yang, M.; Kim, D.H.; Klein, T.R.; Li, Z.; Reese, M.O.; Tremolet De Villers, B.J.; Berry, J.J.; Van Hest, M.F.A.M.; Zhu, K. Highly Efficient Perovskite Solar Modules by Scalable Fabrication and Interconnection Optimization. *ACS Energy Lett.* **2018**, *3*, 322–328. [CrossRef]
7. Higuchi, H.; Negami, T. Largest Highly Efficient 203x203 mm^2 CH3NH3PbI3 Perovskite Solar Modules. *Jpn. J. Appl. Phys.* **2018**, *57*, 1–6. [CrossRef]
8. Di Giacomo, F.; Shanmugam, S.; Fledderus, H.; Bruijnaers, B.J.; Verhees, W.J.H.; Dorenkamper, M.S.; Veenstra, S.C.; Qiu, W.; Gehlhaar, R.; Merckx, T.; et al. Up-Scalable Sheet-to-Sheet Production of High Efficiency Perovskite Module and Solar Cells on 6-in. Substrate Using Slot Die Coating. *Sol. Energy Mater. Sol. Cells* **2018**, *181*, 53–59. [CrossRef]
9. Rong, Y.; Hu, Y.; Mei, A.; Tan, H.; Saidaminov, M.I.; Seok, S.; McGehee, M.D.; Sargent, E.H.; Han, H. Challenges for Commercializing Perovskite Solar Cells. *Science* **2018**, *361*, eaat8235. [CrossRef] [PubMed]
10. Kim, D.H.; Whitaker, J.B.; Li, Z.; van Hest, M.F.A.M.; Zhu, K. Outlook and Challenges of Perovskite Solar Cells toward Terawatt-Scale Photovoltaic Module Technology. *Joule* **2018**, *2*, 1437–1451. [CrossRef]
11. Qiu, L.; Ono, L.K.; Qi, Y. Advances and Challenges to the Commercialization of Organic–Inorganic Halide Perovskite Solar Cell Technology. *Mater. Today Energy* **2018**, *7*, 169–189. [CrossRef]
12. Jena, A.K.; Kulkarni, A.; Miyasaka, T. Halide Perovskite Photovoltaics: Background, Status, and Future Prospects. *Chem. Rev.* **2019**, *119*, 3036–3103. [CrossRef] [PubMed]
13. Christians, J.A.; Habisreutinger, S.N.; Berry, J.J.; Luther, J.M. Stability in Perovskite Photovoltaics: A Paradigm for Newfangled Technologies. *ACS Energy Lett.* **2018**, *3*, 2136–2143. [CrossRef]
14. Asghar, M.I.; Zhang, J.; Wang, H.; Lund, P.D. Device Stability of Perovskite Solar Cells—A Review. *Renew. Sustain. Energy Rev.* **2017**, *77*, 131–146. [CrossRef]
15. Li, Z.; Klein, T.R.; Kim, D.H.; Yang, M.; Berry, J.J.; Van Hest, M.F.A.M.; Zhu, K. Scalable Fabrication of Perovskite Solar Cells. *Nat. Rev. Mater.* **2018**, *3*, 1–20. [CrossRef]
16. Rong, Y.; Ming, Y.; Ji, W.; Li, D.; Mei, A.; Hu, Y.; Han, H. Toward Industrial-Scale Production of Perovskite Solar Cells: Screen Printing, Slot-Die Coating, and Emerging Techniques. *J. Phys. Chem. Lett.* **2018**, *9*, 2707–2713. [CrossRef]
17. Babayigit, A.; Boyen, H.-G.; Conings, B. Environment versus Sustainable Energy: The Case of Lead Halide Perovskite-Based Solar Cells. *MRS Energy Sustain.* **2018**, *5*, 1–15. [CrossRef]
18. Hailegnaw, B.; Kirmayer, S.; Edri, E.; Hodes, G.; Cahen, D. Rain on Methylammonium Lead Iodide Based Perovskites: Possible Environmental Effects of Perovskite Solar Cells. *J. Phys. Chem. Lett.* **2015**, *6*, 1543–1547. [CrossRef]
19. Babayigit, A.; Ethirajan, A.; Muller, M.; Conings, B. Toxicity of Organometal Halide Perovskite Solar Cells. *Nat. Mater.* **2016**, *15*, 247–251. [CrossRef]

20. Babayigit, A.; Duy Thanh, D.; Ethirajan, A.; Manca, J.; Muller, M.; Boyen, H.G.; Conings, B. Assessing the Toxicity of Pb-and Sn-Based Perovskite Solar Cells in Model Organism Danio Rerio. *Sci. Rep.* **2016**, *6*, 1–11. [CrossRef]

21. Bravi, M.; Parisi, M.L.; Tiezzi, E.; Basosi, R. Life Cycle Assessment of Advanced Technologies for Photovoltaic Panels Production. *Int. J. Heat Technol.* **2010**, *28*, 133–140.

22. Parisi, M.L.; Maranghi, S.; Sinicropi, A.; Basosi, R. Development of Dye Sensitized Solar Cells: A Life Cycle Perspective for the Environmental and Market Potential Assessment of a Renewable Energy Technology. *Int. J. Heat Technol.* **2013**, *31*. [CrossRef]

23. Commission of the European Communities. *Green Paper on Integrated Product Policy*; Commission of the European Communities: Brussels, Belgium, 2001.

24. Commission of the European Communities. *Final-Integrated Product Policy-Building on Environmental Life-Cycle Thinking*; Commission of the European Communities: Brussels, Belgium, 2003.

25. Parisi, M.L.; Maranghi, S.; Vesce, L.; Sinicropi, A.; Di Carlo, A.; Basosi, R. Prospective Life Cycle Assessment of Third-Generation Photovoltaics at the Pre-Industrial Scale: A Long-Term Scenario Approach. *Renew. Sustain. Energy Rev.* **2019**. submitted.

26. Gong, J.; Darling, S.B.; You, F. Perovskite Photovoltaics: Life-Cycle Assessment of Energy and Environmental Impacts. *Energy Environ. Sci.* **2015**, *8*, 1953–1968. [CrossRef]

27. Espinosa, N.; Serrano-Luján, L.; Urbina, A.; Krebs, F.C. Solution and Vapour Deposited Lead Perovskite Solar Cells: Ecotoxicity from a Life Cycle Assessment Perspective. *Sol. Energy Mater. Sol. Cells* **2015**, *137*, 303–310. [CrossRef]

28. Serrano-Lujan, L.; Espinosa, N.; Larsen-Olsen, T.T.; Abad, J.; Urbina, A.; Krebs, F.C. Tin- and Lead-Based Perovskite Solar Cells under Scrutiny: An Environmental Perspective. *Adv. Energy Mater.* **2015**, *5*. [CrossRef]

29. Zhang, J.; Gao, X.; Deng, Y.; Li, B.; Yuan, C. Life Cycle Assessment of Titania Perovskite Solar Cell Technology for Sustainable Design and Manufacturing. *ChemSusChem* **2015**, *8*, 3882–3891. [CrossRef] [PubMed]

30. Celik, I.; Song, Z.; Cimaroli, A.J.; Yan, Y.; Heben, M.J.; Apul, D. Life Cycle Assessment (LCA) of Perovskite PV Cells Projected from Lab to Fab. *Sol. Energy Mater. Sol. Cells* **2016**, *156*, 157–169. [CrossRef]

31. Alberola-Borràs, J.-A.; Vidal, R.; Mas-Marzá, E.; Juárez-Pérez, E.J.; Mora-Seró, I.; Guerrero, A. Relative Impacts of Methylammonium Lead Triiodide Perovskite Solar Cells Based on Life Cycle Assessment. *Sol. Energy Mater. Sol. Cells* **2017**, *179*, 169–177. [CrossRef]

32. Lunardi, M.M.; Ho-Baillie, A.W.Y.; Alvarez-Gaitan, J.P.; Moore, S.; Corkish, R. A Life Cycle Assessment of Perovskite/Silicon Tandem Solar Cells. *Prog. Photovolt. Res. Appl.* **2017**, *25*, 679–695. [CrossRef]

33. Celik, I.; Apul, D.; Phillips, A.B.; Song, Z.; Yan, Y.; Ellingson, R.J.; Heben, M.J. Environmental Analysis of Perovskites and Other Relevant Solar Cell Technologies in a Tandem Configuration. *Energy Environ. Sci.* **2017**, *10*, 1874–1884. [CrossRef]

34. International Organization for Standardization. *Environmental Management-Life Cycle Assessment-Principles and Framework*; ISO: Geneva, Switzerland, 2006.

35. International Organization for Standardization. *Environmental Management-Life Cycle Assessment-Requirements and Guidelines*; ISO: Genewa, Switzerland, 2006.

36. European Commission. *International Reference Life Cycle Data System (ILCD) Handbook: Framework and Requirements for Life Cycle Impact Assessment Models and Indicators*; European Commission: Brussles, Belgium, 2010. [CrossRef]

37. *SimaPro Software*, version 8.5.2; PRé Consultant: Amersfoort, The Netherlands, 2018.

38. Hischier, R.; Kunst, H. *Ecoinvent 3.4 Dataset Documentation*; Ecoinvent: Zurich, Switzerland, 2011; pp. 3–6.

39. Guinée, J.B.; Gorrée, M.; Heijungs, R.; Huppes, G.; Kleijn, R.; de Koning, A.; Van Oers, L.; Wegener Sleeswijk, A.; Suh, S.; Udo de Haes, H.A.; et al. *Handbook on Life Cycle Assessment. Operational Guide to the ISO Standards. I: LCA in Perspective. IIa: Guide. IIb: Operational Annex. III: Scientific Background*; Kluwer Academic Publishers: Dordrecht, The Netherlands, 2002.

40. Goedkoop, M.; Spriensma, R. *The Eco-Indicator 99–A Damage Oriented Method for Life Cycle Impact Assessment. Methodology Report*, 3rd ed.; PRé Consultants: Amersfoort, The Netherlands, 2001.

41. Ryberg, M.; Vieira, M.D.M.; Zgola, M.; Bare, J.; Rosenbaum, R. Updated US and Canadian Normalization Factors for TRACI 2.1. *Clean Technol. Environ. Policy* **2014**, *16*, 329–339. [CrossRef]

42. Stocker, T.F.; Qin, D.; Plattner, G.-K.; Tignor, M.; Allen, S.K.; Boshung, J.; Nauels, A.; Xia, Y.; Bex, V.; Midgley, P.M. (Eds.) *Climate Change 2013: The Physical Science Basis*; Cambridge University Press: Cambridge, UK, 2013. [CrossRef]

43. European Commission. *Commission Recommendation of 9 April 2013 on the Use of Common Methods to Measure and Communicate the Life Cycle Environmental Performance of Products and Organisations. Annex II*; European Commision: Brussles, Belgium, 2013.

44. European Commission. *Final Building the Single Market for Green Products Facilitating Better Information on the Environmental Performance of Products and Organisations*; European Commision: Brussles, Belgium, 2013.

45. Frischknecht, R.; Jungbluth, N.; Althaus, H.J.; Bauer, C.; Doka, G.; Dones, R.; Hischier, R.; Hellweg, S.; Humbert, S.; Kollner, T.; et al. *Implementation of Life Cycle Impact Assessment Methods. Ecoinvent Report No. 3, v2.2*; EMPA: Dübendorf, Switzerland, 2010.

46. Rosenbaum, R.K.; Bachmann, T.M.; Gold, L.S.; Huijbregts, M.A.J.; Jolliet, O.; Juraske, R.; Koehler, A.; Larsen, H.F.; MacLeod, M.; Margni, M.; et al. USEtox-The UNEP-SETAC Toxicity Model: Recommended Characterisation Factors for Human Toxicity and Freshwater Ecotoxicity in Life Cycle Impact Assessment. *Int. J. Life Cycle Assess.* **2008**, *13*, 532–546. [CrossRef]

47. Rosenbaum, R.K.; Huijbregts, M.A.J.; Henderson, A.D.; Margni, M.; McKone, T.E.; Van De Meent, D.; Hauschild, M.Z.; Shaked, S.; Li, D.S.; Gold, L.S.; et al. USEtox Human Exposure and Toxicity Factors for Comparative Assessment of Toxic Emissions in Life Cycle Analysis: Sensitivity to Key Chemical Properties. *Int. J. Life Cycle Assess.* **2011**, *16*, 710–727. [CrossRef]

48. Parisi, M.L.; Ferrara, N.; Torsello, L.; Basosi, R. Life Cycle Assessment of Atmospheric Emission Profiles of the Italian Geothermal Power Plants. *J. Clean. Prod.* **2019**, *234*, 881–894. [CrossRef]

Article

Eco-Energetical Life Cycle Assessment of Materials and Components of Photovoltaic Power Plant

Izabela Piasecka [1], Patrycja Bałdowska-Witos [1,*], Katarzyna Piotrowska [2] and Andrzej Tomporowski [1,*]

[1] Faculty of Mechanical Engineering, University of Science and Technology in Bydgoszcz, 85-796 Bydgoszcz, Poland; izabela.piasecka@utp.edu.pl
[2] Faculty of Mechanical Engineering, Lublin University of Technology, 20-618 Lublin, Poland; k.piotrowska@pollub.pl
* Correspondence: patrycja.baldowska-witos@utp.edu.pl (P.B.-W.); a.tomporowski@utp.edu.pl (A.T.)

Received: 12 February 2020; Accepted: 12 March 2020; Published: 16 March 2020

Abstract: During the conversion of solar radiation into electricity, photovoltaic installations do not emit harmful compounds into the environment. However, the stage of production and post-use management of their elements requires large amounts of energy and materials. Therefore, this publication was intended to conduct an eco-energy life cycle analysis of photovoltaic power plant materials and components based on the LCA method. The subject of the study was a 1 MW photovoltaic power plant, located in Poland. Eco-indicator 99, CED and IPCC were used as calculation procedures. Among the analyzed elements of the power plant, the highest level of negative impact on the environment was characterized by the life cycle of photovoltaic panels stored at the landfill after exploitation (the highest demand for energy, materials and CO_2 emissions). Among the materials of the power plant distinguished by the highest harmful effect on health and the quality of the environment stands out: silver, nickel, copper, PA6, lead and cadmium. The use of recycling processes would reduce the negative impact on the environment in the context of the entire life cycle, for most materials and elements. Based on the results obtained, guidelines were proposed for the pro-environmental post-use management of materials and elements of photovoltaic power plants.

Keywords: CED; Eco-indicator 99; IPCC; LCA; photovoltaics panels; recycling; landfill

---

1. Introduction

Climate change has occurred many times in the history of the planet Earth. For the first time, however, the climate is changing faster than before. The natural greenhouse effect, necessary for life, has been intensified in modern times as a result of human activity, and the thermal balance has been significantly shaken. For the protection of the climate, the energy sector is of strategic importance as the largest consumer of energy raw materials and emitter of pollution. A sustainable energy policy should ensure that the social needs of current and future generations are best met by maintaining a balance between energy security, competitiveness of the economy and environmental protection, including the climate [1–4].

Modern civilization has become almost completely dependent on energy. Economic and social analyses indicate that the civilization changes taking place are deepening this relationship (Table 1). Energy in all forms will play an increasingly important role not only in the economic but also in the social sphere [5–7].

Table 1. Energy consumption over the centuries per person per day [7].

Key Event	Historical Period	Approximate Amount of Energy Consumed by One Person [MJ/day]
Prehistoric man (food energy)	Up to 5000 years B.C.	9–10
Man after controlling the fire	Approximately 5000 years B.C.	20
Man after using animals for work	About 3000 years B.C.	50
Man after a technical revolution	End of the 17th century	200
Modern man	XXI century	300–1000

With the current state of technological development, energy is most often obtained by processing energy raw materials, such as coal, natural gas or oil, and to a lesser extent from renewable energy sources (e.g., solar radiation, water, wind, biomass). Energy resources are not evenly distributed everywhere. Some countries do not have them at all or the resources they have at their disposal do not fully meet their energy needs. For this reason, they are forced to obtain the necessary raw materials from regions where they occur in excess. The problem with energy resources is further complicated by the fact that, in the opinion of numerous experts, their resources are limited and are running out [8–10].

One of the reasons that leads to rapid environmental degradation is excessive consumption of energy obtained from conventional sources. Pollution caused by burning fossil fuels is associated with the production of a very large amount of harmful compounds, which include SO_2, NO_x, CO_2, CO, as well as ashes and waste heat. The effects of air pollution by conventional fuel power plants include: human and animal diseases, destruction of vegetation, destruction of building structures (including historic buildings), metal corrosion and increased machine wear, etc. [11–13].

The need to increase the share of renewable energy in the energy balance of each country results from the obligation to reduce CO_2 and other greenhouse gas emissions as a result of the growing greenhouse effect, the need to replace depleting fossil fuel resources with other energy sources, and the desirability of reducing dependence on energy suppliers from other countries [14,15].

Solar installations are becoming increasingly popular around the world. The advantage of photovoltaic cells is undoubtedly that their long-term, trouble-free operation allows for a significant reduction of harmful emissions. However, their production is very energy-intensive, which entails the emission of combustion products. Due to the presence of heavy metals in PV panels, their future recycling may also become a problem. The advantage of this branch of energy is the ubiquity of the sun's rays, environmental friendliness and inexhaustibility. However, disadvantages include daily and annual cyclicity, radiation dispersion and significant costs of the equipment used [16–19].

In the global literature, many analyses can be found, mainly regarding the evaluation of solar panels with particular emphasis on the conditions of the production process. Kumar et al. (2018) indicated that the effect of shape of abrasive and silicon crystal is relevant to the yield and the life cycle of the solar cells over 20+year lifetime. In addition, the production process and the materials and raw materials used are very important for the quality of the solar farm. A detailed analysis of quality and durability was carried out by Kumar et al. (2017), proving that the impact of diamond wire wear impact on surface morphology, roughness and properties of silicon wafer subsurface. Despite numerous publications describing the stages of the production process, no studies have been found on the impact of selected system elements on the condition and development of the natural environment. Therefore, the following hypothesis is worth considering: which of the analyzed elements of the solar plant show the highest level of negative impact on the environment?

The pro-ecological attitude adopted by the authors is aimed at reducing gas emissions due to the operation of PV cells, which must correspond with environmentally friendly technology for producing photovoltaic cells. To this end, the authors made a detailed LCA (Life Cycle Assessment) analysis showing stages throughout the life cycle of the system that have a negative impact on the environment.

Increasing care for nature leads to the development and use of increasingly complex methods that give control of, and the ability to counteract, the human impact on the environment. Therefore, many new ways of assessing the impact of processes, products and industries on the environment have been created. One of them is the method of analysis and assessment in the context of the entire life cycle of products, i.e., their impact from the acquisition of raw materials to development. The Life Cycle Assessment (LCA) method covers the environmental impact of production, operation and post-use management and is in accordance with the principle of sustainable development [20–22].

Each source of energy, even classified as renewable, has a certain impact on the environment. Photovoltaics are widely regarded as a "green", environmentally friendly energy source. During photovoltaic installation exploitation, solar radiation is converted into electricity. This process does not cause emissions of harmful substances into the environment, unlike analogous ones, when using conventional resources (e.g., emissions of CO_2, SO_2, NO_x, dust, etc., as a result of burning coal). The fact that the production and post-use development of plastics and components of photovoltaic power plant is usually overlooked is the need for large material expenditures, for example related to the extraction of raw materials for the production of plant components or chemicals necessary for recycling processes. In addition, the accompanying processes (for example production of PV cells using the Czochralski method) are extremely energy-consuming. During the entire lifecycle of a photovoltaic installation, many compounds and chemicals are emitted that can have a negative impact on the environment, and large amounts of energy are required (especially at the production stage). In view of the above, main target of this study is an ecological and energetical life cycle assessment of materials and components of photovoltaic power plant.

2. Materials and Methods

2.1. Object and Plan of Analysis

The object of this study is a photovoltaic power plant with a capacity of 1 MW, situated in the northern Poland, which produces from 950 to 1100 MWh of electricity per year. As a reference for the purpose of further analyses, it was assumed that the system produced 1,000 MWh per year. The basic elements of the parsed photovoltaic power plant are: supporting structures, photovoltaic panels, cables and straight connectors for electrical installations, container station along with the static inverters (including DC switchgear, DC/AC inverters, AC/LV switchgear, LV/MV transformer, MV switchgear, control and surveillance system, the system of measurement of energy generated).

An LCA study (in accordance with ISO 14000) consists of four stages: determination of goal and scope, life cycle inventory (LCI), life cycle impact assessment (LCIA) and interpretation (Figure 1) [23–25].

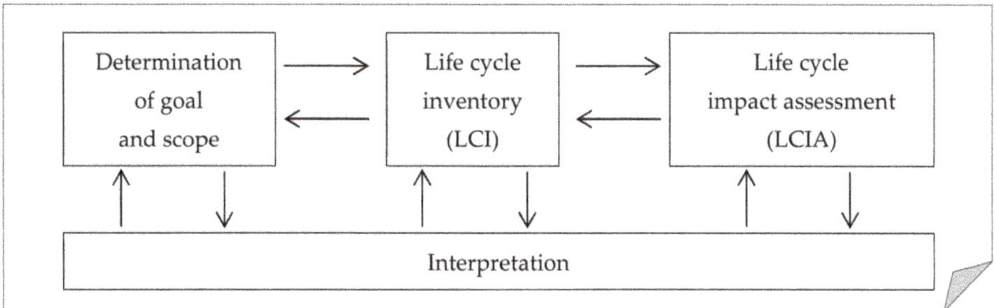

Figure 1. Life Cycle Assessment (LCA) framework.

In accordance with the mentioned ISO standards, the analysis plan consisted of four basic steps. In the first of them, the purpose of the study and its scope were determined, which are described in detail in Section 2.2. The basis for their formulation was the collection of the largest possible amount of data on the studied object. Key data for the analysis were provided by the owner of a 1 MW photovoltaic power plant located in Poland. In addition, detailed information on the production and operation of photovoltaic panels, inverter stations, cables and cabling accessories was obtained from their manufacturers. We also managed to obtain data from a post-use PV panel management company. Details on the second step of the analysis are provided in Section 2.3. The third step involved performing a comprehensive ecological and energetical analysis of the life cycle of the photovoltaic power station under study. SimaPro 8.4 software (PRé Sustainability, LE Amersfoort, Netherlands) was used for this purpose. The basic calculation procedure was the Eco-indicator 99 method, which allows the assessment of the impact of the photovoltaic power plant's life cycle on the environment, including human health, ecosystem quality, and resources. To determine the energy demand and CO_2 emissions at each stage of the life cycle, in addition the CED and the IPCC methods were used. The IPPC method was used for CO_2 emission quantitative assessment to indicate which of the material stages of the life cycle involves the highest level of greenhouse gas emissions (CO_2), the reduction of which is one of the key aims of the European Union countries. The CED method was used to identify the share of a photovoltaic power plant life stages in the energy demand from different sources. In this way, a comprehensive analysis of a photovoltaic power plant cycle, including important areas of sustainable development, that is, CO_2 emission (IPCC method), energy consumption (CED method), human health, ecosystem quality and resources (Eco-indicator 99 method), was provided.

The characteristics of this stage of the analysis are described in Section 2.4 and the results are presented in Sections 3.1–3.5. The last, fourth step was the interpretation of the results of the analysis, which are presented in Sections 2.5 and 4 ("Conclusions").

2.2. Determination of Goal and Scope

This work analyzes several single products connected in a one system—a photovoltaic power station. The analysis, which was conducted as part of this publication, aimed at the numerical determination of the value of the environmental impact related to the life cycle of a 1 MW PV power station. The purpose of the analysis is, most of all, to describe the existing reality (retrospective LCA), but also to model future changes and determine recommendations aimed at developing more pro-environmental solutions (prospective LCA). The procedure will constitute a classic process LCA, the purpose of which will be to determine the extent of the negative environmental impact of the life cycle of the analyzed object [26–28]. For this purpose, selected elements of the solar installation were analyzed: photovoltaic panels, supporting structures, inverter station, electrical installations. The environmental assessment included 11 impact categories: Carcinogens, Resp. Organics, Resp. Inorganics, Climate change, Radiation, Ozone layer, Ecotoxicity, Acidification / eutrophication, Land use, Minerals, Fossil fuels. The research results are divided into four phases, described as: production, exploitation, landfill and recycling. Among the eleven categories available, categories with the highest level of significance were selected for which detailed emissions of compounds into the environment were presented.

Most of the processes performed under the analyzed stages of the life cycle of the photovoltaic power station (production, exploitation, post-use management) take place in Europe. Therefore, the study scope was referenced to the European conditions. The territory of Poland was taken as the geographical area, while the time horizon taken was 20 years (average operation time of photovoltaic systems). Electric energy production was assumed as a function of the photovoltaic power plant. A functional unit was defined as a production of 1000 MWh of electric power by the relevant system in a year. The analysis did not cover the stages of transport, sales, technical tests, and storage. The main reason for this was the lack of appropriate data and large differences in the effects of transport depending on the power plant location.

2.3. Life Cycle Inventory (LCI)

To collect data, special sheets were prepared. Each sheet was assigned to a specific unit process, with a division into process inputs, process performance, and process outputs (Figure 2). Process inputs included main materials, auxiliary materials, and water; process performance involved duration and media consumption; process outputs included main product, waste, and emissions. Data concerning processes and materials less significant from the point of view of environmental impact were obtained from databases included in the SimaPro 8.4 software(PRé Sustainability, LE Amersfoort, Netherlands). Due to confidentiality agreements with companies manufacturing photovoltaic power station elements, any detailed information regarding the design of the analyzed objects and process data are not subject to disclosure in this publication [29,30].

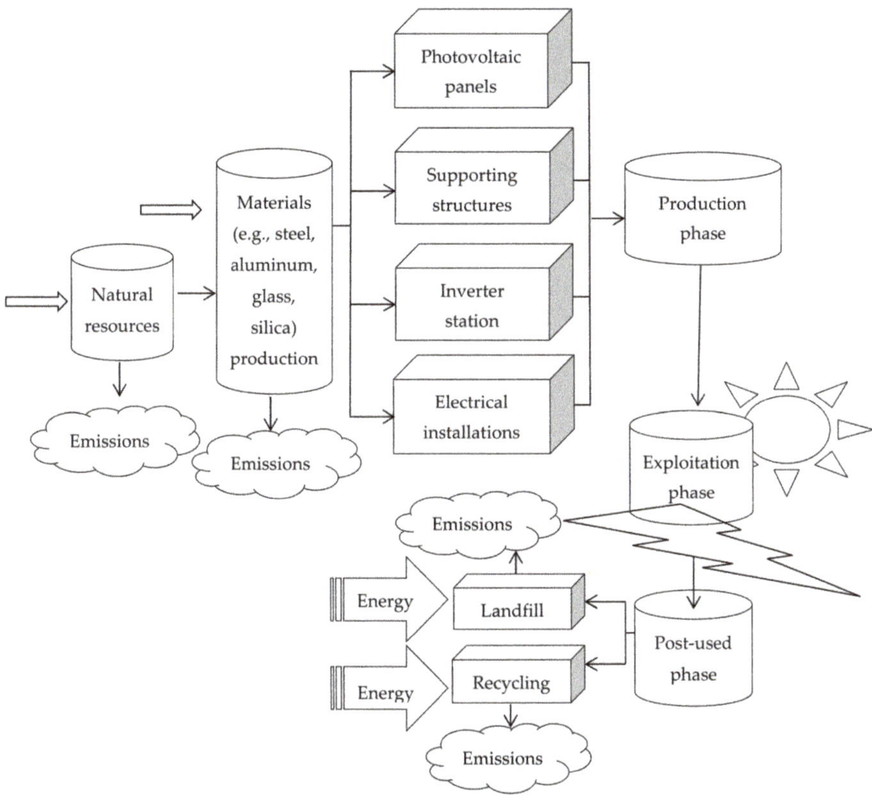

Figure 2. The material life cycle of photovoltaic power plants.

After the data were assigned to the unit processes, they were validated through bilateral energy and mass balance. Models were constructed systematically and filled with data. The input value was equilibrated by the output value. This operation made data aggregation and quantification per functional unit and reference flows possible. By totaling environmental interventions of the same type (inputs of material, energy, waste, emissions, etc.) for all the unit processes, input–output matrices were obtained that were referenced to the reference flows. The next step was to adapt them to a format compatible with the SimaPro 8.4 software. This allowed us to enter data into a calculator and proceed with another stage of the analysis. Information supplied by the manufacturer allowed a precise determination of the values of the materials and energy used in the photovoltaic power station life cycle [31–33].

The relevant power station was equipped with photovoltaic panel support structures made of galvanized steel (mainly due to economic benefits and numerous technical advantages). The support structure in a dual system was placed directly in the ground. Properly selected photovoltaic panels constitute a key element of the entire photovoltaic plant. The construction of the relevant power plant needed a system of 4170 polycrystalline photovoltaic modules with a capacity of 240 W demonstrating a performance up to 17.7%. The manufacturer's declared performance amounts to 91.2% of rated capacity for the first 10 years and 80.7% for a period of another 15 years. Each single module is composed of 60 photovoltaic cells. The panels are made of glass, aluminum, silica, EVA (ethylene-vinyl acetate copolymer), PVB (polyvinyl butyral), cadmium, lead, copper, nickel, selenium, and silver. The modules are connected in series and are inclined at an angle of 35° in the southern direction (this is due to the fact that the angle of the PV panels tilt relative to the horizon depends on the latitude (φ). The location of the analyzed object is 54°N. The angle of the sun above the horizon is H = 90 °-φ; therefore, for the location considered it will be 36°. The highest efficiency is obtained with a perpendicular angle of incidence of sunlight on the panel surface. In Poland, for year-round operation, the optimal angle for placing PV panels is about 30-40° and pointing southwards). To function properly, each photovoltaic system should be equipped with proper wires and cabling accessories. The relevant plant, located on the ground, employed wires with pre-terminated ends (galvanized copper conductor, internal insulation and external sheath made of cross-linked polyolefin), control cables and wires, different types of connectors and splitters (galvanized copper contacts), cable glands (body: PVDF—polyvinylidene fluoride and polyamide PA6; seal: silicone or neoprene) and protection hoses (modified polyamide PA12). In the relevant solution of photovoltaic power station, a central inverter substation was used. Megawatt substation constitutes a comprehensive solution dedicated for solar power plants with a high installed capacity. It contains the electrical equipment necessary to connect photovoltaic power stations to a medium voltage electric power network. The station accommodates two central inverters, optimized transformer, MV switchgear, DC connections for PV modules, and a monitoring system. Made of steel, the insulated container is placed on a concrete base. The entire megawatt substation weighs nearly 20 tons, while the volume of the container amounts to almost 50 m^3 (manufacturer's data).

2.4. Life Cycle Impact Assessment (LCIA)

Life cycle impact assessment was performed with the use of the SimaPro 8.4 calculation software. Cut-off level amounted to 0.1%. LCIA results are presented in Section 3 of this article.

2.4.1. Eco-indicator 99 method

Eco-indicator 99 method was chosen as the base calculation procedure. Eco-indicator 99 belongs to a group methods for modeling the environmental impact of environmental endpoint mechanism. The process of characterization is done for the eleven categories of impact, coming within three larger groups referred to as impact areas or categories of damages. There are the following areas of impact: human health, ecosystem quality, and resources. The results of the impact area indicators are further analyzed through normalization, grouping and weighting into the final Ecolabel. Eco-indicator 99 method offers 11 impact categories with a wide spectrum of analysis areas (Figure 3). The first damage category (carcinogens, resp. organics, resp. inorganics, climate change, radiation, ozone layer) is expressed in DALY (Disability Adjusted Life Years)—the number of years spent in the disease or lost. Assessment of the impact on the environment was performed with the use of a scale from 0 to 1, where 0 stands for a lack of impact on the human health and 1 stands for death. The damage to ecosystem quality is expressed in terms of the percentage of species that have disappeared in a certain area due to the environmental load. Ecotoxicity covers the percentage of all species present in the environment living under toxic stress (PAF—Potentially Affected Fraction). Regarding acidification/eutrophication and land use, the damage to a specific target species (vascular plants) in natural areas is modeled (PDF—Potentially Disappeared Fraction). The damage category covering resource extraction gives a value expressed in MJ surplus energy to indicate the quality of the remaining mineral and fossil resources. The final goal of the grouping and weighting analysis was to obtain environmental factors expressed in environmental points (Pt), constituting aggregated units enabling comparisons of eco-balance sheets. A thousand environmental points are equal to the impact on one's environment, the average European within a year. The value of 1 Pt (eco-point) is representative for one thousandth of the yearly environmental load of one average European inhabitant. It is calculated by dividing the total environmental load in Europe by the number of inhabitants and multiplying it with 1000. Due to the lack of express premises for exclusions, all the impact categories functioning within the Eco-indicator 99 (Eco-indicator 99 (H) V2.06/Europe EI 99 H/A) were subject to analysis [34–37].

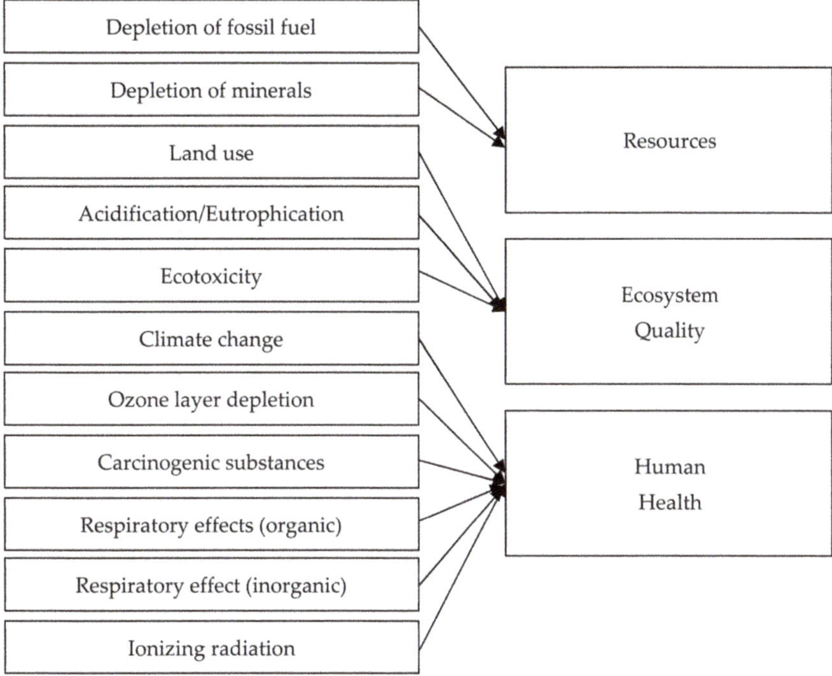

Figure 3. Structure of LCA impact category groupings, Eco-indicator 99 method.

Due to the fact that the purpose of the research was to carry out ecological and energy life cycle analysis of photovoltaic power plant materials and components, in addition to the Eco-indicator 99 method, it was decided to additionally use two other methods—CED and IPCC. As part of Eco-indicator 99, two impact categories ("minerals" and "fossil fuels") refer to energy analyzes (value expressed in MJ surplus energy), but due to the need for expand the scope of their research with an additional assessment of cumulative energy demand in each phase lifecycle. Therefore, the CED method was used, as described in Section 2.4.3. Nowadays, great attention is also paid to the issue of excessive CO2 emissions, which is why it was decided to use another method, IPCC, which makes it possible to assess the impact of greenhouse gases on the increase of the greenhouse effect and obtain results for each stage of the life cycle expressed in kg CO2 eq. As a result of this procedure, the results obtained under the category "climate change" in the Eco-indicator 99 method are more detailed. A more detailed description of the IPCC method is provided in Section 2.4.2.

2.4.2. IPCC Method

The IPCC (Intergovernmental Panel on Climate Change, Global Warming Potential) method has made it possible to perform a quantitative assessment of the impact of particular greenhouse gasses (GHG) for the greenhouse effect, with respect to CO_2. The carbon dioxide indicator in order to assess the impact on the greenhouse effect is equal to 1 (GHG = 1). This research was conducted in accordance with the IPCC standard: IPCC 2007 GWP 100a V1.01 (Intergovernmental Panel on Climate Change, Global Warming Potential, time horizon: 100 years) [38–41].

2.4.3. CED Method

The CED (Cumulative Energy Demand) method allows the determination of the cumulative energy demand. The impact indicators are divided into seven impact categories: two non-renewable (nuclear power, fossil fuels) and five renewable (biomass, water, solar, wind and geothermal energy). The research was conducted in accordance with the CED standard: Cumulative Energy Demand V1.05 [42–44].

2.5. Interpretation

During the analysis, its completeness was checked against the positive result. All the important information and data necessary for interpretation were complete and obtainable directly from the manufacturer, the recycling company, and from the databases of the SimaPro software (PRé Sustainability, LE Amersfoort, Netherlands). Conformity was checked during the analysis. Assumptions, methods, analysis depth, specificity and precision of data for both systems are compliant with the previously assumed goal and scope of analysis. Detailed interpretation of the results obtained is presented in Sections 3 and 4.

3. Results and Discussion

3.1. Eco-Indicator 99

Table 2 presents the results of characterizing the environmental consequences occurring in the life cycle of selected components of the 1 megawatt photovoltaic power plant. Impacts are presented in the 11 categories of impact characteristic of the Eco-indicator 99 method. Two impact models were distinguished: the first one was the life cycle, including landfill disposal as a form of post-disposal management, while the second one was recycling. For all adverse effects in the area of human health, all tested groups showed the highest negative impact on the category of inorganic compounds causing respiratory diseases (e.g., life cycle of photovoltaic panels with landfill: 0.22 DALY; inverter life cycle with storage: 0.14 DALY). The largest quantity is formed at the stage of production of materials and elements, and the maximum share is characterized by sulfur dioxide and nitrogen oxides. These compounds are poisonous to humans and animals and have a harmful effect on plants. Sulfur dioxide is a by-product of burning fossil fuels, which, for example, contributes to atmospheric pollution (smog). In turn, nitrogen oxide is a compound with high biological activity and easily penetrates biological membranes. It is also created, among other things, as a result of burning fossil fuels and industrial processes that can cause smog. For categories affecting environmental deterioration, the ecotoxic compounds category was the most important, for which the maximum level of harmful impact was recorded for photovoltaic panels deposited in landfill: 137,741 PAF·m^2/a. Ecotoxic compounds are substances which, due to their origin, chemical, biological or other properties, constitute or may pose a direct or delayed threat to humans, animals and plants. In the life cycle of a photovoltaic power plant, the largest amount arises from the storage of materials and components at the landfill. A particular threat is copper ion emissions, which in addition to reducing the quality of the environment, can contribute to the formation of diseases of the nervous and digestive systems in humans (for example: mental disorders or liver damage). In the area of processes affecting the depletion of raw material resources, the highest level of harmful impact was recorded in the raw materials category for the life cycle of electrical installations with post-consumer use in the form of landfill (322,646 MJ), and in the fossil fuels category for the life cycle of photovoltaic panels placed in landfill after use (400,584 MJ). The largest amount of fossil fuels is consumed during the plastics and materials plant production phase. The most processes with the highest energy demand include, for e.g., the production of PV cells. Burning of conventional fuels is associated with many hazardous emissions to the atmosphere, water and soil, which are the causes of, for examplem, diseases, increasing the greenhouse effect, ozone layer depletion or increased smog and acid rain. In most of the categories considered, there is a positive impact of the use of recycling processes, to reduce the harmful impact of particular groups of components of the analyzed photovoltaic power plant.

Table 2. Results of characterization of environmental consequences, occurring in the life cycle of selected groups of 1 megawatt photovoltaic power plant.

Impact Category	Photovoltaic Panels		Supporting Structures		Inverter Station		Electrical Installations		Unit
	Landfill	Recycling	Landfill	Recycling	Landfill	Recycling	Landfill	Recycling	
Carcinogens	0.07	-0.01	0.03	0.00	0.02	0.00	0.00	0.00	DALY
Resp. organics	0.00	0.00	0.00	0.00	0.00	0.00	0.00	0.00	DALY
Resp. inorganics	**0.22**	-0.03	0.02	0.01	**0.14**	0.10	0.05	0.05	DALY
Climate change	0.05	-0.03	0.01	0.00	0.02	0.00	0.00	0.00	DALY
Radiation	0.00	0.00	0.00	0.00	0.00	0.00	0.00	0.00	DALY
Ozone layer	0.00	0.00	0.00	0.00	0.00	0.00	0.00	0.00	DALY
Ecotoxicity	**137,741**	-5015	106,619	48,706	60,968	21,959	4326	309	PAF·m^2/a
Acidification/ eutrophication	6650	1234	546	236	4412	3517	1179	1138	PDF·m^2/a
Land use	3605	3458	657	610	4521	4485	1189	1185	PDF·m^2/a
Minerals	68,116	-573	1692	8	10,203	-645	**322,646**	322,644	MJ
Fossil fuels	**400,584**	32,805	25,454	13,088	79,294	24,498	24,074	22,009	MJ

The results of grouping and weighting the environmental after-effects of the existence of selected groups of 1 MW photovoltaic power plants are summarized in Table 3. For the life cycle of photovoltaic panels deposited in landfill, the highest level of harmful impact was recorded in terms of: fossil fuel extraction (9534 Pt), inorganic compounds causing respiratory diseases (5729 Pt), mining of minerals (1621 Pt) and carcinogenic compounds (1822 Pt). Silicon cell production processes are associated with a huge demand for energy, which in the case of the analyzed power station is about 5 million MJ. This energy is most often obtained from non-renewable sources, which causes many negative impacts in relation to human and animal health, and a reduction in the quality of the environment. Another problem is the excessive exploitation of raw material deposits, including silver, used in the electrical contacts of the cells, whose extraction causes the most negative consequences in comparison to other substances and chemical compounds used in the production of PV cells. For the life cycle of supporting structures, including post-use disposal, the categories were: ecotoxic compounds (832 Pt), carcinogenic compounds (827 Pt), and fossil fuel extraction (606 Pt). Supporting structures were made mainly of galvanized steel. The galvanizing process of steel poses a threat to health and the environment. In thermal processes, smoke and zinc vapors (especially particles smaller than 1 μm) can, for example, get into the respiratory system, causing many diseases. In the life cycle of the inverter station including landfill disposal, the highest levels of negative impact were reported in terms of: inorganic compounds causing respiratory disease (3725 Pt), and fossil fuel extraction (1887 Pt). In the cycle of existence of an electrical installation, which after ending its life cycle will be deposited in landfill, the most negative influence on the environment was attributed to mineral extraction (7679 Pt) and inorganic compounds causing respiratory diseases (1327 Pt). An important element found in both inverter station and electrical installation are electrical cables and wires. The main raw material for their production is copper, the extraction and processing of which is associated with very high energy inputs, obtained from conventional sources. As a consequence, many harmful compounds are emitted into the environment, and the resources of raw materials and fuels are depleted. The application of recycling would reduce the harmful impact on the environment of all analyzed groups of photovoltaic elements.

Table 3. Results of grouping and weighting environmental consequences, occurring in the cycle of existence of selected groups of 1 megawatt photovoltaic power plant [unit: Pt].

Impact Category	Photovoltaic Panels		Supporting Structures		Inverter Station		Electrical Installations	
	Landfill	Recycling	Landfill	Recycling	Landfill	Recycling	Landfill	Recycling
Carcinogens	**1822**	-359	**827**	79	557	68	53	1
Resp. organics	7	-8	1	1	4	2	0	0
Resp. inorganics	**5729**	-667	431	219	**3725**	2703	**1327**	1299
Climate change	1431	-810	358	25	502	63	96	88
Radiation	10	9	8	8	12	12	0	0
Ozone layer	1	-2	0	0	0	0	0	0
Ecotoxicity	1074	-39	**832**	380	476	171	34	2
Acidification/ eutrophication	519	96	43	18	344	274	92	89
Land use	281	270	51	48	353	350	93	92
Minerals	**1621**	-14	40	0	243	-15	**7679**	7678
Fossil fuels	**9534**	781	**606**	311	**1887**	583	573	524
Total	22029	-743	3197	1089	8103	4210	9947	9774

The highest total level of harmful impact on the environment is the life cycle of photovoltaic panels ending in landfill storage (22,029 Pt), but in this case, the use of recycling processes significantly reduces their negative impact on the environment. The key reason for this is the abovementioned very high energy demand in the production of PV cells. Reuse of cells recovered in the recycling process is associated with large savings in both energy and materials (e.g., elimination of significant material losses arising during cutting silicon rollers). The lowest total damaging effect was found in the supporting structures (1089 and 3197 Pt) (Figure 4).

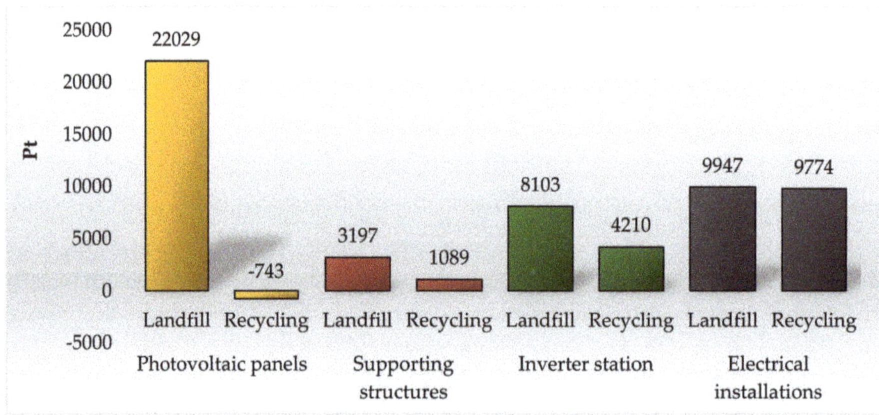

Figure 4. Results of grouping and weighting of environmental impacts occurring during the material life cycle of selected groups of 1 megawatt photovoltaic power plant for all categories of influence [unit: Pt].

The highest level of harmful impact on the environment of 1 MW photovoltaic power plant was observed at the manufacturing stage. The largest category of harmful effects was characterized by the categories of fossil fuel extraction and minerals, while the smallest – compounds causing the increase of the ozone hole. The largest share in the negative impact on the environment was characterized by the production processes of PV panels, which are part of the photovoltaic power plant with the highest demand for energy and materials. The lowest level of negative impact was the exploitation stage (32 Pt total). By comparing the received forms of post-use management, the most negative influence on the environment is the landfill. Carcinogenic compounds and ecotoxic compounds can be classified as the most potent adverse effects. The use of recycling processes would reduce the impact of the life cycle in most of the impact categories analyzed. The fossil fuel mining processes and emissions of inorganic compounds affecting respiratory diseases would be most positively affected (Figure 5).

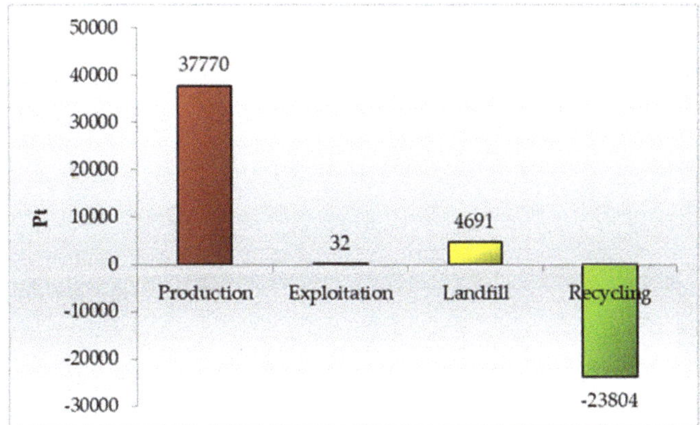

Figure 5. Grouping and weighting results of environmental impacts occurring in the material life cycle stages of a 1 megawatt photovoltaic power plant [unit: Pt].

Post-use disposal of photovoltaic power plant components could be the most important source of oncogenic factors contributing to the development of cancer by mutation of genetic material. The highest level of harmful emissions was recorded for cadmium ions (2547 Pt). Cadmium is an element that easily concentrates in air, water and soil and quickly moves in the soil–plant–human trophic chain. Due to easy absorption and bioaccumulation in living organisms and toxic influence, it is one of the most serious threats to the natural environment and man. For this reason, it is important to minimize the storage of materials and components of solar power plants in landfills. The production phase also has a significant impact on the carcinogenic emissions—538 Pt total—and the highest share of arsenic: 363 Pt. The use of recycling would reduce the harmful impact of carcinogens by a total of 824 Pt, mainly in the range of arsenic (-514 Pt) (Table 4).

Table 4. Results of grouping and weighting environmental after-effects for carcinogenic compounds present in the 1 megawatt photovoltaic power plant [unit: Pt].

Carcinogens	Production	Exploitation	Landfill	Recycling
Cadmium, ion	22.98	0.21	**2547.83**	-19.67
Arsenic, ion	**363.28**	2.44	104.07	**-514.43**
Cadmium	68.65	0.31	0.87	-48.94
Metals, unspecified	43.84	0.00	0.00	-147.56
PAH, polycyclic aromatic hydrocarbons	10.27	0.00	0.00	-57.76
Particulates, < 2.5 μm	13.06	0.03	0.32	0.00
Metallic ions, unspecified	6.58	0.00	0.00	-36.10
Arsenic	9.48	0.03	0.16	0.00
Total	538.14	3.02	2653.26	-824.46

The production stage is distinguished by the highest level of negative impact in organic compounds causing respiratory diseases, altogether 11 Pt, especially in the emissions of non-methane volatile organic compounds (10 Pt). NMVOC is a group of organic compounds that occur as by-products in many industrial processes and are a source of environmental pollution, including, for example: acetone (paints, protective covers, sealants), aliphatic hydrocarbons (paints, glues, sealants, combustion processes), aromatic hydrocarbons (paints, glues, combustion processes), polycyclic aromatic hydrocarbons (paints, polymeric materials, incomplete combustion processes, e.g., car exhaust gas) or chlorine-containing compounds (varnishes, solvents). During storage of the power plant components in landfill, biogenic methane may be the greatest risk: 1 Pt. Recycling processes would reduce the damaging impact of the total lifetime by 17 Pt, including non-methane volatile organic compounds by 16 Pt (Table 5).

Table 5. Results of grouping and weighting of environmental after-effects for organic compounds causing respiratory diseases, occurring in the 1 megawatt photovoltaic power plant [unit: Pt].

Resp. Organics	Production	Exploitation	Landfill	Recycling
NMVOC, non-methane volatile organic compounds	**10.08**	0.00	0.11	**-16.69**
Methane, biogenic	0.01	0.00	0.59	0.00
Hydrocarbons, unspecified	0.32	0.00	0.00	0.00
Methane, fossil	0.20	0.00	0.03	0.00
Methane	0.16	0.00	0.00	-0.31
Ethane	0.20	0.00	0.00	0.00
Hydrocarbons, aromatic	0.03	0.00	0.00	-0.11
Pentane	0.03	0.00	0.00	0.00
Propane	0.02	0.00	0.00	0.00
PAH, polycyclic aromatic hydrocarbons	0.02	0.00	0.00	-0.09
Total	11.10	0.00	0.73	-17.21

The highest level of harmful impact on the environment of respiratory organisms caused by respiratory diseases occurring during the 1 MW photovoltaic power plant cycle was recorded for the production stage–10,837 Pt in total. The highest share was in sulfur dioxide (3194 Pt) and nitrogen oxide (3091 Pt). Recycling would allow a total reduction of harmful effects of 7423 Pt, mainly in the sulfur oxide (-3274 Pt) and particulates in total (-2521 Pt). A key negative health role is played by atmospheric aerosols or particulate matter (PM). They are drops or solid particles of natural or anthropogenic origin (impurities). PM 2.5 (particle size 2.5 μm or smaller) is the most harmful because prolonged exposure to it results in a reduction in life expectancy, while short-term exposure to high concentrations causes an increase in deaths from respiratory and circulatory diseases and increases the risk of emergencies that require hospitalization (for example: worsening of asthma, decreased lung function), because dust enters the blood directly through the lungs. Atmospheric aerosols also contribute to smog (Table 6).

Table 6. Results of grouping and weighting of environmental after-effects for inorganic compounds causing respiratory diseases, occurring in the 1 megawatt photovoltaic power plant [unit: Pt].

Resp. Inorganics	Production	Exploitation	Landfill	Recycling
Sulfur dioxide	**3193.91**	3.42	7.30	0.00
Nitrogen oxides	**3090.90**	2.35	37.28	-1627.29
Sulfur oxides	1453.19	2.01	0.00	**-3273.87**
Particulates, > 2.5 μm, and < 10 μm	1214.04	0.21	3.14	0.00
Particulates	531.42	0.03	0.00	**-2520.56**
Particulates, < 2.5 μm	1023.94	2.53	23.95	0.00
Ammonia	196.78	0.00	0.39	-1.18
Particulates, < 10 μm (stationary)	82.79	0.06	0.00	0.00
Nitrogen dioxide	27.70	0.00	0.00	0.00
Particulates, < 10 μm (mobile)	22.70	0.00	0.00	0.00
Total	10837.39	10.61	72.07	-7422.90

The highest level of emissions of substances causing climate change is the production of materials, and components, which results in a total of 2049 Pt, mainly composed of carbon dioxide (891 Pt). Over the past two centuries there has been a marked acceleration of climate change. The basic factor shaping the speed of these changes is the emission of carbon dioxide and other greenhouse gases, which arise, for example: during the exploitation of fossil fuels, used in the life cycle of a photovoltaic power plant most often for obtaining electricity for various processes, including the energy most intensive—PV cell production. Landfill disposal can result in total negative emissions of 292 Pt, primarily biogenic methane (201 Pt). Recycling, as a form of post-use management, would allow the reduction of dangerous emissions by a total of 2685 Pt, mainly in carbon dioxide (-1947 Pt) (Table 7).

Table 7. Results of grouping and weighting of environmental after-effects for compounds causing climate change, occurring in the 1 megawatt photovoltaic power plant [unit: Pt].

Climate Change	Production	Exploitation	Landfill	Recycling
Carbon dioxide	**890.86**	0.30	0.00	**-1946.84**
Carbon dioxide, fossil	878.77	2.89	13.97	0.00
Methane, biogenic	2.03	0.00	**201.40**	0.00
Tetrafluoromethane, CFC-14	122.12	0.00	0.00	-606.04
Methane, tetrafluoro-, CFC-14	40.81	0.03	65.20	0.00
Carbon dioxide, biogenic	67.46	0.15	10.73	0.00
Methane, fossil	54.93	0.01	0.00	-107.33
Methane	22.61	0.00	0.67	0.00
Carbon monoxide	13.47	0.00	0.00	-25.05
Carbon dioxide, in air	-44.27	-0.04	-0.08	0.00
Total	2048.80	3.35	291.90	-2685.26

Radioactive compounds are characterized by possessing nuclear nuclei with radioactive decay, most commonly associated with alpha particle emission, beta particles, and gamma radiation. The highest number of such elements in the 1 MW photovoltaic power plant cycle was noted for the production phase (total 28 Pt). It is associated with the processes of extracting mineral resources and fossil fuels, during which there are not only emissions of dust and gases containing many harmful substances (e.g., sulfur and nitrogen oxides, chlorine, fluorine, heavy metals), but also radioactive elements such as uranium, thorium, and potassium, and their breakdown products, for example: radium and radon. In this case, these are mainly radon isotopes—^{222}Ra (21 Pt)—and carbon—^{14}C (7 Pt) (Table 8).

Table 8. Results of grouping and weighting of environmental after-effects for radioactive compounds, occurring in the 1 megawatt photovoltaic power plant [unit: Pt].

Radiation	Production	Exploitation	Landfill	Recycling
^{222} Radon	**20.95**	0.02	0.42	0.00
^{14} Carbon	**6.52**	0.00	0.22	0.00
^{137} Cesium	0.56	0.00	0.00	0.00
^{60} Cobalt	0.07	0.00	0.00	0.00
^{134} Cesium	0.05	0.00	0.00	0.00
^{85} Krypton	0.04	0.00	0.00	0.00
Total	28.19	0.02	0.64	0.00

The ozone hole is a phenomenon of a decrease in the concentration of ozone (O_3) in the stratosphere, resulting in a decrease in the level of absorption of ultraviolet radiation reaching the Earth from the Sun. It is, therefore, a threat to living organisms. During the processes of producing materials, and components of the plant under investigation, harmful substances causing the ozone hole to increase in total 2 Pt are formed, mainly bromotrifluoromethane (1,5 Pt). Halon 1301 may have a toxic effect on the central nervous system and other bodily functions (Table 9).

Table 9. Results of grouping and weighting of environmental after-effects for ozone-increasing compounds, occurring in the 1 megawatt photovoltaic power plant [unit: Pt].

Ozone Layer	Production	Exploitation	Landfill	Recycling
Methane, bromotrifluoro-, Halon 1301	**1.48**	0.00	0.01	-1.17
Methane, bromochlorodifluoro-, Halon 1211	0.30	0.00	0.00	0.00
Methane, tetrachloro-, CFC-10	0.15	0.00	0.00	0.00
Ethane, 1,2-dichloro-1,1,2,2-tetrafluoro-, CFC-114	0.06	0.00	0.00	0.00
Methane, chlorodifluoro-, HCFC-22	0.03	0.00	0.00	0.00
Total	2.03	0.00	0.01	-4.17

Comparing all phases of the life cycle, particularly high levels of harmful impact of waste disposal in the form of waste landfills are visible in the category of ecotoxic compounds (a total of 1503 Pt). The most significant level of negative emissions was copper ions (1200 Pt). The high level of emissions of ecotoxic substances is also characterized by the production phase (total 837 Pt), which consists mainly of the harmful effects of nickel (280 Pt) and zinc (263 Pt). Recycling could significantly reduce emissions by a total of -397 Pt, primarily in terms of minimizing the negative impact of nickel (-212 Pt). Nickel is used in the manufacture of many materials and components of a photovoltaic power plant, ranging from steel elements to resistors. The main source of nickel in the environment is the combustion of conventional fuels (especially coal and oil), as well as steel production and electroplating processes. The absorption of nickel into the body is primarily through the respiratory system. Nickel tends to accumulate in the lungs. With wind and rain, it gets into soil and groundwater. It is also one of the components of smog (Table 10).

Table 10. Results of grouping and weighting of environmental after-effects for ecotoxic compounds, occurring in the 1 megawatt photovoltaic power plant [unit: Pt].

Ecotoxicity	Production	Exploitation	Landfill	Recycling
Copper, ion	7.91	0.05	**1199.53**	-8.44
Zinc	**263.02**	0.36	1.04	-20.11
Nickel	**280.00**	0.04	0.51	**-211.57**
Zinc, ion	0.40	0.00	120.31	0.00
Nickel, ion	16.93	0.05	102.76	-8.33
Chromium	128.99	0.02	1.37	-7.28
Lead	61.60	0.04	26.34	30.08
Cadmium, ion	0.10	0.00	51.01	0.00
Metals, unspecified	40.87	0.00	0.00	-161.86
Copper	20.32	0.03	0.17	1.21
Cadmium	12.54	0.07	0.00	-10.48
Total	836.87	0.67	1503.04	-396.76

Acidification of the environment is a phenomenon of progressive decrease in the pH value of its individual components. It can be caused by anthropopressure, for example: by emissions of air pollutants (e.g., SO_2, NO_x, NH_3) as a result of combustion of conventional fuels. Eutrophication, on the other hand, consists of enriching the environment with biophilic elements, mainly phosphorus, which causes an excessive increase in their trophic (biological productivity). Materials and elements of a 1 megawatt photovoltaic power plant can be distinguished primarily by nitrogen oxide (596 Pt) and sulfur dioxide (182 Pt). The total negative impact of the production stage is 975 Pt. Recycling processes would minimize harmful emissions by a total of 502 Pt, including 314 Pt for nitrogen oxide and 187 for sulfur oxide (Table 11).

The highest level of negative impact of land use category was characterized by the production stage (total of 1551 Pt), including primarily the processes related to the transformation to mineral extraction site (338 Pt). A significantly lower level of adverse impact is the potential for landfill disposal—a total of 35 Pt. Extraction of mineral resources is an area of economic activity with one of the highest harmful impacts on human health and the quality of the environment. It is associated with environmental destruction, especially serious in the case of open pit mines. It is characterized by the consumption of huge amounts of water, often causing shortages, and at the same time results in the contamination of surface and groundwater, as well as a reduction in their level by up to several meters (Table 12).

Table 11. Results of grouping and weighting of environmental after-effects for acidifying/eutrophication compounds, occurring in the life cycle of 1 megawatt photovoltaic power plant [unit: Pt].

Acidification/Eutrophication	Production	Exploitation	Landfill	Recycling
Nitrogen oxide	**596.32**	0.45	7.19	**-313.95**
Sulfur dioxide	**182.40**	0.20	0.42	0.00
Ammonia	107.97	0.00	0.22	-0.65
Sulfur oxides	82.99	0.11	0.00	**-186.97**
Nitrogen dioxide	5.34	0.00	0.00	0.00
Total	975.03	0.77	7.83	-501.57

Table 12. Results of grouping and weighting of after-effects of environmental land use processes, occurring in the 1 megawatt photovoltaic power plant [unit: Pt].

Land Use	Production	Exploitation	Landfill	Recycling
Transformation, to mineral extraction site	**338.36**	0.04	13.77	0.00
Occupation, dump site	275.99	0.07	13.15	0.00
Transformation, to arable, non-irrigated	221.31	0.02	0.45	0.00
Transformation, to unknown	142.92	-0.09	0.01	0.00
Transformation, to urban, continuously built	118.14	0.00	0.00	0.00
Occupation, mineral extraction site	111.45	0.02	4.14	0.00
Transformation, to dump site	63.54	0.02	0.03	0.00
Transformation, to industrial area	59.30	0.03	0.02	0.00
Transformation, to water bodies, artificial	47.46	0.01	3.16	0.00
Land use II-III	47.27	0.09	0.00	0.00
Occupation, forest, intensive, normal	41.16	0.10	0.10	0.00
Total	1550.96	0.37	64.41	0.00

Economic development entails an increase in demand for various types of natural resources, resulting in the depletion of non-renewable resources. Although their deposits are limited, their exploitation continues to grow. The stage of production of 1 megawatt photovoltaic power plant, characterized by the highest value of harmful influence in the mineral mining sector, is the production phase (total 9453 Pt), mainly in the field of tin mining (6610 Pt) and copper (2230 Pt). Due to its physical and chemical properties, tin is very important for industry. Its use in the metallurgical industry is the largest. In addition, this element is used for solders alloys. Tin is also used to coat other metals, e.g., steel, with a thin anti-corrosive layer. Recycling would minimize the pervasive effects analyzed, a total of 1889 Pt, mainly in the field of bauxite mining (-1835 Pt) (Table 13).

The highest level of harmful impacts in the category of fossil fuel extraction processes is characterized by a production phase (total of 12,159 Pt), which consists primarily of processes related to extraction of natural gas (5509 Pt) and crude oil (2366 Pt). This is connected to the high energy demand of production processes, in particular PV cells. Recycling as a form of post-disposal management would reduce several adverse effects by a total of 10,095 Pt, mainly in the oil-related processes (-7794 Pt). However, landfill disposal would result in an increase in the unfavorable impact on the life cycle of the tested power plant by 137 Pt (Table 14).

Table 13. Results of grouping and weighting of after-effects of environmental processes related to mineral extraction, occurring in the 1 megawatt photovoltaic power plant [unit: Pt].

Minerals	Production	Exploitation	Landfill	Recycling
Tin, in ground	**6609.87**	0.00	0.00	0.00
Copper, in ground	**2230.23**	8.73	0.00	0.00
Bauxite, in ground	367.14	0.00	0.00	-1834.86
Nickel, in ground	70.80	0.00	0.00	0.00
Nickel, 1.13% in sulfide, Ni 0.76% and Cu 0.76% in crude ore, in ground	68.90	0.00	0.00	0.00
Iron, in ground	50.70	0.00	0.00	-53.67
Nickel, 1.98% in silicates, 1.04% in crude ore, in ground	18.46	0.00	1.13	0.00
Total	9452.86	8.73	1.15	-1888.53

Table 14. Results of grouping and weighting of after-effects of environmental processes related to the extraction of fossil fuels, occurring in the 1 megawatt photovoltaic power plant [unit: Pt].

Fossil Fuels	Production	Exploitation	Landfill	Recycling
Gas, natural, in ground	**5508.88**	0.64	21.28	0.00
Oil, crude, in ground	**2365.64**	0.62	114.85	0.00
Oil, crude, 42.6 MJ per kg, in ground	2166.59	0.96	0.00	-7794.46
Gas, natural, 36.6 MJ per m^3, in ground	822.65	0.00	0.00	-1476.15
Oil, crude, 42.7 MJ per kg, in ground	660.69	0.09	0.00	0.00
Coal, 18 MJ per kg, in ground	204.68	0.03	0.00	-346.74
Gas, natural, 30.3 MJ per kg, in ground	170.19	0.07	0.00	0.00
Gas, natural, 35 MJ per m^3, in ground	140.89	0.72	0.00	-244.92
Oil, crude, 41 MJ per kg, in ground	65.70	0.00	0.00	-232.32
Total	12158.95	3.88	136.56	-10094.60

3.2. IPCC

The analyzed groups of photovoltaic power plant components were also subjected to an IPCC analysis to determine greenhouse gas emissions in kilograms CO_2 equivalents. The results are shown in Figure 6. It follows from that that the largest amount of greenhouse gases is generated in the life cycle of photovoltaic panels that end in landfill storage (269,099 kg CO_2 eq). However, if they are recycled, there is the possibility of significantly reducing the volume of the emissions in question. The lowest greenhouse gas emissions were recorded for post-consumption in recycling cycles (support structures: 3880 Pt, inverter 10,017 Pt, electrical installation: 16,091 Pt). The sources of CO_2 emissions throughout the life cycle include, above all, the burning of fossil fuels to obtain electricity. Its highest demand was observed during the production of photovoltaic cells.

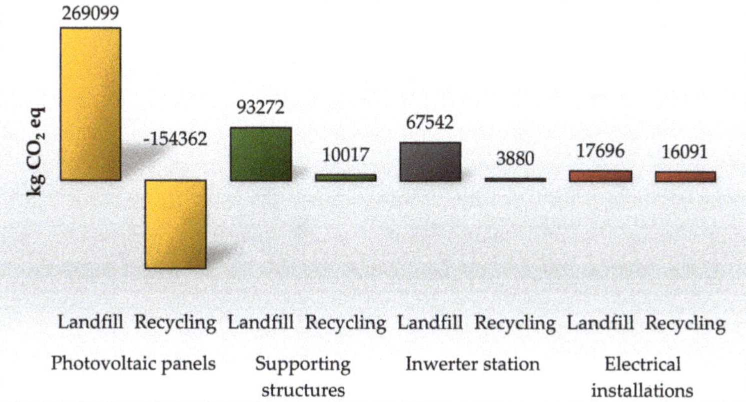

Figure 6. Results of the characterization of environmental consequences for cumulative greenhouse gas (GHG) emissions occurring in the cycle of existence of selected groups of 1 megawatt photovoltaic power plant [unit: kg CO_2 eq].

3.3. CED

The last factor considered was the energy consumption of the life cycle of individual groups of photovoltaic power plants, which was estimated using the CED method. Sustainable development of technical facilities, in addition to having the lowest demand for materials and the least harmful environmental impact of the life cycle, also includes the maximum reduction of energy consumption within its individual stages (which also translates into improving the quality of the environment, especially considering the fact that the main source of energy are conventional fuels). The largest amount of energy is needed to produce photovoltaic panels (e.g., to produce silicon with proper purity), a life cycle that covers the production, operation and management of panels in the form of landfill, consumes nearly 5 million MJ. Similarly, the case of an inverter station represents over 800 thousand MJ, and for supporting structures, a further more than 1 million MJ, and with regard to the electrical installation, almost 250 thousand MJ. The use of recycling processes reduces the energy consumption of all groups of components of the power plant in question. Reuse of photovoltaic cells recovered in recycling processes would minimize energy and material consumption (Figure 7).

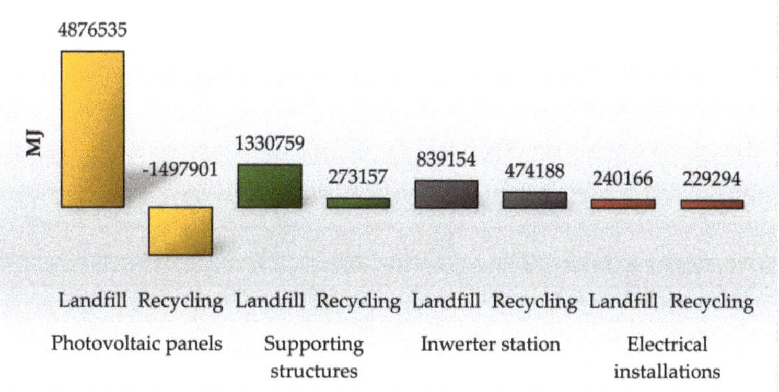

Figure 7. Results of the characterization of environmental consequences, in relation to cumulative energy demand (in MWh), occurring in the cycle of existence of selected groups of 1 megawatt photovoltaic power plant [unit: MJ].

3.4. Recapitulation

Figure 8 shows the results of grouping and weighting the environmental consequences of 1000 kg of selected plastics and materials that are part of the photovoltaic farm components. The highest level of harmful influence on the environment is distinguished by silver (20,512 Pt/1 Mg), nickel (3842 Pt/1 Mg), copper (2363 Pt/1 Mg), PA 6 (656 Pt/1 Mg), lead (638 Pt/1 Mg) and cadmium (586 Pt/1 Mg). The most widely used in photovoltaic panels are: silver, nickel, lead, cadmium, EVA, selenium, silicon, and aluminum; in supporting structures: nickel and steel; in inverter stations: copper, PA6, PVDF, rubber, and steel; and in electrical installations: copper, PA6, PVDF, and rubber.

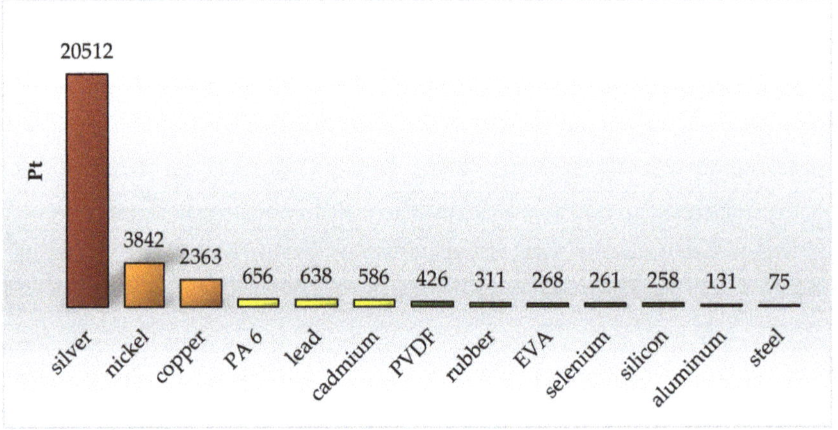

Figure 8. Results of grouping and weighting the environmental consequences of 1,000 kg of selected plastics and materials included in photovoltaic elements for all categories of influence [unit: Pt].

3.5. Other Energy Sources

Using the databases available in the SimaPro software and calculations made previously for a 1 MW photovoltaic power plant, a comparison was made of the environmental impact of the processes of obtaining electricity from photovoltaics with selected, commonly used conventional energy sources and with the structure of the mixed energy characteristic of Poland, mainly based on hard and brown coal (approximately 80%).

The analyzed photovoltaic power station, during its 20-year life cycle, is able to produce about 20,000 MWh of electricity, which value was taken as a reference value. The degree of impact on the surroundings resulting from the combustion of an amount of hard coal, brown coal, and heating oil necessary to obtain the same amount of electricity was analyzed. In addition, an analogous analysis was carried out when 20,000 MWh of energy was obtained in Poland.

Table 15 summarizes the results of the characterization of the environmental after-effects arising from the generation of 20,000 MWh of electricity from the selected energy sources. The highest level of harmful impact of the analyzed energy sources is visible in the area of emissions of compounds having a negative impact on the quality of the environment (category: ecotoxic compounds, compounds causing acidification/eutrophication, land use) and in processes related to the depletion of raw material resources (categories: mineral extraction, extraction of fossil fuels). This is characteristic of energy extraction processes, especially from conventional sources.

Table 15. Results of the characterization of environmental consequences arising as a result of the generation of 20,000 MWh of electricity from selected energy sources.

Impact Category	PV Power Plant (Recycling)	PV Power Plant (Landfill)	Hard Coal	Brown Coal	Natural Gas	Heating Oil	Polish Energy Mix	Unit
Carcinogens	0	0	5	4	0	0	4	DALY
Resp. organics	0	0	0	0	0	0	0	DALY
Resp. inorganics	0	0	13	14	11	5	12	DALY
Climate change	0	0	5	5	6	1	5	DALY
Radiation	0	0	0	0	0	0	0	DALY
Ozone layer	0	0	0	0	0	0	0	DALY
Ecotoxicity	58,446	303,071	717,677	2671,933	218,750	0	1463,288	PAF·m^2/a
Acidification/ Eutrophication	6092	12,623	335,746	284,690	174,296	163,673	300,669	PDF·m^2/a
Land use	8751	8984	192,997	32,729	90,967	0	128,362	PDF·m^2/a
Minerals	319,481	398,914	33,221	18,763	19,079	0	28,350	MJ
Fossil fuels	92,339	520,697	3634,140	307,570	2097,723	10360,976	3844,863	MJ

The results of grouping and weighting environmental consequences arising as a result of generating 20,000 MWh of electricity from selected energy sources are presented in Table 16. The highest value of harmful impact on the environment was noted in the following categories: inorganic compounds causing respiratory diseases (from 3437 to 357,703 Pt), compounds that cause climate change (from 2351 to 163,556 Pt), and the extraction of fossil fuels (from 2198 to 246,591 Pt).

Table 16. Results of grouping and weighting of environmental consequences arising as a result of generating 20,000 MWh of electricity from selected energy sources [unit: Pt].

Impact Category	PV Power Plant (Recycling)	PV Power Plant (Landfill)	Hard Coal	Brown Coal	Natural Gas	Heating Oil	Polish Energy Mix
Carcinogens	−249	3230	123768	102,354	8331	0	104,557
Resp. organics	−6	12	62	24	1265	406	69
Resp. inorganics	3437	10,932	327,703	356,043	286,288	130,653	321,832
Climate change	−627	2351	123,308	121,536	163,556	35,811	119,398
Radiation	28	29	210	121	142	0	297
Ozone layer	−2	2	2	1	13	0	5
Ecotoxicity	456	2364	5598	20,841	1706	0	11,414
Acidification/ eutrophication	475	985	26,188	22,206	13,595	12,766	23,452
Land use	683	701	15,054	2553	7095	0	10,012
Minerals	7604	9494	791	447	454	0	675
Fossil fuels	2198	12,393	86,493	7320	49,926	246,591	91,508
Total	13,997	42,492	709,177	633,445	532,371	426,228	683,220

The highest total level of harmful impact on the environment was from the production of 20,000 MWh of electricity was determined to be from hard coal (709,17 Pt) and brown coal (633,445 Pt). A high degree of negative impact on the environment was also noted for the Polish energy mix (683,220 Pt), due to the fact that it is mainly based on the burning of brown and hard coal. Obtaining energy from solar radiation possessed the lowest level of adverse impact on the environment, with this type of installation exerting an impact from 13,997 (recycling) to 42,492 Pt (storage) throughout its entire life cycle, depending on the form of post-use management. Despite some expenditure of energy and materials in the production and post-use management phase, the use of a renewable energy source, i.e., photovoltaics, causes the least negative environmental consequences compared to conventional energy sources (Figure 9).

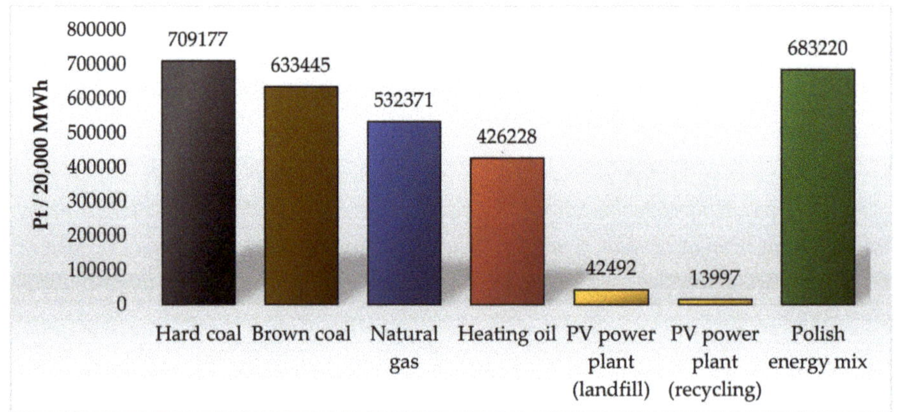

Figure 9. Results of grouping and weighting environmental consequences arising as a result of generating 20,000 MWh of electricity from selected energy sources [unit: Pt].

Additionally, the amount of greenhouse gas emissions resulting from the production of 20,000 MWh of electricity was analyzed from the same energy sources using the IPCC method. The highest level of GHG emissions was found when obtaining energy from natural gas (29,989 Mg CO_2 eq), hard coal (22,872 Mg CO_2 eq), and brown coal (22,241 Mg CO_2 eq). In the case of photovoltaic power plants, the emission level was the lowest and amounted to about 400 Mg CO_2 eq. The obtained results confirm that photovoltaic power plants can be a source of energy enabling the reduction of greenhouse gas emissions, and hence are one of the ways of reducing the greenhouse effect (Figure 10).

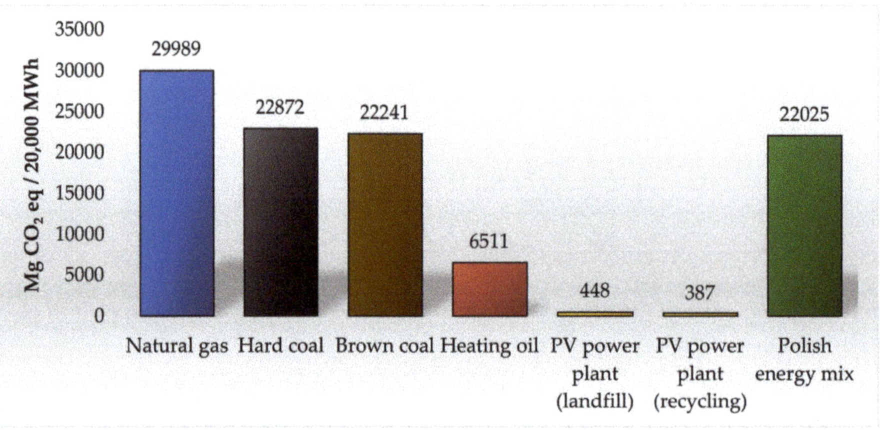

Figure 10. The results of the characterization of the environmental after-effects in relation to cumulated greenhouse gas (GHG) emissions arising from the production of 20,000 MWh of electricity from selected energy sources [unit: Pt].

As a result of these considerations it was found that the total highest level of harmful impact on the environment is the life cycle of post-consumer photovoltaic panels when stored in landfill (22,029 Pt). The largest amount of greenhouse gases generated during the post-consumer life cycle of photovoltaic panels is from storage in landfill (269,099 kg CO_2 eq). The greatest amount of energy absorbed during the life cycle of photovoltaic panels is from the use of this form of storage—nearly 5 million MJ. The use of recycling processes reduces the energy consumption of all groups of components of the power plant in question. Silver, nickel, copper, PA 6, lead, and cadmium are among the materials with the most harmful influence on the environment.

An additional element of the analysis was the comparison of the environmental impact of the processes of obtaining electricity from photovoltaics with selected, most commonly used conventional energy sources and with the structure of the mixed energy characteristic of Poland, which is mainly based on hard and brown coal (approximately 80%). The highest level of harmful impact of the analyzed energy sources is visible in the area of compound emissions: ecotoxic (2,671,933 PAF·m^2/a brown coal), compounds causing acidification/eutrophication (335,746 PAF·m^2/a hard coal), land use (192,997 PAF·m^2/a hard coal) and within processes related to the depletion of raw material resources includes: mineral extraction (398,914 MJ PV power plant (landfill)), and extraction of fossil fuels (10,360,976 MJ heating oil).

4. Conclusions

Renewable energy sources, including solar energy, possess many positive environmental aspects in terms of local, regional and national and, most importantly, global aspects. The beneficial effect of the use of alternative energy sources can be considered in three areas: continuous sustainable development, the environment protection and natural resources, and the timeless and endless nature of the raw materials used [45,46].

Energy security is understood as providing the security of energy supplied to recipients in a particular time and place, and it is one of the priority actions in economic policy. Scattered generation ensures even distribution of heat sources and energy (derived from a wide variety of energy sources, including renewable sources of energy) and have been the subject of considerable interest. They are considered to be important not only for increasing energy security, but also for the reduction of greenhouse gas emissions, particularly carbon dioxide, from burning fossil fuels. Each energy type impacts the quality of the environment, but not all to an identical extent [47–49]. That is why processes related to solar cell production have become so important in environmental assessment, determined by assessing the surface properties, including roughness, thickness and machining [50,51].

In the opinion of society, the use of only conventional energy sources (for example coal or oil), is a threat to human health and the quality of the environment. However, the depletion of non-renewable resources concerns not only traditional ways of obtaining energy, but also alternative ones. As a consequence, we are constantly striving to minimize negative environmental impacts, e.g., by reducing CO_2 emissions [52]. Bearing in mind the future prospects of sustainable development, directions for the environmental assessment of wind farms, small hydropower plants and solar farms have become attractive. This approach makes it possible to strive to meet the growing demand for electricity without burning fossil fuels [53,54]. Thus, this work proved that at the production stage of the elements for a photovoltaic power plant, there is also a demand for raw materials and energy. The same applies to processes related to post-use management. Assuming that the analyzed power station produces annual energy equal to about 1000 MWh, it must work for up to about 2 years to produce an amount of energy equal to the demand for it throughout its entire life cycle (total energy demand for the life cycle with post-use management in the form of storage on the dump is about 2024 MWh—Figure 7). For this reason, it was considered justified to conduct research aimed at ecological and energetical assessment of the life cycle of materials and components of a photovoltaic power plant, and the hypothesis adopted in the work was confirmed.

Based on the results of the research performed and an evaluation of the material stages of the life cycle of the analyzed PV power plant, in terms of pro-environmental, post-production use of materials, materials and elements of photovoltaic power plants, the following suggestions are proposed:

- ameliorating the detrimental effect on the environment of production process (mainly PV panels), which is the most deleterious phase of all phases in environmental life cycle, by introducing the latest technologies, which are less energy absorbent, with lesser usage of materials and lower emissions of harmful particles and substances,

- creating the most pro-environmental algorithm for dealing with plastic materials and elements of photovoltaic power plants after their completion, taking into account, in particular, recycling processes, reducing the energy consumption, material consumption and emissions of harmful substances throughout the life cycle of the power plant,
- employing more environmentally friendly construction materials while at the same time maintaining proper technical, mechanical and qualitative characteristics for specific roles in photovoltaic power plants, in particular limiting the use of materials with the highest levels of negative environmental impact such as silver, nickel, copper, PA 6, lead and cadmium,
- employing construction strategies that allow for easier separation of individual materials, making them easy to identify during post-consumer use,
- development of comprehensive, pro-environmental standards with respect to post-consumption management of materials and elements of photovoltaic power plants,
- popularizing the idea of research and assessment of the impact of renewable energy throughout their life cycle.

Author Contributions: Conceptualization, I.P. and P.B.-W.; methodology, I.P. and P.B.-W.; software, I.P.; validation, I.P., P.B.-W. and K.P.; formal analysis, I.P, P.B.-W, K.P. and A.T..; investigation, I.P. and P.B.-W.; resources, I.P.; data curation, I.P.; writing—original draft preparation, I.P.; writing—review and editing, I.P, P.B.-W, K.P. and A.T.; visualization, I.P.; supervision, I.P, P.B.-W, K.P. and A.T.; project administration, P.B.-W. All authors have read and agreed to the published version of the manuscript.

Funding: This research received no external funding.

Conflicts of Interest: The authors declare no conflict of interest.

References

1. Corcelli, F.; Ripa, M.; Ulgiati, S. End-of-life treatment of crystalline silicon photovoltaic panels. An emergy-based case study. *J. Clean. Prod.* **2017**, *9*, 1129–1142. [CrossRef]
2. Goe, M.; Gaustad, G. Strengthening the case for recycling photovoltaics: An energy payback analysis. *Appl. Energy* **2014**, *5*, 41–48. [CrossRef]
3. McLellan, B. Sustainable Future for Human Security. In *Environment and Resources*; Springer: Singapore, 2018; pp. 37–68. [CrossRef]
4. Singh, R.; Kumar, S. *Green Technologies and Environmental Sustainability*; Springer: Cham, Switzerland, 2017; pp. 157–178. [CrossRef]
5. Andersen, O. *Unintended Consequences of Renewable Energy*; Springer: London, UK, 2013; pp. 81–89. [CrossRef]
6. Coulson, N.E.; Wang, Y.; Lipscomb, C.A. *Energy Efficiency and the Future of Real Estate*; Palgrave Macmillan: New York, NY, USA, 2017; pp. 83–89. [CrossRef]
7. Gronowicz, J. *Ochrona Środowiska w Transporcie Lądowym*; Biblioteka problemów eksploatacji, Instytut Technologii Eksploatacji–PIB: Radom, Poland, 2004; pp. 11–32. ISBN 83-7204-374-4.
8. Granata, G.; Pagnanelli, F.; Moscardini, E.; Havlik, T.; Toro, L. Recycling of photovoltaic panels by physical operations. *Sol. Energy Mater. Sol. Cells* **2014**, *6*, 239–248. [CrossRef]
9. Kruszelnicka, W.; Bałdowska-Witos, P.; Kasner, R.; Flizikowski, J.; Tomporowski, A.; Rudnicki, J. Evaluation of emissivity and environmental safety of biomass grinders drive. *Przem. Chem.* **2019**, *10*, 1494–1498. [CrossRef]
10. Tao, J.; Yu, S. Review on feasible recycling pathways and technologies of solar photovoltaic modules. *Sol. Energy Mater. Sol. Cells* **2015**, *10*, 108–124. [CrossRef]
11. Demirel, Y. *Energy: Production, Conversion, Storage, Conservation and Coupling*; Springer: Cham, Switzerland, 2016; pp. 441–484. [CrossRef]
12. Merkisz, J.; Rymaniak, Ł. The assessment of vehicle exhaust emissions referred to CO_2 based on the investigations of city buses under actual conditions of operation. *Eksploat. Niezawodn.* **2017**, *19*, 522–529. [CrossRef]
13. Wang, Z.; Wu, J.; Liu, C.; Gu, G. *Integrated Assessment Models of Climate Change Economics*; Springer: Singapore, 2017; pp. 1–19. [CrossRef]

14. Dahlquist, E.; Hellstrand, S. *Natural Resources Available Today and in the Future: How to Perform Change Management for Achieving a Sustainable World*; Springer: Cham, Switzerland, 2017; pp. 245–268. [CrossRef]
15. Heshmati, A.; Abolhosseini, S.; Altmann, J. *The Development of Renewable Energy Sources and Its Significance for the Environment*; Springer: Singapore, 2015; pp. 7–29. [CrossRef]
16. Bauer, G.H. *Photovoltaic Solar Energy Conversion*; Springer: Berlin, Germany, 2015; pp. 5–37. [CrossRef]
17. Dincer, I.; Midilli, A.; Kucuk, H. *Progress in Sustainable Energy Technologies: Generating Renewable Energy*; Springer: Cham, Switzerland, 2014; pp. 339–353. [CrossRef]
18. Márquez, F.P.G.; Karyotakis, A.; Papaelias, M. *Renewable Energies*; Springer: Berlin, Germany, 2018; pp. 1–15. [CrossRef]
19. Rigatos, G.G. *Intelligent Renewable Energy Systems: Modelling and Control*; Springer: Cham, Switzerland, 2016; pp. 339–409. [CrossRef]
20. Hossain, J.; Mahmud, A. *Renewable Energy Integration: Challenges and Solutions*; Springer: Singapore, 2014; pp. 69–95. [CrossRef]
21. Toke, D. *Ecological Modernization and Renewable Energy*; Palgrave Macmillan: New York, NY, USA, 2011; pp. 167–179. [CrossRef]
22. Traverso, M.; Asdrubali, F.; Francia, A.; Finkbeiner, M. Towards life cycle sustainability assessment: An implementation to photovoltaic modules. *Int. J. Life Cycle Assess* **2012**, *17*, 1068–1079. [CrossRef]
23. Bałdowska-Witos, P.; Kruszelnicka, W.; Kasner, R.; Rudnicki, J.; Tomporowski, A.; Flizikowski, J. Impact of the plastic bottle production on the natural environment. Part 1. Application of the ReCiPe 2016 assessment method to identify environmental problems. *Przem. Chem.* **2019**, *10*, 1662–1667. [CrossRef]
24. Kulczycka, J.; Lelek, Ł.; Lewandowska, A.; Zarębska, J. Life Cycle Assessment of municipal solid waste management—Comparison of results using different LCA models. *Pol. J. Environ. Stud.* **2015**, *1*, 125–140. [CrossRef]
25. Lelek, Ł.; Kulczycka, J.; Lewandowska, A.; Zarębska, J. Life cycle assessment of energy generation in Poland. *Int. J. Life Cycle Assess* **2016**, *21*, 1–14. [CrossRef]
26. Guineé, J. *Handbook on Life Cycle Assessment: Operational Guide to the ISO Standards*; Springer: Dordrecht, The Netherlands, 2002; pp. 31–108. [CrossRef]
27. Treloar, G.J.; Love, P.E.D.; Faniran, O.O.; Iyer-Raniga, U. A hybrid life cycle assessment method for construction. *Constr. Manag. Econ.* **2000**, *18*, 5–9. [CrossRef]
28. Ulgiati, S.; Raugei, M.; Bargigli, S. Overcoming the in adequacy of single-criterion approaches to Life Cycle Assessment. *Ecol. Modell.* **2006**, *3*, 432–442. [CrossRef]
29. Guinée, J.; Heijungs, R.; Huppes, G.; Zamagni, A.; Masoni, P.; Buonamici, R.; Rydberg, T. Life Cycle Assessment: Past, present, and future. *Environ. Sci. Technol.* **2011**, *1*, 90–96. [CrossRef] [PubMed]
30. Tomporowski, A.; Flizikowski, J.; Wełnowski, J.; Najzarek, Z.; Topoliński, T.; Kruszelnicka, W.; Piasecka, I.; Śmigiel, S. Regeneration of rubber waste using an intelligent grinding system. *Przem. Chem.* **2018**, *10*, 1659–1665. [CrossRef]
31. Klinglmair, M.; Sala, S.; Brandão, M. Assessing resource depletion in LCA: A review of methods and methodological issues. *Int. J. Life Cycle Assess* **2014**, *19*, 580–592. [CrossRef]
32. Kłos, Z. Ecobalancial assessment of chosen packaging processes in food industry. *Int. J. Life Cycle Assess* **2002**, *7*, 309. [CrossRef]
33. Rebitzer, G.; Loerincik, Y.; Jolliet, O. Input-output life cycle assessment: From theory to applications. *Int. J. Life Cycle Assess* **2002**, *7*, 174–176. [CrossRef]
34. Dreyer, L.C.; Niemann, A.L.; Hauschild, M.Z. Comparison of Three Different LCIA Methods: EDIP97, CML2001 and Eco-indicator 99. *Int. J. Life Cycle Assess* **2003**, *8*, 191–200. [CrossRef]
35. Piasecka, I.; Tomporowski, A.; Flizikowski, J.; Kruszelnicka, W.; Kasner, R.; Mroziński, A. Life Cycle Analysis of Ecological Impacts of an Offshore and a Land-Based Wind Power Plant. *Appl. Sci.* **2019**, *9*, 231. [CrossRef]
36. Piotrowska, K.; Kruszelnicka, W.; Bałdowska-Witos, P.; Kasner, R.; Rudnicki, J.; Tomporowski, A.; Flizikowski, J.; Opielak, M. Assessment of the Environmental Impact of a Car Tire throughout Its Life Cycle Using the LCA Method. *Materials* **2019**, *12*, 4177. [CrossRef]
37. Tomporowski, A.; Flizikowski, J.; Kruszelnicka, W.; Piasecka, I.; Kasner, R.; Mroziński, A.; Kovalyshyn, S. Destructiveness of profits and outlays associated with operation of offshore wind electric power plant. Part 1: Identification of a model and its components. *Pol. Marit. Res.* **2018**, *2*, 132–139. [CrossRef]

38. Kendall, A. Time-adjusted global warming potentials for LCA and carbon footprints. *Int. J. Life Cycle Assess* **2012**, *17*, 1042–1049. [CrossRef]

39. Muñoz, I.; Schmidt, J.H. Methane oxidation, biogenic carbon, and the IPCC's emission metrics. Proposal for a consistent greenhouse-gas accounting. *Int. J. Life Cycle Assess* **2016**, *21*, 1069–1075. [CrossRef]

40. Peter, C.; Fiore, A.; Hagemann, U.; Nendel, C.; Xiloyannis, C. Improving the accounting of field emissions in the carbon footprint of agricultural products: A comparison of default IPCC methods with readily available medium-effort modeling approaches. *Int. J. Life Cycle Assess* **2016**, *21*, 791–805. [CrossRef]

41. Stichnothe, H.; Schuchardt, F.; Rahutomo, S. European renewable energy directive: Critical analysis of important default values and methods for calculating greenhouse gas (GHG) emissions. *Int. J. Life Cycle Assess* **2014**, *19*, 1294–1304. [CrossRef]

42. Frischknecht, R.; Wyss, F.; Büsser-Knöpfel, S.; Lützkendorf, T.; Balouktsi, M. Cumulative energy demand in LCA: The energy harvested approach. *Int. J. Life Cycle Assess* **2015**, *20*, 957–969. [CrossRef]

43. Puig, R.; Fullana-Palmer, P.; Baquero, G.; Riba, J.R.; Bala, A. A Cumulative Energy Demand indicator (CED), life cycle based, for industrial waste management decision making. *Waste Manag.* **2013**, *12*, 2789–2797. [CrossRef]

44. Scipioni, A.; Niero, M.; Mazzi, A.; Manzardo, A.; Piubello, S. Significance of the use of non-renewable fossil CED as proxy indicator for screening LCA in the beverage packaging sector. *Int. J. Life Cycle Assess* **2013**, *18*, 673–682. [CrossRef]

45. Tiwari, G.N.; Mishra, R.K. *Advanced Renewable Energy Sources*; Royal Society of Chemistry: Cambridge, UK, 2012; pp. 46–71. ISBN 978-1-84973-380-9.

46. Zacher, L.W. ; *Technology, Society and Sustainability: Selected Concepts, Issues and Cases*; Springer: Basel, Switzerland, 2017; pp. 203–221. [CrossRef]

47. Solmes, L.A. *Energy Efficiency: Real Time Energy Infrastructure Investment and Risk Management*; Springer: Dordrecht, The Netherlands, 2009; pp. 1–34. [CrossRef]

48. Twidell, J.; Weir, T. *Renewable Energy Resources*; Routledge: London, UK, 2015; pp. 151–202. [CrossRef]

49. Yang, M.; Yu, X. *Energy Efficiency: Benefits for Environment and Society*; Springer: London, UK, 2015; pp. 11–42. [CrossRef]

50. Kumar, A.; Melkote, S.N. Diamond Wire Sawing of Solar Silicon Wafers: A Sustainable Manufacturing Alternative to Loose Abrasive Slurry Sawing. *Procedia Manuf.* **2018**, *21*, 549–566. [CrossRef]

51. Kumar, A.; Melkote, S.N.; Kamiński, S.; Arcona, C. Effect of grit shape and crystal structure on damage in diamond wire scribing of silicon. *J. Am. Ceram. Soc.* *100* **2017**, *4*, 1350–1359. [CrossRef]

52. Bai, A.; Popp, J.; Pető, K.; Szőke, I.; Harangi-Rákos, M.; Gabnai, Z. The Significance of Forests and Algae in CO_2 Balance: A Hungarian Case Study. *Sustainability* **2017**, *9*, 857. [CrossRef]

53. Tremeac, B.; Meunier, F. Life cycle analysis of 4.5 MW and 250 W wind turbines. *Renew. Sustain. Energy Rev.* **2009**, *13*, 2104–2110. [CrossRef]

54. Verán-Leigh, D.; Vázquez-Rowe, I. Life cycle assessment of run-of-river hydropower plants in the Peruvian Andes: A policy support perspective. *Int. J. Life Cycle Assess* **2019**, *24*, 1376–1395. [CrossRef]

Article

Influence of Waste Management on the Environmental Footprint of Electricity Produced by Photovoltaic Systems

Sina Herceg *, Sebastián Pinto Bautista and Karl-Anders Weiß

Department of Material Analysis, Fraunhofer Institute for Solar Energy Systems ISE, Heidenhofstr 2, 79110 Freiburg, Germany; sebastian.pinto.bautista@ise.fraunhofer.de (S.P.B.); karl-anders.weiss@ise.fraunhofer.de (K.-A.W.)
* Correspondence: sina.herceg@ise.fraunhofer.de

Received: 30 March 2020; Accepted: 19 April 2020; Published: 1 May 2020

Abstract: PV waste management will gain relevance proportionally to the amounts of waste that are expected to arise with the phasing-out of old installations in the upcoming years and decades. The Life Cycle Assessment (LCA) methodology is used here to analyze the environmental performance of photovoltaic systems and the waste management methods that have been developed recently. Several LCA studies have already been performed for PV technologies, but in most cases these do not include the end of life stage, thus there is still uncertainty about the impacts of recycling on the environmental footprint of PV electricity. The present study offers a more detailed analysis of different end-of-life approaches for the main photovoltaic technologies that are found on the market. The results from the analysis demonstrate that recycling has the potential to improve the environmental profile of PV electricity but at the same time there is room for further improvements in developing dedicated recycling technologies.

Keywords: photovoltaics; waste management; life cycle assessment

---

1. Introduction

1.1. Motivation

Energy transition towards cleaner electricity grids has led to a large deployment of photovoltaic (PV) installations on a global scale. However, until now only small flows of PV waste have had to be handled. This has restrained the consolidation of a specialized waste processing industry due to a lack of profitability. As a consequence, little is known regarding the performance of waste management schemes and the potential ecological benefits that could come along with them. Appropriate waste management is required to mitigate potential ecological impacts of the waste material. If not handled in a proper manner, critical and valuable resources might be lost and toxic materials such as lead or cadmium contained within the modules could leak into the soil and groundwater, representing a threat for biodiversity and human health. While the risk of leakage is low under neutral atmospheric conditions, it rises when the module is exposed to acidic environments as those produced by rainfall in certain locations. Furthermore, PV systems represent a potential source of valuable materials. However, the small flow of End-of-Life (EoL) modules has restrained the profitability of recycling, since dedicated recycling processes demand high volumes of input material to make this activity economically viable [1–3]. The study of waste management will gain increasing relevance in the long term when dealing with big amounts of photovoltaic waste that are already emerging. It is nevertheless difficult to accurately predict these flows of waste since the real lifetime of PV modules still remains uncertain. Most installations have not yet met their expected lifespans, which have been indicated to be around 25–30 years by most manufacturers. Little is known about the modules' performance after this timespan, but a couple of studies indicate that an economically viable operation of PV plants is realistic even after 30–35 years [4–6].

Several environmental assessment studies have proved the environmental benefits of PV technologies, but in most cases the end-of-life stage has been neglected. Likewise, a couple of approaches to processing waste material have been proposed, but their impact on the environmental profile of PV electricity remains uncertain. This study intends to reduce this gap through a comprehensive analysis of different waste management approaches based on recycling. The main objective is to quantify the impacts of EoL management approaches and assess their contribution to the overall environmental footprint of electricity produced by different PV technologies with a special focus on Germany.

1.2. LCA of Waste Management for PV Systems

The most recent Life Cycle Inventories (LCIs) have been published by Wambach et al. [7]. For crystalline Silicon (c-Si) modules, the inventories were built alongside plants designed for the processing of laminated glass, metals or electronic and electric waste. For the specific case of Cadmium-Telluride (CdTe) modules, the inventories included were based on the recycling activities of the company First Solar in Germany. The study concluded that the benefits of recycling were higher than the impacts caused by processing the waste material in all the categories under study with an exception for CdTe in the category 'human toxicity (cancer effects)'. Another study [8] found a reduction of about 4–11% of the modules' environmental burdens and highlighted the influence of transports and electricity use in the burdens of the recycling process. The authors also presented a preliminary study on a new approach for EoL management of thin film modules, which was mostly based on mechanical processes eliminating the need for thermal treatments and reducing the use of chemical agents [9]. The technical feasibility of a new c-Si module recycling technique was studied by Duflou et al. [10] by means of a selective mechanical delamination, performing a comparative LCA with two other existing methods. Others have illustrated and analyzed an innovative process that enables the recovery of other valuable materials present in the modules such as silicon, silver and other metals that cannot be retrieved by traditional means [11,12]. Those studies applied the LCA methodology to a pilot process developed within the project called 'Full Recovery End of Life Photovoltaic' (FRELP), which is composed of a series of mechanical, thermal and chemical treatments. A more detailed analysis on the environmental impact categories according to the different processes involved was given recently [13]. This study shows net benefits from recycling in all the impact categories of analysis, as well as the identification of transportation and thermal treatments as the main burdens within the entire recycling process. A specific analysis focused on waste backsheets in order to evaluate the effects of fluorinated layers on the EoL processing found that fluorine-free backsheets performed better than fluorinated ones for both incineration and pyrolysis; thus, the use of fluoropolymers should be avoided [14]. Other studies evaluating existing waste management techniques and new alternative methods for recycling PV panels have been performed, but without offering the disclosure of inventories or a deep analysis of the ecological footprint of such techniques [15–18].

2. Materials and Methods

The evaluation of the environmental profile of a product or service is the characterization of its ecological footprint; a description of the interactions with its environment, and thus the associated impacts. Several tools have been developed to perform such an evaluation. One of these is the so-called Life Cycle Assessment (LCA), a systematic methodology which determines indicators related to impact categories that quantify the environmental burden of each phase in the lifecycle of the product. Within this paper, a total of six waste management approaches proposed by industry and research programs are evaluated under this methodology. Following the recommendations from [19,20] the impact assessment method for the analysis is the ILCD-2011 and the following impact categories were chosen: (1) Climate change; (2) Human toxicity (non-carcinogenic); (3) Human toxicity (carcinogenic); (4) Particulate matter; (5) Acidification; (6) Freshwater ecotoxicity; (7) Resource depletion.

2.1. Product Systems Description

The system under study comprises the generation of AC electricity produced with a slanted-roof photovoltaic plant installed in Germany and manufactured under the approximated market and technological conditions of 2010–2013, which can be seen in Table 1. This specific period was chosen due to the great increase in deployment of PV installations at the time, also describing the gross composition of the waste to be processed in the future:

Table 1. Module modeling parameters [21].

	Parameter	Mono-Si	Multi-Si
Location of Manufacturing (Market Share)	China (CN)	79.6%	
	Europe (RER)	14.5%	
	Asia Pacific (APAC)	5.9%	
	Plant size (kW_p)	3	3
	Module efficiency	14%	13.60%
	Panel capacity rate (W_p/m^2)	140	136
	Module lifetime (Years)	30	30
	Module weight (kg/m^2)	10	10
	Module Area (m^2)	1.6	1.6

As proposed by [22], the following lifetimes and parameters will be used in this study:

- Life expectancy
  Modules: 30 years (for any type)
  Inverters: 15 years
  Mounting structure: 30 years (for any type of module)
  Cabling: 30 years
  Manufacturing plant: 30 years (for any type of module)
  Recycling plant: 30 years (for any type of module)
- Performance Ratio: Standard value of 0.75
- Degradation rate: Standard linear degradation rate of 0.7% per year
- Conversion efficiency from primary energy to electricity: $\eta G = 30\%$
- Irradiation: Conditions in Germany: 1.055 kWh/m^2 [23]

Figure 1 shows the system boundaries defined for the assessment of electricity production, which include raw materials and energy supply in processing at the initial manufacturing plant. The EoL stage includes burdens from transportation, waste processing and final disposal of the modules. In addition, environmental credits gained from energy and material recovery are given. However, the burdens from the post-recycling refining processes of these materials needed before reintroducing them in the production lines have been left out. The manufacturing of the balance-of-system components (BOS), e.g., mounting structure and the electric installation required for PV systems, is included in the analysis; however, recycling of these has not been considered, since they are treated separately from the PV modules, and thus no credits have been accounted for.

The functional unit (FU) is 1 kWh of electricity produced with a PV system installed in Germany, which allows consistent comparability of the different technologies under study. This FU shall describe the purpose of the system studied and will serve as a reference measure to which the impacts in the different categories will be expressed. The reference flow refers to the size of the system that quantifies the functional unit, which in this case is the 3 kWp plant.

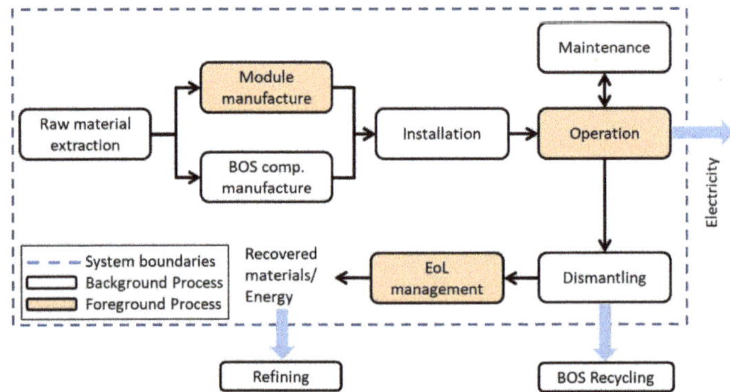

Figure 1. System boundaries of the current study.

2.2. Characterization of Modules and Input Waste

For practical purposes, the mass composition of c-Si input waste was taken from [11]. However, in real life, recycling plants will receive a wide variety of modules, each one with a specific environmental profile. This is because the technological development has diversified the market making it difficult to define a standard composition. The mass fractions assumed for a c-Si module can be found in Table 2.

Table 2. Mass composition of c-Si module input waste.

Module Section	Material	Mass Fraction
Front glass	Glass	70.00%
Frame	Aluminium	18.00%
Junction box	Copper	0.33%
	Plastic junction box	0.67%
Encapsulant	EVA	5.10%
Backsheet	PET	1.50%
Solar cell	Silicon	3.65%
	Aluminium	0.53%
	Copper	0.11%
	Silver	0.05%
others	other metals (tin, lead)	0.05%
	Sealant, potting	0.00%

2.3. Description of the EoL Approaches

The waste management approaches under study are either state-of-the art basic recycling as described, basic recycling with additional benefits from backsheet recovery, or more dedicated recycling strategies as proposed by other research programs (Table 3). These approaches have been subject to research and testing for a long time and are thus likely to define the future scenario. Their main differences consist in the recovered material, the involved processes and the final disposal mechanism that the unrecovered materials undergo.

Basic material recovery refers to those approaches limited to recovering only bulk materials such as aluminum and glass by simple processes in order to comply with the European legal requirements of mass recovery. These approaches leave behind other valuable materials such as wafers and silver due to the added complexity of these processes. Currently, this type of approach is performed mostly in laminated glass, metal and electronic waste recycling plants where PV modules can also be processed.

Figure 2 presents a diagram flow chart of one of the recycling approaches studied as well as how the boundary conditions have been defined.

Table 3. Modelled approaches and respective features.

Appr.	Materials Recovered	Unrecovered Fraction Disposal Method
n°1	Glass, Aluminum, copper	Landfilling
n°2	Glass, Aluminum, copper	Landfilling / Incineration
n°3	Glass, Aluminum, copper	Landf */Inciner **/Inciner. of Backsheet
n°4	Glass, Aluminum, copper	Landf./Inciner./Pyrol. of Backsheet
n°5	Glass, Aluminum, copper, MG-Si, Silver	Landfilling/Incineration
n°6	Glass, Aluminum, copper, MG-Si, Silver	Landfilling/Pyrolysis

* Landfilling. ** Incineration.

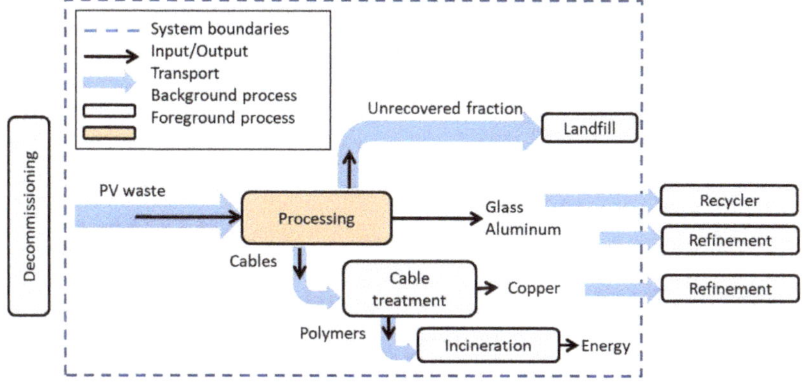

Figure 2. Exemplary process flow chart of a recycling approach (appr. n°1).

Approaches n°1 and n°2 correspond to the most common methods performed nowadays as described by Wambach et al. [7].

Approaches n°3 and n°4 are more specified processes described by Aryan et al. [14] which seek to determine the impacts of (A) fluorine-free and (B) fluorinated backsheets. These approaches consist essentially of the following processes: The initial preprocessing in which the frame, junction box and cables are removed. The cables are used for copper extraction while the plastic cover is sent to an incineration plant. The next stage comprises a series of mechanical steps such as shredding, crushing, sieving and separation of the different fractions. The recovered materials are glass and aluminum, while the unrecovered fraction is sent to a landfill and in some cases is partially incinerated.

In addition to basic bulk recovery, the approaches described as 'dedicated recovery' will be based on the work from Latunussa et al. [11] and correspond to approaches n°5 and n°6, presented in Table 3. The first step of these approaches consists in the dismantling of the modules; this includes removal of frame and wires, which later on receive the same subsequent treatments as described before. The following step is the glass separation by means of heat and mechanical detachment. The recovered glass is later refined, and the clean fraction is reused in other industrial processes, while the contaminated fraction is sent to a landfill. The remaining 'sandwich laminate' is cut into small pieces and sent to an incineration plant. Alternatively, depending on the backsheet composition, pyrolysis could be performed (appr. n°6). The outputs of this process are heat and fly ashes which are sent to a landfill and bottom ashes rich in silicon and other metals that are returned to the recycling plant for further processing. The final processing stage consists in a series of mechanical and chemical treatments such as sieving, leaching, filtration and electrolysis which aim to separate silicon, silver and copper scrap, as well as the remaining fractions of aluminum connectors. These usable fractions are sent to third party processors for further refinement while the sludge and liquid waste also produced within these processes are sent to a landfill.

2.4. Sensitivity Analysis

The initial analysis was performed assuming that recycling was done within the same time frame as the manufacturing of the modules (2010–2013, due to data availability). To determine how changing industrial parameters would affect the environmental performance of recycling technologies, a sensitivity analysis based on technology forecasts and the author's understanding is performed until the year 2040, in order to evaluate the impact of recycling modules produced over the last decade:

- Scenario n°1: Changing electricity mix with higher penetration of renewables. Figure 3 shows the variation in the energy mix as expected in the year 2040 based on actual conditions (Figure 3a) [24] and predictions (Figure 3b) [25].

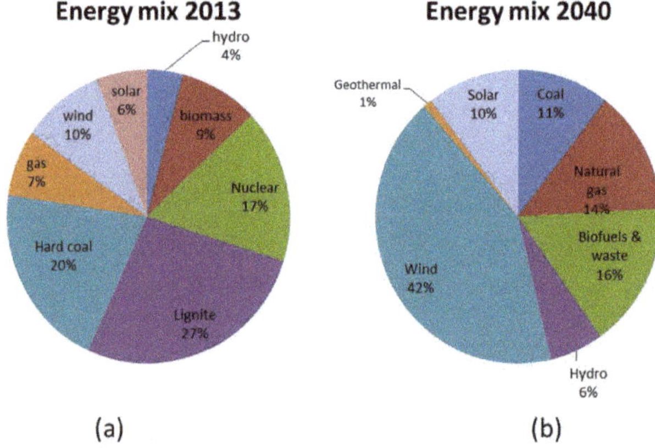

Figure 3. (**a**) Electricity mix for the 2013 baseline conditions [24], (**b**) Electricity mix for 2040 according to predictions from [25].

- Scenario n°2: Changing recycling efficiency due to enhanced processes or bigger flows of material. An energy demand of 60% of the value in the baseline scenario is assumed.
- Scenario n°3: Changing primary material content due to increased recycled fraction. Usually, environmental credits are given for the primary material content existing in the market at the moment the recycling is performed. However, since recycling is assumed to take place after 30 years of installment of the modules, its environmental performance would be better described if the calculations contained estimates of the primary material content in 2040. For the baseline scenario, these values are 48%, 56% and 77%, while for 2040, these are assumed to be 41%, 48% and 65% for aluminum, copper and silver respectively [26,27].
- Scenario n°4: Reduced transportation due to an efficient collection network. It will be assumed that the waste management network is optimized in a way that transportation demands are reduced 50% of the initial estimate.
- Scenario n°5: Year 2040, combined conditions. It is fair to assume that future conditions would be better represented by the combination of the previously described scenarios. To obtain a more accurate perspective of the overall impact in the future the assumptions in the previous scenarios have been taken into account in one single analysis. This scenario is shown for illustrative purposes and entails a high level of uncertainty.

3. Results

3.1. Electricity Production

Figures 4 and 5 show the impact of the different EoL approaches analyzed relative to the impacts of the production of 1 kWh of electricity. For mono-Si and multi-Si the same recycling approaches with the same impacts per kg of waste have been used. Due to different conversion efficiencies, different amounts of waste have to be recycled per kWh of electricity produced. The results are displayed in the negative axis as this describes reductions of the original profile.

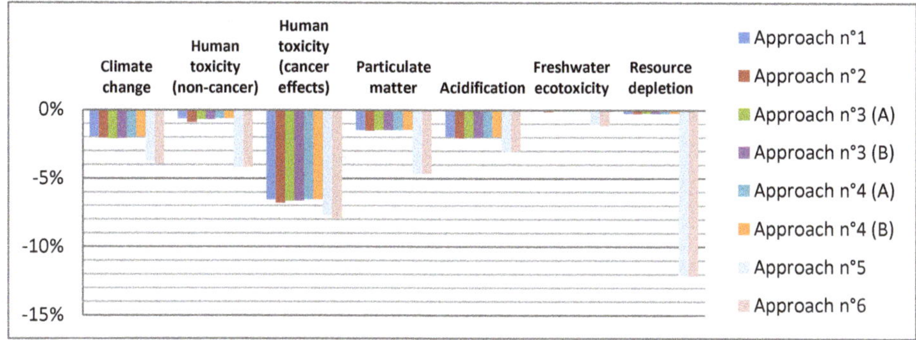

Figure 4. Potential contribution of waste management to the environmental profile of electricity production with mono-Si modules (for numerical values see Appendix A, Table A1).

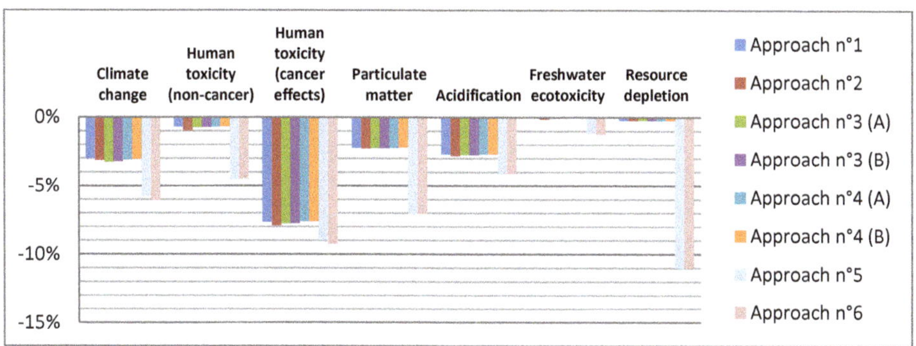

Figure 5. Potential contribution of waste management to the environmental profile of electricity production with multi-Si modules (for numerical values see Appendix A, Table A2).

- Mono-Si modules

Figure 4 shows that the recycling of mono-Si modules has a high capacity of reducing the impacts within the impact category '*human toxicity_(cancer effects)*' for every approach, with potential reductions ranging from −6.6% to −7.9% of the total. '*Resource depletion*' could be reduced by about 12% when dedicated recovery is considered, but it is negligible for any other approach. The lowest effect can be seen for '*freshwater ecotoxicity*', ranging from 0% to −1.1%. It can be seen that for the basic recycling approaches (n°1 to n°4) the environmental benefits are comparable in all the categories. Approaches n°5 and n°6 (dedicated recycling) also perform in a similar way in all impact categories.

- Multi-Si modules

The relative contribution of Multi-Si recycling in every impact category, with an exception of *'resource depletion'*, entails greater potential than the ones from Mono-Si recycling, as presented in Figure 5. For *'climate change'* and *'particulate matter'*, potential reductions of about 3% arise from approaches n°1 to n°4, and of about 6–7% from approaches n°5 or n°6. With potential reductions of up to 1.2%, effects from any approach on *'freshwater toxicity'* are negligible. Within 'resource depletion', impacts from approaches n°1 to n°4 are negligible while those from approaches n°5 and n°6 are 11% respectively.

3.2. Sensitivity Analysis

The results of the sensitivity analysis displayed in Figure 6 are presented in the following way: The 0% line defines the respective baseline approach; values on the positive side represent improvements with regard to the baseline while values on the negative side reflect detriments of performance or lesser benefits obtained.

The values displayed here are proportions of the results presented in Figures 4 and 5. Due to the drastic effects observed in the sensitivity analysis for the impact category *'freshwater ecotoxicity'*, the category has been plotted in an independent graph (Figure 6f). Given that the order of magnitude differs drastically from the other impact categories, a separate display has been chosen. In the baseline impact assessment for approaches n°1 to n°4B, net benefits in this category have been close to zero (Figures 4 and 5). Therefore, small parameter variations as disposed by the sensitivity analysis will easily translate into a large percentage of difference from the baseline. For example, in approach n°4B (Figure 6f) a decline of 1336% of the 0.007% environmental benefits for multi-Si modules (Figure 4) translate into absolute numbers of only 0.1% additional burdens with respect to PV electricity production without EoL management.

From scenario n°1, it can be seen that a higher penetration of renewables lowers the benefits obtained from incineration (appr. n°2) where the recovered energy is assumedly substituting generation in a cleaner grid. This is, however, subject to whether the substituted energy is effectively electricity or just heat. For all other approaches, mostly net environmental benefits can be observed, except for the category *'resource depletion'*, since the input electricity required to run the processes will come from sources with high criticality in this category, e.g., increased silver demand for a greater share of PV. A changing recycling efficiency as analyzed in scenario n°2 entails improvements of between 0.2% and 5.3% for most categories, whereas *'human toxicity (non-cancer)'* benefits the most. A variation in primary material content in PV modules (scenario n°3) leads to fewer environmental benefits from recycling in every category, due to there being less primary material content to be replaced. For the special case of *'freshwater ecotoxicity'*; however (Figure 6f), the horizontal −100% line indicates that the original benefits from recycling face a reduction of 100%. In other words, benefits went down to zero. Values above this line mean that recycling will still lead to environmental benefits, fewer than in the baseline (Figures 4 and 5). Values below this, as is the case in this category, mean that recycling is now leading to additional burdens on the environment. Reducing transport distances as assumed in scenario n°4 has on average the greatest influence, especially in the category *'resource depletion'*, with improvements up to 20%. What can be seen from scenario n°5 is that when combining all the above assumptions, recycling has smaller relative environmental benefits as it would have under current conditions (Figures 4 and 5). Still, recycling would lead to overall environmental benefits of up to 17.2% for 'human toxicity (non-cancer)', with an exception for *'freshwater toxicity'*, where the burdens overlay the benefits.

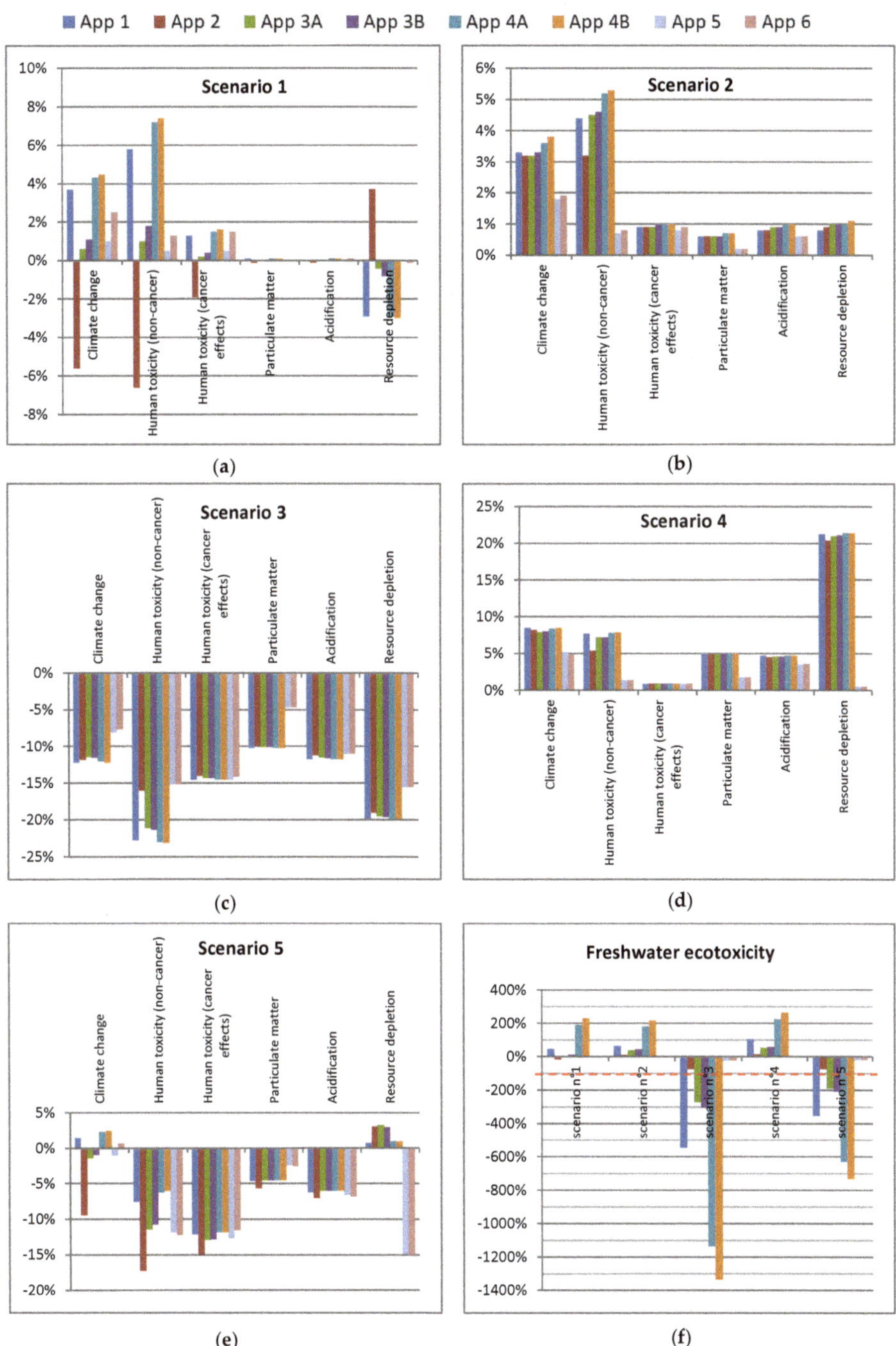

(**a**)

(**b**)

(**c**)

(**d**)

(**e**)

(**f**)

Figure 6. Results from the different scenarios considered in the sensitivity analysis: (**a**) high penetration of renewables, (**b**) increased processing efficiency, (**c**) lower primary material content, (**d**) Optimized collection network, (**e**) combined conditions in 2040, and (**f**) specific plot for *'freshwater ecotoxicity'*.

3.3. Greenhouse Gas Emissions

Potentially avoided greenhouse gas emissions are calculated for each approach as a measure of the global warming potential. For practical purposes, these are displayed as kg CO_2-eq per kilogram of recycled waste, since these units make the calculation of cumulated greenhouse gas over a time period more comprehensible. It can be seen in Figure 7 that the potential to avoid CO_2 emissions is highest for approaches n°5 and n°6, almost twice as high as the CO_2 savings from the basic recycling approaches.

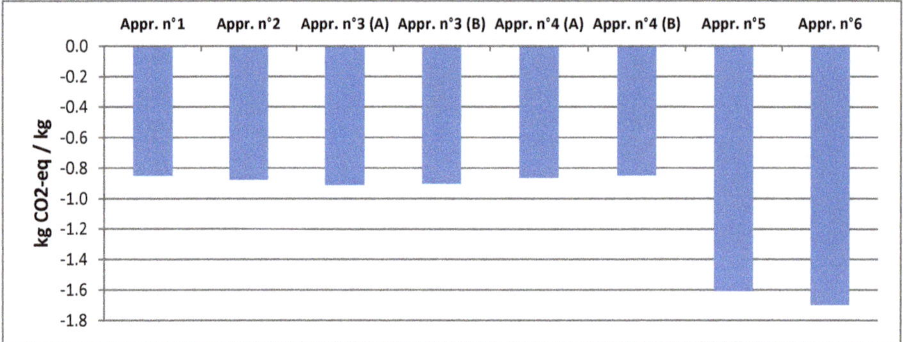

Figure 7. Potentially avoided greenhouse gas production of each approach.

3.4. Energy Payback Time

Table 4 shows the cumulative energy demand (CED) and Energy Payback Time (EPBT) of a 3 kWp plant without the EoL phase as a reference to evaluate the contribution of EoL management. The table further shows the potential reduction in CED and EPBT that arises from the energy savings obtained through the different recycling scenarios. For mono-Si systems the EPBT reduction potential lies between 2.4% and 4.7%. In the case of multi-Si, reductions of about 3.6% up to 7.1% are plausible. This equals an improvement of about one to three months for the EPBT.

Table 4. Energy metrics with and without effects from EoL management.

				Appr. 1	Appr. 2	Appr. 3 (A) Fluor-Free	Appr. 3 (B) Fluor	Appr. 4 (A) Fluor-Free	Appr. 4 (B) Fluor	Appr. 5	Appr. 6
Mono-Si	CED (GJ)	127.3	CED saving (GJ)	−3.00	−3.70	−3.24	−3.20	−3.05	−3.05	−5.93	−5.69
	EPBT (Years)	4.5	EPBT incl. EoL (years)	4.36	4.34	4.35	4.36	4.36	4.36	4.26	4.27
			% Reduction	2.4%	2.9%	2.5%	2.5%	2.4%	2.4%	4.7%	4.5%
Multi-Si	CED (GJ)	88.6	CED saving (GJ)	−3.20	−3.95	−3.45	−3.41	−3.25	−3.25	−6.32	−6.07
	EPBT (Years)	3.1	EPBT incl. EoL (years)	3.00	2.97	2.99	2.99	3.00	3.00	2.89	2.90
			% Reduction	3.6%	4.5%	3.9%	3.8%	3.7%	3.7%	7.1%	6.9%

4. Discussion

The analysis performed within this study shows that dedicated material recovery has a significant influence on reducing the global warming potential of PV electricity. It can be seen that implementing dedicated recovery approaches could contribute, to a great extent, to the reduction of greenhouse gas emissions inherent in module production. The high savings of CO_2 emissions come from the reduced need for primary materials which are now being substituted by those recovered through recycling. Table 5 shows the potentially avoided greenhouse gas emissions in tons of CO_2-eq when combining the results obtained from the analysis with the average amounts of cumulative PV waste as predicted by IRENA [28] until the year 2040 and 2050. Minimum and maximum values are given in consideration of the approaches with the lowest and highest potentials:

Table 5. Potentially avoided greenhouse gas emissions.

Year	Area	PV Waste (Million Tons)	Cumulative Million Ton CO_2-eq	
			MIN	MAX
2040	Germany	2.4	2.04	4.08
	Global	23.5	19.95	39.90
2050	Germany	4.3	3.65	7.30
	Global	69	58.57	117.16

By 2040, the avoided greenhouse gas emissions from recycled PV material in Germany could add up to two to four million tons, which equals around 10% of the possibly avoided average global emissions. By 2050, the same analysis results in possible emissions avoided of around three to seven million tons in Germany, accounting for 6% of the global greenhouse gas emissions that could be avoided by PV recycling. These estimations represent an ideal scenario under the assumption that all the PV waste produced until then will be recycled using any of the studied approaches. To put this in perspective, the average global CO_2 emissions in 2014 were about 5 tons per capita [29].

The results show that current module recycling has moderate to low influence on the EPBT of PV systems. The maximum potential EPBT reduction is found for multi-Si and is of about three months. It is worth mentioning again that the CED used to calculate the EPBT takes into account the energy demand of the whole plant while the energy savings obtained through recycling were calculated only for the treatment of the modules. Additional energy savings and thus higher improvements in the EPBT will be obtained if the whole system undergoes recycling.

This study was conducted for the specific conditions in Germany, which differ greatly from those considered in other studies. The same applies for the boundary conditions, as is the case for some transportation steps and final disposal processes, which were often not taken into account. Additionally, different initial module compositions were used in all of the studies analyzed. For example, the study from [12], where dedicated recovery was initially studied, assumes a module mass of 22 kg, which is 37.5% heavier than the model studied here. These facts make direct comparability impossible. Clarity and transparency when describing the system boundaries become of high relevance for the correct understanding of these differences.

Cost-effectiveness of investments in recycling infrastructure will be heavily influenced by the waste composition and volumes of material to process. Accurate estimation of these parameters is therefore necessary when assessing technical and economic viability for a proper settlement of a recycling industry. Miscalculations, in a business where profits are already low, will most likely lead to economic loss. This can be exemplified with the silver content present in the modules. Reasonably, and due to scarcity and the environmental burden of its mining, research efforts are already aiming at reducing the amount of metal used. However, if the silver content reaches a certain minimum, dedicated recycling could lose economic appeal and, if no other recovery technique has been developed, silver will probably, at a slow rate, end up in landfills.

From the sensitivity analysis it can be seen that the benefits of energy recovery are lower when the production energy comes from renewable sources. Scenario n°1 shows that energy recovery (as in approach n°2) does not have the same impacts in a system based on a cleaner grid as when substituting electricity coming from a more fossil-based grid (as in approach n°1). Additionally, it can be seen that the primary material content is the parameter with the highest influence on the environmental benefits from waste management. It was demonstrated that reducing the primary material content in the modules (scenario n°3) lowers the potential benefits of recycling to a great extent. In other words, the more recycled material is used, the lower the impacts are from module recycling, as can especially be observed in the impact category 'Freshwater ecotoxicity'. With the assumed primary material content of the sensitivity analysis, this effect overcomes the estimated increased benefits obtained from recycling (as observed in scenario 5); however, it is also expected that the impacts of module manufacturing will lower considerably under the same consideration thus improving the overall footprint of electricity.

PV waste management will gain relevance when the impacts of electricity generation become smaller. PV recycling benefits will be more significant for low-impact optimized PV electricity production, for which the benefits of recycling will be found in a similar order of magnitude as the quantified environmental impacts of electricity generation. This can be achieved with better and more efficient industrial manufacturing processes both energy- and material wise.

5. Conclusions

Within this contribution, six different approaches of recycling of PV modules have been compared with respect to their impact on the environmental footprint of electricity production from a standard PV system. As has been shown, dedicated recycling as presented in approach n°5 and n°6 carries the best potential to improve the environmental footprint of PV electricity production. Future analysis should take into account the benefits of recycling of the whole system, including inverter and other BOS components, to further highlight the potential of environmental benefits through recycling. The overall balance shows that the benefits overlay the burdens in all the approaches and for all the impact categories studied, improving the environmental profile of PV and lowering its EPBT. The specific treatment of the backsheet layer, as seen in approaches n°3 and n°4, does not have a major impact on the environmental profile of recycling the modules under study due to their low mass fraction. It should, however, be taken into consideration that certain materials contained, like the fluorine in the backsheet, can become problematic when taking into account the forecasted increase in waste material flow. Efforts in research should focus on reducing the unrecovered fraction that is being landfilled containing polymers, silicon and metals. The ITRPV technology roadmap [30] already contemplates the content reduction for some of these as a countermeasure. While the processing of PV waste until 2040, as studied here, will be characterized by high silver and silicon content, the roadmap estimates significant material content reductions that could make recycling activities unfeasible from the economic point of view driving more material into landfills. Research is thus required into that subject.

The development of an appropriate recycling network will not only bring benefits in environmental aspects but will also have a great impact on the economics and financial balances of the logistic schemes. Such an assessment must also take into account a policy and legislative framework so that the outcomes meet realistic conditions. Additionally, it is very likely that trade of waste between neighboring countries will arise, given the small amount of installed capacity of certain regions which might be more conveniently managed by imports and exports, adding complexity to the analysis.

There is still plenty of room for improvement in PV waste management. Even when the results from the current approaches seem to entail moderate contributions to the ecological footprint of electricity, further development of these and new other techniques will turn recycling into a relevant tool to make photovoltaics an exemplary technology for the energy transition. Additionally, recycling should still be studied as a whole, including the processing of the BOS components as well as the environmental credits that come with it. This will most likely lead to a greater reduction of the electricity production's footprint.

Author Contributions: Conceptualization, S.H. and S.P.B.; methodology, S.H. and S.P.B.; software, S.P.B.; validation, S.H.; formal analysis, S.P.B.; investigation, S.H. and S.P.B.; resources, S.H. and S.P.B.; data curation, S.H.; writing—original draft preparation, S.H. and S.P.B.; writing—review and editing, K.-A.W.; visualization, S.P.B.; supervision, S.H.; project administration, K.-A.W.; funding acquisition, K.-A.W. All authors have read and agreed to the published version of the manuscript.

Funding: This research received no external funding.

Conflicts of Interest: The authors declare no conflict of interest.

Appendix A

Table A1. Potential contribution of waste management to the environmental profile of electricity production with mono-Si modules.

	Climate Change	Human Toxicity (Non-Cancer)	Human Toxicity (Cancer Effects)	Particulate Matter	Acidification	Freshwater Ecotoxicity	Resource Depletion
Appr. n°1	−2.0%	−0.6%	−6.6%	−1.4%	−2.0%	0.0%	−0.2%
Appr. n°2	−2.0%	−0.9%	−6.8%	−1.5%	−2.0%	−0.1%	−0.2%
Appr. n°3 (A)	−2.1%	−0.7%	−6.7%	−1.5%	−2.0%	0.0%	−0.2%
Appr. n°3 (B)	−2.1%	−0.7%	−6.6%	−1.5%	−2.0%	0.0%	−0.2%
Appr. n°4 (A)	−2.0%	−0.6%	−6.5%	−1.4%	−2.0%	0.0%	−0.2%
Appr. n°4 (B)	−2.0%	−0.6%	−6.5%	−1.4%	−2.0%	0.0%	−0.2%
Appr. n°5	−3.7%	−4.2%	−7.7%	−4.6%	−3.0%	−1.1%	−12.1%
Appr. n°6	−3.9%	−4.1%	−7.9%	−4.6%	−3.0%	−1.1%	−12.1%

Table A2. Potential contribution of waste management to the environmental profile of electricity production with multi-Si modules.

	Climate Change	Human Toxicity (Non-Cancer)	Human Toxicity (Cancer Effects)	Particulate Matter	Acidification	Freshwater Ecotoxicity	Resource Depletion
Appr. n°1	−3.0%	−0.7%	−7.6%	−2.2%	−2.7%	0.0%	−0.2%
Appr. n°2	−3.1%	−1.0%	−7.9%	−2.3%	−2.8%	−0.1%	−0.2%
Appr. n°3 (A)	−3.3%	−0.7%	−7.8%	−2.2%	−2.7%	0.0%	−0.2%
Appr. n°3 (B)	−3.2%	−0.7%	−7.7%	−2.2%	−2.7%	0.0%	−0.2%
Appr. n°4 (A)	−3.1%	−0.7%	−7.6%	−2.2%	−2.7%	0.0%	−0.2%
Appr. n°4 (B)	−3.0%	−0.7%	−7.6%	−2.2%	−2.7%	0.0%	−0.2%
Appr. n°5	−5.8%	−4.5%	−9.0%	−7.1%	−4.1%	−1.2%	−11.0%
Appr. n°6	−6.1%	−4.5%	−9.2%	−7.1%	−4.1%	−1.2%	−11.0%

References

1. Choi, J.-K.; Fthenakis, V. Crystalline silicon photovoltaic recycling planning. Macro and micro perspectives. *J. Clean. Prod.* **2014**, *66*, 443–449. [CrossRef]
2. McDonald, N.C.; Pearce, J.M. Producer responsibility and recycling solar photovoltaic modules. *Energy Policy* **2010**, *38*, 7041–7047. [CrossRef]
3. D'Adamo, I.; Miliacca, M. Rosa, Paolo Economic Feasibility for Recycling of Waste Crystalline Silicon Photovoltaic Modules. *Int. J. Photoenergy* **2017**, *2017*, 1–6. [CrossRef]
4. Heinemann, D.; Jürgens, W.; Knecht, R.; Parisi, J. A historical module: The TSG MQ 36/0 solar panel. *Einblicke* **2011**, *54*, 6–11.
5. Jordan, D.C.; Kurtz, S.R. Photovoltaic Degradation Rates—An Analytical Review. *Prog. Photovolt. Res. Appl.* **2013**, *21*. [CrossRef]
6. Virtuani, A.; Caccivio, M.; Annigoni, E.; Friesen, G.; Chianese, D.; Ballif, C.; Sample, T. 35 years of photovoltaics: Analysis of the TISO-10-kW solar plant, lessons learnt in safety and performance—Part 1. *Prog. Photovolt. Res. Appl.* **2019**, 328–339. [CrossRef]
7. Wambach, K. *Life Cycle Inventory of Current Photovoltaic Module Recycling Processes in Europe*; IEA PVPS Task12, Subtask 2, LCA Report IEA-PVPS T12-12:2017; National Renewable Energy Lab. (NREL): Golden, CO, USA, 2017.

8. Held, M. LCA screening of a recycling process for silicon based PV modules. In Proceedings of the Symposium conducted at PV Cycle Conference, Rome, Italy, 12–13 February 2018.

9. Giacchetta, G.; Leporini, M.; Marchetti, B. Evaluation of the environmental benefits of new high value process for the management of the end of life of thin film photovoltaic modules. *J. Clean. Prod.* **2013**, *51*, 214–224. [CrossRef]

10. Duflou, J.; Peeters, J.; Altamirano, D.; Bracquene, E.; Dewulf, W. Demanufacturing photovoltaic panels: Comparison of end-of-life treatment strategies for improved resource recovery. *CIRP Ann. Manuf. Technol.* **2018**, *67*, 29–32. [CrossRef]

11. Latunussa, C.E.L.; Ardente, F.; Blengini, G.A.; Mancini, L. Life Cycle Assessment of an innovative recycling process for crystalline silicon photovoltaic panels. *Sol. Energy Mater. Sol. Cells* **2016**, *156*, 101–111. [CrossRef]

12. Latunussa, C.; Mancini, L.; Blengini, G.; Ardente, F.; Pennington, D. *Analysis of Material Recovery from Photovoltaic Panels*; EUR 27797; Publications Office of the European Union: Luxembourg, 2016. [CrossRef]

13. Ardente, F.; Latunussa, C.; Blengini, G. Resource efficient recovery of critical and precious metals from waste silicon PV panel recycling. *Waste Manag.* **2019**, *91*, 156–167. [CrossRef] [PubMed]

14. Aryan, V.; Danz, P.; Maga, D.; Brucart, M.F. *End-of-Life Pathways for PV Backsheets*; Final Report; Fraunhofer UMSICHT: Oberhausen, Germany, 2017.

15. Komoto, K.; Lee, J.-S. *End-of-Life Management of Photovoltaic Panels: Trends in PV Module Recycling Technologies*; IEA PVPS Task12, Subtask 1, Recycling Report IEA-PVPS T12-10:2018; National Renewable Energy Lab. (NREL): Golden, CO, USA, 2018.

16. Dias, P.; Javimczik, S.; Benevit, M.; Veit, H. Recycling WEEE. Polymer characterization and pyrolysis study for waste of crystalline silicon photovoltaic modules. *Waste Manag.* **2017**, *60*, 716–722. [CrossRef] [PubMed]

17. Dias, P.; Javimczik, S.; Benevit, M.; Veit, H.; Bernardes, A.M. Recycling WEEE. Extraction and concentration of silver from waste crystalline silicon photovoltaic modules. *Waste Manag.* **2016**, *57*, 220–225. [CrossRef] [PubMed]

18. Huang, W.-H.; Shin, W.J.; Wang, L.; Sun, W.-C.; Tao, M. Strategy and technology to recycle wafer-silicon solar modules. *Sol. Energy* **2017**, *144*, 22–31. [CrossRef]

19. Wade, A.; Stolz, P.; Frischknecht, R.; Heath, G.; Sinha, P. The Product Environmental Footprint (PEF) of photovoltaic modules-Lessons learned from the environmental footprint pilot phase on the way to a single market for green products in the European Union. *Prog. Photovolt. Res. Appl.* **2018**, *26*, 553–564. [CrossRef]

20. JRC European Commission. *Recommendations Fir Life Cycle Impact Assessment in the European Context. ILCD Handbook*; JRC European Commission: Ispra, Italy, 2011; ISBN 978-92-79-17451-3.

21. Jungbluth, N.; Stucki, M.; Büsser, S.; Flury, K.; Frischknecht, R. *Life Cycle Inventories of Photovoltaics*; ESU-Services Ltd.: Uster, Switzerland, 2012.

22. Frischknecht, R.; Heath, G.; Raugei, M.; Sinha, P.; de Wild-Scholten, M.; Fthenakis, V.; Kim, H.C.; Alsema, E.; Held, M. *Methodology Guidelines on Life Cycle Assessment of Photovoltaic Electricity*, 3rd ed.; IEA PVPS Task 12, International Energy Agency Photovoltaic Power Systems Programme; Report IEA-PVPS T12-06:2016; IEA: Paris, France, 2016; ISBN 978-3-906042-38-1.

23. Philipps, S.; Warmuth, W. Photovoltaics Report. Fraunhofer Institute for Solar Energy Systems ISE. Available online: https://www.ise.fraunhofer.de/en/publications/studies/photovoltaics-report.html (accessed on 9 March 2020).

24. Fraunhofer ISE Energy Charts. Fraunhofer Institute for Solar Energy Systems ISE. Available online: https://www.energy-charts.de (accessed on 9 March 2020).

25. International Energy Agency. *Energy Policies of IEA Countries*; 2013 Review: OECD; IEA: Paris, France, 2013; ISBN 978-92-64-19076-4.

26. Stolz, P.; Frischknecht, R.; Wambach, K.; Sinha, P.; Heath, G. *Life Cycle Assessment of Current Photovoltaic Module Recycling*; IEA PVPS Task 12; International Energy Agency Power Systems Programme; Report IEA-PVPS T12-13:2018; IEA: Paris, France, 2017.

27. O'Connell, R.; Alexander, C.; Alway, B.; Litosh, S.; Nambiath, S.; Wiebe, J.; Yao, W.; Norton, K.; Li, S.; Aranda, D.; et al. *World Silver Survey 2018*; Produced for The Silver Institute by the GFMS team at Thomson Reuters; The Silver Institute: Washington, DC, USA, 2018; ISBN 978-1-880936-31-3.

28. IRENA and IEA-PVPS. *End-of-Life Management: Solar Photovoltaic Panels*; International Renewable Energy Agency and International Energy Agency Photovoltaic Power Systems; IEA: Paris, France, 2016; ISBN 978-92-95111-99-8.

29. The World Bank. CO2 Emissions (Metric Tons Per Capita) 2018. Available online: https://data.worldbank.org/indicator/EN.ATM.CO2E.PC?end=2014&start=1960 (accessed on 9 March 2020).

30. VDMA Photovoltaic Equipment. *International Technology Roadmap for Photovoltaic (ITRPV)*, 8th ed.; Results including Maturity Report; VDMA Photovoltaic Equipment: Frankfurt, Germany, 2017.

 *energies*

Article

Environmental Assessment of a Coal Power Plant with Carbon Dioxide Capture System Based on the Activated Carbon Adsorption Process: A Case Study of the Czech Republic

Kristína Zakuciová [1,2,*], Jiří Štefanica [1], Ana Carvalho [3] and Vladimír Kočí [2,*]

[1] ÚJV Řež, a. s., Hlavní 130, Řež, 250 68 Husinec, Czech Republic; jiri.stefanica@ujv.cz

[2] Department of Environmental Chemistry, Faculty of Environmental Technology, University of Chemistry and Technology, Prague, Technická 5, 166 28 Praha 6, Czech Republic

[3] CEG-IST, Instituto Superior Técnico, Universidade de Lisboa, 1049-001 Lisbon, Portugal; anacarvalho@tecnico.ulisboa.pt

* Correspondence: kristina.zakuciova@ujv.cz (K.Z.); vlad.koci@vscht.cz (V.K.); Tel.: +420-775-364 032(K.Z.)

Received: 3 April 2020; Accepted: 19 April 2020; Published: 4 May 2020

Abstract: The Czech Republic is introducing new technological concepts for mitigation of greenhouse gases (GHG) in coal-based energy industries. One such technology, in power plants, is post combustion CO_2 capture from flue gases by activated carbon adsorption. A life cycle assessment (LCA) was used as the assessment tool to determine the environmental impacts of the chosen technology. This article focuses on a comparative LCA case study on the technology of temperature-swing adsorption of CO_2 from power plant flue gases, designed for the conditions of the Czech Republic. The LCA study compares the following two alternatives: (1) a reference power unit and (2) a reference power unit with CO_2 adsorption. The most significant changes are observed in the categories of climate change potential, terrestrial acidification, and particulate matter formation. The adsorption process shows rather low environmental impacts, however, the extended LCA approach shows an increase in energy demands for the process and fossil depletion as a result of coal-based national energy mix. The feasibility of the study is completed by the preliminary economical calculation of the payback period for a commercial carbon capture unit.

Keywords: carbon dioxide capture; activated carbon; environmental impacts; life cycle assessment

1. Introduction

In the Czech Republic, around 52.4% of the total gross electricity production (87.6 TWh) is generated from coal, which is approximately 41% of the energy mix [1]. The Czech industry emitted around 120.5 million tons CO_2, with the largest proportion of 98 million tons in 2017 [2]. The major problem of reducing CO_2 emissions and its sequestration lies with the implementation of carbon capture and storage (CCS) systems. It is commonly perceived that the implementation of CCS decreases local CO_2 emissions and, if applied globally, supports mitigation efforts concerning the anthropogenic contribution to climate change. However, CCS technologies can be related to more complex, unexpected, or non-obvious environmental impacts. Thus, there is a common need for a holistic environmental approach that assesses and evaluates new industrial techniques and applications, which can significantly influence the environment. Results and conclusions based on such a holistic approach support the decision making of scientists, environmentalists, and governments concerning the implementation of new techniques and allow the environmental analysis in a wider and more detailed context. Life cycle assessment (LCA) is a tool for the assessment of technologies such as CCS, from both an environmental and sustainable point of view. A LCA uses different database methods and offers several approaches for the optimization of the processes and subsequent calculation of the environmental suitability of the chosen technology.

Considering several scientific references available (summarized in Table 1), there is a need for systematic environmental studies and reports that are built on local and site-specific operational data of the pilot CO_2 capture plants. Most of the available studies deal with averaged environmental data from LCA databases and use hypothetical and mathematical models to describe a specific CO_2 capture system. Current LCA studies focus on the comparison of several sorbents. Most studies specifically target post-combustion capture using monoethanolamine (MEA) sorbent. Comparative LCA analyses of MEA and potassium carbonate solvents have revealed that potassium carbonate solvents contribute to lower environmental impacts [3]. Compared to MEA, this process shows a reduction of emissions and energy cost savings. In [4], Manuilova compared power units with MEA sorbents and without CO_2 capture and corroborated the general conclusions of other research works which reported a decrease in SO_2 and particulate matter and an increase in NOx emissions due to MEA emissions. Additionally, increased levels of smog, water consumption, and water toxicity were also calculated. Koornneef [5] stated that SO_2 and solid particle emissions decreased due to CCS implementation, but NOx, NH_3, and volatile organic compound emissions increased due to the utilization of amine and ammonia-based absorbents for CO_2 capture. Petrakopoulou and Tsatsaronis [6] evaluated the environmental impacts of electricity generated by natural gas and coal power plants. The facts published by others [5,7,8] also revealed that CCS required additional energy consumption, leading to a decrease in power plant efficiency and a greater potential of fossil fuel consumption. One recently studied technology using $CaCO_3$ as the solvent, was the CaO looping system. A comparative LCA study by Clarens et al. [9] showed that CaO looping decreased CO_2 emissions by 73 percent. A suitable alternative to the absorption processes based on amines can be adsorption separation of CO_2. Under certain circumstances, adsorption can exceed CO_2 absorption if the corrosive absorption medium is replaced by a solid sorbent, the absorption media treatment is removed, and the operational costs are decreased due to lower energy consumption during the regeneration step as compared with the regeneration of liquid solvents.

Regarding the activated carbon (AC) adsorption process, our literature survey revealed a study published by Hjaila et al. [10] about LCA of the production of AC from olive waste cakes. This study highlighted the most significant impacts of acidification and eutrophication due to the use of H_3PO_4 and the electricity demand for the AC production process. The software used in this study was SimaPro 7.3 based on the Ecoinvent database and the assessment method was CML 2 Baseline 2000. Another recent study associated with LCA and AC was conducted by Arena et al. [11]. This study evaluated the LCA of the production of AC from coconut shells. The life cycle inventory was based on the Ecoinvent 3.0 database, using CML-2001 as the LCA method and the software GaBi 6.0. The results demonstrated that global warming potential, acidification, and human toxicity represent the most significant environmental impacts. The environmental burdens are mainly associated with the production of electrical energy (based on hard coal) used in the production process of AC.

Currently, in the Czech Republic, CCS is in the stage of technical drafts and optimization of the systems, as well as economic assessment of the optimized solutions. These studies can be subsequently supported by the evaluation of environmental gains and impacts using the LCA approach. The environmental impacts have already been assessed for two technical solutions in the Czech Republic, i.e., a power plant with ammonia scrubbing of CO_2 [12] and a power plant with high-temperature carbonate looping [13]. These processes face various operational issues that could be omitted by using low-temperature solid sorbents, such as zeolites or activated carbon.

Table 1. Summary of life cycle assessment (LCA) studies on carbon capture and storage (CCS) technologies, including methods and results. PC, pulverized coal power plant; NGCC, natural gas combined cycle power plant; IGCC, integrated gasification combined cycle cycle power plant; SC-PC, subcritical pulverized coal; SPC-PC, supercritical pulverized coal; GWP, global warming potential; POP, photochemical oxidation potential; AP, acidifying potential; EP, eutrophication potential; PM10, PM-10 equivalents; ADP, abiotic depletion; ODP, ozone layer depletion; HTP, human toxicity; FAETP, freshwater aquatic ecotoxicity; MAETP, marine aquatic ecotoxicity; TET, terrestrial ecotoxicity.

References	Scope	LCA Software and Database or Method	Significant LCA Impacts
Koornneef et al., 2008	Comparative LCA study of SC-PC and SPC-PC power plants with and without CCS CO_2 capture by MEA absorption	CML 2 baseline 2000 V2.03 SW EcoInvent database v1.3	ADP, GWP, ODP, HTP, FAETP, MAETP, POP, AP, EP
Odeh and Cockerill, 2008	Comparative LCA study of SPC-PC, NGCC, and IGCC power plants with and without CCS CO_2 capture by MEA absorption	SimaPro SW EcoInvent	GWP
Singh et al., 2012	Feasibility study of SC-PC and SPC-PC power plants with and without CO_2 capture, FGD, and SCR. CO_2 capture by MEA absorption	EcoIndicator 99	According to EcoIndicator 99
Hjaila et al., 2013	LCA of AC production from olive waste cakes in Tunisia	Simapro SW CML 2 Baseline 2000 Ecoinvent v 2.2 database	AP, EP
Grant et al., 2014	Comparative life cycle assessment of K_2CO_3 and MEA solvents for CO_2 capture from post combustion flue gases	CML 2001 methodology Human and ecotoxicity based on USETox method	Potassium carbonate solvents improves TET, carcinogen emissions and energy consumption
Manuilova et al., 2014	Case study, Boundary Dam Power Station CO_2 capture by MEA absorption	Operational data by Cenovus Energy, Canadian provinces, USA and other countries	GWP, AP, EP, ODP, HTP, FAETP
Petrakopoulou and Tsatsaronis, 2014	Comparative LCA study of PC and NGCC power plants with and without CCS CO_2 capture by MEA absorption and chemical looping combustion	EcoIndicator 99	According to EcoIndicator 99
Clarens et al., 2016	CaO looping vs. MEA based adsorption	ReCiPe Midpoint for Europe, v 1.04	Reductions in GWP category for CaO looping (73%), conventional MEA (66%), advanced MEA (72%)
Arena et al., 2016	LCA of activated carbon production from coconut shells	GaBi 6.0 software CML-2001databank Ecoinvent 3.0	HTP, AP, GWP

This study aims to analyze an integrated system of Czech brown coal power unit with CO_2 capture based on an AC system, drafted and optimized as part of a national-scale project [14]. The main goal of this study is to identify key environmental impacts and the economic feasibility of the unit with integrated capture of CO_2. The study intends to use operational data from pilot testing of the CO_2 capture method based on adsorption to evaluate environmental impacts on the national conditions. In order to perform a holistic and systematic evaluation, the LCA study was performed in different levels of decision-making processes such as characterization, normalization, Pareto analysis, and input-output analysis. This study was completed by the economical evaluation of such CCU (carbon capture and utilization) unit. In summary, the main contributions of the paper are:

- An extensive literature survey of current LCA studies on various carbon capture technologies;
- Performance of a holistic environmental LCA case study on the unique technology of activated carbon adsorption of CO_2 in Czech energy conditions;
- An LCA case study based on operational data of an operating 250 MW power unit from a national-scale project;
- Performance of robust LCA analyses for the adsorption capture process conducted at four decision making levels (characterization, normalization, Pareto analysis, and input-output analysis);
- An economical calculation of commercial CCU unit with payback period;

- The identification of areas within a carbon capture technological process that can be improved to enhance environmental and economic performance.

The structure of this paper is comprised of the definition of a life cycle approach according to related international standards; the definition of the case study, i.e., description of Czech power unit and adsorption technology; implementation of the LCA methodology for the case study; characterization and interpretation of the environmental results; and finally, an economic evaluation of the case study.

2. Methods

2.1. Environmental Assessment: The Life Cycle Assessment Method

LCA is a tool to evaluate the environmental impacts of products and processes, such as the production of electricity. The LCA method is certified and defined by international standards ISO 14040 [15] as a cradle-to-grave analysis which facilitates a comparison of technological processes regarding their environmental characteristics. This includes all phases of a product´s lifetime. According to the international standards, LCA consists of four steps, i.e., goal and scope definition, inventory analysis, impact assessment, and interpretation which are described as follows:

Step 1 Goal and scope definition

The depth of the analysis is determined by the goal of LCA. This study aimed to create a model and analyze the potential environmental impacts of CO_2 adsorption on activated carbon connected to a 250 MW brown coal power unit. Therefore, two scenarios were considered:

1. Scenario 1 which is the assessment of electricity production by the 250 MW coal power unit.
2. Scenario 2 which is the assessment of the electricity production by the 250 MW coal power unit integrated with the CO_2 adsorption unit.

For the comparison of the LCA results, a compatible functional unit must be defined for each scenario. The functional unit for both scenarios was defined as the power capacity (250 MW) of the power unit. System boundaries included the operational part of the power plant and activated carbon production, emission treatment, CO_2 capture process, and waste generation. System boundaries excluded CO_2 compression, transport, and final storage due to limited information about CO_2 storage in the Czech conditions. Moreover, the environmental assessment included the operational part of the power plant rather than the life cycle of the CO_2. Therefore, the approach used was considered to be "cradle-to-gate."

Step 2 Life cycle inventory (LCI)

LCI starts with data collection and model construction, in compliance with the goal and scope definition, followed by the collection of input-output data and calculation of the resource depletion and emission release during the production process. Operational data for the case study was collected from the pilot project report [14] with detailed technical requirements and descriptions.

Step 3 Life cycle impact assessment (LCIA)

LCIA can be divided into three steps, i.e., characterization, classification, and normalization. For the characterization and classification steps, the impact potentials were calculated. Normalization is an optional step of LCIA. Normalization uses a common reference impact to express results after the characterization and supports the comparison between alternative scenarios by using reference numerical scores. Normalization also gives a basis for comparing different types of impact categories [16]. The additional approach of Pareto analysis was chosen for defining the most significant impact categories. The selected method for the LCA analyses of the study was the ReCiPe 1.08 method. This method combines the problem-oriented approach of the CML method and the damage-oriented approach of EcoIndicator 99. The ReCiPe method is characterized by the following 18 midpoint indicators: ozone depletion (OD), human toxicity (HT), ionizing radiation (IR), photochemical oxidant formation (POF), particulate matter formation (PMF), terrestrial acidification (TA), climate change (CC), terrestrial ecotoxicity (TET), agricultural land occupation (ALO), urban land occupation (ULO), natural land transformation (NLT), marine ecotoxicity (MET), marine eutrophication (ME), fresh water eutrophication (FE), fresh water ecotoxicity (FET), fossil depletion (FD), metal depletion (MD), and water depletion(WD); in addition, there are three endpoint indicators, i.e., human health, ecosystems, and resource surplus costs [17,18].

Step 4 Characterization and interpretation

This step is based on the results of the LCIA phase. The results are defined as a potential environmental impact. The environmental impact is calculated using characterization methods that associate the scale of a pollutant emission to selected characterization factors. The interpretation of the results includes an identification of significant issues, evaluation of completeness, and sensitivity of results. The interpretation phase also includes key conclusions and recommendations [19]. Normalized results are further assessed by the statistical method of Pareto´s rule (80/20 rule), which states that 20 percent from all impact categories cause 80 percent of the total environmental impacts [20,21].

In our case study LCA, for the Pareto analysis, we chose values after normalization for each impact category.

In summary, the full LCA analysis was performed for both scenarios applying the ReCiPe method. Then, the characterization and normalization (according to ReCiPe 1.08, midpoint normalization of the Europe region) steps were performed to interpret the environmental impacts of the chosen scenarios. Additional Pareto analyses with more detailed input-output analyses of the specific processes were performed to identify the most significant processes which influence relevant environmental impacts.

2.2. Economical Assessment and Economical Inventory of the Carbon Capture Unit (CCU)

The economical evaluation of the CCU can have a significant impact on the actual feasibility of the project and contributes to the sustainability assessment of the technology. The calculation predicts the cost for the construction of a newly built commercial CCU and payback period. The economical inventory of the required construction materials was estimated based on previous projects made for the Czech market by the experts of the biggest Czech energetic research group (ÚJV Řež, a. s.). The evaluation was part of a national scale project [14] for the CCU application and connection to the 250 MW power unit.

3. Case Study Definition: Reduction of CO_2 Emissions in the Czech Republic

To understand and define the technological boundaries for comparing the systems, it is important to describe both scenarios from a technical point of view. Scenario 1 defines the basic systems of the reference power plant. Scenario 2 is described by the reference power unit with the adsorption process of CO_2 capture systems.

3.1. Scenario 1: Reference Power Plant

The first scenario considers the concept of the reference case scenario of the real Czech power plant. Mass and energy flows for further LCA analysis are related to the power plant's operation, which consists of three independent power blocks, each with an installed power of 250 MW. Each power block includes a dry bottom boiler, a turbine and its auxiliaries, a generator, a fly-ash separator, a cooling tower, a transformer, and a desulphurization unit. Coal feeding, water management (pipeline, pump, and chemical treatment), stack, auxiliaries for fly-ash handling, and desulphurization are shared systems. Since 1996, several equipment modernizations have been added in the power plant, such as a desulphurization system based on wet-limestone scrubbing. The gypsum, a product of desulphurization, is deposited into an adjusted mining dump site. Moreover, a hydraulic ash removal system has been replaced by deposition of a mixture of ash, gypsum, and wastewater into an adjusted mining dump site. The modernization includes the research and development of suitable CO_2 sorbents for specific conditions. One of the most viable and commercially affordable options seems to be capture by activated carbon [22].

3.2. Scenario 2: Activated Carbon Adsorption for Reference Power Plant

The second scenario considers the same reference power plant with the connected CO_2 adsorption unit. Mass and energy flows for the next LCA analysis include the operational phase of the power plant and the adsorption unit [23]. For the Czech power plant, the pilot adsorption facility was designed by the UJV Rez group and it was a pilot project [14] for the adsorption of operational flue gases of the power unit. The adsorption unit is based on a rotative adsorber (Figure 1). The main advantages are the continuous operation of adsorption, easy regulation of the adsorption process, and being a viable source of activated coal. The pilot facility for CO_2 separation from flue gases by the adsorption was designed for continuous operation in the conditions of real flue gases from the lignite power unit. The concept is based on the rotational adsorption part, where the main functional part is the fixed adsorption bed connected to the motor driven rotor. The adsorption wheel rotates at a predetermined velocity around its own axis and the speed determines the time of the whole adsorption–desorption cycle. The CO_2 separation follows the desulphurization process, where flue gases are purified and cooled by NaOH, and then the oxides SOx and NOx are removed. Cooled and purified flue gases pass through ventilators and through heating to the rotational adsorber for CO_2 adsorption. Purified flue gases without CO_2 are led out of the separation technology to the chimney or cooling tower. In the section of desorption, adsorbed CO_2 is thermally displaced from the carbon sorbent by circulating gas heated by external steam. CO_2 is continually transported for potential compression with 95% purity. The next step is the cooling of the sorbent to the requested temperature for adsorption. The whole process operates continuously by the rotation of the wheel. The pilot case rotative adsorber is pictured in Figure 1 [24]. The primary source of the activated carbon in Czech conditions is assumed to be hard coal from the ČSM mine site (Karvina region) with an annual mining of six million tons of hard coal [2]. The activation of the hard coal is based on two main steps, i.e., carbonization of the raw material in the absence of oxygen and activation of the carbonized product with water vapor. The heat supply necessary for the activation is obtained by combustion of gases produced during activation.

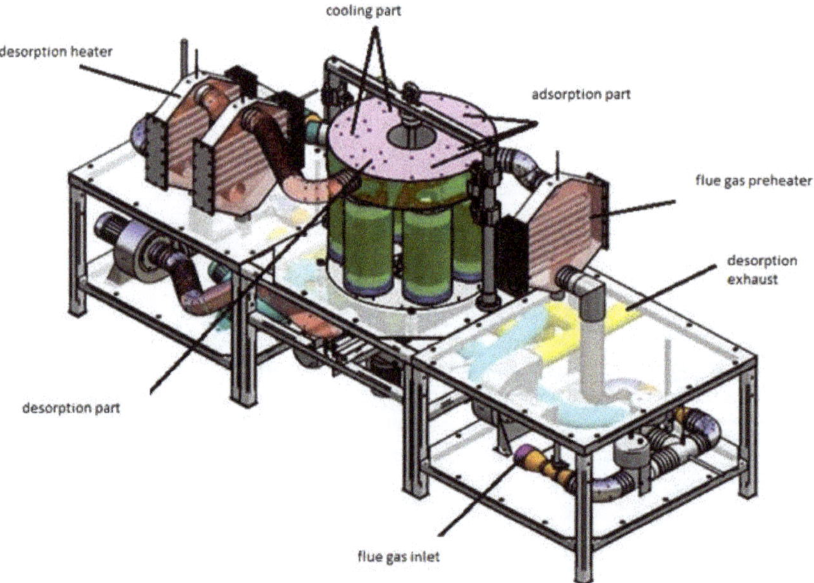

Figure 1. Case rotative adsorber.

3.3. LCA Study: System Boundaries Definition

As stated previously, this study aims to identify the environmental impacts of a power plant with a carbon capture system integration using the designed adsorption method and comparing those impacts with a reference power plant without an adsorption system. The study also focuses on assigning the environmental impacts to the designed CO_2 capture technology itself. The system boundary for Scenario 1 (Figure 2) includes a brown coal production chain from the mining process, transport of fuel to the power unit with power production, combustion, and flue gas treatment processes. Scenario 1 also includes waste and gas emissions production (residuals of flue gas after treatment as nitrogen oxides, carbon dioxide, and sulphur dioxide, released through a stack into the air). Scenario 2(Figure 3) includes a brown coal production chain, power unit operation with power and waste production, adsorption process with all relevant inputs, such as activated carbon production and production of 40% NaOH, and finally, output flows from the adsorption process (captured CO_2, gas emissions and waste products). Although the CCS chain also includes transport and storage of captured CO_2, our specific study does not include any operational data for transport and storage of CO_2, as there is no specific solution of CO_2 transport and storage in the Czech Republic and the distance from an emission source to a storage site with adequate storage capacity and lifetime is unique for every case.

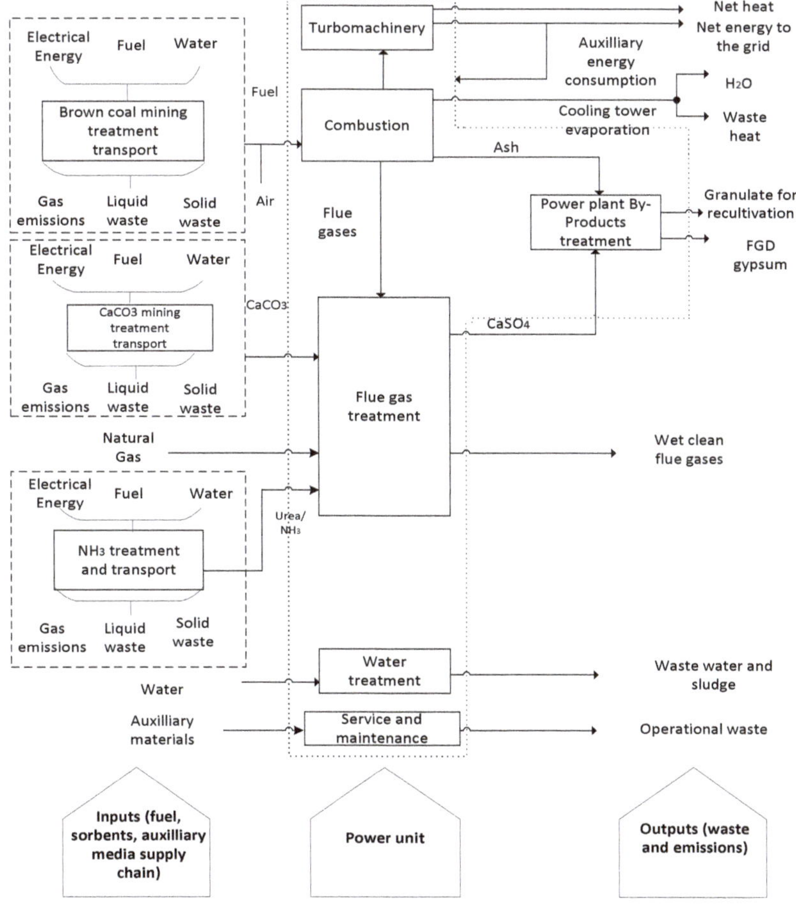

Figure 2. System boundaries for Scenario 1.

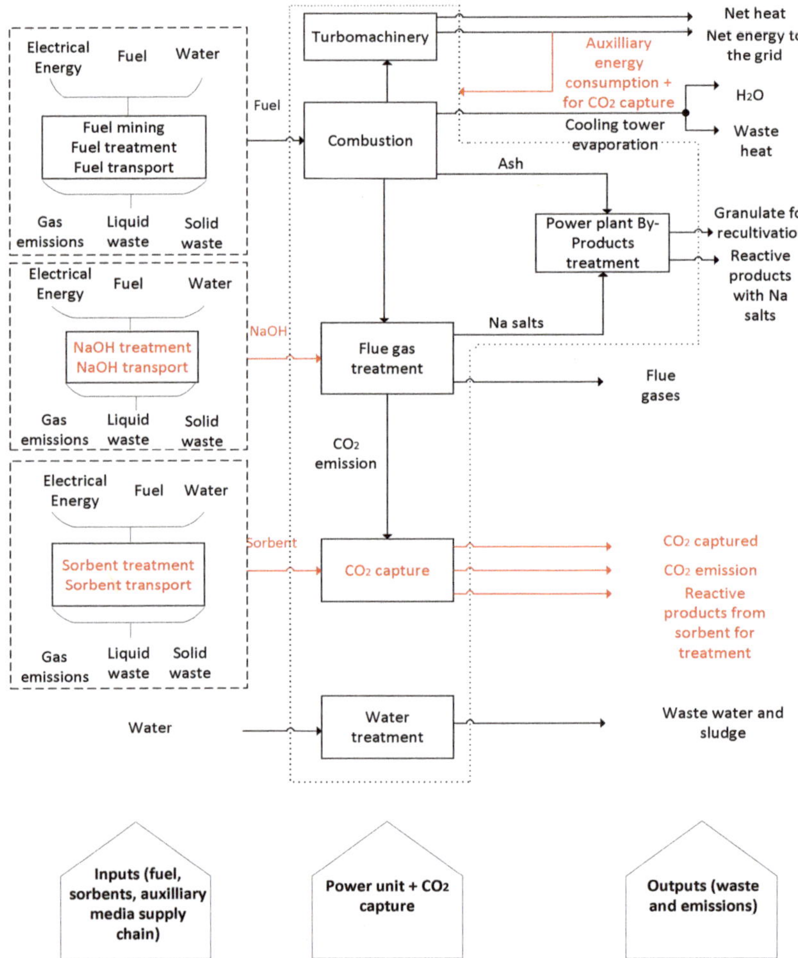

Figure 3. System boundaries for Scenario 2 (differences from Scenario 1 are marked in red).

3.4. Life Cycle Inventory

The input data for the power unit was based on the real power plant operational parameters. Its characteristics are given in Table 2. The heat and mass balances for CO_2 capture technology (Table 3) were evaluated in relation to the power plant characteristics [14] and the results obtained from the pilot testing of the adsorption method.

Table 2. Characteristics of power unit without CO_2 capture.

Parameter	Value	Unit
Nominal power output	250	MW
Brown coal production	214	t/h
Yearly operation	6300	h
Electricity produced	1424	MWh/y
Wet flue gases	766,045	m^3/h

Table 3. Input data for the CO_2 adsorption process.

Parameter	Value	Unit
Consumption of fresh activated carbon	23	kg/h
CO_2	211	t/h
NO_x	159	kg/h
SO_2	119	kg/h
Waste heat	222	MWt
Total energy consumption for CO_2 capture	23.08	MW
Consumption of cooling water	9259	t/h
NaOH consumption	0.305	t/h

In addition to the process of adsorption, the process of activation and carbonization of hard coal is also required to be included in the model. The input data was calculated for the initial batch of 7.6 t of hard coal and 76 t of tar (then activated and transformed into activated carbon). The energy consumption for the hard coal activation was calculated as 1133 MJ. Before the introduction of the flue gas from coal combustion into the CO_2 adsorption stage, it must be secondary treated to decrease the amount of acid gases. To do so, flue gas after coal combustion is washed with a NaOH solution. The composition of the reacted products (output flows) after the reaction with NaOH is described in Table 4.

Table 4. Composition of products after the reaction between flue gas compounds and NaOH.

Compound	Value	Unit
Na_2SO_3	235	kg/h
Na_2SO_4	16.9	kg/h
NaCl	2.54	kg/h
NaF	6.67	kg/h
$NaNO_2$	119	kg/h
$NaNO_3$	147	kg/h

The emission gases released into the air after the CO_2 adsorption stage, the amount of wastewater, and of captured CO_2 are depicted in Table 5.

Table 5. Output data from the CO_2 adsorption process.

Parameter	Value	Unit
SO_2	68.6	kg/h
NOx	13.72	kg/h
CO_2 exhaust gas	18.3	t/h
Captured CO_2	158	t/h
Wastewater	19.73	t/h

Moreover, other conditions and assumptions in the LCA model were taken into consideration with respect to the technical concept of CO_2 adsorption technology and its energy and mass balances as follows:

- The energy required for the 250 MW power unit is determined by brown coal mix production in the conditions of the Czech Republic.
- The reactive product from the NaOH reaction is a non-utilized waste which would be stored at a landfill.
- The wastewater is processed in a wastewater treatment plant and the data for the wastewater treatment plant was taken from the general EU standard dataset thinkstep.

- The primary resource of the activated carbon is hard coal, which is transported by diesel train for an average distance of 1000 km.
- The energy source required for activated carbon activation and carbonization is natural gas.

3.5. Economical Calculation

For the calculation of the CO_2 capture unit construction and connection to the average power plant, the following parameters were assumed:

- Capture effectivity 50%;
- CCU unit would process 0.1% to 0.2% of flue gases produced by an average power plant block;
- CCU unit would capture around 1200 t CO_2/year.

The cost estimation of the construction of CCU is calculated according to the price of the required appliances. Then, the capital expenditure (CapEXP) is multiplied by the coefficient 1.7 (conservative estimate), which refers to the assumption that the construction is built as a fully new technology, and therefore some unexpected expenses could arise.

The operational costs (OpCost) for each item are difficult to estimate. Therefore, 5% of the capital expenditures were used as the operational cost value per year.

Incomes (In) are calculated from the cost of saved CO_2 allowance for each ton of CO_2 and the current market price of CO_2. The authors used a more pessimistic scenario according to the EU commission reference for CO_2 allowance of 25 Euro/t CO_2 and the actual market price which is estimated to be 120 Euro/t CO_2. Then, the payback period (PBT, simple, not discounted) for the commercial CO_2 capture unit is calculated as following:

$$\mathrm{In} = CO_2 \text{ captured} * (\mathrm{CostCO_2}_{\text{ allowance}} + \mathrm{CostCO_2}_{\text{ market}}) \tag{1}$$

$$\mathrm{PBT} = \frac{\mathrm{CapEXP}}{(\mathrm{In} - \mathrm{OpCost})} \tag{2}$$

4. Results

4.1. Life Cycle Impact Assessment

Steps 3 and 4 of LCIA are involved in the Results section. First, each scenario is analyzed separately, and then compared to each other. The results of both considered scenarios are represented in Table 6. Table 6 summarizes the environmental impact categories into the following three groups of results: values in category units, normalization results without any units, and the relative contribution of each impact category to the sum of all categories. The relative contributions are computed from the normalization values.

According to the results, the relative contribution to the sum of impacts in Scenario 1 shows the highest contribution of 46.81% by fossil depletion and in 29.27% by climate change potential. In addition, almost 10% of the environmental impact contribution refers to the terrestrial acidification potential. For Scenario 1, the highest contribution of 77.41% is shown by fossil depletion. All other potential environmental impacts refer to much smaller contributions. For the comparison of both scenarios among the environmental impact categories, the normalization level of the decision-making process was considered. Further Pareto analysis describes the difference between values among the scenarios.

Table 6. LCA results for both scenarios.

Environmental Impact Categories	Scenario 1			Scenario 2		
	Values in Category Units	ReCiPe 1.08, Mid-point Normalization, Europe, excl biogenic carbon	Relative Contribution in %	Values in Category Units	ReCiPe 1.08, Mid-point Normalization, Europe, excl biogenic carbon	Relative Contribution in %
ALO (m^2a)	473	0.11	0.16	473	0.11	0.26
CC, excl biogenic carbon (kg CO_2 eq.)	221,000	19.70	29.27	28,000	2.50	6.14
FD (kg oil eq.)	49,100	31.50	46.81	49,100	31.50	77.41
FET (kg 1,4-DB eq.)	1.24	0.11	0.17	1.25	0.12	0.28
FE (kg P eq.)	0.023	0.05	0.08	0.02	0.05	0.13
HT (kg 1,4-DB eq.)	171	0.29	0.43	171	0.29	0.71
IR (U235 eq.)	314	0.05	0.07	314	0.05	0.12
MET (kg 1,4-DB eq.)	1.46	0.17	0.26	1.46	0.17	0.42
ME (kg N eq.)	7.83	0.78	1.15	2.18	0.22	0.53
MD (kg Fe eq.)	16.20	0.02	0.03	16.20	0.02	0.06
NLT (m^2)	0.014	0.09	0.13	0.01	0.09	0.21
PMF (kg PM10 eq.)	64.70	4.34	6.45	25.40	1.71	4.20
POF (kg NMVOC eq.)	184	3.47	5.16	36.30	0.68	1.68
TA (kg SO_2 eq.)	227	6.61	9.82	109	3.18	7.81
TET (kg 1,4-DB eq.)	0.075	0.01	0.01	0.07	0.01	0.02
ULO (m^2a)	0.016	0.00	0.00	0.02	0.00	0.00
WD (m^3)	6390	0.00	0.00	6400	0.00	0.00
Sum	-	67.30	100	-	40.69	100

4.2. Pareto Analysis of the Scenarios and Processes

The Pareto analysis defines 20% of the potential environmental impact categories that contribute to 80% of the total impact. The environmental impacts were divided into the following two groups of flows: (1) input flows which use, consume, or transform primary soil, land, or resources (agriculture land occupation, natural land transformation, and fossil depletion) and (2) output flows which are considered to be emissions from the considered processes. In the case of the input flows, for both scenarios, all the mentioned environmental categories have equal values and the highest among them has fossil depletion.

Figures 4 and 5 illustrate the most significant environmental categories for output flows by Pareto graphics. On the one hand, Scenario 1 identifies fossil depletion (FD), climate change (CC) potential, and terrestrial acidification (TA) as the highest contributors. On the other hand, Scenario 2 shows that the terrestrial acidification has a higher impact value than CC. For the fossil depletion category, the brown coal mining is shown as having the highest contribution. Therefore, the climate change category is affected mainly by the combustion of brown coal and thermal energy production for the power unit operation.

For the comparison of both scenarios the following graphics (Figures 6–9) represent differences in the significant (CC, TA, PMF, and POF) environmental categories. The graphs show lower impacts in each category for Scenario 2. The most significant difference is seen in CC where the values for Scenario 2 are almost two-thirds lower than those in Scenario 1.

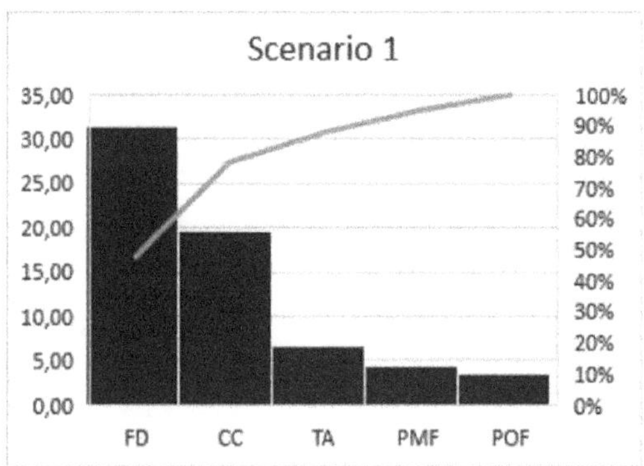

Figure 4. Pareto graph for scenario 1.

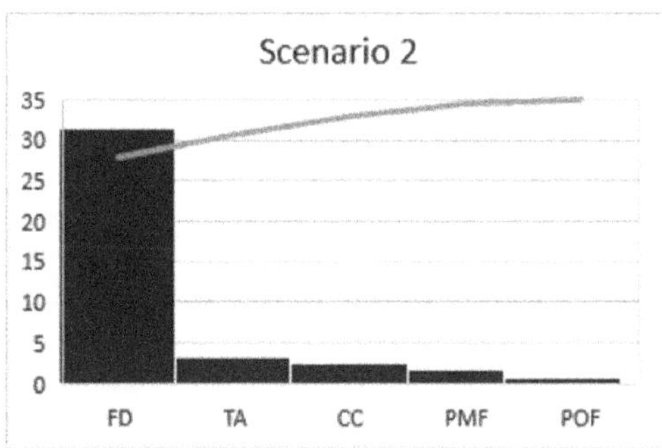

Figure 5. Pareto graph for scenario 2.

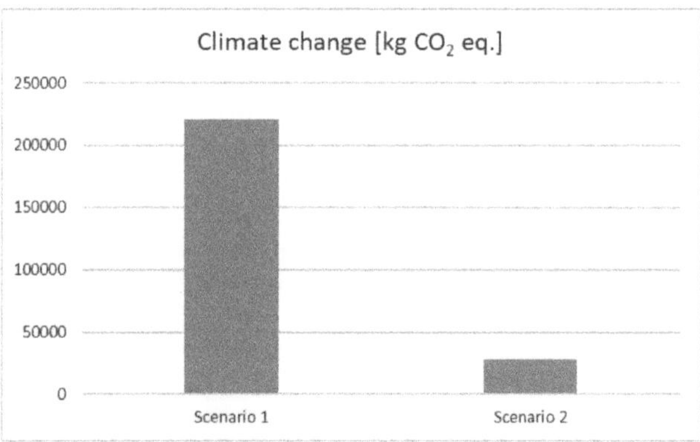

Figure 6. Comparison of scenarios-climate change.

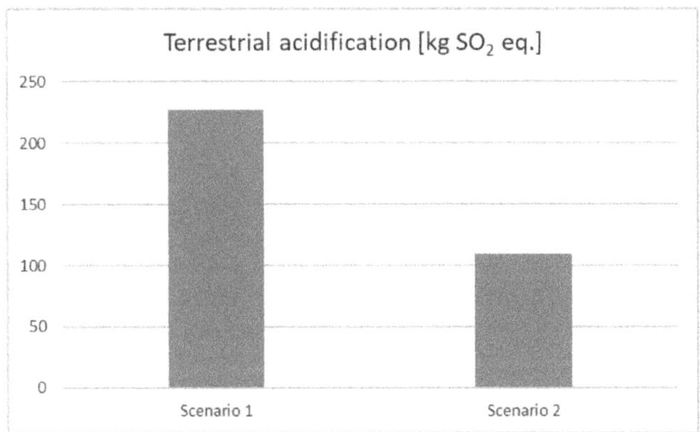

Figure 7. Comparison of scenarios-terrestrial acidification.

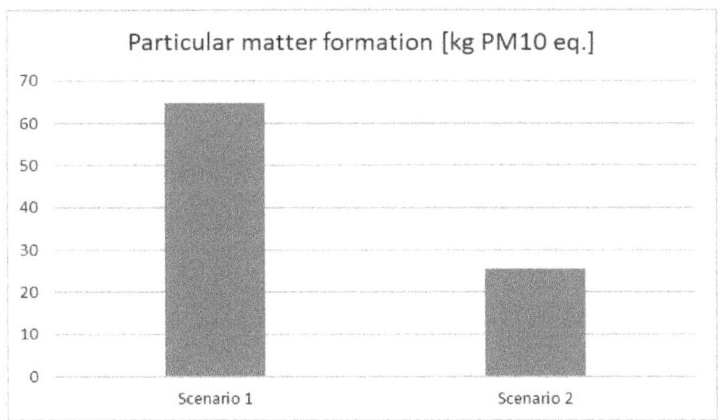

Figure 8. Comparison of scenarios- particular matter formation.

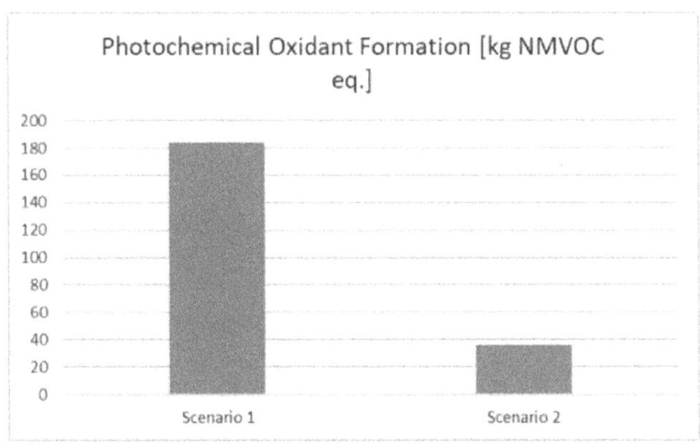

Figure 9. Comparison of scenarios -photochemical oxidant formation.

For Scenario 2, Table 7 shows that the category of fossil depletion is affected mainly by brown coal production and mining. Moreover, terrestrial acidification potential is mainly affected by the CO_2 adsorption process. Brown coal combustion and, consequently, production of thermal energy for the adsorption process are also contributors to the acidification potential. Transport has a minor role for both scenarios

Table 7. Causes of significant environmental impacts in Scenario 2.

Scenario 2	Brown Coal Mining	Transport	CO_2 Adsorption Process	Thermal Energy for Adsorption Process
Terrestrial acidification (kg SO_2 eq.)	15.6	3.68	82.3	7.68
	Brown Coal Mining	Transport	NaOH Production	Water Consumption
Fossil depletion (t oil eq.)	48.10	1.03	0.28	0.59

Environmental Impact Assessment of Activated Carbon Production

The question of the activated carbon production is crucial for the whole environmental impact analysis. Therefore, further detailed input-output analyses of the process of the activated carbon production is required. Table 8 shows the relative contribution of each category for the activated carbon production for the CO_2 adsorption process. The highest contribution refers to the categories of climate change potential and fossil depletion.

Table 8. Environmental assessment of activated carbon production.

Environmental Impact Categories	Values in Category Units	Relative Contribution in %
Climate change (kg CO_2 eq.)	6.98	98.46
Fossil depletion (kg oil eq.)	0.107	1.51
Water depletion (m^3)	0.002	0.03

4.3. Economical Evaluation of the Payback Period of the Pilot CO_2 Capture Unit

Total capital expenditures according to the required components price are listed in Table 9.

Table 9. Total costs of CCU construction.

Cost Estimation for Commercial CCU Unit in EURO	
4x reactor	40,000
Fittings	24,000
Measure appliances	14,000
CO_2 tank (pressurized)	20,000
Electro + regulations	12,000
CO_2 compression	6000
Ventilators	4000
Project	24,000
Cooling	8000
Heating (steam transport)	6000
Others + non predictable	32,000
Construction	40,000
Capital expenditure	434,000
New technology x1,7	737,800

Results in Table 10 for the economic feasibility and payback time period were calculated according to Formulas (1) and (2).

Table 10. Payback period (simple, not discounted) for CCU.

Expenditures	737,800	EUR
Operational costs	36,890	EUR
Captured CO_2	1200	t
Cost of CO_2 allowance	25	EUR/t
Market price of CO_2 produced	120	EUR/t
Income	174,000	EUR
Payback period	5.38	years

5. Discussion

According to the characterization values in Table 6, the most obvious difference between the two scenarios is the climate change category (Figure 6). This result corresponds to the decrease of CO_2 levels by the adsorption process (in Scenario 2), modelled at 75% CO_2 capture from flue gases. The gains in terrestrial acidification potential in Scenario 1 have higher values (227 kg SO_2 eq.), which are mainly influenced by SO_2 emissions (119 kg) released into the air after the flue gas treatment. In Scenario 2, this amount of SO_2 in flue gases is the input into the adsorption process, thus, the values for the acidification category are lower. In addition, Scenario 1 refers to the higher values in the PMF and POF categories. These categories are influenced by fuel combustion emissions.

Studies by Kantová [25] and Vassilev [26] showed that ash from the brown coal combustion consisted of a high volume of particulate matter (93.08% from brown coal) which was volatile and persistent in the atmosphere. Therefore, brown coal combustion and the quality of brown coal are significant factors that influence the level of potential environmental harm. According to Kantová, the key parameters for controlled emissions in the process of combustion are low ash content, low moisture levels, and a constant size of volatile particles. In the context of the Czech integrated register of pollutants and emissions restrictions [27], there are no further chemical descriptions of particulate matter (PM), and thus it is complicated to get parameters of produced PMs that directly affect the environmental category of particulate matter formation. Therefore, the chemical analysis of PM produced from Czech brown coal would be an interesting subject for further research.

The Pareto analysis was chosen to distinguish which processes have the greatest impact, particularly for output flows (emissions) in both scenarios which are harder to decide upon. The input flows, which consider the change of the land and resource depletion, show the highest contribution of fossil depletion potential in both scenarios. In both cases, fossil depletion is related to the processes of brown coal mining and production which require 214 t/h of primary brown coal for the actual operation of the studied power unit. However, the result values for fossil depletion in Scenario 1 does not show any significant difference as compared with Scenario 2, although there is a slightly higher demand for raw material extraction, such as hard coal for activated carbon production. This demand refers to a need for fresh carbon, in the amount of 23 kg/h, which in the evaluation of the whole life cycle means only a small resource demand. The Pareto analysis for output flows (Figures 4 and 5) considers fossil depletion, climate change, and terrestrial acidification as the most significant contributors to the overall environmental impacts. The only difference lays in the degree of priority for each scenario. For both scenarios, fossil depletion has equal significance, as the values of characterization are the same. In Scenario 1, the impact contribution in second place is climate change and, in Scenario 2 it is terrestrial acidification. Climate change in Scenario 1 is mainly caused by the combustion process of the coal and the production of thermal energy for the power unit. For Scenario 2, Table 7 shows that the processes influencing the category of terrestrial acidification are brown coal mining and SO_2 emissions (released into the air after the adsorption process in the amount of 82.3 kg SO_2 eq). The impact of transportation, in both scenarios, is insignificant.

It is also important to mention the category of water consumption. In the relative contribution to environmental impacts, this category does not show any significance. However, in the characterization phase, the consumption of 6400 m^3 of water shows that the energy industry plays a role in water management. This is a reminder that the process must also aim to mitigate excessive water consumption, especially in times of global warming and drought danger.

The next step was to assess the production of the activated carbon. The literature review for the LCA of the activated carbon adsorption process showed that environmental studies have focused on the type of activated carbon production. The case study is considering the production of activated carbon from hard coal as a primary source. Therefore, resource depletion, as a direct connection to the category of fossil depletion potential, is affected by hard coal mining. Tar, as an input flow for the activated carbon production, also contributes to resource depletion. On the site of emissions, CO_2 emissions from the combustion of natural gas (1133 MJ for activation and carbonization of 7.6 t of activated carbon) cause major environmental consequences. Moreover, among all environmental impact categories, climate change has the highest relative contribution (almost 99%). The impact relates to the combustion of natural gas (Table 8).

The results clearly demonstrate that the power unit with the connection of adsorption process leads to decreased environmental impacts, specifically in the categories of climate change, terrestrial acidification, particulate matter formation, and photochemical oxidant formation. The problem is seen in a primary source, i.e., coal extraction, which, in both scenarios, shows relatively high and equal values. The Czech national energy mix is based on brown coal power plants, and therefore the raw materials extraction and resource depletion creates a significant environmental burden. The extraction of hard coal for activated carbon production also contributes to this fact. The case study counts with just an input of 23 kg fresh activated carbon, but the basic batch of hard coal is rather bigger and counts with 7.6 t, which is an additional amount of raw materials extracted from the ground. The production of the activated carbon could be optimized using different resources such as biomass, which could contribute to reduced consumption of the fossil source.

Finally, the economical consideration (Tables 9 and 10) of the newly built CCU shows a payback period of almost six years (relatively fast for such a small CCU). It must be considered that the market price for a ton of CO_2 could be lower, due to lower purity of the CO_2 product. If the market price was one-third lower (40 Euro), incomes would change to 78,000 Euros and the payback period would increase to almost 18 years. We conclude that, even if the process of CO_2 capture was highly effective, the purity of the final product has a significant role in the economy of the whole process. To make the project feasible, there is a technical requirement to solve the purification process of CO_2, leading to an increase of total capital expenditures and of the payback period. Moreover, purified CO_2 as a final product could be more attractive for sectors such as agriculture or the food processing industry, and therefore contribute to the national circular economy.

6. Conclusions

Carbon dioxide capture by activated carbon adsorption seems to be a promising technology from an environmental perspective. The LCA assessment highlights the main environmental impacts that can arise during the life cycle of the technology.

The robust LCA analyses which included characterization, normalization, and Pareto analyses with input-output analyses are approaches that create a precise model to reflect specific conditions of the technology. The LCA helps to identify the key processes that can be improved with respect to their environmental performance. In addition, the economic calculation completes the sustainability assessment of the newly built technology and gives the perspective of the final product (CO_2) utilization. It must be stressed that the designed emissions of the capture method are site specific and reflect the local conditions, for example, the type of power plant, fuel type, natural sources (for capture media production), etc. The presented adsorption method was designed for the purpose of CO_2 capture from subcritical, coal-fired power plants in the Czech Republic. The sum of the environmental impacts with carbon capture is generally lower than the power unit itself. Nevertheless, this study shows that the Czech energy mix (in both scenarios) leads to high levels of CO_2, SO_2 emissions, and solid particulates. As the Czech national energy mix is primarily from brown coal, the depletion of fossils by a primary energy source still remains the main environmental problem, but monitoring of the coal quality, as well as testing the chemical composition of particulate matter, could contribute towards lower potential environmental impacts.

Nevertheless, further research that focuses on various sources (as biomass) for activated carbon production should be considered. Moreover, CCU could become part of the Czech circular economy, if the purification processes and measures of the CO_2 product are wisely chosen and adapted.

Worldwide pressure for low carbon economy transition is forcing the national energy systems to find viable solutions to mitigate the levels of greenhouse gases (GHG). The Czech Republic is slowly shifting towards increased integration of renewable energy systems. However, the coal industry is still the prevailing sector, where a sudden shift could be drastic for the national economy and coverage of the power demand. Therefore, the current systems with optimized CCU could be the solution that would help to overcome the transition process. The implementation of all available tools and knowledge to reach this goal is required and would assist choosing and creation of a reasonable and wise strategy for the sustainable development of the country.

Author Contributions: Conceptualization, K.Z. and A.C.; methodology, K.Z. and V.K.; formal analysis, K.Z.; resources, K.Z.; data curation, K.Z. and J.Š.; writing—original draft preparation, K.Z.; writing—review and editing, K.Z.; supervision, V.K. and A.C.; All authors have read and agreed to the published version of the manuscript.

Funding: This research received no external funding.

Acknowledgments: This work was supported by the Technology Agency of the Czech Republic, project number TH03020169 and project number TA02020205 and by institutional support from the University of Chemistry and Technology Prague.

Conflicts of Interest: The authors declare no conflict of interest.

References

1. European Association for Coal and Lignite. Eurocoal Statistics, Coal in Europe 2017. Available online: www.eurocoal.eu (accessed on 9 February 2019).
2. Czech Statistical Office. Czech Republic in International Comparison—(Selected Indicators)-2017. Available online: www.czso.cz (accessed on 2 February 2019).
3. Grant, T.; Anderson, C.; Hooper, B. Comparative Life Cycle Assessment of Potassium Carbonate and Monoethanolamine Solvents for CO_2 Capture from Post Combustion Flue Gases. *Int. J. Greenh. Gas Control* **2014**, *28*, 35–44. [CrossRef]
4. Manuilova, A.; Koiwanit, J.; Piewkhaow, L.; Wilson, M.; Chan, C.W.; Tontiwachwuthikul, P. Life Cycle Assessment of Post-Combustion CO_2 Capture and CO_2-Enhanced Oil Recovery Based on the Boundary Dam Integrated Carbon Capture and Storage Demonstration Project in Saskatchewan. *Energy Procedia* **2014**, *63*, 7398–7407. [CrossRef]
5. Koornneef, J.; van Keulen, T.; Faaij, A.; Turkenburg, W. Life Cycle Assessment of a Pulverized Coal Power Plant with Post-Combustion Capture, Transport and Storage of CO_2. *Int. J. Greenh. Gas Control* **2008**, *2*, 448–467. [CrossRef]
6. Petrakopoulou, F.; Tsatsaronis, G. Can Carbon Dioxide Capture and Storage from Power Plants Reduce the Environmental Impact of Electricity Generation? *Energy Fuels* **2014**, *28*, 5327–5338. [CrossRef]
7. Odeh, N.A.; Cockerill, T.T. Life Cycle GHG Assessment of Fossil Fuel Power Plants with Carbon Capture and Storage. *Energy Policy* **2008**, *36*, 367–380. [CrossRef]
8. Singh, B.; Strømman, A.H.; Hertwich, E.G. Scenarios for the Environmental Impact of Fossil Fuel Power: Co-Benefits and Trade-Offs of Carbon Capture and Storage. *Energy* **2012**, *45*, 762–770. [CrossRef]
9. Clarens, F.; Espí, J.J.; Giraldi, M.R.; Rovira, M.; Vega, L.F. Life Cycle Assessment of CaO Looping versus Amine-Based Absorption for Capturing CO_2 in a Subcritical Coal Power Plant. *Int. J. Greenh. Gas Control* **2016**, *46*, 18–27. [CrossRef]
10. Hjaila, K.; Baccar, R.; Sarrà, M.; Gasol, C.M.; Blánquez, P. Environmental Impact Associated with Activated Carbon Preparation from Olive-Waste Cake via Life Cycle Assessment. *J. Environ. Manag.* **2013**, *130*, 242–247. [CrossRef] [PubMed]
11. Arena, N.; Lee, J.; Clift, R. Life Cycle Assessment of Activated Carbon Production from Coconut Shells. *J. Clean. Prod.* **2016**, *125*, 68–77. [CrossRef]
12. Štefanica, J.; Smutná, J.; Kočí, V.; Macháč, P.; Pilař, L. Environmental Gains and Impacts of a CCS Technology—Case Study of Post-Combustion CO_2 Separation by Ammonia Absorption. *Energy Procedia* **2016**, *86*, 215–218. [CrossRef]
13. Zakuciová, K.; Lapao Rocha, J.; Koci, V. Life Cycle Assessment Overview of Carbon Capture and Storage Technologies. *Annu. Int. Conf. Sustain. Energy Environ. Sci.* **2016**, 84–90. [CrossRef]
14. Pilar, L.; Slouka, P.; Vlček, Z. *Technical Report of Project TA02020205*; ÚJV Řež, a.s., Řež: Husinec, Czech Republic, 2015.

15. International Organization for Standardisation. *IS/ISO 14044 (2006): Environmental Management-Life Cycle Assessment-Requirements and Guidelines*; Bureau of Indian Standarts: New Delhi, India, 2006.

16. Hauschild, M.Z.; Rosenbaum, R.K.; Olsen, S.I. *Life Cycle Assessment, Theory and Practice*; Springer International Publishing AG: Montpellier, France, 2018. [CrossRef]

17. Goedkoop, M.; Heijungs, R.; Huijbregts, M.; De Schryver, A.; Struijs, J.; Van Zelm, R. *ReCiPe 2008 First Edition (Version 1.08) Report I: Characterisation*; Ministry of foreign Affairs: Rijnstraat, Netherlands, 2013.

18. Jolliet, O.; Margni, M.; Charles, R.; Humbert, S.; Payet, J.; Rebitzer, G.; Rosenbaum, R. Presemmg a New Meth6d IMPACT 2002+: A New Life Cycle Impact Assessment Methodology. *Int. J. Life Cycle Assess* **2003**, *8*, 324–330. [CrossRef]

19. Benini, L.; Mancini, L.; Sala, S.; Schau, E.; Manfredi, S.; Pant, R. *Normalisation Method and Data for Environmental Footprints*; Office of the European Union: Luxembourg, 2014; Volume 113. [CrossRef]

20. Carvalho, A.; Mimoso, A.F.; Mendes, A.N.; Matos, H.A. From a Literature Review to a Framework for Environmental Process Impact Assessment Index. *J. Clean. Prod.* **2014**, *64*, 36–62. [CrossRef]

21. Shen, L.; Patel, M.K. Lca Single Score Analysis of Man-Made Cellulose Fibres. *Lenzing. Ber.* **2010**, *88*, 7.

22. CEZ Group. Power Plant Prunerov. Available online: www.cez.cz (accessed on 9 February 2018).

23. Tzimas, E.; Georgakaki, A.; Peteves, S. Reducing CO_2 Emissions from the European Power Generation Sector-Scenarios to 2050. *Energy Procedia* **2009**, *1*, 4007–4013. [CrossRef]

24. Vávrová, J.; Štefanica, J.; Hájek, P.; Smejkalová, J. *Technical Report of Project TA02020205*; ÚJV Řež, a.s., Řež: Husinec, Czech Republic, 2013.

25. Kantová, N.; Holubčík, M.; Jandačka, J.; Čaja, A. Comparison of Particulate Matters Properties from Combustion of Wood Biomass and Brown Coal. *Procedia Eng.* **2017**, *192*, 416–420. [CrossRef]

26. Vassilev, S.V.; Vassileva, C.G.; Vassilev, V.S. Advantages and Disadvantages of Composition and Properties of Biomass in Comparison with Coal: An Overview. *Fuel* **2015**, *158*, 330–350. [CrossRef]

27. Ministerstvo Zivotniho Prostredi České Republiky. Integrovany Registr Znecistovani. Available online: www.irz.cz (accessed on 8 February 2018).

Article

A Prospective Net Energy and Environmental Life-Cycle Assessment of the UK Electricity Grid

Marco Raugei [1,2,*], **Mashael Kamran** [1,2] **and Allan Hutchinson** [1,2]

1 School of Engineering, Computing and Mathematics, Oxford Brookes University, Wheatley, Oxford OX33 1HX, UK; 19038874@brookes.ac.uk (M.K.); arhutchinson@brookes.ac.uk (A.H.)
2 The Faraday Institution, Didcot OX11 0RA, UK
* Correspondence: marco.raugei@brookes.ac.uk

Received: 30 March 2020; Accepted: 27 April 2020; Published: 2 May 2020

Abstract: National Grid, the UK's largest utility company, has produced a number of energy transition scenarios, among which "2 degrees" is the most aggressive in terms of decarbonization. This paper presents the results of a combined prospective net energy and environmental life cycle assessment of the UK electricity grid, based on such a scenario. The main findings are that the strategy is effective at drastically reducing greenhouse gas emissions (albeit to a reduced degree with respect to the projected share of "zero carbon" generation taken at face value), but it entails a trade-off in terms of depletion of metal resources. The grid's potential toxicity impacts are also expected to remain substantially undiminished with respect to the present. Overall, the analysis indicates that the "2 degrees" scenario is environmentally sound and that it even leads to a modest increase in the net energy delivered to society by the grid (after accounting for the energy investments required to deploy all technologies).

Keywords: LCA; EROI; net energy; energy scenario; energy transition; electricity; grid mix; storage; decarbonization

---

1. Introduction

Energy is an essential commodity for productivity, economic growth and well-being. However, ensuring the delivery of the energy that societies need while preventing major environmental disruption is quickly becoming one of the major challenges of this century. The worldwide increase in energy demand and continuous burning of fossil fuels for energy production since industrialization has led to a rapid increase in carbon dioxide (CO_2) concentration in the atmosphere from 280 ppm (parts per million) to over 400 ppm [1].

Historically, the UK was one of the first countries to experience massive industrialization across all sectors. Before industrialization, most of the UK energy demand was fairly renewable, met by water and wind mills and by burning woody biomass, mostly for agricultural and heating purposes [2]. Subsequently, with the onset of industrialization, the UK population and its appetite for energy grew rapidly. The UK economy shifted from reliance on agriculture to manufacturing, and most manufacturing sectors were based on coal consumption. This propelled UK economic growth in the 18th and 19th century [3] but, as a consequence, it also led to a rapid increase in CO_2 and other greenhouse gas (GHG) emissions to the atmosphere. The burning of coal and peat for manufacturing and heating purposes created a visual smog in major UK cities, too. This led to the first 1956 and 1968 Clean Air Act to reduce the emissions caused by coal [3]. In 2008, the UK legislated the Climate Change Act to reduce its contribution to global warming by cutting down GHG emissions by 80% by 2050, relative to its 1990 level [4].

While the UK was not the only country with legislation already in place aimed at reducing GHG emissions, the Paris Agreement held in December 2015 was the first global agreement to combat climate change signed by 195 countries. It aims to keep the global temperature below 2 °C above

---

pre-industrialization level, while encouraging further efforts to limit global warming to below 1.5 °C to reduce the risk of irreversible climate consequences [5]. To accomplish this target, all parties of the United Nation Framework Convention for Climate Change (UNFCCC) understood the necessary action to limit the CO_2 and other GHG concentration in the atmosphere by decarbonizing their current energy system. In 2019, the UK Committee on Climate Change (CCC) recommended a new national target to fulfil the commitment made when signing the Paris Agreement. This new ambitious target aims to bring all GHG emissions to net zero by 2050 [6].

The current UK energy system comprises of three main energy carriers: oil-derived fuels, natural gas and electricity. Electricity is known to be one of the most desirable energy carriers for a low-carbon future due to its flexibility to operate a wide range of services such as lightning, heating, and transportation, with its transportability over long distances with comparatively little loss, and the efficiency with which it can be converted into useful work at the point of use [7]. Electricity can be generated from a range of primary sources, and in order to meet the climate target, there has recently been an increased deployment of various low-carbon technologies in the UK electricity grid (primarily wind and solar photovoltaics), with an on-going shift away from the combustion of fossil fuels in thermal power plants, and a concomitant increase in the electrification of the heating and transportation sectors. Such trends are expected to be sustained for the coming decades, with a positive overall effect in terms of decarbonization; however, the increased reliance on electricity as the energy carrier of choice for all sectors, paired with the shift to more and more variable renewable energy (VRE) generation, has also led to growing concerns about the future requirement for large-scale energy storage (including lithium-ion batteries) and the possible associated adverse consequences.

Over the past decade, a growing body of literature has analyzed the feasibility and environmental consequences of various energy transition pathways characterized by an increased degree of electrification across all sectors, coupled with a larger reliance on renewable technologies to generate the required electricity. Among such studies that appeared relatively early, a few attracted significant attention by non-specialist media, such as Fthenakis et al.'s seminal "Solar Grand Plan" [8,9], while others seemed to polarize scientific opinion and generated strong controversy [10–14]. More recently, a string of similarly-framed studies respectively focusing on Europe [15], the Americas [16] or the whole world [17,18] have seemingly pointed to a degree of high-level agreement on the general desirability of such transition pathways, albeit with a number of caveats.

Crucially, although key electricity generation technologies such as wind, solar photovoltaics and nuclear are often considered to be "zero carbon" at the point of consumption, the same is not true for their manufacturing, nor are they necessarily 100% environmentally sound in all respects. Life cycle assessment (LCA) is the preferred methodology in order to address all these points while taking into consideration all the life-cycle stages of the various energy supply chains (resource extraction, processing, and delivery, and power plant manufacturing, operation and decommissioning). For instance, a comprehensive LCA of a range of International Energy Agency (IEA) world electricity supply scenarios indicated that, in broad terms, those scenarios that rely more heavily on renewable technologies do indeed tend to stabilize or even reduce global pollution, but entail larger (and sometimes potentially critical) demand for key materials, noticeably among which is copper [19]. Another recent study, whose scope also extended to the whole world, found that those decarbonization strategies relying heavily on wind and solar technologies are comparatively more effective in reducing human health impacts than those relying on carbon sequestration, while the use of bioenergy in the mix raises concerns in terms of land use and associated ecosystem damage [20]. Pehl et al. [21] carried out a thorough life-cycle comparison of all the direct and indirect GHG emissions from various renewable and non-renewable low-carbon technologies, also including seldom-considered factors such as indirect carbon emissions from land use change. Their study produced ranges of results for each technology that reflect the sometimes-large variability that is not only due to technological aspects, but also, critically, system siting (from insolation level for solar photovoltaics, to the rather less obvious methane emissions from biomass degrading in poorly-chosen hydro reservoir sites).

Coming to more UK-centric LCA studies, Stamford and Azapagic [22] compared a range of possible grid mix scenarios up to 2070, and concluded that a low-carbon mix with nuclear and renewables provides the best overall environmental performance, albeit with some increased impacts in terms of terrestrial eco-toxicity, and material resource depletion (elements). Raugei et al. [23] investigated a number of stakeholder-informed photovoltaic-heavy grid mix options, and found that, despite the comparatively low insolation levels that are characteristic of the UK, no such scenarios would be detrimental to the grid performance from a wide range of technical and environmental metrics.

There also remains a question about whether the future electricity grid can continue providing the same level of "net" energy for all societal needs as before, to support continued economic growth and prosperity. The concept of net energy was first described by ecologist Howard T. Odum in 1973 [24] and then reprised by his former student Charles A.S. Hall, who went on to coin the popular term "energy return on investment" (EROI) [25]. Net energy is intended to represent the "true value" of the energy that is delivered to society. In other words, for a society to prosper, the energy delivered to the society must at least be higher than the energy invested in the chain of processes required to convert primary energy resources into useful energy carriers. Carbajales-Dale et al. [26] discuss the importance of carrying out net energy analysis (NEA) to measure the productivity of the energy system and support a sustainable energy transition. Moving towards a low carbon future would require a different re-investment of energy for the electricity grid due to the deployment of new energy generation and storage technologies, and NEA is the way to identify the ensuing changes in the final net energy delivered to society by the electricity grid.

To this date, however, literature is divided as to this important question, with some studies questioning the capability of future renewable-heavy grid mixes to deliver sufficient net energy (e.g., for the case of Australia [27]), and others instead concluding that the net energy set-backs could be minor (e.g., for New York state [28]) to non-existent (e.g., for Chile [29]). Ultimately, the devil is, once again, in the details, and the answer is likely to depend on specific conditions such as the exact grid mix composition, location, degree of gird-level storage, and the demand profile.

This paper presents the results of a joint environmental LCA and NEA of one of the leading energy transition scenarios for the UK, specifically to understand the changes that may be expected in environmental impacts as well as net energy delivery when moving from the current UK grid mix to a future one featuring larger amounts of VRE and associated energy storage (the latter including dedicated stationary solutions as well as mobile—also referred to as "vehicle-to-grid"—schemes).

2. Materials

2.1. Current and Future (Projected) UK Grid Mix

National Grid Electricity System Operator, the UK's largest utility company, produced a study called "Future Energy Scenarios (FES)" [30] analyzing the energy demand and supply of the future electricity and gas networks in Great Britain based on policy support, customer engagement, technological development, economic growth and energy efficiency (National Grid does not provide electricity to Northern Ireland, which instead is served by three interconnectors with the Republic of Ireland. Consequently, while the FES are based on targets set at the national (UK) level, all the data contained therein are actually for Great Britain (GB) only, i.e., excluding Northern Ireland. For the sake of clarity and simplicity, however, in this paper the distinction between GB and UK is dropped in favour of a unified reference to "UK" throughout). Four potential energy pathways to achieve various degrees of decarbonization for the UK are described in the FES report, respectively named: "steady progression", "consumer evolution", "community renewables" and "2 degrees". Out of these, "community renewables" and "2 degrees" are the more aggressive in terms of decarbonization. Both target 80% reduction in GHG emissions by 2050 compared to 1990, based on the 2008 Climate Change Act, and the main difference between the two is that "2 degrees" assumes more "large-scale", centralized electricity generation, while "community renewables" relies more on smaller, decentralized installations. This paper focuses on the "2 degrees" scenario, due to it being the one for which a greater level of detail is provided in the FES report.

The FES "2 degrees" scenario projects electricity generation capacities for all the grid mix technologies and the associated annual electricity generation for the next 30 years. The overall electricity generation is estimated to increase by 35% in 2035 and 72% in 2050, relative to 2019, in response to the increasing demand partly due to the electrification of transport and heating. Figure 1 illustrates the UK grid mix composition in terms of total electricity generated in the year 2019, and Figures 2 and 3 the corresponding projected grid mix compositions in the years 2035 and 2050.

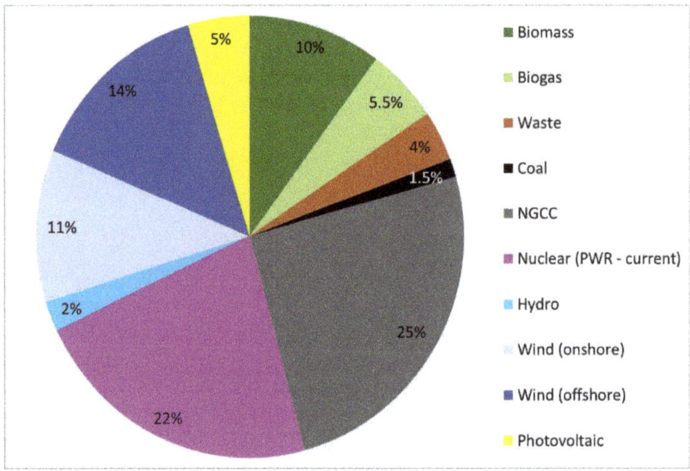

Figure 1. Current (2019) UK grid mix composition, in terms of total electricity generated. NGCC = natural gas combined cycles; PWR = pressure water reactors.

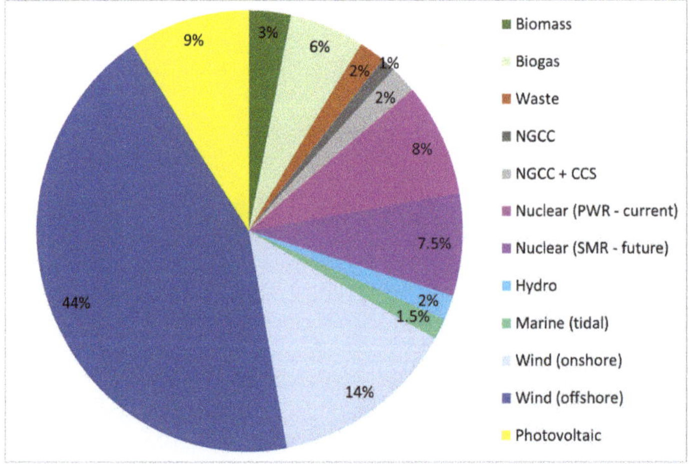

Figure 2. Expected UK grid mix composition in 2035, in terms of total electricity generated (under "2 degrees" scenario assumptions). NGCC = natural gas combined cycles; NGCC + CCS = natural gas combined cycles plus carbon capture and storage; PWR = pressure water reactors; SMR = small modular reactors.

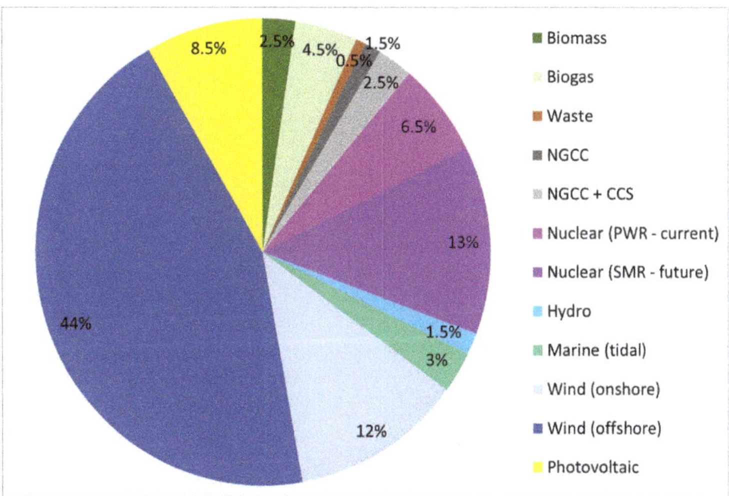

Figure 3. Expected UK grid mix composition in 2050, in terms of total electricity generated (under "2 degrees" scenario assumptions). NGCC = natural gas combined cycles; NGCC + CCS = natural gas combined cycles plus carbon capture and storage; PWR = pressure water reactors; SMR = small modular reactors.

The 2019 grid mix includes a small share of coal-fired electricity, which is expected to be completely phased out within the next few years. Gas-fired generation, currently accounting for 25% of the grid mix, is expected to move towards a peaking plant role in the coming years to provide flexibility in the electricity grid by supplying electricity when the demand is high or when renewable generation is low. While most conventional combined-cycle gas-fired generation is also expected to be phased out (−92% by 2050), a non-negligible share of gas-fired generation is expected to re-emerge from 2030 onwards but equipped with carbon capture and storage. Nuclear generation in 2019 provides 22% of the total electricity generation, but over half of the existing power plants are expected to reach their designated end of life and close down before 2035. However, small modular reactors are then projected to take over, eventually providing around 16 GW of generation capacity, and 13% of total electricity generation, by 2050.

By 2035, 99% of electricity generation is expected to come from "zero carbon" technologies at the point of use (i.e., nuclear, biogas, biomass, waste, hydro, tidal, wind and solar)—cf. blue line in Figure 4. Out of this, a large proportion will be met by renewable technologies (cf. green line in Figure 4), and especially by offshore wind (accounting for approximately 44% of total generation in 2035 and 2050). In fact, most renewable technologies (wind, solar, tidal, and biogas) are projected to increase significantly in output over the next decades, with the sole exceptions of hydro, which is already essentially maxed out in the UK, and biomass, which is expected to drop significantly by 2035. It should be mentioned that in the "2 degrees" scenario tidal generation is expected to account for 76% of total marine generation, with the remaining marine output to be provided by wave technologies; however, due to the lack of data on the latter, in this study, the simplifying assumption was made to consider all marine generation to be tidal.

More specifically, the yellow line in Figure 4 shows that approximately 68% of total generation in 2035 and 2050 is expected to be provided by VRE technologies (i.e., wind, solar and tidal), which, as the phrase implies, are inherently intermittent. As the cumulative share of these technologies increases over the next decades, the grid will thus require increased levels of energy storage to provide the required flexibility to match the energy demand profile.

Storage technologies will therefore increasingly be deployed alongside VRE, to provide such flexibility, as shown by the red line in Figure 4.

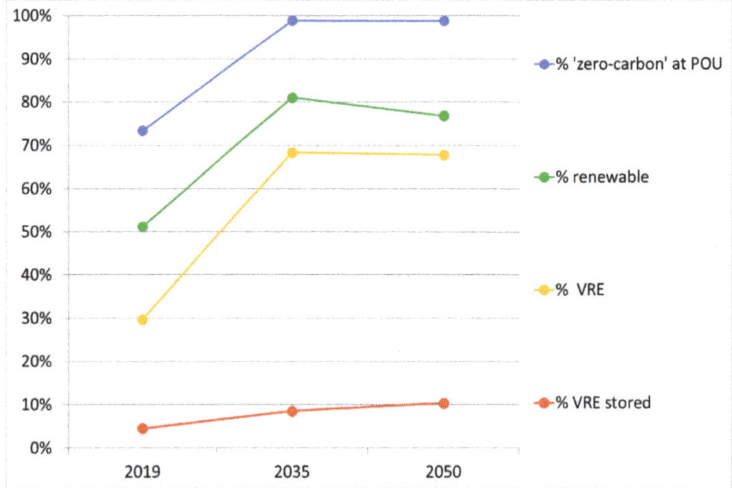

Figure 4. Expected trends (under "2 degrees" scenario assumptions) for, respectively: % of total generation that is considered "zero-carbon" at point of use (i.e., nuclear, biogas, biomass, waste, hydro, tidal, wind and solar); % of total generation that is considered "renewable" (same as above but excluding nuclear); % of total generation that is classified as "Variable Renewable Energy" (VRE; i.e., wind, solar and tidal); % of VRE generation that is sent to energy storage (as opposed to used directly).

National Grid's "2 degrees" scenario considers the deployment of five energy storage technologies: pumped hydro storage (PHS)—in large part relying on existing dammed hydro installations located overseas and tapped via interconnectors, compressed air energy storage (CAES), dedicated lithium-ion batteries (LIB), vehicle-to-grid (V2G) storage provided by the electric vehicle (EV) fleet, and liquid air energy storage (LAES). The first two technologies are generally capable of providing longer-duration storage than the latter three. Specifically, LAES is a fairly new technology that is only projected to ultimately provide 4% of the total storage capacity; due to the very limited information available on its supply chain, in this study the associated storage capacity was instead lumped together with that provided by LIB. The resulting specific assumptions on storage technologies for the future of the UK grid mix are summarized in Table 1.

Table 1. Scenario assumptions on grid-level energy storage.

Technology	Installed Power (GW) 2019	Installed Power (GW) 2035	Installed Power (GW) 2050	Average Daily Storage Duration (h) [1]	Round-Trip Storage Efficiency
PHS	2.7	5.2	5.8	6	80% [31]
CAES	0	1.0	2.0	6	67% [32]
LIB	2.0	8.6	14.0	2	80% [31]
V2G	0	2.7	8.3	2	80% [31]

[1] Own assumption.

Electricity transmission was also included within the boundary of this assessment, albeit limited to the high voltage (HV) network. This was deemed an acceptable simplification, since the vast majority of the electricity generation plants comprising the grid mix at present and in the "2 degrees" scenario are large centralized units which inject HV electricity into the grid (the only exception being rooftop-mounted PV systems). The current HV transmission lines were modelled using the UK-specific life-cycle inventory (LCI) information provided in the Ecoinvent database [33], and then scaled for 2035 and 2050 by simple linear extrapolation based on the projected total gross electricity generation. In terms of energy balance, transmission losses were conservatively set at 5% [34].

Finally, electricity interconnectors with the continent are expected to provide net electricity imports during the demand peaks in winter in the early 2020s. Such imports are, however, expected to decrease significantly in the following years, and to become almost negligible by 2035, eventually leading to very slight net exports by 2050. The choice was therefore made to limit the scope of this analysis to the electricity generated and delivered domestically within the UK, disregarding all electricity exchanged via the interconnectors (with the sole exception of a relatively small share of the future VRE that is assumed to be sent to be stored in PHS facilities located on the continent).

2.2. Electricity Generation and Storage Technologies

2.2.1. Coal

Currently, there are six coal-fired power stations in UK; all are expected to shut down by 2025, except for Drax Power Station, for which there are plans to convert its units to biomass- and gas-fired generation in the near future [35]. The Ecoinvent LCI database "GB" (Great Britain) hard coal electricity production model was adopted for the life cycle inventory of coal-fired electricity generation in the UK.

2.2.2. Natural Gas Combined Cycles

UK natural gas combined cycle (NGCC) inventory was based on the corresponding Ecoinvent GB database model. The model output was reduced by 19% to account for the difference in assumed vs. reported natural gas energy input per kWh of electricity output at the power plant stage, as well as the actual average heating value of the UK gas feedstock [36].

2.2.3. Natural Gas Combined Cycles Plus Carbon Capture and Storage (Future Technology)

At present, there is no detailed inventory information available for natural gas combined cycles with carbon capture and storage (NGCC-CCS). Literature data on the carbon capture process were adopted for 90% CO_2 capture, and incorporated in the adjusted model for conventional NGCC plants. The electricity output of the plant was reduced by 15% to account for the CO_2 capture and compression process (in addition to the initial output adjustment described in Section 2.2.2) [37,38]. The life cycle inventory for the transportation of CO_2 captured and stored is excluded from the study due to the uncertainty involved in the location of the plant and possible storage size.

Appendix A—Tables A1 and A2 provide the construction and operational inventory quantities and the adjusted emissions per kWh of electricity generated.

2.2.4. Biomass

The biomass used for electricity generation in the UK mainly consists of wood pellets from North America forestry, domestic wood chips from UK forests, and a small percentage of domestic residues [39,40]. According to the digest of UK energy statistics (DUKES) 2019 report, based on the 2018 renewable flow-chart, 45% of the biomass share was domestically sourced and 55% was imported [36]. The GB heat and power co-generation model from the Ecoinvent database was selected to represent biomass electricity generation. However, this model does not include mixed inputs of woody biomass feedstocks, and it was therefore modified to account for such. Specifically, the Ecoinvent "RER" (regional European) wood chip model was used to represent the share of wood chips used (since there is not one available specifically for the UK), and two additional processes were added to the model to account for the wood pellet imports. The first process accounts for wood pellet production, and the second one for the transportation of the wood pellets from North America by freight ship. The transport distance was assumed to be 5300 km [41] and the mass of the dry pellets was multiplied by a factor of 1.2 to account for the average moisture content in the transported pellets [42].

2.2.5. Biogas

Biogas is a mixture of gases (mainly methane and carbon dioxide, but with multiple trace gases such as ammonia and hydrogen sulphide) produced by the anaerobic degradation ("digestion") process of organic waste from landfills and, to a lesser extent, agricultural sludge. The biogas is fed into a combined heat and power plant station (CHP), where it is combusted to generate electricity and heat [43]. The Ecoinvent GB biogas-fired heat and power co-generation model was selected to assess biogas electricity generation, and all energy and environmental impacts of the CHP multi-output process were allocated on an energy content basis.

2.2.6. Waste

Electricity generation from waste incineration is also a co-product of a multi-output process. According to ISO recommendations [44], in this case system expansion was adopted in preference to allocation, since waste incineration with energy recovery represents an almost textbook example in which: (i) a primary function is clearly identified (i.e., getting rid of the waste), and (ii) a comparable alternative process exists which delivers only one of the two outputs (i.e., an incinerator without energy recovery). Consequently, the energy recovery process and associated electricity generation was calculated to have negligible emissions assigned to it, since the only additional up-front inputs required vs. the incinerator without energy recovery are those for the boiler and turbine system, while the use-phase emissions at the stack are virtually the same.

2.2.7. Nuclear

There are 15 operating nuclear reactors in UK, 14 of which are advanced gas-cooled reactors (AGR) which are expected to shut down before 2035, and one is a large pressurized water reactor (PWR) which was initially also expected to shut down in 2035, but whose operation may be extended for 20 more years [45]. Out of the currently operating nuclear reactor technologies in the UK, life cycle information in Ecoinvent was only available for PWRs, and therefore the latter was used to model the life cycle inventory associated to nuclear electricity generation

2.2.8. Nuclear (Small Modular Reactors—Future Technology)

Small Modular Reactors (SMR) are factory-built nuclear reactors of less than 300 MWe installed power, inspired by the current large nuclear power plants [46]. They are also known as integral PWR since their main components, such as the stream generator, reactor and pressurizer, are all located in one vessel. SMRs offer the opportunity to add nuclear generating capacity with a smaller capital cost and thus reduce construction risks. They can be categorized in two groups: (1) Generation III water-cooled SMR based on existing large nuclear plants but on a smaller scale, and (2) Generation IV SMRs based on the use of novel fuels and coolants, which can provide other services such as heat for industrial processes [47]. Generation IV small modular reactors are not expected to achieve commercial maturity until 2030 onwards [48], while Generation III SMR are considered to be more mature technologies as they are based on the existing large nuclear plants concept. According to the world nuclear association there are currently two potential SMR projects, one with NuScale and other with Rolls Royce both based on light-water pressurized SMR designs [45].

At present there is no existing LCI database model for SMR technologies, therefore, in this study the Ecoinvent model for PWRs was adopted as a basis for the life cycle inventory of light-water pressurised SMR. The latter are the scaled-down version of large PWR which utilize the same working concepts, but instead of having pumps and coolant loops for directing the flow of water, they utilize natural circulation to direct the cool water to the reactor core after going through the steam generator to turn the turbine to generate electricity [49]. The main components of both systems are considered to be the same, and both are expected to have the same lifetime of 60 years. The Ecoinvent model was adjusted to account for the reduction in efficiency and improvement of the capacity factor (CF) for SMRs compared to large PWRs, which lead to an overall reduced output (−15% in relative terms) [48] (The Capacity Factor is defined as the ratio of the actual power generated by a system, averaged over its service lifetime, to its nominal installed power).

2.2.9. Hydroelectric

Hydropower electricity generation in UK consist of 24% run-of-river and 76% reservoir [36]. The Ecoinvent model for GB run-of-river hydroelectricity model was adopted, and the Ecoinvent "DE" (Germany) model for hydro-reservoirs was used as a proxy, as the database does not contain a corresponding GB model.

2.2.10. Marine Tidal (Future Technology)

Tidal energy is generated through the rise and fall of tides, due to the interaction of gravitational pull of moon and to a lesser extend the sun on the ocean and the rotation of the Earth [50]. There are three types of tidal technologies: lagoon, barrage and stream turbines. National Grid's FES 2019 scenarios assume that the target tidal capacities will be met primarily using tidal lagoons, and secondarily stream turbines. However, there is mounting uncertainty on the future development of tidal technologies, with on the one hand, the recent cancellation of one large tidal lagoon project [51], and on the other hand, new upcoming developments and installed projections on stream turbine [52]. In this study, the assumption was therefore made that the electricity generated by tidal will be harnessed by tidal stream turbines.

There is no model in the Ecoinvent database for this technology. Therefore, all life-cycle inventory and technical information for use in this study was sourced from published scientific literature [53] on an OpenHydro tidal stream turbine. The inventory information includes energy inputs for the installation, manufacturing and maintenance of the system and the material inputs for the construction of the device, power cabling and foundation. The system as described was expected to have a lifetime of 20 years and was rated at 2 MW. The average CF for the stream turbine tidal plant was taken as 5.5% from the DUKES 2019 report [36], and all energy and material inputs were duly scaled to 1 kWh of electricity generated over the life-time of the system.

The resulting foreground inventory information is provided in Appendix A-Table A3.

2.2.11. On-Shore Wind

The Ecoinvent electricity production model for GB 1–3 MW onshore wind turbines was used to represent the total onshore wind electricity generation in UK. The model assumes a 20-year lifetime for all moving components and a 40-year lifetime for all the stationary components of the wind installation [33], and a 26% CF.

2.2.12. Off-Shore Wind

The Ecoinvent electricity production model for GB 1–3 MW offshore wind turbine was used to represent the total offshore wind electricity generation in UK. The model assumes the same lifetimes as for onshore wind turbines, and a 30% CF.

2.2.13. Solar Photovoltaic

National Grid's "2 degree" scenario considers distribution-connected and micro-connected solar capacity and accordingly in this study the assumption was made that most of the solar photovoltaic (PV) generation will come from roof-top mounted systems; additionally, the latest Fraunhofer Institute for Solar Energy report [54] confirms that multi-crystalline silicon (mc-Si) continues to be the leading PV technology by far in terms of global annual production. In order to limit the complexity of the model, a single Ecoinvent process (GB roof-top mounted mc-Si PV) was therefore adopted as the basis for the assessment of solar PV electricity generation in UK. However, since PV systems are still on a continuously and rapidly improving trend, the model was adjusted to reflect the current and expected future mc-Si module efficiencies, respectively reported at 17% in 2019 [54], and estimated at 20% in 2035 and 25% in 2050 [55]. This information was used to adjust the area of solar panels required to produce 1 kWp of installed power in the model.

An average insolation of 1000 kWh/(m^2·yr) was then assumed [56], which, combined with a performance ratio (PR) of 80% [57], led to a calculated CF of 9.1%. Finally, the expected lifetime of the PV modules was kept at 30 years until 2035 [57], and then increased to 35 years for 2050 [55].

2.2.14. Pumped Hydro Storage (PHS)

Pumped hydro storage (PHS) uses electricity to pump water into the high-elevation reservoirs during high generation and low demand, and then releasing the water to generate electricity at peak demand. Since PHS for the UK is projected to utilize pre-existing hydro reservoir systems (mainly located overseas and accessed via the interconnectors), which were built for the primary function of generating hydroelectricity, and since the electricity used for pumping the water uphill would otherwise have to be curtailed, the life-cycle impacts associated with PHS were taken to be zero, thus avoiding any double-counting.

2.2.15. Compressed Air Energy Storage (CAES)

Compressed Air Energy Storage (CAES) systems store excess electricity by compressing air to high pressure in underground reservoirs such as pre-existing salt mines. The stored air is then heated and expanded to drive a turbine to generate electricity when the electricity demand is high and the generation is low [32]. Currently there are two CAES systems operating worldwide, one is in Huntorf, Germany since 1969 and the other is in McIntosh, United States since 1991 [58]. Both work by burning natural gas to provide heat for the expansion of air to drive the turbine generator.

However, the FES 2019 "2 degrees" scenario expects CAES systems to be deployed from 2030 onwards, and therefore in this study the assumption was made that by that time the UK's CAES installed capacity will be of the more advanced adiabatic type (A-CAES). This type of CAES works by retaining and storing the heat generated during the compression of the air using a thermal energy storage (TES) system, and then reusing the stored heat for the expansion process instead of burning natural gas. There has been a lot of on-going research on A-CAES over the last decade, including the planned EU-based research Project "ADELE" [59], the Storelectric project planning to build large-scale A-CAES in Holland [60], and a recently completed demonstration project by Hydrostor in Toronto Island, Canada [61].

Life cycle inventory (LCI) information on A-CAES is not available in the Ecoinvent database. The technical data were therefore adopted from available literature; specifically, the maximum number of storage cycles was taken to be 10,000 [62], and the cycle efficiency of the plant was taken to be 67% [32]. It was also assumed that the compressed air will be stored in pre-existing underground caverns, with a cumulative energy storage capacity of 6 GWh deployed by 2035 (cf. Table 1). The information on material and energy inputs for plant construction, compression unit, heat expander and thermal energy storage system was adopted from published literature [63] and rescaled linearly in terms of storage capacity.

The inventory information for the input quantities per kWh of electricity storage capacity is provided in Appendix A-Table A4.

2.2.16. Lithium-Ion Battery Storage (LIB)

Dedicated grid-level lithium-ion battery (LIB) storage was modelled on the basis of the Ecoinvent model for lithium manganese oxide (LMO) technology. LMO is among the most mature options for LIBs, and although it lags behind some of the other cathode formulations in terms of energy density [64,65], this was deemed relatively unimportant for dedicated stationary applications, and counterbalanced by its comparatively long cycle life, its overall stability, and its reliance on abundant and eco-friendly materials [64]. The maximum number of charge-discharge cycles was assumed to be 7000 [66].

2.2.17. Vehicle-to-Grid Storage (V2G)

For each year of analysis, vehicle-to-grid (V2G) storage schemes rely on the Li-ion batteries already installed in the existing electric vehicle (EV) fleet, when connected to the network of charging points, to provide short-duration storage (i.e., frequency response services) for grid support. In order to avoid incurring in double-counting, energy storage made available through V2G is therefore regarded to have zero impacts assigned to it, since the primary function of vehicle batteries is to provide electricity storage for transportation.

3. Methods

3.1. Life Cycle Assessment (LCA)

From a methodological perspective, the work presented here was conducted in strict adherence to the current International Organization for Standardization norms on LCA [44,67].

The functional unit (FU) of this study was set as 1 kWh of electricity delivered by the UK grid as a whole, including energy storage and transmission.

The main data source used for the life cycle inventory (LCI) analysis was the reputable and widely-used Ecoinvent version 3.5 database [33], complemented where appropriate and required by a range of other literature sources as described in detail in Section 2.2.

As concerns life cycle impact assessment (LCIA), the choice was made to focus on a set of key impact categories, individually discussed in Sections 3.1.1–3.1.5, and all evaluated at "mid-point" using the widely-adopted and well-regarded CML method [68]. Normalization and weighting were not conducted because, while potentially facilitating the interpretation of the results by a less technical audience, they inevitably remain the most arbitrary steps in any LCA, and the choice of the weighting factors is to a large extent political, with very little if any scientific relevance. In fact, because of this, according to ISO, normalization and weighting are always optional steps and are discouraged for any "comparative assertion intended to be disclosed to the public" [44].

Finally, from a practical point, the whole analysis was carried out using the latest release of the dedicated LCA software package GaBi [69].

3.1.1. Global Warming Potential (GWP)

Global Warming Potential is calculated applying IPCC-derived characterization factors to gaseous emissions, on the basis of their respective equivalent warming potential relative to carbon dioxide (CO_2), with a time horizon of 100 years.

Two alternative accounting rules are possible with regard to biogenic carbon emissions, i.e., those that arise from the combustion of biomass (wood chips and pellets, and biogas derived from the anaerobic degradation of organic matter). The argument for excluding them from the calculation of the grid's GWP is that the same amount of carbon (on a molar basis) was previously absorbed from the atmosphere during the biomass growth phase (including the trees used for the wood chips and pellets, as well as those used for paper and cardboard and all the agricultural and food crops that are eventually biodegraded to produce biogas), thereby effectively "closing the loop" and resulting in net zero carbon emissions. Of course, this calculation only applies to C emitted as CO_2 (i.e., under complete combustion conditions), otherwise each mole of C that is absorbed as CO_2 and later emitted as CH_4 would result in a net contribution to GWP, as quantified by Equation (1):

$$\text{Net GWP per mole of biogenic C} = MM_{CH4} \, (CF_{CH4}\text{-}1) \, [g \, CO_2\text{-eq}]$$

where:

$$MM_{CH4} = \text{molar mass of } CH_4 \, [g/mol]$$
$$CF_{CH4} = \text{GWP characterization factor for } CH_4 \, [g \, CO_2\text{-eq} / g \, CH_4]$$

(1)

Even so, the argument for excluding biogenic C from the accounting only holds fully if it can be proven that the totality of such biomass was in fact grown sustainably (e.g., from well-managed short-rotation forestry that leads to net zero standing biomass change over time). In reality, this condition can rarely be expected to be met completely. Specifically, wood chips coming from domestic forestry residues may often be closer to being net zero C than wood pellets imported from overseas and paper and cardboard coming from a wide range of international sources. As a result, the real-world net carbon emissions of biogenic feedstocks will always be higher than zero, with considerable uncertainty on the exact figures, depending on often hard-to-ascertain factors such as feedstock origin and supply chain practices.

To reflect such uncertainty, in this work the choice was made to calculate and report GWP under both assumptions, i.e., respectively including and excluding biogenic carbon emissions, and to discuss the results in the light of the considerations made above.

3.1.2. Acidification Potential (GWP)

Acidification Potential is calculated applying stoichiometry-derived characterization factors to airborne emissions, on the basis of their respective acidification potential relative to sulphur dioxide (SO_2).

3.1.3. Human Toxicity Potential (HTP)

Human Toxicity Potential is calculated applying characterization factors to all emissions (to air, water and soil), on the basis of their respective toxicity potential relative to 1,4-dichlorobenzene (1,4-DB).

It is noteworthy that HTP results are inevitably affected by a larger margin of uncertainty than those for all other impact categories, due to the intrinsic methodological difficulty of comparing and combining the individual toxicity potentials of a wide and diverse range of organic and inorganic emissions into a single indicator. The uncertainty is especially large in the case of metal emissions [70].

3.1.4. Abiotic Depletion Potential (ADP)

Abiotic Depletion Potential is calculated applying characterization factors to all non-living LCI inputs from the geosphere, expressing their respective scarcity relative to the element antimony (Sb), based on estimates of ultimate reserves and current extraction rates. In order to avoid partially duplicating the information provided by the nr-CED indicator (cf. Section 3.1.5), and potentially obfuscating the impact arising from the depletion of non-energy resources, this indicator is calculated excluding the contributions of all energy inputs (such as fossil fuels and uranium).

It should be noted that abiotic depletion is an impact category that is still frequently the object of methodological discussion, and alternative approaches exist to the quantification of the associated impact, often reflecting differences in problem definition [71–73]. Additionally, ADP's specific dependence on the estimate of ultimate reserves makes it susceptible to obsolescence, especially when using this indicator to assess a depletion-related impact taking place several decades into the future [74].

3.1.5. Non-Renewable Cumulative Energy Demand (nr-CED)

Non-renewable Cumulative Energy Demand is calculated by applying characterization factors to all LCI inputs, based on the total non-renewable primary energy directly and indirectly harvested from the environment for their provision, and expressed in terms of Joules of crude oil equivalent [75].

3.2. Net Energy Analysis (NEA)

Net Energy Analysis (NEA) provides a different and complementary viewpoint to LCA [23,26,28,29,76], and it is carried out here using the same underlying inventory analysis, thereby providing an internally coherent platform for the calculation and discussion of the results. Specifically, while LCA draws the boundary of the system under analysis so as to account for all natural resources used as inputs (and all emissions to the environment as outputs), NEA is only concerned with those energy inputs that are already available as energy carriers within the technosphere, and which are deliberately "invested" in the system (Inv) for the purpose of harvesting more primary energy (PE) from nature and delivering (the term "returning" is also often used) a usable energy carrier (EC) to society. This is illustrated in Figure 5 with a simplified diagram of a generalized energy supply chain.

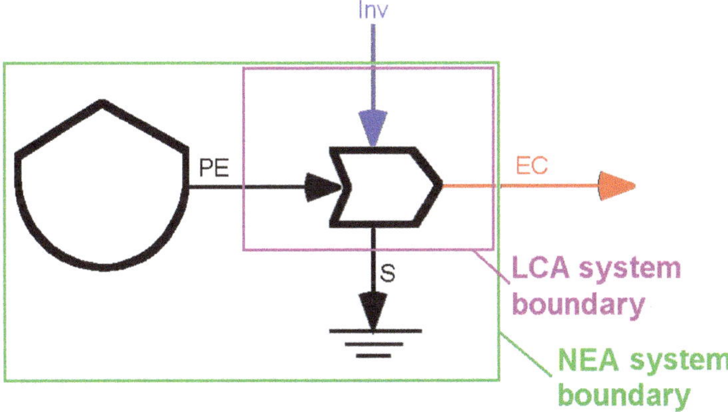

Figure 5. Simplified diagram of a generalized energy supply chain, illustrating the different boundaries set by life cycle assessment (LCA) and net energy analysis (NEA). PE = Primary Energy; Inv = energy "investment"; EC = Energy Carrier (the energy "return"); S = energy dissipated to the environment.

For the purpose of consistency, all energy "investments" (Inv) to the energy system are accounted for in terms of the total (i.e., renewable plus non-renewable) primary energy directly and indirectly harvested from the environment for their provision (expressed in MJ of oil equivalents). This corresponds to including in the analysis the whole supply chain for "Inv", like in LCA.

The energy "returned" by the same system in the form of a usable energy carrier may then alternatively be accounted for either on the basis of the actual energy in the carrier (EC), or on the basis of its equivalent primary energy ($EC_{PE\text{-}eq}$, measured in MJ of oil equivalents).

When EC is electricity, its equivalent primary energy may be calculated according to an LCA substitution logic, i.e., by calculating how much primary energy is directly and indirectly harvested (in total) from the environment to produce one unit of electricity using the current grid mix. Expressing the electricity "return" in terms of equivalent primary energy thus makes the NEA of any of the individual electricity technologies (e.g., natural gas combined cycles, wind, PV, etc.) inextricably linked to the specific grid mix into which it is embedded. However, doing so also has the following advantages:

(i) It enables the calculation of an "energy return ratio", often referred to in literature as Energy Return on Investment (EROI) [25,77], which features consistent units of primary energy equivalents at both the numerator and denominator. This simplifies the interpretation, since otherwise, "if the numerator and denominator are not measured by the same rule, one loses the intuitively appealing interpretation that EROI > 1 is the absolute minimum requirement a resource must meet in order to constitute a net energy source" [78]. In order to clarify when EROI is calculated using units of equivalent primary energy at the numerator, the notation $EROI_{PE\text{-}eq}$ is used in this work—cf. Equation (2).

$$EROI_{PE\text{-}eq} = EC_{PE\text{-}eq}/Inv \tag{2}$$

(ii) It enables the methodologically consistent calculation of the Net-to-Gross (NTG) energy return ratio, defined as per Equation (3). NTG thus provides a clear and easy-to-interpret indication of how much of the "gross" energy delivered to society by the UK grid is a "net" energy output that remains available for all societal uses, vs. how much needs to be re-invested to keep the grid itself operational.

$$NTG = (EC_{PE\text{-}eq} - Inv)/EC_{PE\text{-}eq} \tag{3}$$

By combining Equations (2) and (3) and simply re-arranging the terms, NTG may also be expressed as in Equation (4), which leads to what has often been referred to as the "Net energy cliff" [28,79–81] (Figure 6).

$$NTG = 1 - 1/EROI_{PE\text{-}eq} \tag{4}$$

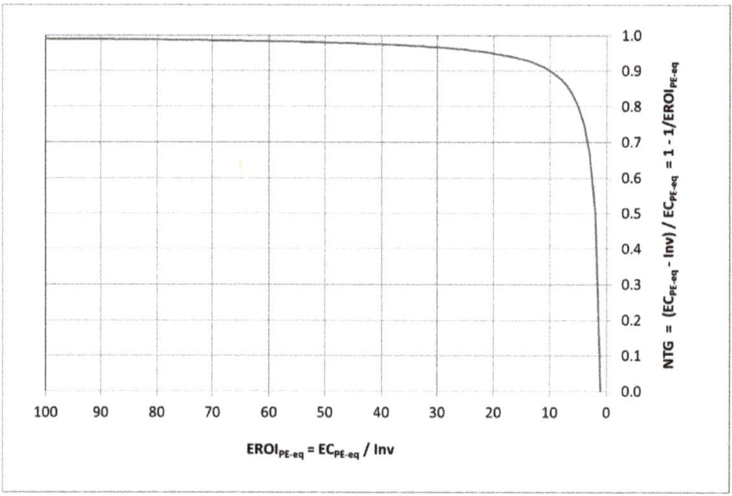

Figure 6. "Net energy cliff" diagram displaying the relation between Energy Return on Investment ($EROI_{PE\text{-}eq}$) and Net-to-Gross ratio (NTG).

The interpretation of the "Net energy cliff" is that once the $EROI_{PE\text{-}eq}$ of an energy system starts dropping below approximately 10, the share of its gross energy output that remains available for other societal uses (after subtracting the primary energy required to sustain the operation of the energy system itself) starts being reduced significantly for each additional reduction in $EROI_{PE\text{-}eq}$; in other words, for $EROI_{PE\text{-}eq} < 10$, NTG quickly "falls off a cliff". Conversely, for all values of $EROI_{PE\text{-}eq} > 10$, NTG remains safely > 0.9, which means that beyond such "threshold" there is less and less significant competitive advantage to a system characterized by increasingly larger $EROI_{PE\text{-}eq}$.

4. Results and Discussion

4.1. Life Cycle Impact Assessment Results

4.1.1. Global Warming Potential

Figures 7 and 8 both illustrate the expected life-cycle GHG emissions associated to the future UK grid mix in 2035 and 2050, expressed per unit of electricity delivered, and relative to the emissions in 2019. The difference between the two figures is that the former includes the contribution of biogenic carbon, while the latter excludes it.

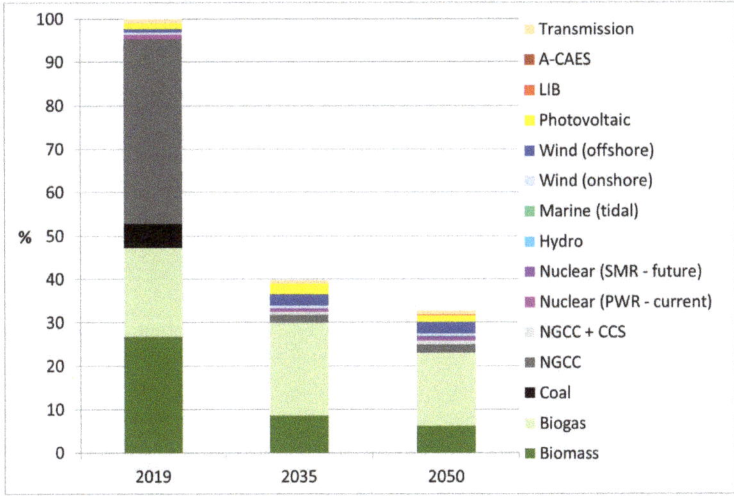

Figure 7. Global warming potential (including biogenic carbon) per unit of electricity delivered by the UK grid mix (including energy storage and transmission), expressed as relative to the 2019 value (100% = 269 g CO_2-eq/kWh).

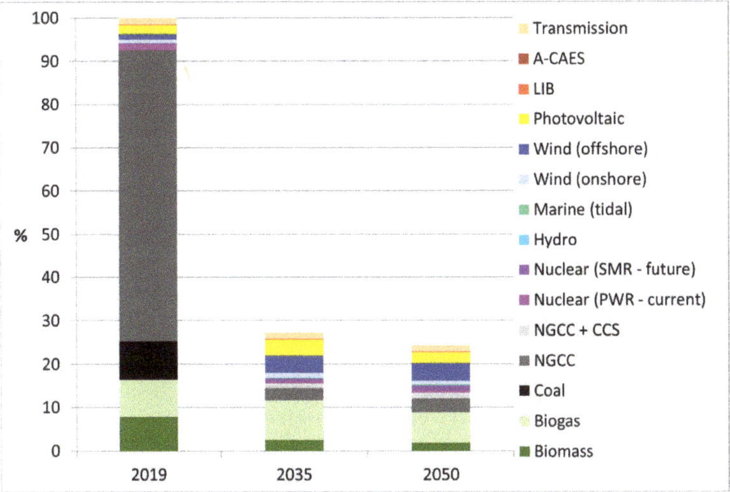

Figure 8. Global warming potential (excluding biogenic carbon) per unit of electricity delivered by the UK grid mix (including energy storage and transmission), expressed as relative to the 2019 value (100% = 170 g CO_2-eq/kWh).

As expected, the two technologies that are most significantly affected by the different assumptions on how to account for biogenic C are biomass- and biogas-fired electricity generation. Additionally, it is noteworthy that even when excluding biogenic C emissions (Figure 8), the life-cycle GHG emissions caused by biogas-fired electricity are expected to still amount to 9% of the grid mix total in 2035, despite such technology delivering only 5.7% of the total electricity generation (cf. Figure 2). This points to biogas being almost twice as carbon intensive as the grid mix as a whole at that point in time, and definitely nowhere near as deserving of the "zero carbon" designation as the leading renewable technology for the UK, i.e., wind, which is responsible for just 5% of total GHG emissions while generating 58% of the total electricity. On the one hand, this might lead to question of whether retaining, and if fact increasing, biogas-fired electricity generation in the future grid mix even makes sense at all. On the other hand, though, one must also consider what the alternative would be, i.e., what would happen to the biogas that is produced by anaerobic degradation of organic matter in landfills and in sewage sludge if it were not used as a feedstock for electricity generation. Clearly, releasing it directly to the atmosphere would not be an option, as biogas is rich in methane and would cause even more global warming. Therefore, the focus really shifts from the energy sector to the waste management sector, and points to a need to reduce the reliance on all landfills (municipal, industrial and agricultural) to the maximum extent possible, and instead incentivize the reuse, recycling and thermovalorization of waste flows (as applicable, and broadly in that order of merit), as well as composting (the latter providing the added benefit of returning valuable nutrients to the soil).

Unsurprisingly, the planned decommissioning of coal- and gas-fired power plants is shown to have the largest effect on the decarbonization of the grid mix. It is also noteworthy that the contribution of energy storage technologies (A-CAES and LIB) to the total carbon budget is absolutely negligible, even out to 2050 when 10% of the total yearly VRE generation is assumed to be routed into storage. This is a reassuring result that dispels any potential concerns about the negative effect that the requirement for energy storage might have on the planned decarbonization of the grid.

Considering the total grid mix results as a whole, from 2019 to 2035, GHG emissions may be expected to drop by 60% (i.e., from 263 to 106 g CO_2-eq/kWh, when including biogenic C emissions), or even as much as 73% (i.e., from 170 to 46 g CO_2-eq/kWh, when excluding biogenic C emissions). Further, but less significant, emission reductions are then expected when extending the analysis to 2050. As discussed in Section 3.1, the real total net GHG emissions lie somewhere in between those reported respectively in Figures 7 and 8, but are likely to be somewhat closer to the latter. Be that as it may, it is worth noting that in neither case does the reduction in the total life-cycle grid mix emissions approach 99%, as one might naively assume when considering the share of electricity output that is generated by technologies that are considered "zero carbon at point of use" (cf. blue line in Figure 4). This should not be interpreted as a damning result per se, but rather as a stark reminder of the importance of always duly including all life cycle stages in the analysis.

4.1.2. Acidification Potential

When looking at the results for acidic emissions (Figure 9), two things are readily apparent: (i) the total reductions in potential impact that may be expected for 2035 and even 2050 are much less significant than in the case of global warming potential (respectively, only −8% and −28% with respect to 2019); and (ii) biogas-fired electricity is by far the worst offender in the mix (>80% of total acidic emissions). The latter result is striking, but it is broadly confirmed by other independent studies [82,83], and it may be understood if one considers that the anaerobic degradation of biomass produces significant levels of ammonia and hydrogen sulfide alongside methane, and that those two gases are readily oxidized to, respectively, NO_2 (then hydrated to nitric acid) and SO_2 (then hydrated to sulphuric acid).

While these AP results may look somewhat discouraging, it should also be recognized that there is potentially ample scope for improvement if the biogas is "upgraded" to biomethane by subjecting it to water scrubbing and membrane separation prior to feeding it to the thermal power plant, which would essentially rid it of most ammonia and hydrogen sulfide and render it almost indistinguishable from natural gas [84]. While not yet common practice in the UK, such pre-treatment of the biogas feedstock may become more widespread in the future and contribute to curbing the acidic emissions of biogas-fired electricity, and hence of the whole grid mix, by 2035 or 2050.

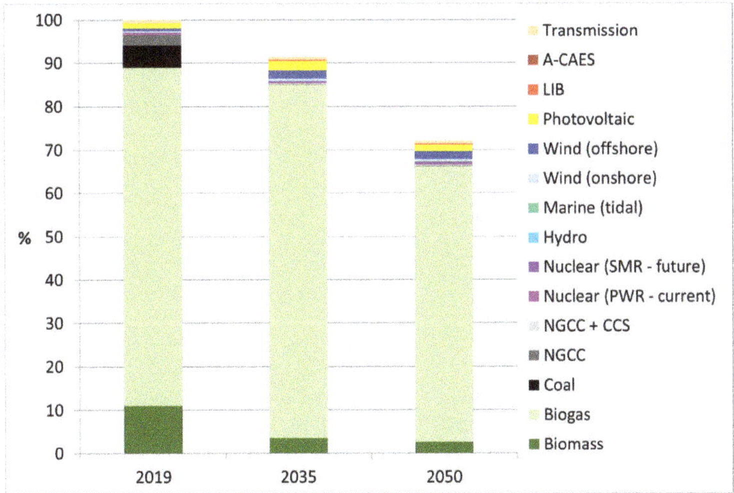

Figure 9. Acidification potential per unit of electricity delivered by the UK grid mix (including energy storage and transmission), expressed as relative to the 2019 value (100% = 2.1 g SO_2-eq/kWh).

4.1.3. Human Toxicity Potential

When moving to consider the life-cycle environmental impacts of the grid mix in terms of human toxicity (Figure 10), a reversal of the trend is observed for the first time, whereby the total impact increases going from 2019 to 2035 (+10%), only to then decrease again slightly in 2050 (+8% relative to 2019). Because of the unavoidably large uncertainty associated with the quantification of HTP (cf. Section 3.1.3), such relatively small changes are probably at the limit of what ought to be considered statistically significant, and another, possibly more scientifically valid way of looking at the results is that the overall human toxicity potential of the grid mix is expected to remain broadly stable for the next three decades.

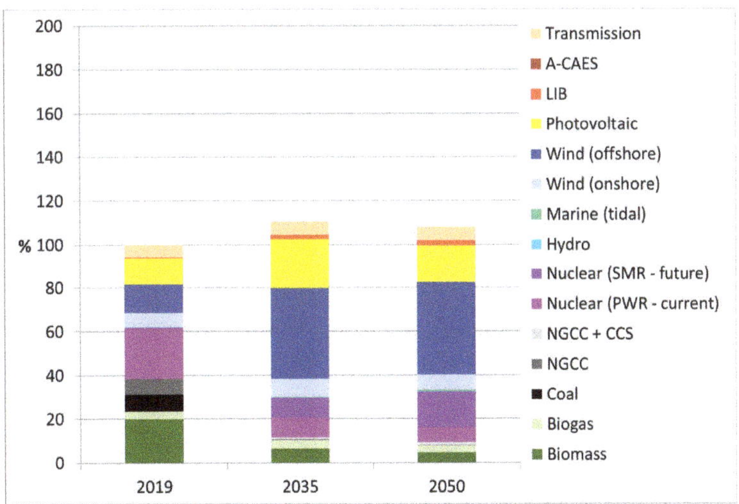

Figure 10. Human toxicity potential per unit of electricity delivered by the UK grid mix (including energy storage and transmission), expressed as relative to the 2019 value (100% = 83 g 1,4-DB-eq/kWh).

What is interesting, nonetheless, is that the relative contributions of the various technologies to the total impact are significantly different than for GWP or AP, and for the first time even those technologies that are conventionally regarded as the "greenest" (i.e., wind and solar PV) end up being responsible for sizeable shares of the total impact. These results are due to a combination of these technologies' comparatively large demand for heavy metals (mainly Cu, Al and Ni) per functional unit (also corroborated by previous independent studies [19,85]), and the toxic emissions associated to the respective metal supply chains (mainly at the mining and beneficiation stages) [33,86]. For the same reasons, electricity transmission lines and LIBs are also non-negligible contributors to this impact category. The HTP of nuclear electricity is likewise significant in the mix, almost entirely due to the emissions arising from the uranium supply chain.

4.1.4. Abiotic Depletion Potential (Elements)

The abiotic resource depletion results illustrated in Figure 11 ("elements", i.e., excluding energy resources such as fossil fuels and uranium) are even more striking, in that they point to a significant increase of the total grid mix impact: initially +76% in 2035 and then down to +55% in 2050 (both values relative to 2019). Even more so than for HTP, these results are mainly driven by wind and PV's increased demand for metals (mainly Cu) per unit of electricity delivered, and LIB and transmission lines once again play a non-negligible role.

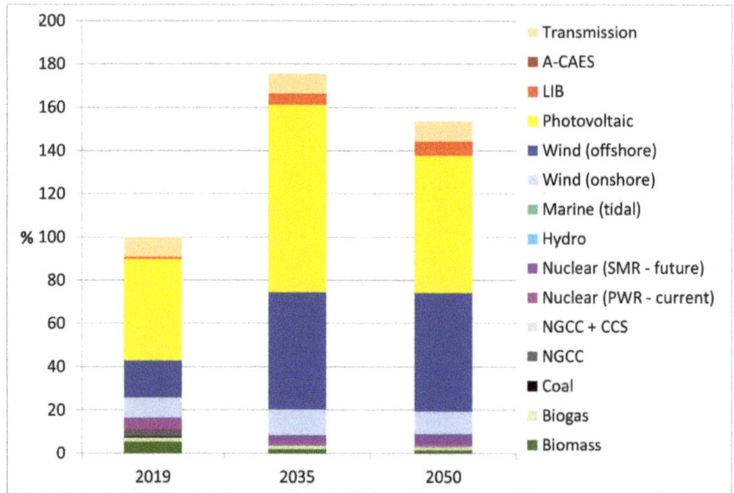

Figure 11. Abiotic depletion potential (elements) per unit of electricity delivered by the UK grid mix (including energy storage and transmission), expressed as relative to the 2019 value (100% = $3.6 \cdot 10^{-4}$ g Sb-eq/kWh).

4.1.5. Non-Renewable Cumulative Energy Demand

The analysis of the grid's overall demand for non-renewable primary energy (Figure 12) is straightforward to interpret, with those thermal technologies relying on non-renewable energy feedstocks (i.e., coal, natural gas and uranium) being responsible for the vast majority of the impact. The continued reliance on nuclear power as a significant contributor to the future mix (15.8% in 2035 and up then to 19.5% in 2050—cf. Figures 2 and 3) also poses a major constraint on the achievable improvement at grid level (only −24% in 2035 and −40% in 2050), which contrasts with the more significant gains in terms of decarbonization (cf. Figures 7 and 8).

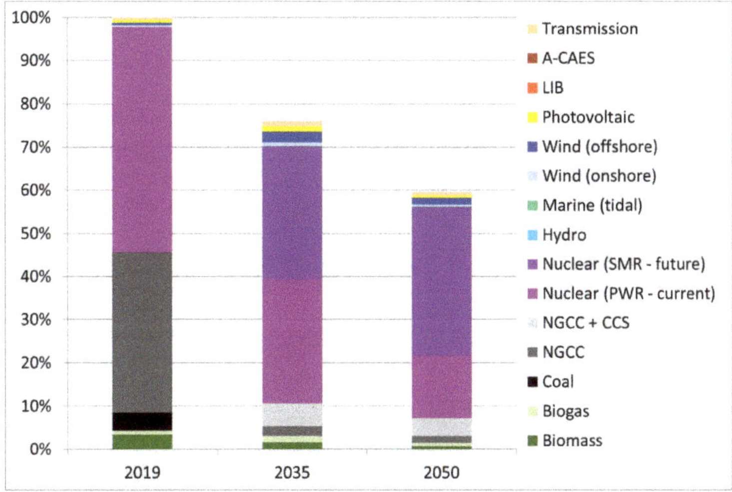

Figure 12. Non-renewable Cumulative Energy Demand (nr-CED) per unit of electricity delivered by the UK grid mix (including energy storage and transmission), expressed as relative to the 2019 value (100% = 6.0 MJ oil-eq/kWh).

4.2. Net Energy Analysis Results

Figure 13 illustrates the total primary energy that must be invested to operate the various energy supply chains that "feed" the UK grid, per unit of final electricity delivered by the grid itself. It is important to recall that, as explained in Section 3.2, these figures exclude the primary energy resources that are directly harvested from nature and converted into electricity (e.g., the coal itself that is extracted, pulverized and delivered to the coal-fired power plants; the woody biomass is harvested, compressed into pellets and delivered to the biomass-fired power plants; the solar energy is captured by the PV panels; and so on and so forth). What these results show is that, over time, the UK grid mix will require less and less commercial energy to be diverted from other societal uses and (re)invested to support its operation, per unit of electricity delivered.

In terms of the break-down of the total energy investment by technology, the results for the three years considered in the analysis broadly reflect the changing grid mix composition, with a few notable observations to be made, as follows.

(i) The energy investment for biomass-fired electricity is disproportionally large (over 30% of the total in 2019, vs. a share of only 10% of electricity generated). The planned reduced reliance on biomass as an energy resource for electricity generation in the future (3% of the grid mix in 2035 and 2% in 2050—cf. Figures 2 and 3) therefore seems justified from a net energy perspective.

(ii) At the opposite end of the scale, the expected future energy investment for wind electricity (approximately 30% of the total in 2035 and 2050) appears to be well justified in view of the associated energy returns (the share of total electricity generated by wind approaches 60% in the same years—cf. Figures 2 and 3).

(iii) The energy investment required for energy storage technologies (A-CAES and LIB) remains very small, even when such technologies are relied upon to store 10% of the VRE generated. This is a reassuring result that indicates that, from an energy balance perspective, there likely will not be any disruptive consequences when moving to a future grid mix which will rely heavily on intermittent renewable generators and will therefore require energy storage.

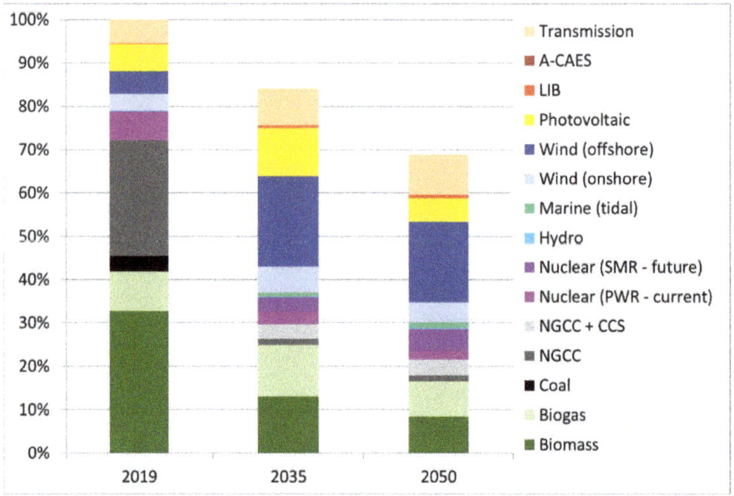

Figure 13. Total energy investment per unit of electricity delivered by the UK grid mix (results include the effects of energy storage and transmission), expressed as relative to the 2019 value (100% = 0.65 MJ oil-eq/kWh).

Figure 14 then illustrates the same overall results, but in terms of the Net-to-Gross primary energy return ratio. NTG is shown to creep steadily upwards on the "net energy cliff", from 2019 to 2050, once again indicating that the planned energy transition according to National Grid's "2 degrees" scenario is sound from a net energy perspective. This is a significant result, which should allay some of the often-voiced concerns about potentially reduced future availability of net energy.

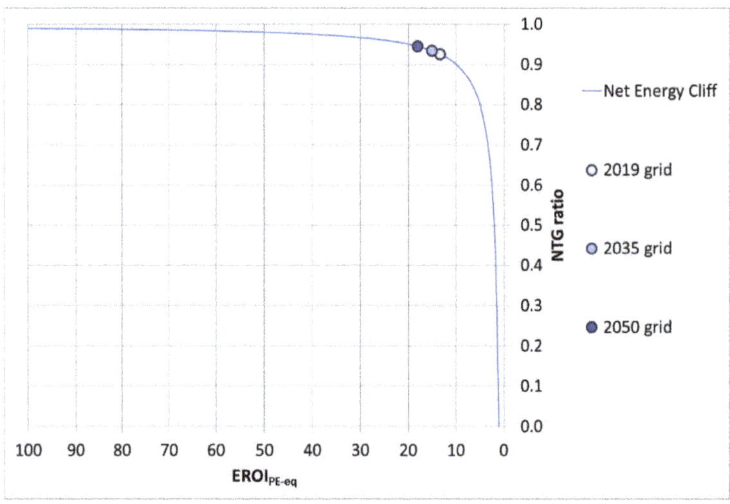

Figure 14. Evolution of the Net-to-Gross (NTG) primary energy return ratio of the UK grid mix (results include the effects of energy storage and transmission).

Having said all this, it is also important to underline that the present analysis was carried out using an "integrative" approach [87], which is characteristic of LCA and which "discounts" (i.e., virtually spreads out) all inputs and emissions associated to each individual energy system over its respective full life cycle. In other words, each individual energy supply chain is treated as a "black box" from a temporal perspective, irrespective of exactly when, during its life cycle, a particular (energy) input is supplied or a specific emission occurs. Instead, in reality, most of the material and energy inputs required for renewable technologies such as wind and PV take place up front during their manufacturing phase, while the associated energy "returns" are reaped over the course of approximately three decades (the typical lifetime of these systems). When these technologies are deployed quickly and on a massive scale, this type of temporal mismatch could, in some instances, result in a temporary reduction in the initial year-by-year availability of net energy, as discussed elsewhere in literature [88]. While this fact needs to be duly taken into account in terms of energy policy planning, it is fair to argue that, in light of the overall positive longer-term NTG trend identified here, it should not be seen as reason to dismiss the fundamental soundness of the whole long-term strategy, but instead as a further call for the application of a "sower's strategy" (whereby today's energy "seeds" are planted to reap the energy "crops" of tomorrow) [89].

5. Conclusions

This thorough life-cycle analysis of National Grid's "2 degrees" future energy scenario has produced a wide range of quantitative results which, when analyzed together, allow for a comprehensive and balanced appraisal of its strengths and weaknesses. To summarize, the following considerations may be made, which capture the key findings and provide policy guidance.

Firstly, these life-cycle carbon budget calculations have confirmed that the "2 degrees" strategy would be very effective at decarbonizing UK electricity, albeit not to the radical extent that might have been inferred by just taking at face value the share of electricity that is projected to be generated by "zero carbon" technologies (nuclear, wind, tidal, hydro, biogas, biomass and gas with CCS).

Secondly, the analysis has shown that biogas and, to a lesser extent, biomass are not especially benign energy resources for electricity production, despite their intuitively reassuring "bio" designation. Both are negatively affected by low energy return on investment, and biomass-fired electricity is responsible for very large acidic emissions. However, while reducing the reliance on biomass would be relatively easy to achieve by just curbing imports of wood pellets from North America, biogas-fired electricity may prove difficult to phase out, since biogas is produced in landfills and agricultural sludge, and using it to produce electricity may be the lesser evil in terms of global warming. Cleaning up biogas electricity generation by either upgrading the biogas feedstock to biomethane or applying drastic scrubbing at the power plant stack output would however still be recommendable.

Thirdly, the results have indicated that moving to the large-scale penetration of variable renewable energy (VRE) generation entails some trade-offs in terms of the depletion of metal resources and potential associated life-cycle toxicity impacts. This is partly due to wind and PV's demand for technology-specific metals and semi-metals, but also in large part simply determined by their less spatially concentrated nature, which calls for more copper transmission lines per unit of electricity delivered.

Fourthly, the planned continued (and even increased, after 2030) reliance on nuclear generators puts a hard cap on the achievable gains in terms of decreased dependence on non-renewable (and imported) energy resources. While the ultimate availability of nuclear ore reserves may not pose a major issue for a long time still, this fact does have immediate negative consequences in terms of the UK's energy sovereignty (since no such ores are available domestically).

Importantly, the deployment of energy storage technologies has shown not to cause any major set-backs in terms of any of the impact categories (with the possible partial exception of human toxicity and abiotic depletion). This is a welcome and reassuring result in and of itself, and it may also be interpreted to indicate that perhaps an even more aggressive roll-out of wind and PV might be attempted (the intermittency of which could be mitigated with even more energy storage), which would lead to a reduced demand for new small modular nuclear reactors.

However, a margin of uncertainty remains on the possible residual requirement for some curtailment of VRE, despite the substantial energy storage capacity that is assumed to be deployed in the "2 degrees" scenario. Such uncertainty could potentially be reduced by employing a high-temporal-resolution modelling approach whereby the hourly electricity supply and demand profiles are extrapolated on the basis of historical datasets.

Additionally, further research is already under way, whereby the interlinkages between the electricity and transport sectors will be explicitly modelled dynamically, with specific focus on the twin roles played by lithium-ion batteries (LIBs). On the one hand, LIBs are expected to be used as on-board storage in electric vehicles (EVs), which may be privately-owned or used for transport-as-a-service (TaaS), and in both cases potentially provide vehicle-to-grid (V2G) storage capacity. On the other hand, LIBs will also be used for dedicated stationary grid-level storage, in which case they may be produced using a combination of virgin and recycled materials, or even re-purposed after their first use in EVs (i.e., a second life application). The ensuing coupled mass-flow model will enable a more detailed and robust assessment of the energy and environmental impacts of energy storage on the future electricity grid mix, by improving on the current blanket assumptions about the future deployment of LIB and V2G storage.

Finally, this analysis has produced significant, and to some extent, possibly even ground-breaking results by dispelling the often-voiced concerns that a massive transition to renewable energy technologies must necessarily entail a worrisome reduction in the net energy that is made available to society. In fact, it was shown that the evolution of the UK grid under "2 degree" scenario conditions would even result in a modest increase of its net-to-gross energy return ratio.

All in all, this study's results should be taken as a sobering reminder that there is more to "greening the grid" than meets the eye, and that forecasting the complex effects of any energy policy strategy calls for a holistic life-cycle approach.

Author Contributions: Conceptualization, M.R.; methodology, M.R.; software, M.R. and M.K.; validation, M.R., M.K. and A.H.; formal analysis, M.R.; investigation, M.R. and M.K.; resources, M.R. and M.K.; data curation, M.K.; writing—original draft preparation, M.R. and M.K.; writing—review and editing, M.R. and A.H.; visualization, M.R.; supervision, M.R. and A.H.; project administration, A.H.; funding acquisition, A.H. All authors have read and agreed to the published version of the manuscript.

Funding: This work was supported by the Faraday Institution [grant number FIRO05].

Conflicts of Interest: The authors declare no conflict of interest.

Appendix A

This appendix contains foreground life-cycle inventory tables for those electricity generation and storage technologies for which no pre-existing data were available in the Ecoinvent database, and which had to be modelled from scratch on the basis of literature information.

Table A1. Foreground inventory of life-cycle inputs for carbon capture and storage (CCS) technology, complementing natural gas combined cycle (NGCC) power plants. All values are per kWh of electricity generated.

Item	Quantity	Unit
Activated Carbon	$3.2 \cdot 10^{-5}$	kg
Concrete	$2.1 \cdot 10^{-7}$	kg
Electricity [1] (for CO_2 compression)	$4.7 \cdot 10^{-2}$	kWh
Monoethanolamine (MEA)	$1.8 \cdot 10^{-4}$	kg
Polyethylene, high density (HDPE)	$7.1 \cdot 10^{-7}$	kg
Sodium hydroxide (NaOH)	$5.5 \cdot 10^{-5}$	kg
Steel (low alloyed)	$7.7 \cdot 10^{-5}$	kg

[1] Accounted for by deduction from plant output.

Table A2. Foreground inventory of use-phase emissions per kWh of electricity generated by the NGCC + CCS system.

Item	Quantity	Unit
Carbon dioxide (CO_2)	47	g
Nitrogen oxides (NO_x)	$1.7{\cdot}10^{-1}$	g
Sulphur dioxide (SO_2)	$3.8{\cdot}10^{-3}$	g
Particulate matter (PM)	$2.2{\cdot}10^{-3}$	g
Formaldehyde (HCHO)	$1.1{\cdot}10^{-1}$	g
Acetaldehyde (CH_3-CHO)	$7.0{\cdot}10^{-2}$	g
Ammonia (NH_3)	$1.5{\cdot}10^{-2}$	g
Monoethanolamine (MEA)	$2.6{\cdot}10^{-2}$	g

Table A3. Foreground inventory of life-cycle inputs for stream turbine tidal electricity generation. All values are per kWh of electricity generated.

Item	Quantity	Unit
Cast Iron	$1.5{\cdot}10^{-6}$	kg
Cement	$2.5{\cdot}10^{-5}$	kg
Copper	$3.2{\cdot}10^{-6}$	kg
Electricity (for plant construction)	$1.9{\cdot}10^{-2}$	kWh
Glass fibre reinforced plastics (GRP)	$9.4{\cdot}10^{-6}$	kg
Polyethylene (PE)	$4.7{\cdot}10^{-7}$	kg
Steel (low alloyed)	$1.6{\cdot}10^{-4}$	kg

Table A4. Foreground inventory of life-cycle inputs for adiabatic compressed air energy storage (A-CAES) technology, including plant construction, compressors, thermal energy storage (TES) and heat expanders. All values are per MWh of electricity storage capacity.

Item	Quantity	Unit
Aluminium	$4.4{\cdot}10^{-1}$	kg
Cast Iron	48	kg
Concrete	$5.2{\cdot}10^2$	kg
Copper	4.0	kg
Diesel (burnt in building machines)	$9.1{\cdot}10^2$	MJ
Electricity (for plant construction)	18	kWh
Foam Glass	3.2	kg
Heavy fuel oil (burnt in industrial machines)	$9.1{\cdot}10^2$	MJ
Insulation (rock wool)	19	kg
Limestone	4.6	kg
Lubricating oil	$2.5{\cdot}10^3$	kg
Polypropylene (PP)	$6.3{\cdot}10^{-1}$	kg
Sand-lime brick	24	kg
Steel (high alloyed)	91	kg
Steel (low alloyed)	$1.3{\cdot}10^2$	kg
Steel (unalloyed)	$1.1{\cdot}10^2$	kg

References

1. Ritchie, H.; Roser, M. CO2 and Greenhouse Gas Emissions. Our World in Data. Available online: https://ourworldindata.org/co2-and-other-greenhouse-gas-emissions#global-warming-to-date (accessed on 19 March 2020).
2. Keay, M. *Energy: The Long View*; Oxford Institute for Energy Studies: Oxford, UK, 2007.

3.   Agbugba, G.; Okoye, G.; Giva, M.; Marlow, J. The Decoupling of Economic Growth from Carbon Emissions: UK Evidence. Office for National Statistics. Available online: https://www.ons.gov.uk/economy/nationalaccounts/uk sectoraccounts/compendium/economicreview/october2019/thedecouplingofeconomicgrowthfromcarbonemis sionsukevidence (accessed on 19 March 2020).

4.   Committee on Climate Change. UK Climate Action Following the Paris Agreement. 2016. Available online: https://www.theccc.org.uk/publication/uk-action-following-paris/ (accessed on 19 March 2020).

5.   Dejuán, Ó.; Lenzen, M.; Cadarso, M.Á. (Eds.) *Environmental and Economic Impacts of Decarbonization: Input-Output Studies on the Consequences of the 2015 Paris Agreements*; Routledge: London, UK, 2017; p. 402. ISBN 978-1-31-522593-7. [CrossRef]

6.   UK Climate Change Act. 2008. Available online: http://www.legislation.gov.uk/ukpga/2008/27/part/1/crossh eading/the-target-for-2050 (accessed on 19 March 2020).

7.   Brown, T.W.; Bischof-Niemz, T.; Blok, K.; Breyer, C.; Lund, H.; Mathiesen, B.V. Response to 'Burden of proof: A comprehensive review of the feasibility of 100% renewable-electricity systems'. *Renew. Sust. Energy Rev.* **2018**, *92*, 834–847. [CrossRef]

8.   Fthenakis, V.; Mason, J.E.; Zweibel, K. The technical, geographical, and economic feasibility for solar energy to supply the energy needs of the US. *Energy Policy* **2009**, *37*, 387–399. [CrossRef]

9.   Zweibel, K.; Mason, J.E.; Fthenakis, V. A Solar Grand Plan. *Scientific American.* 2008. Available online: https://www.scientificamerican.com/article/a-solar-grand-plan/ (accessed on 17 April 2020).

10. Jacobson, M.Z.; Delucchi, M.A.; Cameron, M.A.; Frew, B.A. Low-cost solution to the grid reliability problem with 100% penetration of intermittent wind, water, and solar for all purposes. *Proc. Natl. Acad. Sci. USA* **2015**, *112*, 15060–15065. [CrossRef] [PubMed]

11. Bistline, J.E.; Blanford, G.J. More than one arrow in the quiver. *Proc. Natl. Acad. Sci. USA* **2016**, *113*, 3988. [CrossRef] [PubMed]

12. Jacobson, M.Z.; Delucchi, M.A.; Cameron, M.A.; Frew, B.A. Reply to Bistline and Blanford: Letter reaffirms conclusions and highlights flaws in previous research. *Proc. Natl. Acad. Sci. USA* **2016**, *113*, 3989–3990. [CrossRef]

13. Clack, C.T.M.; Qvist, S.A.; Apt, J.; Bazilian, M.; Brandt, A.R.; Caldeira, K.; Davis, S.J.; Diakov, V.; Handschy, M.A.; Hines, P.D.H.; et al. Evaluation of a proposal for reliable low-cost grid power with 100% wind, water, and solar. *Proc. Natl. Acad. Sci. USA* **2017**, *114*, 6722–6727. [CrossRef]

14. Jacobson, M.Z.; Delucchi, M.A.; Cameron, M.A.; Frew, B.A. The United States can keep the grid stable at low cost with 100% clean, renewable energy in all sectors despite inaccurate claims. *Proc. Natl. Acad. Sci. USA* **2017**, *114*, 5021–5023. [CrossRef]

15. Child, M.; Kemfert, C.; Bogdanov, D.; Breyer, C. Flexible electricity generation, grid exchange and storage for the transition to a 100% renewable energy system in Europe. *Renew. Energy* **2019**, *139*, 80–101. [CrossRef]

16. Jacobson, M.Z.; Cameron, M.A.; Hennessy, E.M.; Petkov, I.; Meyer, C.B.; Gambhir, T.K.; Maki, A.T.; Pfleeger, K.; Clonts, H.; McEvoy, A.L.; et al. 100% clean and renewable Wind, Water, and Sunlight (WWS) all-sector energy roadmaps for 53 towns and cities in North America. *Sustain. Cities Soc.* **2018**, *42*, 22–37. [CrossRef]

17. Jacobson, M.Z.; Delucchi, M.A.; Cameron, M.A.; Hennessy, E.M. Matching demand with supply at low cost in 139 countries among 20 world regions with 100% intermittent wind, water, and sunlight (WWS) for all purposes. *Renew. Energy* **2018**, *123*, 236–248. [CrossRef]

18. Bogdanov, D.; Farfan, J.; Sadovskaia, K.; Aghahosseini, A.; Child, M.; Gulagi, A.; Oyewo, A.S.; de Souza Noel Simas Barbosa, L.; Breyer, C. Radical transformation pathway towards sustainable electricity via evolutionary steps. *Nat. Commun.* **2019**, *10*, 1077. [CrossRef] [PubMed]

19. Hertwich, E.G.; Gibon, T.; Bouman, E.A.; Arvesen, A.; Suh, S.; Heath, G.A.; Bergesen, J.D.; Ramirez, A.; Vega, M.I.; Shi, L. Integrated life-cycle assessment of electricity-supply scenarios confirms global environmental benefit of low-carbon technologies. *Proc. Natl. Acad. Sci. USA* **2015**, *112*, 6277–6282. [CrossRef] [PubMed]

20. Luderer, G.; Pehl, M.; Arvesen, A.; Gibon, T.; Bodirsky, B.L.; de Boer, H.S.; Fricko, O.; Hejazi, M.; Humpenöder, F.; Iyer, G.; et al. Environmental co-benefits and adverse side-effects of alternative power sector decarbonization strategies. *Nat. Commun.* **2019**, *10*, 5229. [CrossRef]

21. Pehl, M.; Arvesen, A.; Humpenöder, F.; Popp, A.; Hertwich, E.G.; Luderer, G. Understanding future emissions from low-carbon power systems by integration of life-cycle assessment and integrated energy modelling. *Nat. Energy* **2017**, *2*, 939–945. [CrossRef]

22. Stamford, L.; Azapagic, A. Life cycle sustainability assessment of UK electricity scenarios to 2070. *Energy Sustain. Dev.* **2014**, *23*, 194–211. [CrossRef]

23. Raugei, M.; Leccisi, E.; Azzopardi, B.; Jones, C.; Gilbert, P.; Zhang, L.; Zhou, Y.; Mander, S.; Mancarella, P. A multi-disciplinary analysis of UK grid mix scenarios with large-scale PV deployment. *Energy Policy* **2018**, *114*, 51–62. [CrossRef]

24. Odum, H.T. Energy, ecology, and economics. *Ambio* **1973**, *2*, 220–227. Available online: https://www.jstor.org/stable/4312030 (accessed on 19 March 2020).

25. Hall, C.; Lavine, M.; Sloane, J. Efficiency of Energy Delivery Systems: I. An Economic and Energy Analysis. *Environ. Manag.* **1979**, *3*, 493–504. [CrossRef]

26. Carbajales-Dale, M.; Barnhart, C.; Brandt, A.; Benson, S. A better currency for investing in a sustainable future. *Nat. Clim. Chang.* **2014**, *4*, 524–527. [CrossRef]

27. Trainer, T. Estimating the EROI of whole systems for 100% renewable electricity supply capable of dealing with intermittency. *Energy Policy* **2018**, *119*, 648–653. [CrossRef]

28. Murphy, D.J.; Raugei, M. The Energy Transition in New York: A Greenhouse Gas, Net Energy and Life-Cycle Energy Analysis. *Energy Technol.* **2020**. [CrossRef]

29. Raugei, M.; Leccisi, E.; Fthenakis, V.; Moragas, R.E.; Simsek, Y. Net energy analysis and life cycle energy assessment of electricity supply in Chile: Present status and future scenarios. *Energy* **2018**, *162*, 659–668. [CrossRef]

30. National Grid ESO. Future Energy Scenarios. 2019. Available online: http://fes.nationalgrid.com/media/1409/fes-2019.pdf (accessed on 19 March 2020).

31. U.S. Grid Energy Storage Factsheet. University Of Michigan—Center for Sustainable Systems, 2018. Available online: http://css.umich.edu/factsheets/us-grid-energy-storage-factsheet (accessed on 19 March 2020).

32. Kaldmeyer, C.; Boysen, C.; Tuschy, I. Compressed Air Energy Storage in the German Energy System—Status Quo & Perspectives. *Energy Procedia* **2016**, *99*, 298–313. [CrossRef]

33. Ecoinvent Life Cycle Inventory Database. Available online: https://www.ecoinvent.org/database/database.html (accessed on 19 March 2020).

34. Transmission Losses. National Grid ESO, 2019. Available online: https://www.nationalgrideso.com/document/144711/download (accessed on 19 March 2020).

35. Coal Countdown. Power Stations of the UK. Available online: http://www.powerstations.uk/coal-countdown/ (accessed on 26 March 2020).

36. Digest of UK Energy Statistics. National Statistics, 2019. Available online: https://www.gov.uk/government/statistics/renewable-sources-of-energy-chapter-6-digest-of-united-kingdom-energy-statistics-dukes (accessed on 25 March 2020).

37. Fadeyi, S.; Arafat, H.A.; Abu-Zahra, M.R.M. Life Cycle Assessment of Natural Gas Combined Cycle Integrated with CO_2 Post Combustion Capture Using Chemical Solvent. *Int. J. Greenh. Gas Control* **2013**, *19*, 441–452. [CrossRef]

38. Singh, B.; Strømman, A.H.; Hertwich, E. Life Cycle Assessment of Natural Gas Combined Cycle Power Plant with Post-Combustion Carbon Capture, Transport And Storage. *Int. J. Greenh. Gas Control* **2011**, *5*, 457–466. [CrossRef]

39. Life Cycle Impacts of Biomass Electricity in 2020. Department Of Energy and Climate Change, 2014. Available online: https://assets.publishing.service.gov.uk/government/uploads/system/uploads/attachment_data/file/349024/BEAC_Report_290814.pdf (accessed on 25 March 2020).

40. A Burning Issue: Biomass Is the Biggest Source of Renewable Energy Consumed in the UK. Office for National Statistics, 2019. Available online: https://www.ons.gov.uk/economy/environmentalaccounts/articles/aburningissuebiomassisthebiggestsourceofrenewableenergyconsumedintheuk/2019-08-30 (accessed on 25 March 2020).

41. International Container Shipping—Online Freight Marketplace. Sea Rates. Available online: https://www.searates.com/ (accessed on 25 March 2020).

42. Woody Biomass. University of Strathclyde Glasgow. Available online: http://www.esru.strath.ac.uk/EandE/Web_sites/07-08/Biomass_feasibility/overview/woody.html (accessed on 25 March 2020).

43. Biogas. Planet-Biogas. Available online: http://www.planet-biogas.co.uk/info/ (accessed on 25 March 2020).

44. Environmental Management. *Life Cycle Assessment. Principles and Framework*; Standard ISO 14044; International Organization for Standardization: Geneva, Switzerland, 2006. Available online: https://www.iso.org/standard/38498.html (accessed on 27 March 2020).
45. Nuclear Power in the United Kingdom. World Nuclear Association, 2020. Available online: https://www.world-nuclear.org/information-library/country-profiles/countries-t-z/united-kingdom.aspx (accessed on 25 March 2020).
46. Pannier, C.; Skoda, R. Comparison of Small Modular Reactor and Large Nuclear Reactor Fuel Cost. *EPE* **2014**, *6*, 82–94. [CrossRef]
47. Advanced Nuclear Technologies. Department for Business, Energy & Industrial Strategy, 2019. Available online: https://www.gov.uk/government/publications/advanced-nuclear-technologies/advanced-nuclear-technologies (accessed on 25 March 2020).
48. Carless, T.S.; Griffin, W.M.; Fishbeck, P.S. The Environmental Competitiveness of Small Modular Reactors: A Life Cycle Study. *Energy* **2016**, *114*, 84–99. [CrossRef]
49. Godsey, K. Life Cycle Assessment of Small Modular Reactors Using U.S. Nuclear Fuel Cycle. Clemson University, 2019. Available online: https://tigerprints.clemson.edu/all_theses/3235 (accessed on 25 March 2020).
50. Hammons, T.J. Tidal power. *Proc. IEEE* **1993**, *81*, 419–433. [CrossRef]
51. Bid to Resurrect Swansea Bay Tidal Lagoon. BBC News, 2019. Available online: https://www.bbc.co.uk/news/uk-wales-50656253 (accessed on 25 March 2020).
52. Noonan, M. Tidal Stream: Opportunities for Collaborative Action. Offshore Renewable Energy Catapult, 2019. Available online: https://s3-eu-west-1.amazonaws.com/media.newore.catapult/app/uploads/2019/05/14111713/Tidal-Stream-Opportunities-for-Collaborative-Action-ORE-Catapult.pdf (accessed on 25 March 2020).
53. Walker, S.; Howell, R.; Hodgson, P.; Griffin, A. Tidal energy machines: A comparative life cycle assessment study. *Proc. Inst. Mech. Eng. Part M J. Eng. Marit. Environ.* **2013**, *229*, 124–140. [CrossRef]
54. Photovoltaics Report. Fraunhofer Institute for Solar Energy Systems, 2019. Available online: https://www.ise.fraunhofer.de/en/publications/studies/photovoltaics-report.html (accessed on 20 March 2020).
55. Frischknecht, R.; Itten, R.; Wyss, F.; Blanc, I.; Heath, G.; Raugei, M.; Sinha, P.; Wade, A. Life Cycle Assessment of Future Photovoltaic Electricity Production from Residential-Scale Systems Operated in Europe. Report T12-05:2015. *International Energy Agency*. 2015. Available online: http://www.iea-pvps.org (accessed on 20 March 2020).
56. Global Solar Atlas. Available online: https://globalsolaratlas.info/ (accessed on 20 March 2020).
57. Frischknecht, R.; Heath, G.; Raugei, M.; Sinha, P.; de Wild-Scholten, M.; Fthenakis, V.; Kim, H.C.; Alsema, E.; Held, M. *Methodology Guidelines on Life Cycle Assessment of Photovoltaic Electricity*, 3rd ed.; Report T12-08:2016; International Energy Agency: Paris, France, 2016; Available online: http://www.iea-pvps.org (accessed on 20 March 2020).
58. Budt, M.; Wolf, D.; Span, R.; Yan, J. A review on compressed air energy storage: Basic principles, past milestones and recent developments. *Appl. Energy* **2016**, *170*, 250–268. [CrossRef]
59. ADELE—Adiabatic Compressed-Air Energy Storage for Electricity Supply. RWE Power, 2010. Available online: http://www.rwe.com/web/cms/mediablob/en/391748/data/235554/1/rwe-power-ag/press/company/Brochure-ADELE.pdf (accessed on 20 March 2020).
60. A De-Carbonised and Cost-Effective Energy System. Storelectric, 2018. Available online: https://www.storelectric.com/wp-content/uploads/2018/03/A-De-Carbonised-and-Cost-Effective-Energy-System.pdf (accessed on 20 March 2020).
61. Hydrostor and NRStor Announce Completion of World's First Commercial Advanced-CAES Facility. Available online: https://www.globenewswire.com/news-release/2019/11/25/1952039/0/en/Hydrostor-and-NRStor-Announce-Completion-of-World-s-First-Commercial-Advanced-CAES-Facility.html (accessed on 20 March 2020).
62. Venkataramani, G.; Parankusam, P.; Ramalingam, V.; Wang, J. A review on compressed air energy storage—A pathway for smart grid and polygeneration. *Renew. Sustain. Energy Rev.* **2016**, *62*, 895–907. [CrossRef]
63. Bouman, E.A.; Øberg, M.M.; Hertwich, E.G. Environmental impacts of balancing offshore wind power with compressed air energy storage (CAES). *Energy* **2016**, *95*, 91–98. [CrossRef]
64. Zubi, G.; Dufo-López, R.; Carvalho, M.; Pasaoglu, G. The lithium-ion battery: State of the art and future perspectives. *Renew. Sustain. Energy Rev.* **2018**, *89*, 292. [CrossRef]

65. Placke, T.; Kloepsch, R.; Dühnen, S.; Winter, M. Lithium ion, lithium metal, and alternative rechargeable battery technologies: The odyssey for high energy density. *J. Solid State Electrochem.* **2017**, *21*, 1939–1964. [CrossRef]

66. Xu, B.; Oudalov, A.; Ulbig, A.; Andersson, G.; Kirschen, D.S. Modeling of Lithium-Ion Battery Degradation for Cell Life Assessment. *IEEE Trans. Smart Grid* **2016**, *9*, 1131–1140. [CrossRef]

67. *Environmental Management. Life Cycle Assessment. Principles and Framework*; Standard ISO 14040. International Organization for Standardization: Geneva, Switzerland, 2006. Available online: https://www.iso.org/standard/37456.html (accessed on 27 March 2020).

68. CML-IA Characterisation Factors. Institute of Environmental Sciences (CML), University of Leiden. Available online: https://www.universiteitleiden.nl/en/research/research-output/science/cml-ia-characterisation-factors (accessed on 12 March 2020).

69. GaBi LCA Software. ThinkStep. Available online: https://www.thinkstep.com/software/gabi-software (accessed on 19 March 2020).

70. Declaration of Apeldoorn on LCIA of Non-Ferrous Metals. Available online: https://www.lifecycleinitiative.org/wp-content/uploads/2013/01/Declaration_Apeldoorn_final.pdf (accessed on 17 March 2020).

71. Guinée, J.B.; Gorrée, M.; Heijungs, R.; Huppes, G.; Kleijn, R.; de Koning, A.; van Oers, L.; Wegener Sleeswijk, A.; Suh, S.; Udo de Haes, H.A.; et al. *Handbook on Life Cycle Assessment. Operational Guide to the ISO Standards*; Kluwer Academic Publishers: Dordrecht, The Netherlands, 2002; p. 692. ISBN 1-4020-0228-9.

72. Schulze, R.; Guinée, J.; van Oersa, L.; Alvarenga, R.; Dewulf, J.; Drielsma, J. Abiotic resource use in life cycle impact assessment—Part I—Towards a common perspective. *Resour. Conserv. Recycl.* **2020**, *154*, 104596. [CrossRef]

73. Schulze, R.; Guinée, J.; van Oersa, L.; Alvarenga, R.; Dewulf, J.; Drielsma, J. Abiotic resource use in life cycle impact assessment—Part II—Linking perspectives and modelling concepts. *Resour. Conserv. Recycl.* **2020**, *154*, 104595. [CrossRef]

74. van Oers, L.; Guinée, J.; Heijungs, R. Abiotic resource depletion potentials (ADPs) for elements revisited—Updating ultimate reserve estimates and introducing time series for production data. *Int. J. Life Cycle Assess.* **2019**, *25*, 294–308. [CrossRef]

75. Frischknecht, R.; Wyss, F.; Büsser Knöpfel, S.; Lützkendorf, T.; Balouktsi, M. Cumulative energy demand in LCA: The energy harvested approach. *Int. J. Life Cycle Assess.* **2015**, *20*, 957–969. [CrossRef]

76. Murphy, D.J.; Carbajales-Dale, M.; Moeller, D. Comparing Apples to Apples: Why the Net Energy Analysis Community Needs to Adopt the Life-Cycle Analysis Framework. *Energies* **2016**, *9*, 917. [CrossRef]

77. Hall, C.A.S.; Cleveland, C.J. Petroleum Drilling and Production in the United States: Yield per Effort and Net Energy Analysis. *Science* **1981**, *211*, 576–579. [CrossRef] [PubMed]

78. Arvesen, A.; Hertwich, E.G. More caution is needed when using life cycle assessment to determine energy return on investment (EROI). *Energy Policy* **2015**, *76*, 1–6. [CrossRef]

79. Murphy, D.J.; Hall, C.A.S. Year in review—EROI or energy return on (energy) invested. *Ann. N. Y. Acad. Sci.* **2010**, *1185*, 102–118. [CrossRef]

80. Murphy, D.J. The implications of the declining energy return on investment of oil production. *Philos. Trans. R. Soc. A* **2014**, *372*. [CrossRef]

81. Raugei, M.; Leccisi, E. A comprehensive assessment of the energy performance of the full range of electricity generation technologies deployed in the United Kingdom. *Energy Policy* **2016**, *90*, 46–59. [CrossRef]

82. Nielsen, M.; Nielsen, O.-K.; Plejdrup, M. *Danish Emission Inventories for Stationary Combustion Plants*; Danish Centre for Environment and Energy: Aarhus, Denmark, 2014; p. 188. ISBN 978-87-7156-073-2.

83. Paolini, V.; Petracchini, F.; Segreto, M.; Tomassetti, L.; Naja, N.; Cecinato, A. Environmental impact of biogas: A short review of current knowledge. *J. Environ. Sci. Health A* **2018**, *53*, 899–906. [CrossRef]

84. Outlook for Biogas and Biomethane—Prospects for Organic Growth. World Energy Outlook Special Report. International Energy Agency, 2020. Available online: https://www.iea.org/reports/outlook-for-biogas-and-biomethane-prospects-for-organic-growth (accessed on 23 March 2020).

85. Kleijn, R.; van der Voet, E.; Jan Kramer, G.; van Oers, L.; van der Giesen, C. Metal requirements of low-carbon power generation. *Energy* **2011**, *36*, 5640–5648. [CrossRef]

86. Beylot, A.; Villeneuve, J. Accounting for the environmental impacts of sulfidic tailings storage in the Life Cycle Assessment of copper production: A case study. *J. Clean. Prod.* **2017**, *153*, 139–145. [CrossRef]

87. Jones, C.; Gilbert, P.; Raugei, M.; Leccisi, E.; Mander, S. An Approach to Prospective Consequential LCA and Net Energy Analysis of Distributed Electricity Generation. *Energy Policy* **2017**, *100*, 350–358. [CrossRef]
88. Carbajales-Dale, M.; Barnhart, C.; Benson, S.M. Can we afford storage? A dynamic net energy analysis of renewable electricity generation supported by energy storage. *Energy Environ. Sci.* **2014**, *7*, 1538–1544. [CrossRef]
89. Sgouridis, S.; Csala, D.; Bardi, U. The sower's way: Quantifying the narrowing net-energy pathways to a global energy transition. *Environ. Res. Lett.* **2016**, *11*, 094009. [CrossRef]

Article

Identification of Suitable Areas for Biomass Power Plant Construction through Environmental Impact Assessment of Forest Harvesting Residues Transportation

Maria Pergola [1], Angelo Rita [2], Alfonso Tortora [2], Maria Castellaneta [2], Marco Borghetti [2], Antonio Sergio De Franchi [2], Antonio Lapolla [2], Nicola Moretti [2], Giovanni Pecora [2], Domenico Pierangeli [2], Luigi Todaro [2] and Francesco Ripullone [2,*]

[1] Ages s.r.l. s—Academic Spin-off, University of Basilicata, Viale dell'Ateneo Lucano, 10-85100 Potenza, Italy; mariateresa_pergola@virgilio.it

[2] School of Agricultural, Forestry, Food and Environmental Sciences, University of Basilicata, viale dell'Ateneo Lucano, 10. I-85100 Potenza, Italy; angelo.rita@unibas.it (A.R.); alfonso.tortora@unibas.it (A.T.); maria.castellaneta@unibas.it (M.C.); marco.borghetti@unibas.it (M.B.); sergio.defranchi@unibas.it (A.S.D.F.); antonio.lapolla@unibas.it (A.L.); nicola.moretti@unibas.it (N.M.); giov.pecora@gmail.com (G.P.); domenico.pierangeli@unibas.it (D.P.); luigi.todaro@unibas.it (L.T.)

* Correspondence: francesco.ripullone@unibas.it; Tel.: +39-097-120-5354

Received: 3 April 2020; Accepted: 26 May 2020; Published: 28 May 2020

Abstract: In accordance with European objectives, the Basilicata region intends to promote the use of energy systems and heat generators powered by lignocellulosic biomass, so the present study aimed to investigate the availability of logging residues and most suitable areas for the construction of bioenergy production plants. The life cycle assessment (LCA) methodology was employed to conduct an environmental impact assessment of the biomass distribution and its transport, and spatial LCA was used to evaluate the impact of regional transport. One cubic meter kilometer (m^3 km^{-1}) was used as the functional unit and a small lorry was considered for the transport. The results showed that the available harvesting residues amounted to 36,000 m^3 and their loading environmental impact accounted for 349 mPt m^{-3}. The impacts of transport (4.01 mPt m^{-3}) ranged from 3.4 to 144,400 mPt km^{-1} forest parcel^{-1}, mainly affecting human health (95%) and, second, the ecosystem quality (5%). Three possible sites for bioenergy plant location were identified considering the environmental impact distribution due to feedstock transport. Findings from this research show the importance of considering the LCA of biomass acquisition in site selection and can fill the knowledge gaps in the available literature about spatial LCA.

Keywords: bioenergy; life cycle assessment; geographic information system (GIS); harvesting residues

---

1. Introduction

Harvesting residues are the biomass left on fields after wood harvesting (tops, branches, and little non-marketable trunks) [1]. On average, 10% to 15% of this biomass is left on site as forest residues following harvesting operations [2] because it is expensive to harvest and transport and there are few markets for this wood material. Occasionally, some of the larger logging wood is removed as firewood for domestic consumption [3].

At the same time, the use of wood biomass is believed to be an important component of renewable energies, particularly for producing thermal energy or joint thermal and electrical energy with a view to creating smart energy cities. Bailey et al. [4], Perez-Verdin et al. [5], and Moon et al. [6] argued that the use of wood biomass residues to produce energy or fuel can encourage the rise of regional economies and the creation of new employment opportunities. In the last decade, there has been increased awareness in using this residual biomass as a raw material for renewable energy as a response to the standards for renewable fuels and energy markets [7]. This is because the production of forest biomass energy has the potential to reduce carbon emissions when replacing fossil fuels, although several authors have reported contrasting evidence [8]; retrieve waste that would otherwise be disposed of in landfills or be incinerated; create jobs (especially in rural areas); and supply local and sustainable energy for communities, reducing their dependence on the international fuel market, as affirmed by Shabani et al. [9], Saidur et al. [10], and Ahtikoski et al. [11]. However, these residues are often not fully utilized due to the lack of demand within the immediate vicinity of the processing plant. Furthermore, transporting residues to an area with high demand is considered uneconomical [12], and significant costs are associated with the supply of forest residues from the forest. Transport also constitutes one of the major sources of air pollution, in particular, due to emissions of sulfur dioxide (SO_2), nitrogen oxides (NOx), volatile non-methane organic compounds (NMVOC), primary particulate matter (PM 2.5), and carbon monoxide (CO). The latter are produced during fuel combustion, but other non-exhaust emissions of particulates, are produced during road and rail transport due to the abrasion of brakes, wheels, etc. [13]. Thus, the inclusion of some collection sites is out of the question due to long distances, harsh topographic conditions, or ecological restrictions. In any case, it is advised that 30% of the harvesting wood residues are left in place in order to restore the fertility of the forest soil [14–17].

When analyzing the whole life cycle of the product, Law and Harmon [18] and Schulze et al. [19] highlighted the fact that some aspects of production could worsen rather than contribute to mitigating climate change such as long-distance lumber transport. Farkavcova et al. [20] stated that in Europe, transport represents 22% of the total emissions, and that these emissions are constantly increasing. In addition, urban transport is responsible for around 25% of the CO_2 emissions produced by all transport [21]. Referring to the forestry sector, cradle-to-grave life cycle assessment (LCA) studies have shown that transport significantly contributes to the overall results by representing 60%–70% of the overall environmental impact [20,22]. In this regard, LCA is very useful, since it is a useful tool for evaluating all environmental impacts linked with a product, process, or activity as well as the consumption and emissions of resources [23]. LCA is constantly evolving, and its application to bioenergy systems has been a key factor for the development of the process in the last few years. In the literature, bioenergy production chains have been evaluated from an environmental and energy point of view by several authors [24–30]. Particular attention has been paid to saving greenhouse gas (GHG) emissions and energy balances for the production of liquid biofuels [31]. Some reviews have considered electricity, while only one study has included heat in addition to generating electricity [24]. Referring to transportation systems, an LCA study includes the identification of direct, indirect, and supply chain emissions affecting the system. In particular, direct emissions refer to energy consumption and emissions associated with vehicle movement, namely, air emissions (CO_2, CO, SO_x, NO_x, PM, etc.) from diesel combustion. In LCA transport, fuel use and the related produced emissions are called direct emissions because they are associated with the direct objective of the system to ease the movement of residues. Indirect processes are those that must exist for the direct process to exist, in this case, vehicles, infrastructure, and energy production services; vehicle production; infrastructure construction, management, and maintenance; and fuel and electricity production. Additionally, these indirect processes depend on a supply chain to produce materials, services, and other activities, probably far from where the vehicle acts [32]. Similarly, the direct energy input represents the energy effectively used to sustain a process (fuel and oil consumption of machineries, and energy consumption of humans during the work), while the indirect energy input stands for the energy stored in the materials used in the process (the energy value for the production of machinery and tools) [33,34].

Wood biomass residues are geographically allocated, with alterations in space–time availability. Therefore, energy and environmental evaluation requires a decision support system for efficient planning [35,36]. To plan a biomass facility, a preliminary and precise database including the distribution of residues and the seasonal variation (peak period and decreased availability period) is essential. Logistics such as the harvesting, storage, and transport of residues are spatially interconnected and require accurate planning. A geographic information system (GIS) is an important territorial decision-making tool that allows for a precise evaluation of distributed resources for renewable energy [37–39]. The joint use of LCA and GIS, also known as spatial LCA [35], can be useful for estimating the biomass potential in a region, and enhancing the results of environmental impact assessment by counting spatial variations and considering power plant design.

According to the European commitment of the last few years, which is aimed at solving the international economic and environmental problems linked to the climate and energy supply, the Basilicata region has highlighted the importance of the agricultural and forestry sector in the development and diffusion of renewable agro-energy sources [40]. The regional commitment aims at strengthening the financial instruments to support research and experimentation as well as the involvement of interested companies in implementing pilot projects in the regional territory. In 2012, the potential supply of forest wood biomass for bioenergy production in the Basilicata region was estimated to be around 500,000 tons per year [41]. Of this quantity, the same authors identified a mean annual production of about 22,000 tons of residual biomass of forestry origin and an average annual production of about 400,000 tons of residual biomass of agricultural origin as well as an average annual unitary production of dry biomass from dedicated crops, consisting of approximately 60,000 tons on private fields and approximately 57,000 tons on public fields [41]. Thus, the regional administration has assigned a strategic role to the energy sector to relaunch the territories with the aim of creating new and qualified job opportunities and environmentally friendly development [40]. Hence, the present research is part of the "Smart Basilicata Research and Development" project. This project aims to develop innovative techniques for the management of wood biomass including their use for energy purposes. The aim of the present research was the selection of co- and trigeneration plant construction sites and the optimal residual forest woody material collection areas in the regional territory after first investigating the availability and amount of harvesting residues through an analysis of forest management plans (FMPs) still in force. GIS was used as a decision-making spatial tool for the accurate assessment of spatially logging residues for lignocellulosic bioenergy production. Additionally, the LCA methodology was employed for an environmental impact assessment of the biomass load and transport and spatial LCA was used to investigate the distribution of the impact on the regional territory.

2. Materials and Methods

2.1. Case Study Description

The study was performed in the Basilicata region, one of the most forested areas in southern Italy (356,426 hectares) in which the forest sector is governed by Regional Law No. 42 of the 30th November 1998 "Rules on Forestry" [42]. The objective of forest management planning in this region is to apply sustainable forest management guidelines, which are carried out through FMPs. In the present study, only wood biomass produced directly from forest plans was taken into consideration (i.e., the logging residues estimated as percentages of forest utilization), according to the guidelines developed by the Basilicata region for the reduction of FMPs [42].

Considering the forestry sector and despite the benefits of forest management, in the Basilicata region, as in other Mediterranean areas and South-East Europe, the seasonality in the demand for wood products such as firewood, the substantial investments required to purchase woodland lots by forest companies, and the high cost of transactions due to the slowness of the administrative and authorization procedures, has resulted in an excessive bureaucracy [43,44], which significantly reduces the gross operating margins of companies. All this does not incentivize the purchase and consequent management of woods, especially in terms of public ownership. This leads to the abandonment of forests and sometimes to the degradation phenomena that affects the capacity of forest ecosystems [45].

Therefore, in Basilicata, there are currently 83 FMPs, 35 of which are still in force. Consequently, the study was performed in these latter municipalities (Figure 1).

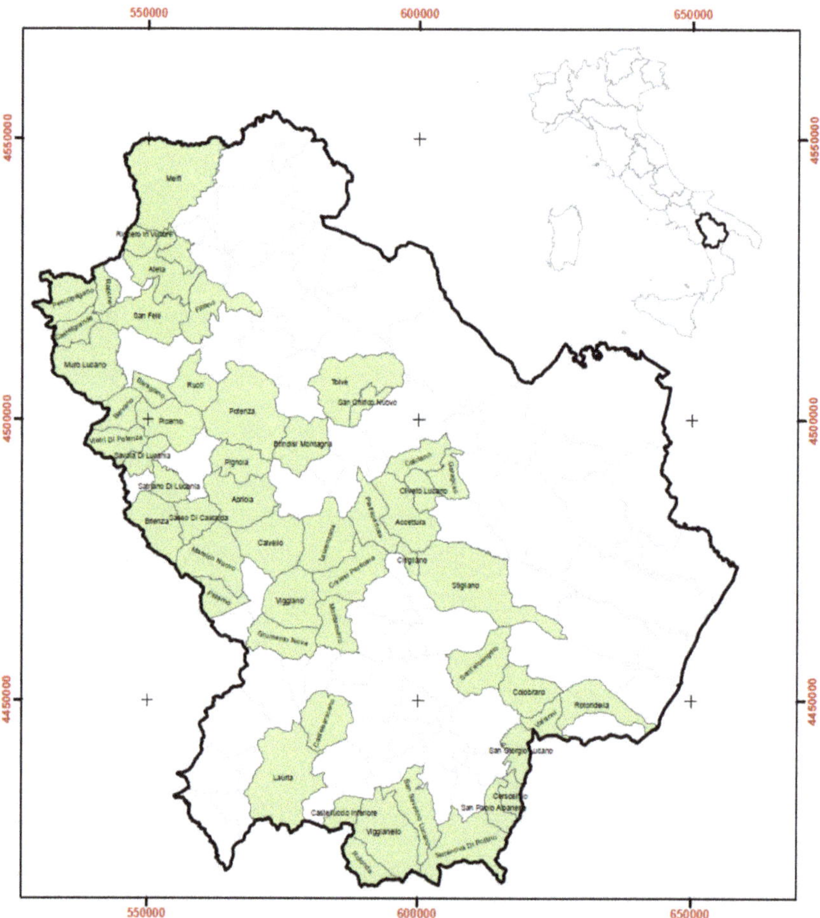

Figure 1. Study area: municipalities of the Basilicata region in which forest management plans (FMPs) are still in force.

2.2. The Life Cycle Assessment (LCA) Approach

The present study was designed to investigate the optimization of biomass supply distances in order to designate suitable sites for the implementation of cogeneration or trigeneration plants powered by biomass in the Basilicata region. An LCA analysis was performed according to the ISO 14040/44 (2006) [23,46].

The objective of the present analysis was to address the movement of the harvesting residues, by road, from the production forest parcels in the regional territory. Therefore, the system boundaries include the environmental impacts during all phases of the transport (transport operation and infrastructure), from raw material extraction to their use and, finally, disposal. Moreover, through the LCA methodology, all important emissions were quantified as well as their related environmental and health impacts and the issue of the resource consumption combined with transport.

In order to estimate the environmental impact of transport services and correlate transport datasets with other product life cycles, environmental loads are determined using the functional unit of one cubic meter kilometer (m^3 km^{-1}). A cubic meter kilometer is defined as the transport of a cubic meter of harvesting waste from a given transport service over a kilometer [47].

Data on the available harvesting residues and their distribution in the regional territory were gathered from an analysis of the 35 FMPs still in force. In particular, these residues were estimated as percentages of forest utilization, according to the guidelines developed by the Basilicata region for the redaction of FMPs. These percentages reflect the share of residues in the mean annual cut, as indicated by Cozzi et al. [48]. In order to ensure sustainable harvesting levels (max 60% of the annual increment for high forest and 90% for coppice forest), the percentage of forest utilization was between 5% and 50% of the total wood mass in the examined FMPs, and the relative percentage of residues available ranged from 9% to 20%. Data related to harvesting residues (tops, branches, and little non-marketable trunks usually left on sites) and load (forestry machines used, duration of the operation and fuel consumed) were taken from information reported in Pergola et al. [49], while data on the transportation were extrapolated from SimaPro's Life Cycle Inventory (LCI) databases and, in particular, from databases of scientific relevance and accuracy such as Ecoinvent 3 [50]. Since one of the objectives of this research was to investigate the regional distribution of the environmental impacts of the transport of woody residues, the item "Small lorry transport, Euro 0, 1, 2, 3, 4 mix, 7.5 ton total weight, 3.3 ton max payload RER S" was employed, whose emissions calculation was obtained from the literature [51] based on measurements [52].

Environmental assessment was performed using the SimaPro 8.0.4.30 (copyright PRé Consultants bv 2014) software by means of the LCA Eco-indicator 99 endpoint method, in which "environment" was defined as being affected by three types of damage [53]:

(1) Human health, which encompasses the number and duration of illnesses and years of life lost because of early death from environmental impacts. These latter comprise global warming, ozone layer reduction, carcinogenic and respiratory effects, etc. The measurement unit is the disability-adjusted life year (DALY);
(2) Ecosystem quality, which encompasses the effect on animal and plant biodiversity, particularly related to issues linked to acidification, ecotoxicity, and land use including the reduction of agricultural resources such as sand and gravel. Its indicator unit is the potentially disappeared fraction (PDF) of species for a given area and for a precise period (PDF m^{-2} $year^{-1}$); and
(3) Resources, which include the excess energy needed in the future to extract lower-quality mineral and fossil resources. Its indicator unit is the surplus of MJ.

In addition, the impact assessment was performed following the endpoint approach, which expresses the total environmental impact in a single score using the point (Pt) or millipoint (mPt) as the standard unit [53].

2.3. The Geographic Information System (GIS) Analysis

The ability to analyze the environment and understand all the factors that characterize it is a prerequisite for carrying out a study on a territory's suitability, and the main tools are often represented by the GIS. The latter is itself a system of tools designed to acquire, extract, archive, manipulate, analyze, manage, visualize, and present all types of geographically referenced data, which are data from the real world [54].

In particular, geo-referenced data of the total harvesting residues load and transport impacts were imported into a project in the GIS software package, together with maps of the main road network, main electricity grid, borders of the Basilicata region and municipalities, and main protected areas, which were freely available through the RSDI Basilicata portal (http://rsdi.regione.basilicata.it/web/guest/mappe-in-linea). The residential areas were taken from the digitalized map (1:25,000) of the Istituto Geografico Militare (IGM). To compute the spatial distribution of the transport impact, we multiplied the harvesting residues transport impacts per distance from the parcel, where buffers were generated using the buffer tool with distances of 1, 5, and 10 km. We highlighted that longer distances from the parcels do not maximize the cost-effectiveness of the residual transport. Finally, layers were then arranged to produce the final maps using ArcMap® 10.4.1 software by Esri (Copyright © Esri). A detailed workflow of the processes is provided in the Supplementary Materials (Figure S1).

3. Results and Discussion

3.1. Harvesting Residue Availability

Table 1 reports the results for each municipality of the analysis of 35 FMPs, referring to the availability of harvesting residues for bioenergy production (expressed in m^3) and environmental impacts (expressed in mPt) relative to harvesting residues loading and harvesting transport per kilometer.

Table 1. Harvesting residues available for bioenergy production and environmental impacts per studied municipality.

AREA	Municipalities	Harvesting Residues Available for Bioenergy (m^3)	Impacts of Harvesting Residues Loading (mPt)	Impacts of Harvesting Residues Transport (mPt km^{-1})
	Atella	368	128,250	1475
	Balvano	787	274,482	3157
	Baragiano	126	43,795	504
	Castelgrande	233	81,105	933
	Melfi	746	260,138	2992
NORTH	Muro Lucano	1132	394,702	4540
	Rapone	2317	807,670	9291
	Rionero	231	80,602	927
	Ruoti	1584	552,146	6351
	Satriano di Lucania	253	88,271	1015
	Savoia di Lucania	89	31,113	358
	Total North	7867	2,742,273	31,545
	Abriola	4430	1,544,438	17,766
	Brienza	2618	912,635	10,498
	Grumento Nova	623	217,092	2497
	Marsico Nuovo	1948	678,995	7811
	Paterno	738	257,324	2960
	Accettura	1560	543,811	6256
CENTER	Brindisi di Montagna	224	78,074	898
	CorletoPerticara	2439	850,363	9782
	Garaguso	234	81,503	938
	Laurenzana	2246	782,865	9005
	Oliveto Lucano	689	240,222	2763
	San Chirico Nuovo	94	32,834	378
	Tolve	508	176,939	2035
	Total Center	18,351	6,397,095	73,587
	Castel Saraceno	544	189,744	2183
	Cersosimo	311	108,491	1248
	Chiaromonte	135	47,223	543
	Colobraro	256	89,372	1028
	Lauria	151	52,474	604
SOUTH	Rotondella	46	15,993	184
	San Giorgio Lucano	873	304,436	3502
	San Paolo Albanese	1270	442,786	5093
	Sant'Arcangelo	49	17,057	196
	Terranova di Pollino	1875	653,603	7518
	Viggianello	4287	1,494,381	17,190
	Total South	9798	3,415,561	39,290
	TOTAL	36,015	12,554,929	144,421

Referring to the availability of harvesting residues, the analysis of the 35 FMPs showed that they amounted to about 36,000 m^3, and the Basilicata region could be split into three areas: the largest supply basin was located in the central part (about 18,300 m^3), followed by the south (about 7900 m^3), and then the north (about 9798 m^3). Harvesting residues per forestry parcel ranged from 0.85 to 668 m^3.

In line with the "Smart Basilicata Research and Development" project, investigating the availability and quantity of residual forest woody materials is useful for understanding the feasibility of cogeneration or trigeneration plants powered by biomass, which could be used to produce thermal and electrical energy to create district energy systems. The latter is a growing phenomenon in many cities around the world [55] and, as stated in Perea-Moreno et al. [56], the introduction of such schemes into urban networks has important benefits such as the availability of an open energy supply grid, greater use of renewable energy sources, less reliance on imported resources and fossil fuels, greater leverage over energy supply, and the development of energy supply [57].

At the same time, harvesting residues are widespread in the regional territory, as shown in Figure 2, so there is a need to transport and concentrate them in specific areas for subsequent bioenergetic purposes.

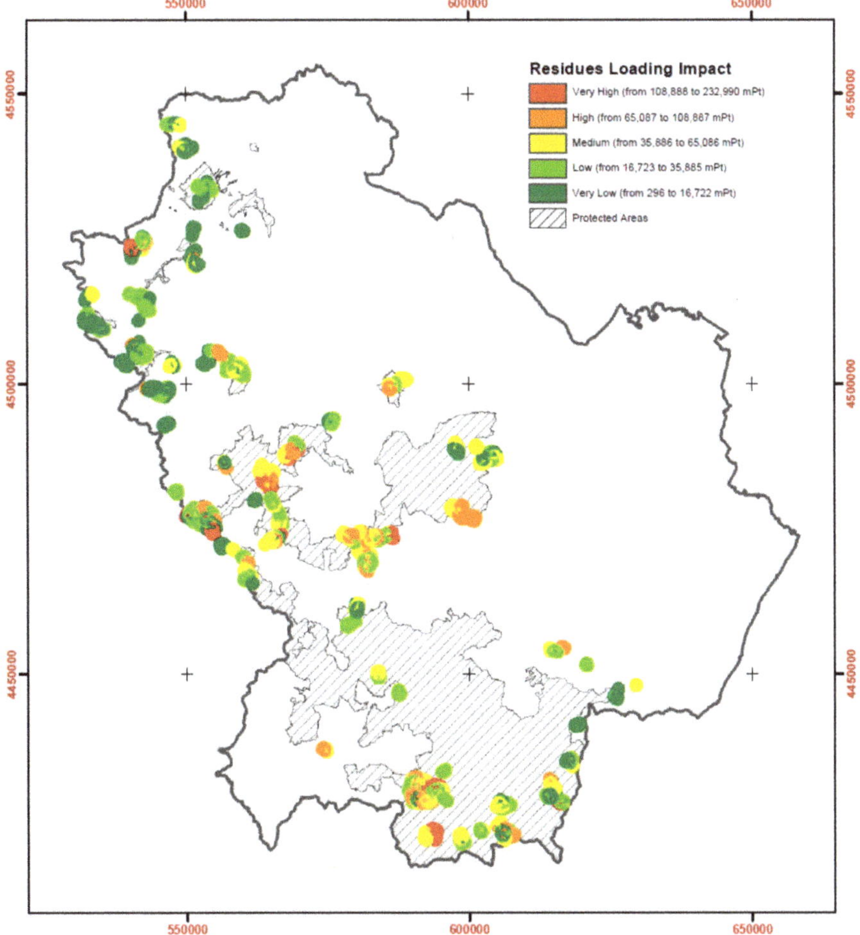

Figure 2. Regional distribution of harvesting residues and relative load impacts per impact class.

3.2. Loading Impacts

Harvesting residues load impacts were, on average, equal to 349 mPt m^{-3} and ranged from about 300 to 233,000 mPt forest parcel^{-1}. The main impacts concern human health (284 mPt m^{-3}), followed by resource depletion (59 mPt m^{-3}), and ecosystem quality (6 mPt m^{-3}). Human health damage (in total, equal to 0.173 DALY) was mainly due to the fuel consumed in the various loading operations and the impact categories most responsible for this damage were climate change (75%) and radiation (24%). Resource depletion (in total, equal to 7344 MJ surplus) was essentially affected by the materials and processes necessary for the construction of forest machines. Referring to ecosystem quality (in total, equal to 10,090 PDFm^{-2}year^{-1}), the most important impact categories were air acidification and water eutrophication, representing 73% of this damage (Table 2).

Table 2. Characterization of the total harvesting residues load impacts.

Impact Categories	Unit	Total	Diesel	Forest Machinery
Carcinogens	DALY	1.2×10^{-3}	1.8×10^{-4}	1.0×10^{-3}
Respiratory organics	DALY	5.3×10^{-4}	4.3×10^{-4}	1.1×10^{-4}
Respiratory inorganics	DALY	1.3×10^{-1}	1.1×10^{-1}	2.2×10^{-2}
Climate change	DALY	4.2×10^{-2}	3.4×10^{-2}	7.3×10^{-3}
Radiation	DALY	1.8×10^{-5}	1.5×10^{-5}	3.3×10^{-6}
Ozone layer	DALY	1.1×10^{-5}	9.6×10^{-6}	1.6×10^{-6}
Ecotoxicity	PDFm^{-2} year^{-1}	4.5×10^{3}	1.2×10^{3}	3.4×10^{3}
Acidification/Eutrophication	PDFm^{-2} year^{-1}	1.2×10^{4}	1.2×10^{4}	5.7×10^{2}
Land use	PDFm^{-2} year^{-1}	-2.8×10^{3}	-7.0×10^{2}	-2.1×10^{3}
Minerals	MJ surplus	7.3×10^{3}	2.8×10^{2}	7.1×10^{3}

Many of the analyzed forest parcels fell in protected areas and in particular, in two national parks (Appennino Lucano-Val d'Agri-Lagonegrese and Pollino National Parks), in two regional parks (Gallipoli Cognato Piccole Dolomiti Lucane and Vulture), and in several Natura 2000 Network sites (Figure 2), so the load operations should be performed with caution to ensure their conservation for present and future generations [51]. In particular, these areas of relevant naturalistic and ecological value are subjected to a specific rule of protection and management to preserve animal and plant species; safeguard anthropological, historical values, and agro-forestry–pastoral and traditional activities; promote education, training, and scientific research activities; defend and replenish the hydraulic and hydro-geological balance; and promote the enhancement and testing of compatible production activities [58].

3.3. Transport Impacts

LCA analysis showed that the transport of 1 m^3 of harvesting residues caused environmental damage equal to 4.01 mPt km^{-1}, which, in total, for the whole regional territory, corresponded to 144,421 mPt km^{-1} (Table 1). Needless to say, the greatest impacts were recorded in the municipalities with the greatest amount of residues to transport and, therefore, in the middle area (73,587 mPt m^{-3}) (Table 1). Similar to loading, the greatest impacts were on human health (3.81 mPt m^{-3} km^{-1}), but in this case, were followed by ecosystem quality (0.195 mPt m^{-3} km^{-1}).

Human health was mainly affected by climate change (59%) and respiratory inorganics (39%); in particular, the latter refers to "winter" smog caused by inorganic substance emissions. At the same time, eutrophication/acidification was the impact category with the greatest negative effects on the quality of the environment, representing 97% of the total impact of ecosystem quality (Table 3).

Table 3. Total transport damage characterization per impact categories.

Impact Categories	Unit	Values per KM
Carcinogens	DALY	1.8×10^{-5}
Respiratory organics	DALY	2.8×10^{-6}
Respiratory inorganics	DALY	4.5×10^{-4}
Climate change	DALY	6.9×10^{-4}
Radiation	DALY	5.9×10^{-8}
Ozone layer	DALY	5.5×10^{-9}
Ecotoxicity	$PDFm^{-2}year^{-1}$	5.6×10^{-0}
Acidification/Eutrophication	$PDFm^{-2} year^{-1}$	1.6×10^{2}
Land use	$PDFm^{-2} year^{-1}$	0.0×10^{0}
Minerals	MJ surplus	2.5×10^{-1}

Farkavcova et al. [20] stated that transport from the forest to the production site caused significant environmental impacts and, more precisely, mainly caused the consumption of fuel fossil resources, the abiotic depletion of the non-fossil resources, and the potential reduction of the ozone layer. In addition, the environmental impacts of transport are particularly due to the consumption or partial combustion of non-renewable fossil fuels as well as trace elements in fuel and tire abrasion. According to Handler et al. [59] and Sonne [60], the transport of biofuels or raw materials for bioenergy is potentially the major source of environmental impacts in the total supply chain. All of this was confirmed by Murphy et al. [22], and in accordance with these observations, several studies have reported that forest biomass transport accounts for most of the energy consumption and environmental impacts in forest biomass systems [61–63]. The results of these studies have shown that the transport of biomass significantly contributes to both the energy demand and GHG emissions, representing 70%–78% of the overall energy needs and 68%–75% of GHG emissions.

Other comparisons of this research and the results of other LCA studies were not possible since they only focused on the global warming potential (GWP) or energy demand, and therefore did not provide a complete picture; Farkavcova et al. [20] rightly advised that the whole set of indicators should be considered. Moreover, as stated by Murphy et al. [22], comparisons of results are complicated by discrepancies in system boundaries, geographic areas, and the employed characterization methods. According to Heinimann [64], LCA studies neglecting embodied burdens of road infrastructure and forest machines, called "truncated LCAs", always result in an underestimation of environmental impacts or an overestimation of environmental performance, respectively.

3.4. The Territorial Distribution of Impacts

Harvesting residues were widespread in the regional territory and, consequently, their loading and transportation for bioenergy production also had widespread impacts. Figure 3 shows the distribution of the total environmental impact for each forest parcel when transporting residues within 10 km from the source. Since there were many forest parcels (327), for a better representation of the environmental impacts, the calculation was simplified and considered the calculation of the impacts for 1, 5, and 10 km, and not for each kilometer, to best represent the environmental impacts at a territorial level. Red areas represent areas with the greatest environmental impacts, given by the sum of the various "impact rays" calculated for each forest plot. Therefore, the total impact ranged from a minimum of 3.41 mPt to a maximum of about 276,000 mPt (Figure 3).

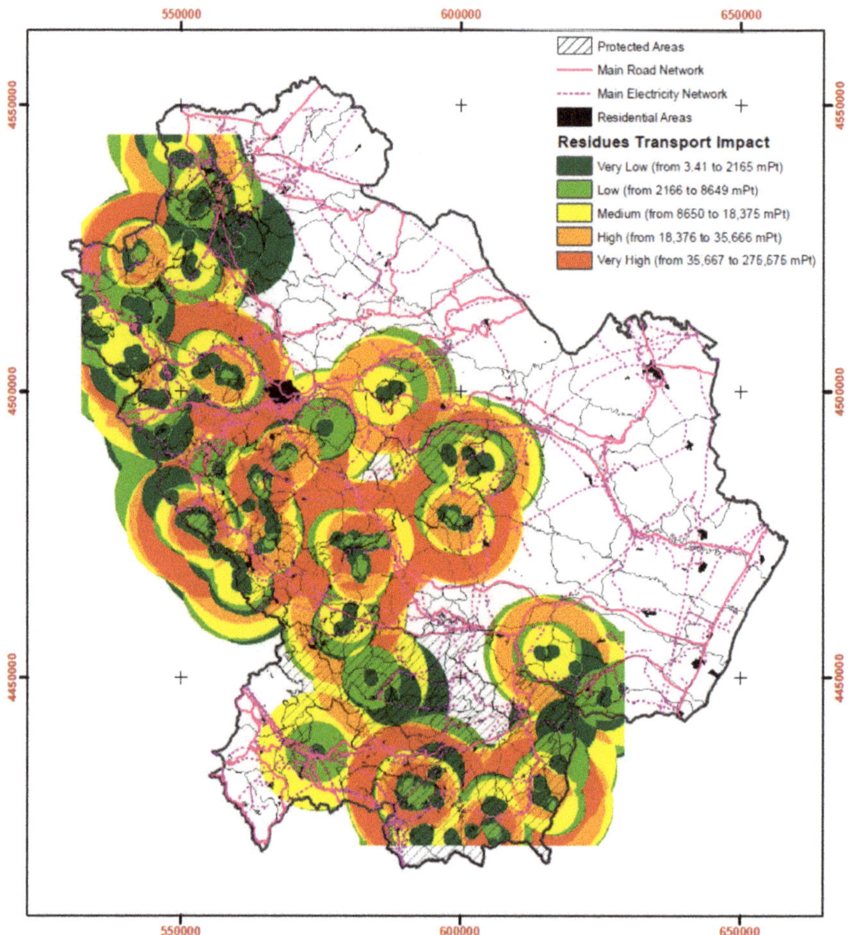

Figure 3. Territorial distribution of harvesting residues transport impacts.

To our best knowledge, this paper is one of the few studies that wish to represent transport environmental impacts territorially, in order to understand how they are distributed when more displacements are involved. In conducting a complete environmental analysis, the present study considered a set of indicators, rather than a single category (e.g., GWP). Other LCA studies of the biomass supply chain [20,22,24] have referred to the movement of lumber from the forest to a bioenergy plant, while the present study, through additionally considering environmental impacts, tried to give indications for the regional administration on which areas may be the most suitable locations for a bioenergy production plant (i.e., without compromising the environment and human health).

Indeed, one of the objectives of the "Smart Basilicata Research and Development" project is the development of cogeneration or trigeneration plants powered by lignocellulosic biomass in the regional territory. As stated by Zubaryeva et al. [65], the site should be readily reachable by transport, close to service points, and achievable for the best planning of energy transport lines. In addition, the plant should be established at an acceptable distance from residential areas, natural reserves, and protected areas to diminish the potential negative impacts of plant operation and waste disposal [35].

According to Hiloidhari et al. [35], the site selection of a biomass power plant based on GIS can be carried out through two methods: (i) suitability analysis and (ii) optimization analysis. The former allows users to recognize the most appropriate site for a power plant among many candidate sites based on user-defined constraints and support criteria. On the other hand, the best analysis accounts for the relationship between biomass and power plants to determine the optimization locations of power facilities with minimal transport and distribution costs [66].

In the present study, we tried to select the most suitable sites considering the map of the impacts; the proximity to the main road and electricity networks as well as residential areas; and the presence/absence of protected areas. Therefore, as reported in Figure 4, three possible sites were identified (one for each area of the Basilicata region):

(1) North Area near Muro Lucano municipality, since it is one of the northern municipalities with the most harvesting residues, has low impacts due to the transport, is not located in a protected area, and is well-served in terms of main roads and the electricity network;

(2) Central Area near Marsico Nuovo municipality, given that it is outside the different protected areas that characterize this area of Basilicata, does not have very high transport-related impact levels, and is well-served in terms of roads and the electricity network; and

(3) South Area near Viggianello, one of the municipalities with the highest amount of forest residues (4300 m^3), obviously outside the Pollino National Park.

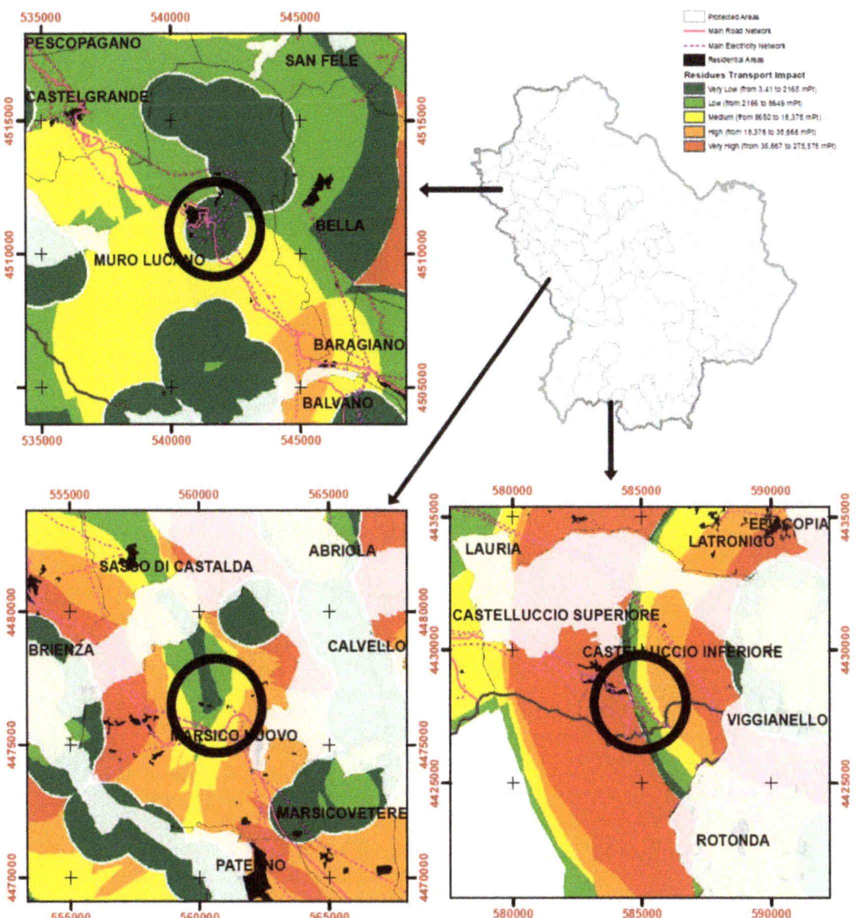

Figure 4. The three sites suitable for biomass power plant construction *.

The optimization of biomass power plant location may be carried out through the modeling of the location–allocation or the modeling of the supply area including or not including the impacts, but, as stated by Cozzi et al. [48], regardless of the applied method, the selection of an appropriate biomass center should take into account several aspects (energy, environmental, and economic) to be in line with the three pillars of sustainability (economic viability, environmental protection, and social equity). From a landscape perspective, it would be effective to place the plant in urbanized areas with similar structures, in order to avoid areas typified by agricultural and forestry aspects, keeping in mind the costs for the transport of biomass, since lower costs are obtained in areas close to the biomass processing plant [48]. Furthermore, it must be considered that the studied forest particles mainly fall in protected areas, which are rural though heavily frequented areas, where the concentration of emissions during traffic congestion could enhance by 100 times [21], further damaging human health and the quality of ecosystems.

4. Conclusions

The present study aimed to identify suitable sites for locating cogeneration/trigeneration plants powered by lignocellulosic biomass in the Basilicata region based on GIS–LCA information, after investigating the quantity of harvesting residues and environmental impacts of their loading and transport. This research allows us to take further steps forward in our knowledge about spatial LCA with regard to bioenergy production. Indeed, traditional LCAs are inadequate to identify the spatial dimensions of environmental impacts, however, this becomes feasible when they are carried out applying a GIS framework. In this study, we first assessed the environmental impacts per kilometer ($4.01\ \mathrm{mPt\ m^{-3}}$), and then built a map of cumulative impacts over a radius of 10 km for the different analyzed forest parcels, in order to identify areas with major and minor impacts. In this way, we were able to identify three areas to locate biomass plants after considering the main road network, electricity network, proximity to residential areas, and excluding protected areas. The present study represents a replicable example of how it is important to consider the environmental impact distribution of feedstock transport and not only those of bioenergy facility construction in the site selection of a biomass power plant.

Finally, we emphasize these essential aspects: only biomass residues from locally performed forest harvesting operations, or wood residues from local saw milling activities should be used for bioenergy production, and each project for bioenergy production should be preceded by a careful assessment of the potential impact of biomass removal on soil fertility and forest ecosystem biodiversity.

Supplementary Materials: The following are available online at http://www.mdpi.com/1996-1073/13/11/2699/s1, Figure S1: Workflow of the processes for the GIS analysis.

Author Contributions: Conceptualization, M.P., F.R.; Methodology, M.P., A.R.; Software, M.P., A.R., A.T.; Validation, F.R., M.C., A.R.; Formal Analysis, M.P., M.C., A.R.; Investigation, M.C., G.P.; Resources, M.C.; Data Curation, M.C., M.P.; Writing—Original Draft Preparation, M.P., A.R.; Writing—Review & Editing, A.R., F.R., M.B., D.P., L.T.; Visualization, A.S.D.F., N.M., G.P., D.P., L.T., A.L.; Supervision, F.R.; Project Administration, F.R., D.P., M.B., N.M.; Funding Acquisition, F.R., D.P., M.B., N.M. All authors have read and agreed to the published version of the manuscript.

Funding: This research was funded by the "Smart Basilicata" project (Notice MIUR n. 84/Ric 2012, PON 2007–2013 of 2 March 2012).

Acknowledgments: This research was carried out in the framework of the project "Smart Basilicata", which was approved by the Italian Ministry of Education, University and Research (Notice MIUR n. 84/Ric 2012, PON 2007–2013 of 2 March 2012) and was co-funded by the Cohesion Fund 2007–2013 of the Basilicata Regional authority. "Ufficio Foreste e Tutela del Territorio" of the Basilicata region is duly acknowledged for their kind permission to use data extracted from the "Forest Management Plans". The PhD program in 'Agricultural, Forest and Food Sciences' at the University of Basilicata supported O.P.

Conflicts of Interest: The authors declare no conflict of interest.

References

1. Moon, D.; Kitagawa, N.; Genchi, Y. CO_2 emissions and economic impacts of using logging residues and mill residues in Maniwa Japan. *For. Policy Econ.* **2015**, *50*, 163–171. [CrossRef]
2. Kuiper, L.; Oldenburger, J. The Harvest of Forest Residues in EUROPE. Probos. Report on bus ticket no. *D15a*. 2006. Available online: https://www.probos.nl/biomassa-upstream/pdf/reportBUSD15a.pdf (accessed on 20 January 2020).
3. Hoyne, S.; Thomas, A. Forest Residues: Harvesting, Storage and Fuel Value. COFORD (National Council for Forest Research and Development). 2001. Available online: http://www.coford.ie/media/coford/content/publications/projectreports/residues.pdf (accessed on 10 February 2020).
4. Bailey, C.; Dyer, J.F.; Teeter, L. Assessing the rural development potential of lignocellulosic biofuels in Alabama. *Biomass Bioenergy* **2011**, *35*, 1408–1417. [CrossRef]
5. Perez-Verdin, G.; Grebner, D.L.; Munn, I.A.; Sun, C.; Grado, S.C. Economic impacts of woody biomass utilization for bioenergy in Mississippi. *Prod. J.* **2008**, *58*, 75–83.
6. Moon, D.; Isa, A.; Yagishita, T.; Minowa, T. The regional economic impacts on the development of wood chip utilization in Maniwa city. *J. Wood Sci.* **2013**, *59*, 321–330. [CrossRef]
7. Rothe, A.; Moroni, M.; Neyland, M.; Wilnhammer, M. Current and potential use of forest biomass forenergy in Tasmania. *Biomass Bioenergy* **2015**, *80*, 162–172. [CrossRef]
8. Norton, M.; Baldi, A.; Buda, V.; Carli, B.; Cudlin, P.; Jones, M.B.; Korhola, A.; Michalski, R.; Novo, F.; Oszlányi, J.; et al. Serious mismatches continue between science and policy in forest bioenergy. *GCB Bioenergy* **2019**, *11*, 1256–1263. [CrossRef]
9. Shabani, N.; Akhtari, S.; Sowlati, T. Value chain optimization of forest biomass for bioenergy production: A review. *Renew. Sustain. Energy Rev.* **2013**, *23*, 299–311. [CrossRef]
10. Saidur, R.; Abdelaziz, E.A.; Demirbas, A.; Hossain, M.S.; Mekhilef, S. A review on biomass as a fuel for boilers. *Renew. Sustain. Energy Rev.* **2011**, *15*, 2262–2289. [CrossRef]
11. Ahtikoski, A.; Heikkilä, J.; Alenius, V.; Siren, M. Economic viability of utilizing biomass energy from young stands—The case of Finland. *Biomass Bioenergy* **2008**, *32*, 988–996. [CrossRef]
12. FAO. Bioenergy and Food Security Rapid Appraisal (BEFS RA). User Manual. Forest Harvesting and Wood Processing Residues. 2014. Available online: http://www.fao.org/sustainable-forest-management/toolbox/tools/tool-detail/en/c/410267/ (accessed on 25 January 2020).
13. Merchan, A.L.; Léonard, A.; Limbourg, S.; Mostert, M. Life cycle externalities versus external costs: The case of inland freight transport in Belgium. *Transp. Res. Part D Transp. Environ.* **2019**, *67*, 576–595. [CrossRef]
14. Pokharel, R.; Grala, R.K.; Grebner, D.L.; Grado, S.C. Factors affecting utilization of woody residues for bioenergy production in the southern United States. *Biomass Bioenergy* **2017**, *105*, 278–287. [CrossRef]
15. Fletcher, R.J.; Robertson, B.A.; Evans, J.; Doran, P.J.; Alavalapati, J.R.R.; Schemske, D.W. Biodiversity conservation in the era of biofuels: Risks and opportunities. *Front. Ecol. Environ.* **2011**, *9*, 161–168. [CrossRef]
16. Perez-Verdin, G.; Grebner, D.L.; Sun, C.; Munn, I.A.; Schultz, E.B.; Matney, T.G. Woody biomass availability for bioethanol conversion in Mississippi. *Biomass Bioenergy* **2009**, *33*, 492–503. [CrossRef]
17. Panichelli, L.; Gnansounou, E. Estimating greenhouse gas emissions from indirect land-use change in biofuels production: Concepts and exploratory analysis for soybean-based biodiesel production. *J. Sci. Ind. Res. India* **2008**, *67*, 1017–1030.
18. Law, B.E.; Harmon, M.E. Forest sector carbon management, measurement and verification, and discussion of policy related to climate change. *Carbon Manag.* **2011**, *2*, 73–84. Available online: https://content.sierraclub.org/ourwildamerica/sites/content.sierraclub.org.ourwildamerica/files/documents/Law%20and%20Harmon%202011.pdf (accessed on 15 February 2020). [CrossRef]
19. Schulze, E.D.; Körner, C.; Law, B.E.; Haberl, H.; Luyssaert, S. Large-scale bioenergy from additional harvest of forest biomass is neither sustainable nor greenhouse gas neutral. *GCB Bioenergy* **2012**, *4*, 611–616. [CrossRef]
20. Farkavcova, V.G.; Rieckhof, R.; Guenther, E. Expanding knowledge on environmental impacts of transport processes for more sustainable supply chain decisions: A case study using life cycle assessment. *Transp. Res. D Transp. Environ.* **2018**, *61*, 68–83. [CrossRef]

21. Pashkevich, A.; Beliakova, M.; Ivanov, A.; Purju, A. Development of Interactive Monitoring System for Urban Environmental Impact Assessment of Transport System. 16th Conference on Reliability and Statistics in Transportation and Communication, Riga, Latvia, 19–22 October 2016. *Procedia Eng.* **2017**, *178*, 42–52. [CrossRef]

22. Murphy, F.; Devlin, G.; McDonnell, K. Forest biomass supply chains in Ireland: A life cycle assessment of GHG emissions and primary energy balances. *Appl. Energy* **2014**, *116*, 1–8. [CrossRef]

23. ISO 14040. *Environmental Management, Life Cycle Assessment—Principles and Framework*; International Organization for Standardization (ISO): Geneva, Switzerland, 2006.

24. Muench, S.; Guenther, E. A systematic review of bioenergy life cycle assessments. *Appl. Energy* **2013**, *112*, 257–273. [CrossRef]

25. Gold, S.; Seuring, S. Supply chain and logistics issues of bio-energy production. *J. Clean. Prod.* **2011**, *19*, 32–42. [CrossRef]

26. Whittaker, C.; Mortimer, N.; Murphy, R.; Matthews, R. Energy and green house gas balance of the use of forest residues for bioenergy production in the UK. *Biomass Bioenergy* **2011**, *35*, 4581–4594. [CrossRef]

27. Butnar, I.; Rodrigo, J.; Gasol, C.M.; Castells, F. Life-cycle assessment of electricity from biomass: Case studies of two bio crops in Spain. *Biomass Bioenergy* **2010**, *34*, 1780–1788. [CrossRef]

28. Fantozzi, F.; Buratti, C. Life cycle assessment of biomass chains: Wood pellet from short rotation coppice using data measured on a real plant. *Biomass Bioenergy* **2010**, *34*, 1796–1804. [CrossRef]

29. Gasol, C.M.; Gabarrell, X.; Anton, A.; Rigola, M.; Carrasco, J.; Ciria, P.; Rieradevall, J. LCA of poplar bioenergy system compared with Brassica carinata energy crop and natural gas in regional scenario. *Biomass Bioenergy* **2009**, *33*, 119–129. [CrossRef]

30. Wihersaari, M. Greenhouse gas emissions from final harvest fuel chip production in Finland. *Biomass Bioenergy* **2005**, *28*, 435–443. [CrossRef]

31. González-García, S.; Dias, A.C.; Clermidy, S.; Benoist, A.; Maurel, V.B.; Gasol, C.M.; Gabarrell, X.; Arroja, L. Comparative environmental and energy profiles of potential bioenergy production chains in Southern Europe. *J. Clean. Prod.* **2014**, *76*, 42–54. [CrossRef]

32. Chester, M.; Matute, J.; Bunje, P.; Eisenstein, W.; Pincetl, S.; Zoe, E.; Cepeda, C. Life-Cycle Assessment for Transportation Decision-making. UCLA Center for Sustainable Urban Systems. 2010. Available online: https://www.transitwiki.org/TransitWiki/images/7/73/Life-cycle_assessment_fortransportation_decision-making.pdf (accessed on 30 January 2020).

33. Ignea, G.; Ghaffayian, M.R.; Borz, S.A. Impact of operational factors on fossil energy inputs in motor-manual tree felling and processing: Results of two case studies. *Ann. Res.* **2017**, *60*, 161–172. [CrossRef]

34. Picchio, R.; Maesano, M.; Savelli, S.; Marchi, E. Productivity and Energy balance in conversion of a Quercus cerris L. Coppice stand into high forest in Central Italy. *Croat. J. For. Eng.* **2009**, *30*, 15–26.

35. Hiloidhari, M.; Baruah, D.C.; Singh, A.; Kataki, S.; Medhi, K.; Kumari, S.; Ramachandra, T.V.; Jenkins, B.M.; Thakur, I.S. Emerging role of Geographical Information System (GIS), Life Cycle Assessment (LCA) and spatial LCA (GIS-LCA) in sustainable bioenergy planning. *Bioresource Technol.* **2017**, *242*, 218–226. [CrossRef]

36. Sacchelli, S.; Meob, I.D.; Palett, A. Bioenergy production and forest multifunctionality: A trade-off analysis using multi scale GIS model in a case study in Italy. *Appl. Energy.* **2013**, *104*, 10–20. [CrossRef]

37. Angelis-Dimakis, A.; Biberacher, M.; Dominguez, J.; Fiorese, G.; Gadocha, S.; Gnansounou, E.; Guariso, G.; Kartalidis, A.; Panichelli, L.; Pinedo, I.; et al. Methods and tools to evaluate the availability of renewable energy sources. *Renew. Sustain. Energy Rev.* **2011**, *15*, 1182–1200. [CrossRef]

38. Yue, C.; Wang, S. GIS-based evaluation of multifarious local renewable energy sources: A case study of the Chigu area of southwestern Taiwan. *Energy Policy* **2006**, *34*, 730–742. [CrossRef]

39. Ramachandra, T.V.; Krishna, S.V.; Shruthi, B.V. Decision support system to assess regional biomass energy potential. *Int. J. Green Energy* **2005**, *1*, 407–428. [CrossRef]

40. Consiglio Regionale. *Piano di Indirizzo Energetico Ambientale Regionale*; Consiglio Regionale: Potenza, Italy, 2010; Available online: https://www.regione.basilicata.it/giunta/files/docs/DOCUMENT_FILE_543546.pdf (accessed on 20 December 2019).

41. Catullo, G. Sviluppo Agro-Energetico. Benefici Economici e Ambientali Contro la Crisi delle Risorse Fossili. Ufficio Stampa Consiglio regionale della Basilicata. 2012. Available online: https://consiglio.basilicata.it/archivio-news/files/docs/32/36/07/DOCUMENT_FILE_323607.pdf (accessed on 15 January 2020).

42. Consiglio Regionale della Basilicata. Regional Law n.42 of 30 November 1998 "Rules on Forestry". 1998. Available online: https://consiglio.basilicata.it//consiglionew/site/Consiglio/detail.jsp?sec=107173&otype=1150&id=182264&anno=1998 (accessed on 15 December 2019).

43. Rauch, P.; Borz, S.A. Reengineering the Romanian Timber Supply Chain from a Process Management Perspective. *Croat. J. For. Eng.* **2020**, *41*, 85–94. [CrossRef]

44. Rauch, P.; Wolfsmayr, U.J.; Borz, S.A.; Triplat, M.; Krajnc, N.; Kolck, M.; Oberwimmer, R.; Ketikidis, C.; Vasiljevic, A.; Stauder, M.; et al. SWOT analysis and strategy development for forest fuel supply chains in South East Europe. *Forest Policy Econ.* **2015**, *61*, 87–94. [CrossRef]

45. Müllerová, J.; Hédl, R.; Szabó, P. Coppice abandonment and its implications for species diversity in forest vegetation. *For. Ecol. Manag.* **2015**, *343*, 88–100. [CrossRef]

46. ISO 14044. *Environmental Management, Life Cycle Assessment–Requirements and Guidelines*; International Organization for Standardization(ISO): Geneva, Switzerland, 2006.

47. Spielmann, M.; Scholz, R. Life Cycle Inventories of Transport Services: Background Data for Freight Transport (10pp). *Int. J. Life Cycle Ass.* **2005**, *10*, 85–94. [CrossRef]

48. Cozzi, M.; Di Napoli, F.; Viccaro, M.; Romano, S. Use of Forest Residues for Building Forest Biomass Supply Chains: Technical and Economic Analysis of the Production Process. *Forests* **2013**, *4*, 1121–1140. [CrossRef]

49. Pergola, M.; Gialdini, A.; Celano, G.; Basile, M.; Caniani, D.; Cozzi, M.; Gentilesca, T.; Mancini, I.M.; Pastore, V.; Romano, S.; et al. An environmental and economic analysis of the wood-pellet chain: Two case studies in Southern Italy. *Int. J. Life Cycle Ass.* **2018**, *23*, 1675–1684. [CrossRef]

50. Ecoinvent Version 3. 2013. Available online: http://www.ecoinvent.org/database/database.html (accessed on 14 February 2020).

51. HBEFA (Handbook Emission Factors for Road Transport). Umweltbundesamt Wien: Handbuch Emissionsfaktoren des Straßenverkehrs, Version 3.1, Berlin, Bern, Vienna/Germany, Switzerland, Austria. 2010. Available online: http://www.hbefa.net (accessed on 22 January 2020).

52. Fraunhofer. Documentation for Truck Transport Processes. 2012. Available online: http://www.gabi-software.com/fileadmin/gabi/documentation5/Documentation_GaBi_Transport_Processes_Truck.pdf (accessed on 15 November 2019).

53. Ministry of Housing, Spatial Planning and the Environment. *Eco-Indicator 99 Manual for Designers. A damage Oriented Method for Life Cycle Impact Assessment*; Ministry of Housing, Spatial Planning and the Environment: The Hague, The Netherlands, 2000; Available online: https://www.pre-sustainability.com/download/EI99_Manual.pdf (accessed on 15 December 2019).

54. Clarke, K.C. Advances in geographic information systems. *Comput. Environ. Urban Systems* **1986**, *10*, 175–184. [CrossRef]

55. Fuchs, M.; Teichmann, J.; Lauster, M.; Remmen, P.; Streblow, R.; Müller, D. Work flow automation for combined modeling of buildings and district energy systems. *Energy* **2016**, *117*, 478–484. [CrossRef]

56. Perea-Moreno, A.J.; Perea-Moreno, M.A.; Hernandez-Escobedo, Q.; Manzano-Agugliaro, F. Towards forest sustainability in Mediterranean countries using biomass as fuel for heating. *J. Clean. Prod.* **2017**, *156*, 624–634. [CrossRef]

57. Allegrini, J.; Orehounig, K.; Mavromatidis, G.; Ruesch, F.; Dorer, V.; Evins, R. A review of modelling approaches and tools for the simulation of district-scale energy systems. *Renew. Sustain. Energy Rev.* **2015**, *52*, 1391–1404. [CrossRef]

58. Legge 6 Dicembre 1991, n.394. Legge Quadro Sulle Aree Protette. Available online: http://www.parks.it/federparchi/leggi/394.html (accessed on 16 February 2020).

59. Handler, R.M.; Shonnard, D.R.; Lautala, P.; Abbas, D.; Srivastava, A. Environmental impacts of round wood supply chain options in Michigan: Life-cycle assessment of harvest and transport stages. *J. Clean. Prod.* **2014**, *76*, 64–73. [CrossRef]

60. Sonne, E. Greenhouse gas emissions from forestry operations: A life cycle assessment. *J. Environ. Qual.* **2006**, *35*, 1439–1450. [CrossRef]

61. Jäppinen, E.; Korpinen, O.J.; Ranta, T. The Effects of Local Biomass Availability and Possibilities for Truck and Train Transportation on the Greenhouse Gas Emissions of a Small-Diameter Energy Wood Supply Chain. *Bio Energy Res.* **2013**, *6*, 166–177. [CrossRef]

62. Gustavsson, L.; Eriksson, L.; Sathre, R. Costs and CO_2 benefits of recovering, refining and transporting logging residues for fossil fuel replacement. *Appl. Energy* **2011**, *88*, 192–197. [CrossRef]

63. Lindholm, E.L.; Berg, S.; Hansson, P.A. Energy efficiency and the environmental impact of harvesting stumps and logging residues. *Eur. J. Forest. Res.* **2010**, *129*, 1223–1235. [CrossRef]
64. Heinimann, H. Life Cycle Assessment (LCA) in Forestry—State and Perspectives. *Croat. J. For. Eng.* **2012**, *33*, 357–372.
65. Zubaryeva, A.; Zaccarelli, N.; Giudice, C.D.; Zurlini, G. Spatially explicit assessment of local biomass availability for distributed biogas production via anaerobic co-digestion—Mediterranean case study. *Renew. Energy* **2012**, *39*, 261–270. [CrossRef]
66. Shi, X.; Elmore, A.; Li, X.; Gorence, N.J.; Jin, H.; Zhang, X.; Wang, F. Using spatial information technologies to select sites for biomass power plants: A case study in Guangdong Province, China. *Biomass Bioenergy* **2008**, *32*, 35–43. [CrossRef]

 *energies*

Article

Consequential Life Cycle Assessment of Swine Manure Management within a Thermal Gasification Scenario

Mahmoud Sharara [1], **Daesoo Kim** [2], **Sammy Sadaka** [3,*] **and Greg Thoma** [2]

[1] Biological & Agricultural Engineering Department, North Carolina State University, Raleigh, NC 27695, USA; m_sharara@ncsu.edu

[2] Department of Chemical Engineering, University of Arkansas, Fayetteville, AR 72701, USA; dskim@uark.edu (D.K.); gthoma@uark.edu (G.T.)

[3] Department of Biological & Agricultural Engineering, University of Arkansas, Little Rock, AR 72204, USA

* Correspondence: ssadaka@uaex.edu; Tel.: +1-501-671-2298

Received: 20 September 2019; Accepted: 23 October 2019; Published: 25 October 2019

Abstract: Sustainable swine manure management is critical to reducing adverse environmental impacts on surrounding ecosystems, particularly in regions of intensive production. Conventional swine manure management practices contribute to agricultural greenhouse gas (GHG) emissions and aquatic eutrophication. There is a lack of full-scale research of the thermochemical conversion of solid-separated swine manure. This study utilizes a consequential life cycle assessment (CLCA) to investigate the environmental impacts of the thermal gasification of swine manure solids as a manure management strategy. CLCA is a modeling tool for a comprehensive estimation of the environmental impacts attributable to a production system. The present study evaluates merely the gasification scenario as it includes manure drying, syngas production, and biochar field application. The assessment revealed that liquid storage of manure had the highest contribution of 57.5% to GHG emissions for the entire proposed manure management scenario. Solid-liquid separation decreased GHG emissions from the manure liquid fraction. Swine manure solids separation, drying, and gasification resulted in a net energy expenditure of 12.3 MJ for each functional unit (treatment of 1 metric ton of manure slurry). Land application of manure slurry mixed with biochar residue could potentially be credited with 5.9 kg CO_2-eq in avoided GHG emissions, and 135 MJ of avoided fossil fuel energy. Manure drying had the highest share of fossil fuel energy use. Increasing thermochemical conversion efficiency was shown to decrease overall energy use significantly. Improvements in drying technology efficiency, or the use of solar or waste-heat streams as energy sources, can significantly improve the potential environmental impacts of manure solids gasification.

Keywords: life cycle assessment; environmental impact; greenhouse gas; gasification; swine manure management

---

1. Introduction

Agricultural systems are significant contributors to global climate change and ecosystems degradation [1]. Local, regional, and global agreements are increasingly mandating legislative and regulatory actions to restrict emissions to mitigate the short-term and long-term environmental degradation. However, both legislative and regulatory efforts to reduce environmental impacts, especially greenhouse gas (GHG) emissions, will eventually put a more significant burden on agricultural and industrial sectors as well as increase the cost of production. Livestock production, in particular, has been recognized as a significant source of GHG emissions and a driver of both freshwater and marine water eutrophication [2,3]. Therefore, the vulnerability of livestock production and the agriculture sector to climate change further incentivizes the search for and adoption of sustainable agricultural practices [4].

In livestock production, manure management is a significant source of direct GHG emissions, such as methane (CH_4) and nitrous oxide (N_2O) [5]. Land application is the most common practice to handle swine manure to use available nutrients for crop production. However, applying swine manure to crop and grass fields where nutrients are available more than agronomic crop needs or where fields have historically received large volumes of manure application increase environmental risks to surrounding ecosystems. Liquid manure management systems, relevant to swine production, are also a significant source of gaseous emissions. Liquid manure storage promotes anaerobic conditions, which transform organic matter into CH_4 and ammonia (NH_3). Besides, uncontrolled anaerobic and aerobic conditions initiate nitrification-denitrification processes, which convert a share of manure nitrogen to N_2O, which is a potent GHG. Solid-liquid separation of swine manure has been recognized as an emission mitigation strategy. However, increased N_2O and CH_4 emissions have been reported during storage of manure-separated solids [6]. Transforming separated solids into a gas fuel (syngas) and a stable nutrient-rich co-product (biochar), via gasification, can potentially reduce emissions associated with manure-separated solids and generate value-added products. Furthermore, gasification-derived biochars have been shown to have adsorbing characteristics for various organic contaminants such as *p*-Cresol [7,8].

Evaluating emissions and impacts associated with this conversion strategy, i.e., gasification of swine manure solids, can facilitate adoption and expand the set of technologies available to livestock producers for manure management. Life cycle assessment (LCA) is an essential tool to assist decision-makers by evaluating the environmental performance of proposed management strategies. According to ISO standard 14040 [9], LCA considers the various input and output flow, and the corresponding environmental burdens, resulting from production, consumption, and disposal of associated product systems. Energy recovery from swine manure incineration was found to be a promising pathway to reduce GHG emissions associated with manure management [10]. Several LCA studies have reported on the performance on anaerobic digestion as manure management and energy recovery as a sole feedstock or in combination with other biomass streams [11–13]. Wu et al. [10] performed an LCA comparing GHG emissions between land application and gasification as manure management practices. The study showed that gasification has high potential to reduce GHG emissions due to the environmental benefits of syngas production and biochar application to crop field. Biochar also has been used for carbon sequestration, soil amendment and biomass waste management [14].

In a comprehensive review of swine manure conversion technologies, Sharara and Sadaka [15] highlighted the scarcity of research studies that investigated swine manure solids gasification and pyrolysis. Accordingly, it was recommended to develop an LCA of swine manure management systems. Therefore, the objective of this study is to evaluate potential environmental impacts of a manure management scenario that utilizes thermal gasification of swine manure solids as a disposal/energy retrieval strategy using consequential life cycle assessment (CLCA) methodology.

2. Methods

2.1. LCA Goal and Scope

The goal of this CLCA is to determine the impacts associated with swine manure management using gasification for manure solids. Figure 1 shows a schematic diagram of the proposed swine manure management (SMM) scenario. The scope of this CLCA covers manure management activities associated with 1000 kg of flushed swine manure at 5% dry matter content, without accounting for animal maintenance (feed, drinking water, climate control - all assumed to be unaffected by treatment), until the land application of both the liquid fraction (slurry) and the solid fraction (biochar). The functional unit (FU) is the treatment of one metric ton (1000 kg) of swine manure slurry, at 5% DM, via gasification and land application. The thermal gasification of manure solids produces three co-products: syngas, heat, and biochar. All three co-products were modeled as displacing existing processes in the system with surpluses used to displace their alternatives beyond the system boundary. A share of the produced syngas is consumed as fuel in a boiler, while the excess syngas is considered a replacement for natural gas. The heat generated during the thermal gasification is used for drying, and biochar is land applied as a fertilizer replacement.

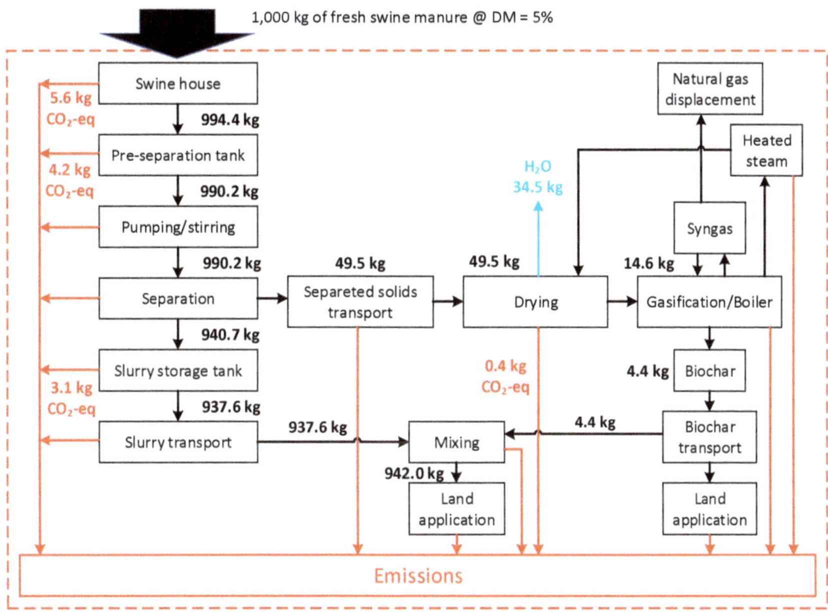

Figure 1. The scope of proposed swine manure management scenario illustrating mass balances and emissions flows (Note: black arrows represent main product flow; red arrows represent direct and indirect greenhouse gas (GHG) emissions. Blue arrow represents water evaporation. Numbers in red are direct emissions of GHG in kg CO_2-eq).

The excreted manure (urine and feces) is flushed from shallow pits under the house. The flushed manure is stored in a holding pond, and stirred, before being pumped into the separation stage. Manure separation is accomplished using a screw press separator. This class of size-separators utilizes a tapered screw and a fine-mesh screen (0.75–3 mm) to fractionate the manure into solids-rich and liquids-rich fractions. The solids-rich fraction is transported to a thermochemical conversion facility that contains a dryer, a gasifier, and a gas boiler. The manure gasification is accomplished in an atmospheric, fluidized-bed gasifier to produce syngas, which is subsequently fired (burned) in the gas boiler for heat, in addition to biochar. The liquid-fraction (slurry) is stored, then transported to an agricultural field for land application. In this model, the emissions associated with land application of biochar and slurry are presented in detail separately first, then combined to estimate the total impacts of biochar-slurry land application. The total impacts represent the summation of the impacts from each substrate without including any synergistic effects. SimaPro© 8.5.2 software (PRé Consultants, The Netherlands) and ecoinvent v3.4 database [16] with IMPACT World+ midpoint method [17] were used to model the impacts. We modified the characterization factors in IMPACT World+ by adopting the most recent Intergovernmental Panel on Climate Change (IPCC, v1.02) 100-year global warming potentials (GWP 100a): CH_4 biogenic: 27.8 kg CO_2-eq, CH_4 fossil: 30.5 kg CO_2-eq and N_2O: 265 kg CO_2-eq [18]. The inventory of mass, energy, and emission flows, in each stage, is presented in the following sections.

2.2. Life-Cycle Inventory Assessment

2.2.1. Swine House

According to manure characteristics standard [19], the amount of total solids in as-excreted swine manure is between 5% and 10% by weight for grow-finish pigs. Manure can be collected from barns through flushing, scrapping, or using pull-plug systems that rely on gravity. Collection systems significantly impact the concentration of solids in the collected manure. In this study, the composition of swine manure solids was taken from first-hand analyses of manure solids sampled from the gravity-collected slurry in a feeder-finisher farm in Washington County, Arkansas. Table 1 shows the characteristics of swine manure solids. The accumulated manure is collected every two weeks by gravity using a pull-plug system (no energy or mechanical power is needed for drainage). During the 2-week storage, various biogenic emissions, namely NH_3, N_2O, CO_2, and CH_4, are released due to aerobic and anaerobic activities in the manure substrate. NH_3 and N_2O emissions as nitrogen during this period were estimated at 16% and 0.5% of total manure nitrogen [20]. Manure-related GHG emissions, i.e., CH_4 and N_2O were estimated using the IPCC Tier 1 approach for GHG emissions in livestock [1].

Table 1. Characteristics of swine manure under the current study.

Characteristics	Value	Units	Characteristics	Value	Units
Dry matter (DM)	5.0	%	P (Ash)	20.9	%
Volatile matter (DM)	81.6	%	K (Ash)	21.1	%
Ash (DM)	18.4	%	Na (Ash)	10.3	%
C (DM)	50.8	%	Ca (Ash)	20.5	%
Total-N (DM)	4.3	%	Mg (Ash)	12.1	%
O (DM)	21.0	%	Cu (Ash)	0.01	%
H (DM)	6.9	%	Zn (Ash)	0.04	%
S (DM)	1.3	%			

2.2.2. Pre-Separation Tank/Stirring and Mixing

Pre-separation storage was modeled as an opened storage tank. The projected NH_3 loss is 2% of the total N in the manure [21]. IPCC guidelines for GHG emissions were used to estimate the N_2O and CH_4 emissions in the pre-separation tank. For the agitation/mixing step, Wesnaes et al. [21] and Nguyen et al. [22] reported that the energy requirements for pumping and stirring 1000 kg of manure slurry to be 0.5 kWh and 1.2 kWh, respectively. Thus, the total energy consumption associated with this stage (stirring and pumping) was taken as 1.7 kWh.

2.2.3. Mechanical Separation and Separated Solids Transport

Moller et al. [23] estimated the power required for manure solid-liquid separation using a mechanical screen press to be 0.50 kWh per metric ton. Therefore, the energy required for separation was modeled as 0.5 kWh. The U.S. electricity mix was used to model the impacts of electric power utilization. No air emissions or water contamination are associated with manure during this stage.

The separation indices reported for screw presses [24] were used to determine the amount and composition of separation products. In that study, the solids content in the original slurry varied from 1.8% to 6.3%. As shown in Table 2, the separation index (%) is defined as the mass of a given compound in the solid fraction to the mass of that compound in the original (unseparated) slurry. The mass of the separated solids fraction (containing both solids and slurry) ranges between 5.0% and 7.3% [25,26]; the lower value (5.0%) was used in this study. The separated solid fraction (TS = 28% *weight basis*) were assumed to be transported 500 meters (0.5 km) from the separation platform to the drying and conversion facility. Emissions associated with this step were estimated per shipping unit, i.e., ton-kilometer (tkm) using ecoinvent (v3.4) inventory (transport, tractor, and trailer, agricultural {GLO}| market for | Conseq, U) for farm operations [16].

Table 2. Separation indices for mechanical screw press separation [23].

	Raw Pig Slurry	Separation Index (%)			
	DM (%)	Volume	Dry Matter	N-Total	P-Total
Mean [1]	4.7	5.75	35	11	20.3
Standard Deviation	(2.01)	(1.50)	(16)	(10)	(16.5)

[1] Based on data collected from ref. [23].

2.2.4. Drying

Gasifying organic material requires that the moisture content should be 15% or lower [24,27]. Therefore, the separated solids fraction must be dried first. Drying can be accomplished passively by relying on solar heating and natural air circulation, or through the mechanical circulation of heated air through the wet mixture using blowers or fans. Ideally, for such a system to be efficient, the material is moved inside the dryer to ensure quick and uniform dryness. Passive drying of swine manure solids can be a source of objectionable odors and can reduce the organic carbon content of the material. Therefore, a heated air-drying technique was modeled for this study. The thermal energy required to remove 1 kg of moisture from manure was reported to be 2.3 MJ [23]. Hospido *et al.* [28] evaluated different scenarios for utilizing solid sludge from urban wastewater treatment plants (WWTP) using 1,000 kg of dried sludge as the unit basis. In their study, the electricity and heat consumption associated with sludge drying were 118 kWh and 1,638 kWh per 1 metric ton of dried sludge, respectively.

During drying, as much as 20% of the manure-N was reported to volatilize, typically as NH_3 [29]. Also, C loss during drying was reported to be around 4%. In the municipal sludge drying process model, 44.3 g of volatile organic compound (VOC) emissions were reported per ton of dried sludge. The same emissions factor was used to model the manure emissions in this study.

2.2.5. Gasification/Boiler

In this thermochemical conversion process, the dry manure solids are transformed at temperatures between 600 and 800 °C to gas (referred to as producer gas, or syngas) in addition to biochar, and a small amount of condensable material (tar) is produced. Gasification utilizes air, or another oxidizing agent, to partially oxidize the biomass C into CO and CO_2. However, given the scarcity of the data on the gasification of swine manure solids, the dataset used in this study (Table 3) was compiled from available studies on swine manure solids and feedstock, such as, poultry litter, sewage sludge, cattle feedlot manure, that have similar characteristics as swine manure solids, i.e., high ash, and nitrogen content. The primary product, syngas, is combusted in a steam boiler to generate steam that is used to satisfy heating needs on the farm, e.g., the drying manure solids, and heating the farrowing crates. The syngas displaces natural gas demand and, consequently, the impacts associated with natural gas production. To account for the summer season when the heating is not necessary on the farm, we deducted 25% of the heat production to be claimed as a credit in the computational modeling.

Table 3. A gasification process model for 1 kg of dry (15% moisture) swine manure solids.

Parameters	Values	Source
Air requirements (kg_{air} kg^{-1})	2.54	Calculated from composition, ER = 0.20
Manure solids HHV [†] ($MJ\ kg^{-1}$)	19–20	[30,31]
Cold-gas efficiency (%)	50–80	[32]
Boiler thermal efficiency (%)	78	[33]
Char yield ($g\ kg^{-1}$)	300–490	[32,33]
Electricity req. ($kWh\ kg^{-1}$)	0.339	[34]
Thermal energy req. ($MJ\ kg^{-1}$)	1.2	[35]

[†] HHV: higher heating value.

The cold-gas efficiency in Table 3 is the chemical energy retained in the syngas as a share of the total chemical energy in the feedstock, without considering the gas sensible heat. In this case, however, since the syngas is used to replace a heat source (natural gas), both the sensible and chemical energies in the syngas were considered. Accordingly, the conversion efficiency increases, with thermal gas efficiency (HGE) taken to be between 60% and 90%. A 70% HGE was used in this model. Accordingly, the amount of heat generated (MJ) due to gasification was calculated after subtracting the thermal energy required for the process (calculated using the pyrolysis enthalpy in Table 3 taken here to be 1.0 MJ kg^{-1}). Gasification also yields a biochar fraction, which is utilized as a soil amendment. The biochar produced was assumed to be a nitrogen-free co-product since all nitrogen typically devolatilizer during gasification as N-species. P and K were assumed to be sequestered entirely in the biochar fraction. Table 4 presents the emission associated with the current study via the gasification facility for the swine manure solids.

Table 4. Emissions to the air resulting from the gasification process (per 1 kg of dry matter of swine manure solids).

Substance	Value (g)	Substance	Value (g)
CO_2	1458.5	VOCs	0.016
NOx	1.1400	HF	0.0005
CO	0.1460	Hg	0.0001
SO_2	0.076	As	0.0001
HCl	0.047	Ni	0.0000
PM	0.018	Cd	0.0000

2.2.6. Biochar Transport and Land Application

The environmental emissions associated with biochar transport to the field were considered with a transportation distance assumed to be 10 km (6.2 miles), which is slightly more than the upper bound on average manure hauling distances, i.e., between 1.6 and 6.4 km (1 and 4 miles). Biochar land application is beneficial both as a fertilizer/soil conditioner and as a carbon sequestration option [36]. In this study, the benefits of biochar application to the soil were determined as the avoided synthetic fertilizers due to the presence of P and K in the biochar. Nutrient credits were assigned for P and K in biochar as 80% equivalency of commercial fertilizer, according to the Wisconsin study [37]. Additional benefits of biochar application include improved water holding capacity and reduced N_2O emissions. However, due to the scarcity of quantifiable data on these benefits and the strong dependence on the crop, soil and climate conditions, these additional benefits are not considered in this study, i.e., no GHG emissions from the biochar field application were considered. The amount of avoided P_2O_5 and K_2O fertilizers, and sequestered CO_2 were determined, to be 0.71 kg P_2O_5, 0.66 kg K_2O, and 3.43 kg CO_2 per ton of functional unit.

2.2.7. Post-Separation Tank (Liquid-Fraction)

The separated slurry is stored in an exposed tank until it is transported to a field for application. During storage, the organic fraction of this slurry transforms, resulting in GHG emissions. The following section describes the computations for the various emissions. IPCC guidelines [38] were used to estimate CH_4 emissions in the swine house, and the estimating method used for the pre-separation tank was also used here to estimate emissions during post-separation storage of the liquid slurry. It should be noted that the volatile solids loading (VS) in this storage step is much lower than in the pre-separation tank, i.e., 21.5 kg per functional unit. Similarly, NH_3 and N_2O emissions were estimated using emission factors outlined for pre-separation tank emissions.

2.2.8. Liquid Fraction Transport, Mixing with Biochar, and Land Application

The liquid fraction transportation distance to the application field was assumed to be equal to that for biochar, 10 km. This distance has been used before in a similar study to model the impacts of dairy cow slurry digestion and land application [39]. The energy requirement for slurry and biochar mixing is 1.2 kWh ton^{-1}, and the land application energy requirements and emissions were modeled using the vacuum spreader model available in the ecoinvent v3.4 database (Ecoinvent Centre, 2019). The impacts of slurry application (without the biochar) are presented in the following section. Table 5 below presents the summary of inputs and emissions for the functional unit as well as nutrient credits associated with land application of liquid slurry and biochar. The avoided N fertilizer value was calculated from N availability in a liquid slurry, using Delin et al. [39], as 52% equivalency of commercial nitrogen fertilizer.

Table 5. Summary of emissions, energy, and transportation requirements as well as an avoided burden for the functional unit.

1. Swine house			
NH_3 emissions (kg)	0.418	CO_2 emissions, biogenic (kg)	2.82
N_2O emissions (kg)	0.017	CH_4 emissions (kg)	2.36
2. Pre-separation storage tank			
NH_3 emissions (kg)	0.052	CO_2 emissions, biogenic (kg)	2.14
N_2O emissions (kg)	0.017	CH_4 emissions (kg)	1.94
3. Stirring and pumping			
Electrical power requirements (kWh)	1.7		
4. Mechanical separation			
Electrical power requirement (kWh)	0.5		
5. Solids transportation			
Transportation (tkm)	0.025		
6. Drying (solid fraction)			
Heat requirements (MJ)	86.1	VOC emissions (kg)	0.00064
Electricity requirements (kWh)	1.72	NH_3 emissions (kg)	0.43
Water requirement (m^3)	0.222		

Table 5. *Cont.*

7. Gasification-Boiler (solid fraction)			
Electricity requirements (kWh)	4.94	HCl emissions (g)	0.47
Air needed (kg)	37.05	PM emissions (g)	0.18
Thermal energy requirement (MJ)	17.5		
Generated heat (MJ)	140.7	VOCs emissions (g)	0.16
Char Produced (kg)	4.38	HF emissions (g)	0.005
CO_2 emissions (kg)	1.458	Hg emissions (g)	0.001
NO_x emissions (g)	11.38	As emissions (g)	0.0009
CO emissions (g)	1.46	Ni emissions (g)	0.0006
SO_2 emissions (g)	0.76	Cd emissions (g)	0.0001
8. biochar transportation (solid fraction)			
Transportation (tkm)	0.044		
9. Post-land application for the biochar (solid fraction)			
Avoided P_2O_5 (kg)	0.71	P leaching (kg)	0.039
Avoided K_2O (kg)	0.66		
10. Post-separation storage tank (liquid fraction)			
NH_3 emissions (kg)	0.033	CO_2 emissions, biogenic (kg)	1.78
N_2O emissions (kg)	0.011	CH_4 emissions (kg)	1.24
11. Slurry transportation (liquid fraction)			
Transportation (tkm)	9.395		
12. Mixing and land application (solid and liquid fractions)			
Mixing (kWh)	1.2		
13. Post-land application for the slurry (liquid fraction)			
NH_3 emissions (kg)	0.204	Avoided P_2O_5 (kg)	2.62
N_2O emissions (kg)	0.021	Avoided K_2O (kg)	1.13
CO_2 emissions, biogenic (kg)	27.2	NO_3 leaching (kg)	1.8
Avoided N fertilizer (kg)	0.398	P leaching (kg)	0.15

NH_3 devolatilization resulting from manure land application is among the primary sources of N emissions in the agricultural sector. Rates of NH_3 emissions vary significantly with variability in manure slurry characteristics, soil type, and weather conditions. Misselbrook et al. [40] studied the influence of manure type (cattle, and pig manure), and land type (arable and grassland) on NH_3 emissions. They reported NH_3 emissions between 6.0 and 21.5% of the total ammoniacal nitrogen (TAN) in the pig manure. Sommer and Hutchings [41] reported NH_3 emissions to be 5% of total NH_4 in an applied slurry with trail hose application, and 8-10% of total NH_4 under broad spreading. According to literature, an estimated 39% of TAN in swine slurry devolatilizes as NH_3 during spring season land application [42]. In this study, NH_3 devolatilization was modeled as 20% of TAN in the slurry. The TAN was taken from Buckley *et al.* [43] to be 75% of the total N in the swine manure slurry (S.D. = 17%).

According to the Intergovernmental Panel on Climate Change (IPCC) guidelines [18], the emission factor for N_2O resulting from organic amendments application (EF_{N2O}) is 0.01 kg N_2O-N kg N^{-1}. Rochette et al. [44] estimated the cumulative C loss (as CO_2) due to swine slurry application to spring maize plots to be 63% of the original slurry C. In this study, the N and P leaching through the soil profile was assumed as 35% and 10% of manure N and P, respectively [22].

3. Results and Discussions

3.1. Impact Assessment

Table 6 presents a summary of the cumulative potential environmental impacts of the swine manure management scenario according to selected categories. Positive impact characterization values indicate an added environmental burden, while negative values represent avoided burden. Detailed descriptions are addressed in the following sections.

Table 6. Characterization of impacts for the proposed swine manure management scenario.

Impact Category	Unit	Proposed Swine Manure Management
Global warming (GWP 100-year)	kg CO_2-eq *	166
Fossil energy use	MJ	−57.9
Water use	m^3	−0.015
Marine eutrophication	kg N-eq	0.470
Aquatic eutrophication	kg PO_4^--eq	0.551

* eq Represents equivalent.

3.2. Global Warming Potential (GWP)

The proposed manure management scenario has net emissions of 166 kg CO_2-eq emitted per ton of swine manure slurry treatment. A detailed representation of the contribution of each stage to the cumulative GHG emissions is shown in Figure 2. Emissions during manure storage under slatted floors in the house and during external storage represented the majority of the GHG emissions, with the two stages contributing 42.1% and 35.1% respectively of the total emissions. This significant contribution is attributed to the high levels of N_2O and CH_4 emissions during these two steps, with both gases having a significantly higher impact on global warming potential. Similarly, the third-largest contributing stage to GHG emissions is post-separation slurry storage, i.e., 22.4% of scenarios of GWP.

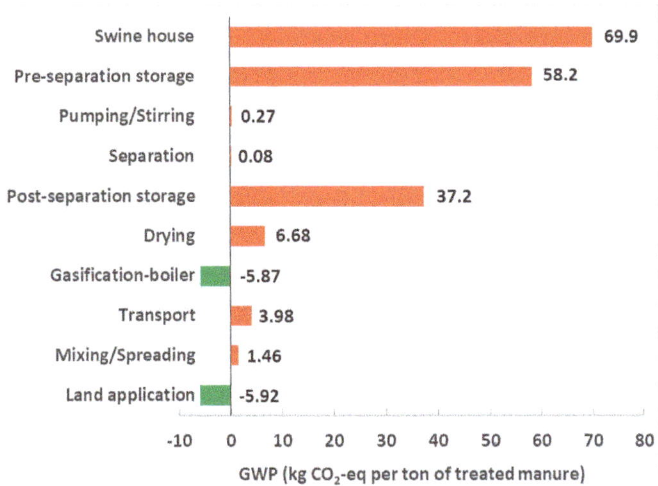

Figure 2. The net contribution of each stage to the cumulative global warming potential over 100 years (GWP 100a) in units of kg CO_2 equivalent (kg CO_2-eq). We used modified Intergovernmental Panel on Climate Change (IPCC, V1.02) [18] GWP factors embedded in the IMPACT World+ method [16].

Manure solids gasification and syngas combustion (in a boiler), represented as one coupled process (Figure 2), contribute - 3.54% of the total GWP. The net negative contribution here indicates that the avoided GHG emissions by syngas combustion, instead of natural gas, completely offset the combined emissions from syngas combustion and those associated with gasifier electricity consumption. Even though the low hot gas efficiency, 70%, and the low boiler efficiency, 78%, was used in this model, the overall ratio of avoided natural gas use resulted in a net negative GWP.

GWP for drying manure solids, 6.68 kg CO_2-eq, represented 4.02% of overall GWP emissions. Despite being an energy-intensive process, the low GWP contribution here for drying is attributed to the fact that the process heat is recycled from the gasification-boiler output heat, which reduces the overall energy requirement for drying and thus the impact. The following stages: pumping, stirring, separation, and transportation cumulatively contributed 3.49% of the total GHG emissions. Land application of liquid slurry and biochar, which is a co-product of thermal gasification, contributed net negative GHG emissions (- 5.92 kg CO_2-eq) due to the credit of displacing synthetic fertilizer. One thing to note is that CO_2 emission during land application is accounted for as biogenic CO_2 emission. The land application represents a 3.57% reduction of total GHG emissions.

3.3. Fossil Fuel Use

Cumulative fossil fuel energy use in this scenario was - 58.0 MJ per functional unit. Figure 3 details the individual contribution of manure management stages to overall fuel consumption. Manure storage steps, from an energy perspective, were all-passive and therefore had no fossil fuel expenditure or saving. The maximum energy burden was associated with the drying stage, which represented 62.0% of total fossil fuel energy input, followed by the slurry transportation stage, which represented 24.6% of total fossil fuel energy input. The gasification-boiler stage was attributed with the net negative energy use of - 95.7 MJ, by offsetting natural gas firing to produce the credited amount of thermal energy. The energy demand for the drying process, 107 MJ, represents the electricity demand in the dryer, which cannot be met through the gasification-boiler stage supply. The primary energy saving in this scenario, - 135 MJ, was attributed to the consequences of slurry-biochar land application. This savings is from the avoided synthetic fertilizers and the fossil fuel energy used in their production. For illustration, production of 1 kg N fertilizer requires 88.0 MJ of energy using global unit process of ecoinvent v3.4 database [16]. Similarly, production of 1 kg of P_2O_5 and K_2O require 20.1 and 18.4 MJ of fossil fuel energy in their production.

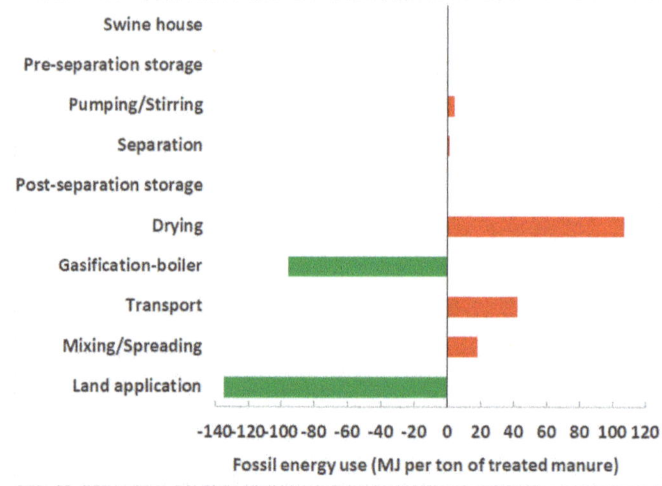

Figure 3. The net contribution of each stage to the cumulative fossil fuel energy use (MJ).

3.4. Water Depletion

This category indicates the total water use from different water sources: lakes, rivers, and wells. In this study, total water depletion was a process credit, i.e., avoided water depletion of 0.015 m^3 per functional unit. This credit is an indirect water-saving resulting from displacing synthetic fertilizers with the slurry-biochar mixture. The difference between land application impacts on water depletion, - 0.112 m^3, and total impact, 0.111 m^3, is attributed to all the energy-positive stages in the scenario. However, the savings accrued by displaced fertilizers outweighed the combined water depletion potential for these stages.

3.5. Marine Eutrophication

This mid-point impact category expresses the amounts of nutrients emitted, expressed in units of kg N equivalent, which potentially reach marine water causing eutrophic conditions. The studied scenario had a net positive (a burden) of marine eutrophication, 88.5% of which is attributed to the slurry-biochar land application. This results from nitrate (NO_3) leaching, and NH_3 emissions following land application. Considering the full lifecycle, 80% of marine eutrophication potential is attributed to NO_3 leaching, while the remainder is due to NH_3 emissions. The eutrophying effect of NH_3 occurs through the formation of acid rains that deposit back in water bodies causing N enrichment. Swine houses and pre-separation storage are together responsible for 5.7% of total marine eutrophication potential due to their NH_3 emissions.

3.6. Freshwater Eutrophication

Given that P is the limiting nutrient for most freshwater bodies, introducing P to rivers and lakes results in eutrophying conditions. In this study, 98.5% of total freshwater eutrophication potential is attributed to the impacts of slurry-biochar application. The leaching of 10% of P from the slurry is responsible for this impact.

3.7. Model Sensitivity to Thermochemical Conversion Parameters

To improve understanding of the implications of the proposed thermochemical conversion system (drying-gasification-boiler) on swine manure treatment, the conversion parameters, i.e., hot-gas efficiency (HGE) and boiler efficiency was varied to represent two additional alternatives. The first set represents low-efficiency conditions: HGE and boiler efficiencies at 60% and 68%, respectively. The second, a high-efficiency scenario, shows HGE and boiler efficiency at 80% and 88%, respectively. Figure 4 shows the impacts of the performance levels on the gasification-boiler stage. A 10% increase in the performance of both the gasifier and the boiler yielded a decrease in the GWP for this stage by 2.6 kg CO_2-eq (from 2.4 kg CO_2-eq to -0.2 kg CO_2-eq), and a corresponding increase in fossil fuel energy saving by 40.5 MJ (from -11.9 MJ to - 52.4 MJ). A 10% drop in the efficiencies increased GWP, from 2.4 to 4.6 kg CO_2-eq, and a change from a fossil fuel energy use of -11.9 MJ to an energy expenditure of 23.5 MJ. The non-linear response in the efficiency scenarios is because the overall efficiency for the gasification-boiler is the product of the conversion and the boiler efficiencies. No noticeable changes were observed in the other impact categories with changes in the efficiencies.

For the full treatment system, increasing the thermochemical conversion efficiency by 10% relative to the baseline led to a 1.5% decrease in GWP and an increase of the fossil fuel savings of 52.4 MJ. These findings suggest that the range of sensitivity for the thermal conversion system has a marginal impact on the GWP for the entire management scenario. It is worth noting, however, that the combined GWP for the separation, drying, and gasification-boiler stages, 0.89 kg CO_2-eq, is lower than the difference in GWP between pre-separation storage, 58.2 kg CO_2-eq, and post-separation storage, 37.3 kg CO_2-eq. The separation-drying-gasification-boiler combination can be considered an emission reduction measure for manure storage. From an energy perspective, the gasification system has a beneficial impact on the total energy use in manure management, notwithstanding high energy requirements for drying. Improvements to thermal conversion efficiency (gasification and syngas firing) combined with improvements to the drying technology can significantly improve the overall environmental performance for swine manure management via thermochemical conversion.

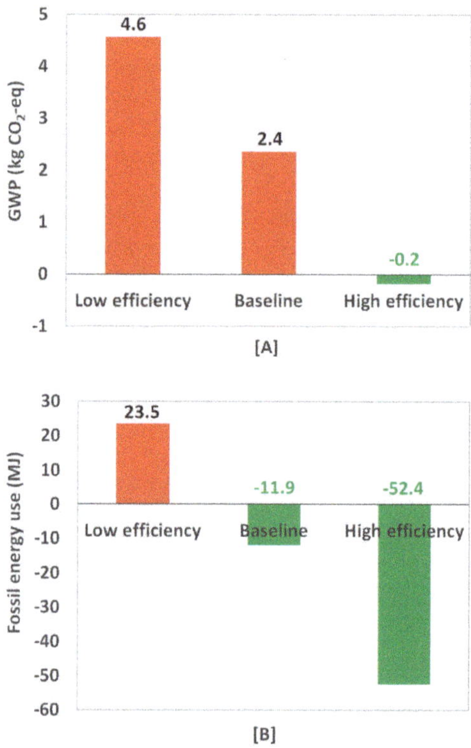

Figure 4. Impacts of gasification-boiler performance on (A) global warming potential (GWP 100a), and (B) fossil fuel energy use (MJ).

4. Implications of the Study

The findings in this investigation contribute to the ongoing discussion on manure management best practices. Given the swine manure composition and the management techniques practiced on the farm, the thermochemical conversion is a challenging technique to dispose of wet swine manure. Improvements to the solid-liquid separation system that reduces moisture content in solid fraction can potentially improve the environmental process of the proposed system.

From GHG emissions and energy use perspectives, the land application step of swine manure management appears beneficial due to the credits from replacing synthetic fertilizer consumption. However, in regions of intensive swine production where manure land application regulations are strict, thermochemical conversion can be an alternative approach to utilize manure. Adopting innovative sludge drying technologies, i.e., biodrying technology [16], can significantly reduce the drying energy demand, and consequently, the GHG emissions. Also, more studies towards a better understanding of biochar agronomic value could potentially help in incentivizing the thermochemical conversion of swine manure solids.

5. Conclusions

- Swine manure liquid storage (before and after solid-liquid separation) contributed 57.5% of the GHG emissions for the entire proposed manure management scenario.

- Swine manure solids separation, drying, and gasification resulted in a net energy expenditure of 12.3 MJ for each functional unit (treatment of 1 metric ton of manure slurry).

- The high energy demand associated with manure drying represented greater energy requirement than the energy produced from the gasification/boiler stage.

- Land application of slurry-biochar mixture is credited with 5.9 kg CO_2-eq in avoided GHG emissions, and 135 MJ of avoided fossil fuel energy use resulting from avoided synthetic fertilizer production.

- Improvements to drying and thermochemical conversion efficiencies may further significantly reduce fossil fuel use in thermochemical manure management.

Author Contributions: All authors contributed equally and significantly to the data collection, manuscript writing, and editing.

Funding: USDA National Institute of Food and Agriculture. Agriculture and Food Research Initiative Competitive Grant No. 2011-68002-30208.

Acknowledgments: Acknowledgment is due to the Agriculture and Food Research Initiative for their support of the Competitive Grant No. 2011-68002-30208 from the USDA National Institute of Food and Agriculture. Their financial support is much appreciated.

Conflicts of Interest: The authors declare no conflict of interest.

References

1. Dong, H.; Mangino, J.; McAllister, T.; Have, D. Emissions from livestock and manure management. In *IPCC Guidelines for National Greenhouse Gas Inventories 4*; 2006; Available online: https://www.alice.cnptia.embrapa.br/handle/doc/1024926 (accessed on 17 October 2019).

2. Bockstaller, C.; Vertès, F.; Fiorelli, J.; Rochette, P.; Aarts, H. Tools for evaluating and regulating nitrogen impacts in livestock farming systems. *Adv. Anim. Biosci.* **2014**, *5*, 49–54. [CrossRef]

3. Reddy, P.P. Agriculture as a Source of GHGs. In *Climate Resilient Agriculture for Ensuring Food Security*; Springer: New Delhi, India, 2015; pp. 27–42.

4. Kurukulasuriya, P.; Rosenthal, S. Climate change and agriculture: A review of impacts and adaptations. *World Bank* **2013**, *1*, 91. Available online: https://openknowledge.worldbank.org/bitstream/handle/10986/16616/787390WP0Clima0ure0377348B00PUBLIC0.pdf?sequence=1&isAllowed=y (accessed on 17 October 2019).

5. EPA 430-R-14-003. *Inventory of U.S. Greenhouse Gas Emissions and Sinks: 1990-2012*; Environmental Protection Agency (EPA): Washington, DC, USA, 2014.

6. Fangueiro, D.; Coutinho, J.; Chadwick, D.; Moreira, N.; Trindade, H. Effect of Cattle Slurry Separation on Greenhouse Gas and Ammonia Emissions during Storage. *J. Environ. Qual.* **2008**, *37*, 2322–2331. [CrossRef] [PubMed]

7. Fitzgerald, S.; Kolar, P.; Classen, J.; Boyette, M.; Das, L. Swine manure char as an adsorbent for mitigation of p cresol. *Environ. Prog. Sustain. Energy* **2015**, *34*, 125–131. [CrossRef]

8. Lentz, Z.A.; Classen, J.; Kolar, P. Thermochemical Conversion: A Prospective Swine Manure Solution for North Carolina. *Trans. ASABE* **2017**, *60*, 591–600. [CrossRef]

9. *ISO 14040 Environmental Management–Life Cycle Assessment–Principles and Framework*; International Organization for Standardization: London, UK; British Standards Institution: Hutchings, UK, 2006.

10. Wu, H.; Hanna, M.A.; Jones, D.D. Life cycle assessment of greenhouse gas emissions of feedlot manure management practices: Land application versus gasification. *Biomass Bioenergy* **2013**, *54*, 260–266. [CrossRef]

11. Lijó, L.; González-García, S.; Bacenetti, J.; Fiala, M.; Feijoo, G.; Lema, J.M.; Moreira, M.T. Life Cycle Assessment of electricity production in Italy from anaerobic co-digestion of pig slurry and energy crops. *Renew. Energy* **2014**, *68*, 625–635. [CrossRef]

12. Rodriguez-Verde, I.; Regueiro, L.; Carballa, M.; Hospido, A.; Lema, J.M. Assessing anaerobic co-digestion of pig manure with agroindustrial wastes: The link between environmental impacts and operational parameters. *Sci. Total Environ.* **2014**, *497*, 475–483. [CrossRef]

13. Pehme, S.; Veromann, E.; Hamelin, L. Environmental performance of manure co-digestion with natural and cultivated grass—A consequential life cycle assessment. *J. Clean. Prod.* **2017**, *162*, 1135–1143. [CrossRef]

14. Roberts, K.G.; Gloy, B.A.; Joseph, S.; Scott, N.R.; Lehmann, J. Life cycle assessment of biochar systems: Estimating the energetic, economic, and climate change potential. *Environ. Sci. Technol.* **2010**, *44*, 827–833. [CrossRef]

15. Sharara, M.A.; Sadaka, S.S. Opportunities and Barriers to Bioenergy Conversion Techniques and Their Potential Implementation on Swine Manure. *Energies* **2018**, *11*, 957. [CrossRef]

16. Ecoinvent Centre. The Ecoinvent Database. Available online: https://www.ecoinvent.org/database/database.html (accessed on 11 October 2019).

17. Bulle, C.; Margni, M.; Patouillard, L.; Boulay, A.; Bourgault, G.; De Bruille, V.; Cao, V.; Hauschild, M.; Henderson, A.; Humbert, S.; et al. IMPACT World+: A globally regionalized life cycle impact assessment method. *Int. J. Life Cycle Assess.* **2019**, *24*, 1653–1674. [CrossRef]

18. IPCC. *Climate Change 2013: The Physical Science Basis. Contribution of Working Group I to the Fifth Assessment Report of the Intergovernmental Panel on Climate Change*; Stocker, T.F., Qin, D., Plattner, G., Allen, S.K., Xia, Y., Eds.; Cambridge University Press: Cambridge, UK; New York, NY, USA, 2013; p. 1535. Available online: http://www.climatechange2013.org/images/report/WG1AR5_Frontmatter_FINAL.pdf (accessed on 17 October 2019).

19. ASABE ASAE D384.2. *Manure Production and Characteristics*; American Society of Agricultural and Biological Engineers: St. Joseph, MI, USA, 2005.

20. Mikkelsen, M.; Gyldenkærne, S.; Poulsen, H.; Olesen, J.; Sommer, S. *Emissions of Ammonia, Nitrous Oxide and Methane from Danish Agriculture 1985–2002: Methodology and Estimates (no. 231)*; National Environmental Research Institute (NERI), Ministry of Environment: Frontlinien, Denmark, 2006.

21. Wesnaes, M.; Wenzel, H.; Petersen, B.M. *Life Cycle Assessment of Slurry Management Technologies*; Danish Ministry of the Environment/Danish Environmental Protection Agency: Copenhagen, Denmark, 2009.

22. Nguyen, T.L.T.; Hermansen, J.E.; Mogensen, L. Fossil energy and GHG saving potentials of pig farming in the EU. *Energy Policy* **2010**, *38*, 2561–2571. [CrossRef]

23. Moller, H.B.; Lund, I.; Sommer, S.G. Solid–liquid separation of livestock slurry: Efficiency and cost. *Bioresour. Technol.* **2000**, *74*, 223–229. [CrossRef]

24. Day, D.; Funk, T.; Hatfield, J.; Stewart, B. *Processing Manure: Physical, Chemical and Biological Treatment, in Animal Waste Utilization: Effective Use of Manure as a Soil Resource*; CRC Press: Boca Raton, FL, USA, 1998.

25. Sadaka, S.; Ahn, H. Evaluation of a biodrying process for beef, swine, and poultry manures mixed separately with corn stover. *Appl. Eng. Agric.* **2012**, *28*, 457–463. [CrossRef]

26. Hjorth, M.; Christensen, K.V.; Christensen, M.L.; Sommer, S.G. Solid–liquid separation of animal slurry in theory and practice. A review. *Agron. Sustain. Dev.* **2010**, *30*, 153–180. [CrossRef]

27. Sharara, M.A.; Sadaka, S.; Costello, T.A.; VanDevender, K. Influence of aeration rate on the physio-chemical characteristics of biodried dairy manure-wheat straw mixture. *Appl. Eng. Agric.* **2012**, *28*, 407–415. [CrossRef]

28. Hospido, A.; Teresa Moreira, M.; Fernandez-Couto, M.; Feijoo, G. Environmental Performance of a Municipal Wastewater Treatment Plant. *Int. J. Life Cycle Assess.* **2004**, *9*, 261–271. [CrossRef]

29. Jensen, L.S. Animal Manure Residue Upgrading and Nutrient Recovery in Biofertilisers. Animal Manure Recycling: Treatment and Management, 271–294. Available online: https://openknowledge.worldbank.org/handle/10986/16616 (accessed on 27 January 2015).

30. Ro, K.S.; Cantrell, K.B.; Hunt, P.G. High-temperature pyrolysis of blended animal manures for producing renewable energy and value-added biochar. *Ind. Eng. Chem. Res.* **2010**, *49*, 10125–10131. [CrossRef]

31. Wnetrzak, R.; Kwapinski, W.; Peters, K.; Sommer, S.G.; Jensen, L.S.; Leahy, J.J. The influence of the pig manure separation system on the energy production potentials. *Bioresour. Technol.* **2013**, *136*, 502–508. [CrossRef]

32. Arena, U. Process and technological aspects of municipal solid waste gasification. A review. *Waste Manag.* **2012**, *32*, 625–639. [CrossRef] [PubMed]

33. Pa, A.; Bi, X.T.; Sokhansanj, S. A life cycle evaluation of wood pellet gasification for district heating in British Columbia. *Bioresour. Technol.* **2011**, *102*, 6167–6177. [CrossRef] [PubMed]

34. Zaman, A.U. Life cycle assessment of pyrolysis–gasification as an emerging municipal solid waste treatment technology. *Int. J. Environ. Sci. Technol.* **2013**, *10*, 1029–1038. [CrossRef]

35. Daugaard, D.E.; Brown, R.C. Enthalpy for pyrolysis for several types of biomass. *Energy Fuels* **2003**, *17*, 934–939. [CrossRef]

36. Evans, M.R.; Jackson, B.E.; Popp, M.; Sadaka, S. Chemical Properties of Biochar Materials Manufactured from Agricultural Products Common to the Southeast United States. *HortTechnology* **2017**, *27*, 16–23. [CrossRef]

37. Laboski, C.A.; Peters, J.B.; Bundy, L.G. *Nutrient Application Guidelines for Field, Vegetable, and Fruit Crops in Wisconsin*; Division of Cooperative Extension of the University of Wisconsin-Extension: Madison, WI, USA, 2006.

38. Pehme, S. *Life Cycle Inventory & Assessment Report: Dairy Cow Slurry Biogas with Grass as an External C Source*; Baltic Manure Report; Estonia, 2013; Available online: http://www.balticmanure.eu/download/Reports/wp5_report_estonia_web.pdf (accessed on 17 October 2019).

39. Delin, S.; Stenberg, B.; Nyberg, A.; Brohede, L. Potential methods for estimating nitrogen fertilizer value of organic residues. *Soil Use Manag.* **2012**, *28*, 283–291. [CrossRef]

40. Misselbrook, T.H.; Nicholson, F.A.; Chambers, B.J. Predicting ammonia losses following the application of livestock manure to land. *Bioresour. Technol.* **2005**, *96*, 159–168. [CrossRef]

41. Sommer, S.G.; Hutchings, N.J. Ammonia emission from field applied manure and its reduction. *Eur. J. Agron.* **2001**, *15*, 1–15. [CrossRef]

42. Hutchings, N.J.; ten Hoeve, M.; Jensen, R.; Bruun, S.; Søtoft, L.F. Modelling the potential of slurry management technologies to reduce the constraints of environmental legislation on pig production. *J. Environ. Manag.* **2013**, *130*, 447–456. [CrossRef]

43. Buckley, K.E.; Mohr, R.M.; Therrien, M.C. Yield and quality of oat in response to varying rates of swine slurry. *Can. J. Plant Sci.* **2010**, *90*, 645–653. [CrossRef]

44. Rochette, P.; Angers, D.A.; Chantigny, M.H.; Bertrand, N.; Côté, D. Carbon dioxide and nitrous oxide emissions following fall and spring applications of pig slurry to an agricultural soil. *Soil Sci. Soc. Am. J.* **2004**, *68*, 1410–1420. [CrossRef]

 *energies*

Article

The Role of Biorefinery Co-Products, Market Proximity and Feedstock Environmental Footprint in Meeting Biofuel Policy Goals for Winter Barley-to-Ethanol

Sabrina Spatari [1,2,*], Alexander Stadel [1], Paul R. Adler [3], Saurajyoti Kar [1], William J. Parton [4], Kevin B. Hicks [5], Andrew J. McAloon [5] and Patrick L. Gurian [1]

[1] Department of Civil, Architectural, and Environmental Engineering, Drexel University, Philadelphia, PA 19104, USA; ajs83@drexel.edu (A.S.); sk3586@drexel.edu (S.K.); plg28@drexel.edu (P.L.G.)
[2] Civil and Environmental Engineering, Technion, Israel Institute of Technology, 3200003 Haifa, Israel
[3] Pasture Systems and Watershed Management Research Unit, The United States Department of Agriculture-The Agricultural Research Service, University Park, PA 16802, USA; paul.adler@usda.gov
[4] Natural Resource Ecology Laboratory, Colorado State University, Fort Collins, CO 80523, USA; william.parton@colostate.edu
[5] Sustainable Biofuels and CoProducts Research Unit, The United States Department of Agriculture-The Agricultural Research Service, Wyndmoor, PA 19038, USA; kevin.hicks@ars.usda.gov (K.B.H.); andrew.mcaloon@ars.usda.gov (A.J.M.)
* Correspondence: ssabrina@technion.ac.il or spatari@drexel.edu

Received: 12 March 2020; Accepted: 25 April 2020; Published: 3 May 2020

Abstract: Renewable fuel standards for biofuels have been written into policy in the U.S. to reduce the greenhouse gas (GHG) intensity of transportation energy supply. Biofuel feedstocks sourced from within a regional market have the potential to also address sustainability goals. The U.S. Mid-Atlantic region could meet the advanced fuel designation specified in the Renewable Fuel Standard (RFS2), which requires a 50% reduction in GHG emissions relative to a gasoline baseline fuel, through ethanol produced from winter barley (*Hordeum vulgare* L.). We estimate technology configurations and winter barley grown on available winter fallow agricultural land in six Mid-Atlantic states. Using spatially weighted stochastic GHG emission estimates for winter barley supply from 374 counties and biorefinery data from a commercial dry-grind facility design with multiple co-products, we conclude that winter barley would meet RFS2 goals even with the U.S. EPA's indirect land use change estimates. Using a conservative threshold for soil GHG emissions sourced from barley produced on winter fallow lands in the U.S. MidAtlantic, a biorefinery located near densely populated metropolitan areas in the Eastern U.S. seaboard could economically meet the requirements of an advanced biofuel with the co-production of CO_2 for the soft drink industry.

Keywords: biofuel policy; life cycle assessment; GHG mitigation; energy security; indirect land use change

---

1. Introduction

Much has been written, debated, and recast over what constitutes a "green" or environmentally sustainable biofuel. The development of renewable and low carbon fuel standards has shaped the way in which life cycle assessment (LCA) tools are used to judge a fuel's "greenness," specifically for addressing greenhouse gas (GHG) mitigation [1]. For example, LCA methods dictate choice of temporal, spatial, and process boundary selection [2,3], and treatment of co-products [4,5]. Crop-based fuels such as ethanol derived from corn have come under scrutiny for disrupting food markets and for the risks they pose towards indirect deforestation and carbon emissions, known as indirect land use change (iLUC) [6,7]. This is because the starch-based crops used to produce ethanol via wet or dry milling processes are grown on prime agricultural lands that compete with food, and thereby disrupt food prices. However, recent studies have shown the penalty of iLUC to be small yet complex when grains like corn and soybean are traded in global markets [8]. Ways of lessening or circumventing the iLUC CO_2 "penalty" include using waste or growing energy crops, whether starch or lignocellulosic, on marginal lands, or growing those crops when land has been historically fallow. Winter double-cropping can make use of fallow land because in certain agricultural regions it is seldom economical to grow winter crops (e.g., winter wheat) for food or feed, and historic records show that farmers have left the land fallow.

If a winter crop used for energy does not disrupt summer crop yield in a rotation, the indirect CO_2 penalty may be minimized. Furthermore, when winter double crops are grown using best management practices (BMP) such as conservation tillage and with optimized nitrogen fertilizer application, they may provide water quality benefits through reducing the NO_3 leaching and runoff from agricultural fields [9] that contribute to seasonal hypoxic zones [10]. Jayasundara et al. [11] reported improved fertilizer uptake efficiency and reduced gaseous and leached losses of NO_3–N through incorporating BMPs in corn–soybean–winter wheat rotations in Ontario, Canada. Therefore, winter small grains such as barley could, in the interim while lignocellulose-based fuel technologies mature, be promising agricultural feedstocks for low carbon fuels that promote energy security and rural economic development—ingredients necessary for deeming a fuel "sustainable."

A suite of biofuels derived from non-corn starch-based feedstocks converted to ethanol that meet a 50% reduction in life cycle GHG emissions relative to gasoline could meet the advanced fuels designation under the Renewable Fuel Standard (RFS2) of the Energy Independence and Security (EISA) Act of 2007 [12]. Barley-to-ethanol produced under select operating conditions was proposed by the U.S. Environmental Protection Agency (EPA) as a renewable and advanced biofuel [13] as was grain sorghum-to-ethanol [14]. In the U.S. Mid-Atlantic region and surrounding states that could support winter cropping (Virginia, North Carolina, Maryland, Delaware, Pennsylvania, and Kentucky), barley (*Hordeum vulgare* L.) can be grown as a winter double-crop. Other double-crops like camelina (*Camelina sativa* L.) and field pennycress (*Thlaspi arvense* L.) oilseeds have been proposed as feedstock for biofuel that can provide ecosystem services if integrated into corn–soybean crop rotations [15]. For example, Krohn and Fripp [16] estimate that biodiesel made from spring camelina grown as a double crop achieved a 40% to 60% reduction in life cycle GHGs, assuming no iLUC effects. Tabatabaie et al. examined soil parameters affecting camelina as a winter crop for biodiesel production in Western U.S. states to estimate multiple life cycle impact assessment (LCIA) metrics [17].

Starch-based cereal grains such as barley [18], wheat (*Triticum aestivum* L.) [19], and sorghum (*Sorghum bicolor* (L.) Moench) [20] can be converted to ethanol using commercial technology. In most cases, results from technoeconomic analyses (TEA) have concluded that stable or low feedstock costs along with rising fossil fuel prices are needed to render grain alcohol competitive. Additionally, value-added co-products can improve biorefinery economics. Environmental life cycle assessments have been undertaken for agricultural commodities like barley grain [21] and straw [22], and for ethanol produced from grain sorghum [23], wheat [24], and cellulosic feedstocks [25] to understand the life cycle impacts related to global warming, eutrophication and acidification potential. To date, LCA of winter barley grain as a feedstock to produce ethanol has not been studied.

Malca and Freire [24] examined uncertainties in life cycle GHG emissions for wheat-to-ethanol in Europe, and showed that the highest and most variable GHG emissions related first to soil organic carbon (SOC), and second to nitrous oxide emissions during crop production. The contribution of SOC to the life cycle of crops strongly depends on the land use history of the growing region as noted in previous biofuel research [23,26–28]; thus, it is best modeled using Tier 3 IPCC (Intergovernmental Panel on Climate Change) biogeochemical or process models to predict the consequences of SOC and N_2O emissions when winter crops are introduced into a rotation. Tier 3 methods have been used to predict soil N_2O emissions to understand policy compliance for barley in advanced biofuel and low carbon fuel standard policy frameworks [29]. The biogeochemical models used in Tier 3 approaches require many detailed input parameters that cannot be known precisely. To address this uncertainty, Gao et al. [29] explored the use of a "margin of safety" approach in which emissions estimates are based not on the central tendency of model results, but on plausible upper bounds, such that one can have confidence that actual emissions will in all likelihood be lower than the estimated value (see Springborn et al. [30] for a discussion of the theoretical basis of margin of safety approaches).

The overall objective of this paper is to evaluate the environmental conditions for meeting advanced fuel designation for biofuels produced within the vicinity of high population centers on the U.S. East Coast using regionally available winter barley and commercial dry-grind technology. Using a life cycle modeling approach, we investigate alternative co-product scenarios and policy implications for meeting the advanced fuel designation in the Mid-Atlantic region of the U.S. using winter barley in corn (*Zea mays* L.)–soybean (*Glycine max* L.) rotations through examination of attributional and consequential LCA approaches and uncertainty. Prior research has used LCA with Monte Carlo simulation to understand the sources and effects of uncertain model parameters on GHG emissions in grain [24,31] and lignocellulosic feedstocks and technology [32] used to produce ethanol transport fuel. Moreover, LCA studies have demonstrated the importance of biomass logistics and scale [33–35], spatial aspects [36], and uncertainty [37] in feedstock supply to biorefineries on life cycle environmental impacts. We investigate the role of biorefinery proximity to end-markets for co-products in relation to meeting RFS2 advanced biofuel policy requirements, while considering the uncertainty in soil GHG emissions in the winter barley-to-ethanol (WBE) life cycle using Monte Carlo methods. We compare WBE to grain sorghum-to-ethanol, also a starch-based fuel substitute for gasoline to compare each pathway's compliance with RFS2.

2. Materials and Methods

2.1. System Boundary Definition

We use life cycle assessment (LCA) following ISO standards [38] to evaluate scenarios for meeting the advanced fuel designation under RFS2 policy in the U.S. Mid-Atlantic region. We focus on two LCIA metrics, the 100-year global warming potential (GWP) to examine RFS2 policy compliance and cumulative energy demand (CED), a metric that measures renewable and nonrenewable energy inputs in a product life cycle and is relevant to evaluating proposed renewable fuel pathways as studied in prior bioenergy systems [39,40]. Criteria specified by the U.S. EPA for meeting advanced fuel standing require that a non-corn feedstock fuel's life cycle GHG emissions be 50% lower than the 2005 gasoline baseline, which is set to 93 g CO_2e MJ^{-1}. We evaluate winter barley grown in six states (Figure 1) that could supply the Mid-Atlantic market under four biorefinery scenarios (Table 1) that include up to four co-products. The system boundary (Figure 2) includes all processes for growing, harvesting, and transporting winter barley grain to the biorefinery, conversion to ethanol and co-products, transporting, and distributing the fuel, and consuming the fuel. The functional unit defined is 1 MJ of denatured ethanol product (E98). We evaluate GHG emissions using the 100-year global warming potential (GWP) where CO_2, CH_4, and N_2O have a GWP of 1, 25, and 298 g CO_2 equivalents (CO_2e), respectively, based on the IPCC's fourth Assessment Report (AR4) [41].

Winter barley is typically planted in late October to early November and harvested in late May to mid-June. In the Mid-Atlantic region, the most widely deployed acres are fallow during winter of the two-year corn and soybean rotation. Farmers in the Chesapeake Bay watershed have grown winter cover and double crops through incentive programs to reduce nutrient leaching for many years. Moreover, agricultural crops, including winter double crops and lignocellulosic feedstocks have been proposed as strategies for reducing nutrient and sediment runoff from farm fields, thereby improving the water quality of the Chesapeake Bay [42]. Thus, winter double-crops such as winter barley may be environmentally and socioeconomically attractive to growers in the Mid-Atlantic region since adding them to the agricultural landscape could improve water quality and create economic opportunities for growers, while also meeting EISA policy objectives.

The technology used to convert winter barley to ethanol [18] is similar to commercial corn dry-grind milling [43,44], which produces a fuel (ethanol), a protein co-product (barley protein), and in the case of winter barley, energetic and chemical commodities from the barley hulls and fermentative CO_2, respectively. The ethanol biorefinery for this study was based on a 245 million liter per year facility that was under construction between 2010 and 2011 by the company Osage Bio Energy in Hopewell, Virginia, near the Richmond metropolitan area. In addition to fuel ethanol, the biorefinery was designed to co-produce barley protein meal, process steam from barley hulls, and food-grade fermentative CO_2 for the beverage industry, all value-added products that improve biorefinery economics and life cycle environmental performance (Table 1).

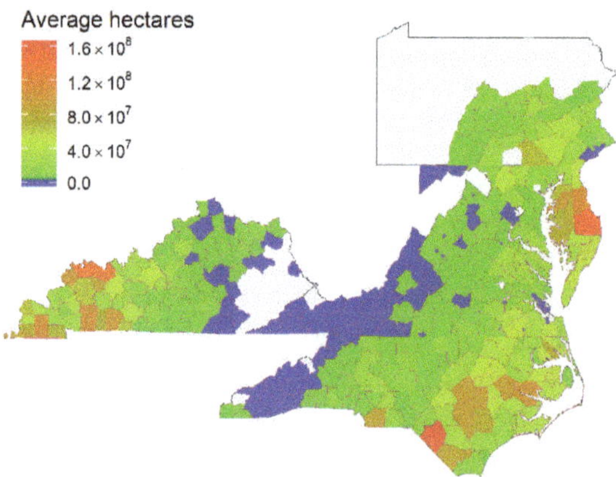

Figure 1. Winter fallow land for potential barley production in the U.S. Mid-Atlantic region and Kentucky.

Table 1. Winter barley-to-ethanol (WBE) production scenarios with alternative co-product crediting examined through attributional LCA (life cycle assessment) boundary.

Scenario	Major Assumptions
WBE1	Biorefinery includes co-product crediting for: fermentative CO_2 capture for the beverage industry; barley protein meal; onsite steam production; barley hull waste pelleting for cofiring with coal for power generation.
WBE2	Biorefinery includes co-product crediting for: barley protein meal; onsite steam production; barley hull waste pelleting for cofiring with coal for power generation.
WBE3	Biorefinery includes the production of co-products (barley protein meal, onsite steam production, and barley hull waste pelleting for coal cofiring). A co-product credit is only assigned to the barley protein meal.
WBE4	Biorefinery includes the production of three co-products (barley protein meal, onsite steam production, and barley hull waste pelleting for coal cofiring). No credits are applied for the avoided co-products.

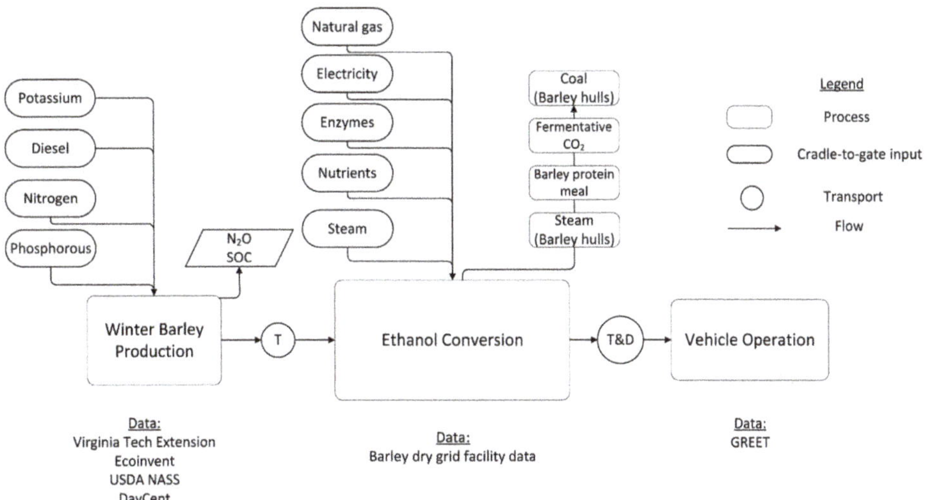

Figure 2. System boundary for the life cycle of winter barley-to-ethanol.

2.2. Life Cycle Inventory Analysis

Data for constructing the life cycle inventory (LCI) model include crop production, transportation of the feedstock to the biorefinery, fuel conversion, transportation and distribution of denatured ethanol, and the combustive emissions for using the fuel. The LCI model was constructed using SimaPro 8.4 [45] with agronomic inputs specified by Thomason [46], nitrous oxide emissions and changes in soil organic carbon (SOC) generated using the DayCent model [47], process–guarantee data from a starch-based biorefinery completed in Hopewell, Virginia, in 2011, and vehicle in-use emissions from the GREET 1 model [48]. Uncertainty in the most sensitive parameter inputs were considered and integrated into the LCI model using Monte Carlo simulation.

2.2.1. Feedstock Production

To support a 245 million liter ethanol facility in the Mid-Atlantic U.S. winter barley can be grown on available annual winter fallow land ranging from 158,000 to 195,000 ha in Virginia, North Carolina, Maryland, and Delaware, and thus support 0.68 million dry metric tons year^{-1} (31.3 million bushels year^{-1}) of winter barley. We also consider available winter fallow land in Pennsylvania and Kentucky (Figure 1). The winter crop is planted in October (following corn) and harvested in mid and late June (prior to soybean) in a two-year rotation using reduced tillage methods. U.S. Department of Agriculture (USDA) crop acreage data from 2008 [49] were evaluated to estimate fallow land available within a 80 and 160 km radius of the proposed Hopewell, Virginia, biorefinery. Between 0.74 and 1.47 million dry metric tons of winter barley could be grown on winter fallow lands within the existing corn, soybean, hay, and converted tobacco croplands.

The LCI includes addition of N, P, and K nutrients, and energy for field preparation and feedstock harvesting. Preharvest nutrient additions described by Thomason [46] specify a winter barley yield of 5380 kg ha^{-1} and application of fertilizers in the following proportions: nitrogen (N), preplant urea (12 kg N ha^{-1}) and spring liquid urea ammonium nitrate (70 kg N ha^{-1}); phosphate (P), diammonium phosphate (23 kg N ha^{-1}, 58 kg P ha^{-1}); potassium, (52 kg K ha^{-1}); and lime (0.74 metric tons ha^{-1}). Lubricant, oil, and diesel fuel inputs are applied in preharvest (12.7 L diesel ha^{-1}) and in harvest (16.9 L diesel ha^{-1}) portions. Addition of herbicides are low and deemed to be outside of the cutoff criteria for the system boundary and their respective emissions were not analyzed.

2.2.2. Winter Barley Soil GHG Emissions

We estimate N_2O and SOC GHG emissions resulting from the introduction of winter barley on existing winter fallow land in six states (Virginia, North Carolina, Maryland, Delaware, Pennsylvania, and Kentucky). These states represent a feedstock supply radius larger than the 80 to 160 km radius of Hopewell, Virginia. The goal of selecting a wider winter barley growing region was to capture a larger extent of soil and climate region, which influences N_2O emissions and SOC. N_2O is emitted from agricultural soils directly to air through the process of nitrification/denitrification, and also indirectly following leaching into nearby water sources and reaction of NOx, NH_3, and water. With the exception of a few studies [23,27,50,51], much prior grain-to-alcohol LCA literature (e.g., Hsu et al. [31] and Wang et al. [52]) used Tier 1 IPCC [53] methods to estimate direct and indirect N_2O emissions. Tier 1 approaches to N_2O and SOC estimation do not distinguish among different soil types, which would show variation in denitrification, and they do not integrate land use history, which impacts SOC.

We use the biogeochemical model DayCent, a Tier 3 IPCC method, to estimate incremental N_2O emissions and SOC associated with the incorporation of winter barley into the corn–soybean rotation. We estimate changes to N_2O emissions on 374 counties over 20 years resulting from integrating winter barley into a four-year cycle that alternates between wheat (W), fallow (F), barley (B), fallow (F) winter periods. The barley is planted following corn and harvested prior to soybean planting and growth during summer months. The incremental changes to the four-year barley rotation are calculated by taking the difference in N_2O and SOC emissions between the barley case (WFBF) and the no-barley case (WFFF). Total N_2O emissions are the sum of direct and indirect sources. Direct N_2O is an output from DayCent, while indirect N_2O is determined by applying IPCC [53] methods to DayCent output of volatile (NH_3, NO) and leached (NO_3) nitrogen. Reduced tillage methods are common in this region and we assume all introduced winter barley land is under this type of soil management. Land under some type of tillage would have higher rates of soil carbon loss and consequently lower gains after winter barley is added to the rotation. Barley straw is not harvested. If the barley straw were harvested, the rate of soil carbon gain after winter barley addition to the rotation would also be lower. Currently, the straw market in Virginia is small (15–20% of acres, down from 35% during high demand from the construction industry), which supports the assumption of leaving the straw on the land, but large in Pennsylvania (99% of acreage with demand from animal bedding, landscape mulch, and mushroom compost). We estimate SOC as the average over 20 years. The rate and time of nitrogen fertilizer application can affect N_2O emissions. Nitrogen fertilizer rate and time of application for winter barley are based on data from university experiment stations in the Mid-Atlantic region. This rate of N application is about 0.454 kg N bu^{-1} ± 40%; our analysis assumes 0.454 kg N bu^{-1} expected barley yield. Farms using a higher rate of N application would have higher losses of N_2O. We apply 25% of N in the fall at planting and 75% in the spring with a one-time application, except in Virginia where spring N is applied in two equal applications. As noted above, the base crop rotation is assumed to be alternating years of corn followed by soybeans, with wheat planted after corn once every four years. We use a geographically weighted estimate of N_2O and change in soil organic carbon (SOC).

2.3. Barley Transport and Ethanol Conversion

The barley feedstock is transported from within a 120 km average one-way distance from farms to the biorefinery; we assume a 120 km one-way biomass transport distance by a 40-ton truck, and account for the return trip. A dry-grind process like the technology for converting corn to ethanol was modified to adjust for yield of ethanol from the starches and protein in winter barley. The conversion of barley to ethanol uses a cocktail of amylases, beta-glucanases, and beta-glucosidases to generate fermentable sugars from both kernel starch and beta-glucans, which are then converted to ethanol at high efficiency [18]. Steam is co-generated and recycled into the biorefinery operations, producing a co-product credit, which is assumed to displace power generated from the local electricity grid and natural gas heat needed for steam production. Carbon dioxide from fermentation can also be captured, liquefied, and sold as a food-grade CO_2 co-product at a significant additional investment. Distiller's dried grains and soulbles (DDGS), described here as barley protein meal, are co-produced and are assumed to displace soybean meal on an equivalent protein value basis. The barley protein meal was multiplied by the ratio of protein fraction between barley and soybean in a ratio of 33:48. Finally, the barley hulls are pelleted and sold as fuel to a neighboring coal-powered utility, to displace an energy equivalent quantity of coal. Proprietary data provided by Osage Bio Energy are used to model the fuel conversion process using ecoinvent [54] data with SimaPro 8.4 LCA software [45]. Inputs to fuel conversion include electricity and natural gas for process energy, chemical reactants, enzymes, and process and cooling water. We assume a facility lifespan of 10 years, and thus account for 10% of the physical infrastructure GHG emissions amortized per year. In total, the facility converts 680,000 dry metric tons of barley grain to 245 million liters of denatured ethanol annually.

Using system expansion described in ISO 14040 [38], the barley-to-ethanol product system was credited with avoided GHG emissions from co-products. We evaluate four co-product scenarios (Table 1) that include the following: the most optimistic case (WBE1) where animal protein feed, barley hull residues, and fermentative CO_2 displace products on the market; a scenario (WBE2) that does not include investment in the fermentative CO_2 product, given that it involves a high capital investment, but includes the animal protein and fossil energy displacement due to barley hull residues; a scenario (WBE3) that only includes animal protein meal displacement; and a scenario (WBE4) that includes production of co-products in scenario WBE2, but does not include displacement credits to demonstrate the most conservative attributional LCA operating conditions. The three scenarios are defined to reflect the effect of uncertainty in avoided product (0% to 100%), substitution of liquefied CO_2, DDGS, barley hull incineration to displace coal at an industrial coal furnace, and onsite steam production through combined heat and power (CHP) production. Table 2 summarizes the inputs of feedstock, chemicals, and energy on a functional unit basis (1 MJ).

Table 2. Cradle-to-gate input of feedstock, chemicals, and energy and co-products in the biorefinery. All data expressed per 1 MJ fuel produced.

Input	Quantity	Unit
Barley (feedstock)	59	kg
Enzymes	126	g
Liquid ammonia	65	g
Urea	65	g
Hydrous ammonia	122	g
Sulfuric acid	35	g
Inorganic chemicals	2.1	g
Lime	72.2	g
Yeast	0.71	g
Make-up water	167	L
Electrical utilities	7	kWh
Electricity for CO_2 capture (WBE1 Scenario only)	3	kWh
Externally sourced steam	11	kg
Onsite steam	40	kg
Natural gas (drying)	112	MJ
Natural gas (onsite steam)	121	MJ
Co-products:		
Barley hulls (coal and onsite steam displacement)	75	MJ
Barley protein meal (dry basis)	19	kg
CO_2	14	kg

2.4. Ethanol Transport, Distribution and Use

The LCI model accounts for transportation and distribution (29 km) of the denatured ethanol fuel (E98; 98% ethanol and 2% denaturant) separately from the barley hull (80 km), protein meal co-products (420 km), and fermentative CO_2 (80 km) by 40 ton trucks. The E98 fuel is assumed to be sold into U.S. East Coast markets, in place of ethanol imported from the Midwest, and therefore reducing transportation and distribution costs, which at high volume can range from 0.29 to 0.62 USD L^{-1} (1.30–2.80 USD gallon^{-1}) according to Wakeley et al. [55]. We model the emissions from combustion of the E98 during motor vehicle operation using the GREET 1 [48]. GREET 1 is used to model the Osage E98 fuel using a dedicated 2010 model-year ethanol vehicle (25 MPG, gasoline equivalent), with 2% (by volume) gasoline as denaturant, which accounts for upstream gasoline emissions in addition to the ethanol and vehicle in-use emissions.

2.5. LCA Model Uncertainty and Indirect Land Use Change Effects

We develop probability distributions from DayCent predictions of N_2O emissions and SOC change from the 30 year and 374 county data set. DayCent model simulations of N_2O emissions and SOC change are averaged over 30 years, and the county averages are fit to a normal distribution (see Supplemental Materials for assessment of goodness-of-fit). The overall mean is based on a weighted average of the 30 year county means with weights based on the estimated winter barley cropland area within the 374 county dataset. Figures S1–S3 (in Supplemental Materials) show the histograms and QQ–plots for N_2O, SOC change, and net soil GHG emissions. Monte Carlo sampling with 10,000 iterations is used to estimate life cycle GHG emissions by summing stochastic soil GHG emissions and foreground GHG emissions from all other life cycle inputs. A similar Monte Carlo sampling approach to uncertainty estimation is implemented in [56]. Table S5 summarizes statistics for N_2O emissions, and SOC change and life cycle GHGs for WBE1 to WBE4 over a 99 percent confidence interval. Scenarios described in Table 1 consider only the spatially weighted uncertainty from N_2O emissions and SOC change at the 99th percentile.

The LCA model we develop for winter barley assumes that the feedstock is grown on existing fallow land that does not compete with food crops for prime agricultural land. Therefore, our boundary definition examines the incremental effects of winter barley placed into the agricultural landscape of the region supplying a Mid-Atlantic biorefinery. This idealized boundary assumes no land competition and therefore does not include an estimate of iLUC CO_2 impacts on the fuel life cycle. There are two significant limitations to this assumption. The first assumption is that the barley crops attain the yields necessary to sustain the biorefinery feedstock demand without delaying the summer crop rotation. This assumption may not hold with the longer growing season in the northern part of the feedstock geographic boundary we examine. The second assumption is that there is no competition with other crops (e.g., winter wheat) for the fallow land in the growing region. Our assumptions may not hold given that the introduction of a new barley market could cause shifts in other commodities. Therefore, to understand possible iLUC effects of barley expansion, we compare our analysis with that conducted by the U.S. EPA [13], which used consequential LCA to estimate direct and indirect life cycle GHG emissions from expansion in domestic and international agricultural sectors using the Forestry and Agricultural Sector Optimization Model with Greenhouse Gases (FASOM-GHG) [57,58] and Food and Agricultural Policy Research Institute Center for Agricultural and Rural Development (FAPRI–CARD) [59] models, respectively. The approach used by the EPA includes co-product displacement crediting that is embedded within the economic models they use, but also aggregates spring and winter barley produced in different U.S. growing regions. We discuss our and the EPA's barley-to-ethanol LCA models along with sorghum-to-ethanol, another small grain whose conversion to ethanol also qualifies as a renewable fuel under RFS2, and the differences and limitations of applying EPA consequential LCA results to winter barley due to aggregating spring and winter barley. Table S5 (Supplemental Materials) summarizes the iLUC factors for the EPA cases for winter barley and sorghum.

3. Results and Discussion

3.1. Greenhouse Gas Emissions for Winter Barley-to-Ethanol

Winter barley-to-ethanol LCA results are presented in Table 3 and Figure 3. Table 3 reports the 99 percentile estimates for soil GHG emissions as these values incorporate a "margin of safety" that allows for confidence in Tier 3 estimates despite the uncertainties inherent in the modeling approach. Figure 3, which can more readily convey ranges, reports central tendencies, upper bounds, and lower bounds, in keeping with governmental guidance for reporting the results of benefits–cost assessments [60].

Life cycle GHG emissions for the winter barley biorefinery configurations show that WBE can meet the EPA's advanced fuel standing only when co-product credits are included (Table 3). The scenarios presented in Table 3 assume no variability in life cycle model parameters, but use the 99th percentile, incorporating a significant margin of safety, from the 374-county geographically weighted range of N_2O and SOC emissions. Barley production accounts for most life cycle emissions, mainly attributable to N fertilizer manufacturing. The CED fossil energy metric (Table S1) follows a similar pattern for all winter barley scenarios in that the case with fermentative CO_2 capture (WBE1) and full co-product crediting for displacing fossil energy with energy recovery from combusted barley hulls yields requires the least fossil energy input per unit of ethanol produced, 0.55 MJ MJ^{-1}. Moreover, the scenario without fermentative CO_2 capture but with energy recovery from barley hulls (WBE2), also requires less fossil energy input relative to each unit of ethanol produced 0.71 MJ MJ^{-1}. Scenarios WBE3, with only crediting of barley protein meal (1.05 MJ MJ^{-1}), and WBE4, with no co-product crediting (1.18 MJ MJ^{-1}), with the WBE4 scenario approaching the fossil energy well-to-wheel performance of gasoline, 1.2 ± 0.1 MJ MJ^{-1} [61]. This finding underscores the importance of co-products as noted in prior biofuel LCA studies [62] in reducing the environmental and resource demand impacts of biorefineries and it further demonstrates the importance of market proximity for the case of fermentative CO_2 capture.

Co-products greatly improve the carbon balance of WBE. Scenario WBE1 includes the maximum co-product credits possible. With fermentative CO_2 capture, 80% of the high volume of biogenic CO_2 can be captured and liquefied onsite to be sold as a food-grade CO_2 product. This CO_2 condensation recovery step requires an electrical input of 170 kWh ton^{-1} CO_2, but this incremental energy input favors both biorefinery economics and the GHG budget. The Osage biorefinery was designed to purchase steam produced from waste heat from nearby/co-located industrial facilities. The credit from steam amounts to approximately 6 g CO_2e MJ^{-1} (128 million kg CO_2e year^{-1}) of avoided steam that would otherwise be produced from natural gas. In addition to onsite steam, 57,700 metric tons year^{-1} of barley hulls could substitute for coal use in a nearby industrial coal furnace. At 17 MJ kg^{-1}, these hulls account for an additional credit of 973 million MJ of heat. The barley protein meal (BPM) co-product is evaluated as a protein substitute for soybean meal processed at an oil mill, and thus a soybean meal credit is applied to the LC GHG emissions profile (Table 3). The avoided soybean meal (48% protein) is adjusted to the fraction of protein in the BPM (33% protein). About 225,000 metric tons of barley protein meal are produced each year, at 10% moisture. Thus, the 202,000 metric tons year^{-1} of barley protein meal is assumed to displace and avoid production of 139,000 metric tons year^{-1} soybean meal.

The results (Table 3) document the expected life cycle GHG emissions for each phase of ethanol production. We assume the E98 fuel combusted in use emits the CO_2 sequestered during growth of harvested barley grain, and small quantities of CH_4 and N_2O, which are factored into life cycle GHG emissions. Fertilizer addition during crop production accounts for high emissions and fossil energy input (Table S1). However, the nitrous oxide reduction benefits of barley as a winter crop offsets much of these emissions. Gao et al. [29] discuss differences between the Tier 3 approach we apply and Tier 1 approaches, which are significant, particularly in regulatory standards that credit incremental improvement (reduction) in GHG emissions, like California's LCFS. Avoided product credits reduce the field to wheel GHG emissions of the winter barley to ethanol process by displacing steam generation, protein meal production, and adding a barley hull fuel to an industrial coal boiler.

Table 3. Life cycle greenhouse gas (GHG) emissions (g CO_2e MJ^{-1} fuel produced) for three winter barley-to-ethanol scenarios. Nitrous oxide and soil organic carbon (SOC) change reflect the 99th percentile of stochastic simulations from Figure 3.

	(WBE1)	(WBE2)	(WBE3)	(WBE4)
Feedstock Production, Collection, and Transport:				
Fertilizer (N)	11.6	11.6	11.6	11.6
Nitrous Oxide	7.0	7.0	7.0	7.0
SOC change	−5.9	−5.9	−5.9	−5.9
Farm operations (diesel)	2.2	2.2	2.2	2.2
Fertilizer (P)	2.1	2.1	2.1	2.1
Feedstock transport	2.1	2.1	2.1	2.1
Fertilizer (K_2O)	0.8	0.8	0.8	0.8
Lime	0.1	0.1	0.1	0.1
Ethanol Production and Distribution:				
Natural Gas	32.3	32.3	32.3	32.3
Electricity	12.3	9.3	9.3	9.3
Chemicals and enzymes	2.3	2.3	2.3	2.3
Co-product transport	1.7	1.6	1.6	1.6
Barley protein meal credit	−13.2	−13.2	−13.2	0
Barley hull coal substitution credit	−19.9	−19.9	0	0
Onsite steam credit	−5.9	−5.9	0	0
Liquid CO_2 recovery credit	−21.8	0	0	0
Fuel transport and distribution	0.4	0.4	0.4	0.4
Vehicle operation	2.0	2.0	2.0	2.0
Total	10	29	55	68

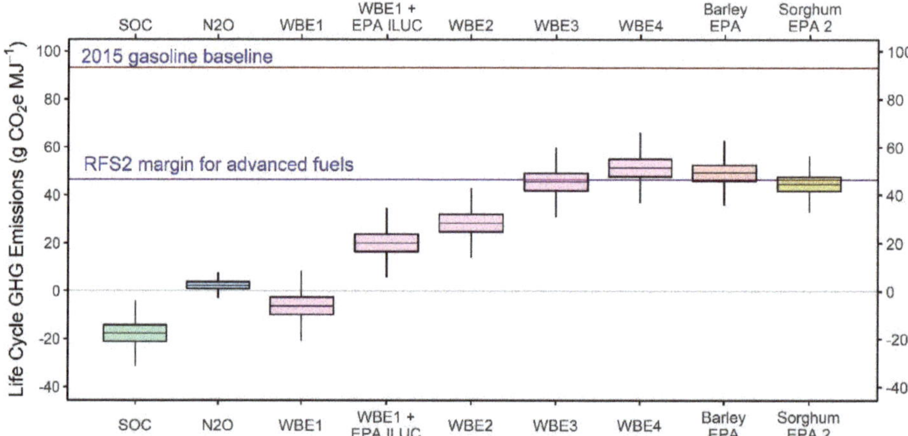

Figure 3. Stochastic greenhouse gas emissions for nitrous oxide emissions (N_2O) and soil organic carbon (SOC) change contributions to life cycle GHG emissions of winter barley-to-ethanol, net life cycle GHG emissions for WBE, and U.S. EPA barley- and sorghum-to-ethanol pathways.

We summarize the life cycle GHG emissions disaggregated by major input and co-product credits for the four WBE pathways (Table 3). The biorefinery has a net global warming potential that ranges from 10 to 68 g CO_2 eq MJ^{-1} for the maximum co-product crediting case to the no co-product crediting case. Co-product credits are mainly attributable to the avoided energy from a coal and fossil fuel powered grid needed to produce the avoided products. By converting all components of the feedstock into marketable co-products that can displace existing products on the market that rely on fossil energy throughout their production cycles (e.g., coal, food-grade CO_2, etc.), the barley-to-ethanol product reduces life cycle GHG emissions through avoided GHG emissions. Capture of biogenic CO_2 adds a credit of 21.8 g CO_2e MJ^{-1}, albeit with an increase in capital costs. In-use combustion emissions are low (2 g CO_2e MJ^{-1}), as are the emissions from feedstock transportation (2.1 g CO_2e MJ^{-1}) and fuel transportation and distribution (0.4 g CO_2e MJ^{-1}).

3.2. Implications of Stochastic Life Cycle Greenhouse Gas Emissions of Winter Barley and Other Starch Pathways for Renewable Fuel Policy

Results presented in Table 3 summarize net life cycle emissions with contributions from the most uncertain life cycle processes set to the 99th percentile, i.e., a high margin of safety. Stochastic distributions (Figure 3) with uncertain contributions from the geographically weighted N_2O and SOC emissions show compliance with advanced fuel designation at the 99th percentile for WBE1 and WBE2, and at the 50th percentile, the level applied in the RFS2 rule, for WBE1, WBE2, and WBE3. Assuming no iLUC from winter barley feedstocks in the growing region we evaluate, a biorefinery that invests in three major co-products would meet advanced fuel status. Comparing our stochastic WBE scenarios with U.S. EPA pathways (Figure 3) for ethanol from barley and sorghum produced in dry milling plants with DDGS co-products, EPA's barley pathway falls below the advanced fuel threshold if the biorefinery uses grid electricity and natural gas for thermal energy needs; however, as with the sorghum-to-ethanol pathway, they specify conditions for meeting advanced biofuel designation if using biogas for thermal energy needs and additionally using barley hulls for thermal energy needs and off-grid electricity.

The EPA uses a consequential LCA framework, projecting expected expansion of barley and sorghum to year 2022 relative to baseline production scenarios for each crop. The EPA barley case assumes expansion of mostly spring barley land, change in import/export due to increase in domestic production, increase in domestic use of barley feedstock, and replacement of livestock feed with DDGS byproduct. These conditions render a midpoint barley-to-ethanol iLUC and life cycle GHG emissions of 25 and 49 g CO_2e MJ^{-1}, respectively, placing barley above the advanced fuel threshold unless a barley biorefinery uses renewable thermal energy and co-produces DDGS. Although the EPA assumes winter barley will comprise 227 million L (60 million gal) compared to 303 million L (80 million gal) of spring barley, approximately 95% of land expansion is from spring barley. In the EPA's 2022 scenario, winter barley increases modestly, but most of it is used for biofuel (Table S2), unlike spring barley, which increases more modestly in the year 2022 scenario with a smaller portion being used for biofuel. Although the EPA's scenarios project little expansion of barley in Virginia, where winter barley would be grown (Table S4), it is difficult to assess the extent of winter barley's impact alone on iLUC given that the EPA analysis aggregates the effects of iLUC, including when a commodity shifts in use, as it does for winter barley, from serving animal feed markets to producing biofuels, and make-up quantities may then lead to iLUC.

Although the EPA sorghum case specifies a higher midpoint iLUC (28.4 g CO_2e MJ^{-1}) compared to the barley case (24.6 g CO_2e MJ^{-1}), with co-produced DDGS and use of renewable thermal energy, it meets the advanced fuel designation. Moreover, while both EPA pathways for sorghum and barley suggest conditions for meeting advanced fuel designation through incorporating renewable energy supply to the biorefinery, we argue here that both geographic growing conditions, in the case of certain winter fallow lands (Figure 1) and co-products that economically favor local markets, can render the winter barley crop suitable in meeting the advanced fuel designation. Even if applying the EPA's iLUC emission to WBE1 (Figure 3), we estimate that a biorefinery that uses winter barley and co-produces fermentative CO_2 for the local soft drink beverage industry would meet the advanced fuel designation and be economically beneficial for the region's growers and local markets.

4. Conclusions

Biorefinery co-product credits are essential for effectively meeting EISA and RFS2 policy. They are also critical to biorefinery economics as they increase revenue. For ethanol derived from winter barley to qualify as an advanced fuel under the RFS2, a life cycle assessment would need to show a 50% reduction in life cycle greenhouse gas (GHG) emissions relative to gasoline. Relative to the baseline gasoline fuel, an advanced fuel would need to emit no more than 48 g CO_2 e MJ^{-1} to meet this criterion. Our results for winter barley with co-product crediting demonstrate that winter barley would meet RFS2 advanced fuel requirements even when a substantial margin of safety is allowed and even when fermentative CO_2 is not captured. Without any co-products WBE would not meet the standard since the GHG emissions at the 50th percentile are above the threshold (WBE4). Results from U.S. EPA models aggregate the emissions from all types of barley production and thus include aggregation-level errors even if a biorefinery can demonstrate sole-sourcing of winter barley. Our analysis demonstrates that winter barley feedstocks would effectively meet national energy policy goals for East Coast U.S. population centers and potentially for other regions of the country where climatic and agronomic conditions are similar. Additional research on the socioeconomic and water quality impacts using LCIA metrics like eutrophication potential in agricultural growing regions of the Mid-Atlantic would need to be undertaken to fully understand the sustainability benefits of adopting those crops for biofuel production. Based on our analysis, winter barley demonstrates beneficial GHG emission reduction outcomes for the Mid-Atlantic U.S.

Supplementary Materials: The following are available online at http://www.mdpi.com/1996-1073/13/9/2236/s1, Table S1: Fossil Energy Input in Cumulative Energy Demand (MJ/MJ); Table S2: Projected Changes in Barley Production in the United States (Millions of Bushels); Table S3: Projected Changes in Barley Uses in the United States, U.S. EPA's Barley Biofuel Scenario; Table S4: Projected Change in Barley Production (million bushels) by State; Figure S1: A (**a**) histogram and a (**b**) Q–Q plot of spatially varying soil N_2O emissions (gCO$_2$e MJ^{-1}) for the studied counties. Total 284 counties are identified with cropland available for growing winter barley. Visual study of the plots suggest we assume data to be normally distributed; Figure S2: A (**a**) histogram and a (**b**) Q–Q plot of spatially changing soil SOC (gCO$_2$e MJ^{-1}) for the studied counties. The positive value represents net soil carbon increase from WB. A total 284 counties are identified with cropland available for growing winter barley. Visual study of the plots suggest we assume data to be normally distributed; Figure S3: A (**a**) histogram and a (**b**) Q–Q plot of spatially varying total (N_2O + SOC) emissions (gCO$_2$e MJ^{-1}) for the studied counties. Visual study of the plots suggest we assume data to be normally distributed. Negative gCO$_2$e MJ^{-1} represents carbon sequestration to soil. A total 284 counties are identified with cropland available for growing winter barley.

Author Contributions: Conceptualization, S.S.; methodology, S.S., A.E., P.R.A. and W.J.P.; software, S.S., A.E. and S.K.; formal analysis, S.S. and A.S.; investigation, A.J.M.; data curation, S.S., A.E., P.R.A. and W.J.P.; writing—original draft preparation, S.S. and A.S.; writing—review and editing, S.S., P.R.A., S.K., K.B.H. and P.L.G. All authors have read and agreed to the published version of the manuscript.

Funding: This research was funded by Osage Bio Energy of Richmond, Virginia.

Acknowledgments: The authors thank Winnie Yee (USDA–ARS), Wade Thomason (Virginia Polytechnic Institute and State University), and Gregory W. Roth (the Pennsylvania State University) for providing data and comments on the preparation of this analysis.

Conflicts of Interest: The authors declare no conflict of interest. The funders provided material balance data for the operation of a dry-grind barley-to-ethanol facility but had no role in the design of the study; in the collection, analyses, or interpretation of data; in the writing of the manuscript, or in the decision to publish the results.

References

1. Yeh, S.; Sperling, D. Low carbon fuel policy and analysis. *Energy Policy* **2013**, *56*, 1–4. [CrossRef]
2. McKechnie, J.; Pourbafrani, M.; Saville, B.A.; MacLean, H.L. Exploring impacts of process technology development and regional factors on life cycle greenhouse gas emissions of corn stover ethanol. *Renew. Energy* **2015**, *76*, 726–734. [CrossRef]
3. Ahlgren, S.; Björklund, A.; Ekman, A.; Karlsson, H.; Berlin, J.; Börjesson, P.; Ekvall, T.; Finnveden, G.; Janssen, M.; Strid, I. Review of methodological choices in LCA of biorefinery systems-key issues and recommendations. *Biofuels Bioprod. Biorefin.* **2015**, *9*, 606–619. [CrossRef]
4. Cai, H.; Han, J.; Wang, M.; Davis, R.; Biddy, M.; Tan, E. Life-cycle analysis of integrated biorefineries with co-production of biofuels and bio-based chemicals: Co-product handling methods and implications. *Biofuels Bioprod. Biorefin.* **2018**, *12*, 815–833. [CrossRef]

5. Riazi, B.; Zhang, J.; Yee, W.; Ngo, H.; Spatari, S. Life Cycle Environmental and Cost Implications of Isostearic Acid Production for Pharmaceutical and Personal Care Products. *ACS Sustain. Chem. Eng.* **2019**, *7*, 15247–15258. [CrossRef]

6. Searchinger, T.; Heimlich, R.; Houghton, R.A.; Dong, F.; Elobeid, A.; Fabiosa, J.; Tokgoz, S.; Hayes, D.; Yu, T.-H. Use of U.S. Croplands for Biofuels Increases Greenhouse Gases Through Emissions from Land Use Change. *Science* **2008**, *319*, 1238–1240. [CrossRef]

7. Creutzig, F.; Ravindranath, N.H.; Berndes, G.; Bolwig, S.; Bright, R.; Cherubini, F.; Chum, H.; Corbera, E.; Delucchi, M.; Faaij, A.; et al. Bioenergy and climate change mitigation: An assessment. *GCB Bioenergy* **2015**, *7*, 916–944. [CrossRef]

8. Taheripour, F.; Tyner, W.E. US biofuel production and policy: Implications for land use changes in Malaysia and Indonesia. *Biotechnol. Biofuels* **2020**, *13*, 11. [CrossRef]

9. Bryant, R.B.; Veith, T.L.; Kleinman, P.J.A.; Gburek, W.J. Cannonsville Reservoir and Town Brook watersheds: Documenting conservation efforts to protect New York City's drinking water. *J. Soil Water Conserv.* **2008**, *63*, 339–344. [CrossRef]

10. Dzombak, D. Nutirent Control in Large-Scale U.S. Watersheds: The Chesapeake Bay and Northern Gulf of Mexico. *Bridge* **2011**, *41*, 13–22.

11. Jayasundara, S.; Wagner-Riddle, C.; Parkin, G.; Bertoldi, P.; Warland, J.; Kay, B.; Voroney, P. Minimizing nitrogen losses from a corn–soybean–winter wheat rotation with best management practices. *Nutr. Cycl. Agroecosyst.* **2007**, *79*, 141–159. [CrossRef]

12. U.S. Congress. *Energy Independence and Security Act*; 12 December 2007; U.S. Congress: Washington, DC, USA, 2017.

13. USEPA. *Notice of Data Availability Concerning Renewable Fuels Produced from Barley under the RFS Program*; EPA-HQ-OAR-2013-0178; 23 July 2013; U.S. Environmental Protection Agency: Washington, DC, USA, 2013; pp. 44075–44089.

14. USEPA. *Supplemental Determination for Renewable Fuels Produced Under the Final RFS2 Program From Grain Sorghum*; EPA-HQ-OAR-2011-0542; FRL-9760-2; 17 December 2012; Federal Register: Washington, DC, USA, 2012; pp. 74592–74607.

15. Sindelar, A.J.; Schmer, M.R.; Gesch, R.W.; Forcella, F.; Eberle, C.A.; Thom, M.D.; Archer, D.W. Winter oilseed production for biofuel in the US Corn Belt: Opportunities and limitations. *GCB Bioenergy* **2017**, *9*, 508–524. [CrossRef]

16. Krohn, B.J.; Fripp, M. A life cycle assessment of biodiesel derived from the "niche filling" energy crop camelina in the USA. *Appl. Energy* **2012**, *92*, 92–98. [CrossRef]

17. Tabatabaie, S.M.H.; Tahami, H.; Murthy, G.S. A regional life cycle assessment and economic analysis of camelina biodiesel production in the Pacific Northwestern US. *J. Clean. Prod.* **2018**, *172*, 2389–2400. [CrossRef]

18. Nghiem, N.P.; Ramírez, E.C.; McAloon, A.J.; Yee, W.; Johnston, D.B.; Hicks, K.B. Economic analysis of fuel ethanol production from winter hulled barley by the EDGE (Enhanced Dry Grind Enzymatic) process. *Bioresour. Technol.* **2011**, *102*, 6696–6701. [CrossRef] [PubMed]

19. Joelsson, E.; Erdei, B.; Galbe, M.; Wallberg, O. Techno-economic evaluation of integrated first- and second-generation ethanol production from grain and straw. *Biotechnol. Biofuels* **2016**, *9*, 1. [CrossRef] [PubMed]

20. Nghiem, N.P.; Montanti, J.J.D.B. Sorghum as a renewable feedstock for production of fuels and industrial chemicals. *AIMS Bioeng.* **2016**, *3*, 75–91. [CrossRef]

21. Fallahpour, F.; Aminghafouri, A.; Ghalegolab Behbahani, A.; Bannayan, M. The environmental impact assessment of wheat and barley production by using life cycle assessment (LCA) methodology. *Environ. Dev. Sustain.* **2012**, *14*, 979–992. [CrossRef]

22. Garcia-Garcia, G.; Rahimifard, S. Life-cycle environmental impacts of barley straw valorisation. *Resour. Conserv. Recycl.* **2019**, *149*, 1–11. [CrossRef]

23. Adler, P.R.; Spatari, S.; D'Ottone, F.; Vazquez, D.; Peterson, L.; Del Grosso, S.J.; Baethgen, W.E.; Parton, W.J. Legacy effects of individual crops affect N_2O emissions accounting within crop rotations. *GCB Bioenergy* **2018**, *10*, 123–136. [CrossRef]

24. Malça, J.; Freire, F. Addressing land use change and uncertainty in the life-cycle assessment of wheat-based bioethanol. *Energy* **2012**, *45*, 519–527. [CrossRef]

25. Fertitta-Roberts, C.; Spatari, S.; Grantz, D.A.; Jenerette, G.D. Trade-offs across productivity, GHG intensity, and pollutant loads from second-generation sorghum bioenergy. *GCB Bioenergy* **2017**, *9*, 1764–1779. [CrossRef]

26. Adler, P.R.; Mitchell, J.G.; Pourhashem, G.; Spatari, S.; Del Grosso, S.J.; Parton, W.J. Integrating biorefinery and farm biogeochemical cycles offsets fossil energy and mitigates soil carbon losses. *Ecol. Appl.* **2015**, *25*, 1142–1156. [CrossRef] [PubMed]

27. Adler, P.R.; Del Grosso, S.J.; Parton, W.J. Life-cycle Assessment of Net Greenhouse Gas Flux for Bioenergy Cropping Systems. *Ecol. Appl.* **2007**, *13*, 675–691. [CrossRef] [PubMed]

28. Adler, P.R.; Del Grosso, S.J.; Inman, D.; Jenkins, R.E.; Spatari, S.; Zhang, Y.M. Mitigation Opportunities for Life-Cycle Greenhouse Gas Emissions during Feedstock Production across Heterogeneous Landscapes. *Manag. Agric. Greenh. Gases Coord. Agric. Res. Through Gracenet Address Our Chang. Clim.* **2012**, 203–219. [CrossRef]

29. Gao, S.; Gurian, P.L.; Adler, P.R.; Spatari, S.; Gurung, R.; Kar, S.; Ogle, S.M.; Parton, W.J.; Del Grosso, S.J. Framework for improved confidence in modeled nitrous oxide estimates for biofuel regulatory standards. *Mitig. Adapt. Strateg. Glob. Chang.* **2018**, *23*, 1281–1301. [CrossRef]

30. Springborn, M.; Yeo, B.-L.; Lee, J.; Six, J. Crediting uncertain ecosystem services in a market. *J. Environ. Econ. Manag.* **2013**, *66*, 554–572. [CrossRef]

31. Hsu, D.D.; Inman, D.; Heath, G.A.; Wolfram, E.J.; Mann, M.K.; Aden, A. Life cycle environmental impacts of selected U.S. ethanol production and use pathways in 2022. *Environ. Sci. Technol.* **2010**, *44*, 5289–5297. [CrossRef]

32. Spatari, S.; Bagley, D.; MacLean, H. Life cycle evaluation of emerging lignocellulosic ethanol conversion technologies. *Bioresour. Technol.* **2010**, *101*, 654–667. [CrossRef]

33. Kylili, A.; Christoforou, E.; Fokaides, P.A. Environmental evaluation of biomass pelleting using life cycle assessment. *Biomass Bioenergy* **2016**, *84*, 107–117. [CrossRef]

34. Liu, W.; Wang, J.; Richard, T.L.; Hartley, D.S.; Spatari, S.; Volk, T.A. Economic and life cycle assessments of biomass utilization for bioenergy products. *Biofuels Bioprod. Biorefin.* **2017**, *11*, 633–647. [CrossRef]

35. Sorunmu, Y.; Billen, P.; Elangovan, S.E.; Santosa, D.; Spatari, S. Life-Cycle Assessment of Alternative Pyrolysis-Based Transport Fuels: Implications of Upgrading Technology, Scale, and Hydrogen Requirement. *ACS Sustain. Chem. Eng.* **2018**, *6*, 10001–10010. [CrossRef]

36. Longato, D.; Gaglio, M.; Boschetti, M.; Gissi, E. Bioenergy and ecosystem services trade-offs and synergies in marginal agricultural lands: A remote-sensing-based assessment method. *J. Clean. Prod.* **2019**, *237*, 117672. [CrossRef]

37. Nguyen, L.; Cafferty, K.; Searcy, E.; Spatari, S. Uncertainties in Life Cycle Greenhouse Gas Emissions from Advanced Biomass Feedstock Logistics Supply Chains in Kansas. *Energies* **2014**, *7*, 7125–7146. [CrossRef]

38. ISO. *ISO 14044: Environmental Management—Life Cycle Assessment—Requirements and Guidelines*; ISO 14044:2006(E); International Organization for Standardization: Geneva, Switzerland, 2006.

39. Gaglio, M.; Tamburini, E.; Lucchesi, F.; Aschonitis, V.; Atti, A.; Castaldelli, G.; Fano, E.A. Life Cycle Assessment of Maize-Germ Oil Production and the Use of Bioenergy to Mitigate Environmental Impacts: A Gate-To-Gate Case Study. *Resources* **2019**, *8*, 60. [CrossRef]

40. Pourhashem, G.; Spatari, S.; Boateng, A.A.; McAloon, A.J.; Mullen, C.A. Life Cycle Environmental and Economic Tradeoffs of Using Fast Pyrolysis Products for Power Generation. *Energy Fuels* **2013**, *27*, 2578–2587. [CrossRef]

41. Forster, P.; Ramaswamy, V.; Artaxo, P.; Berntsen, T.; Betts, R.; Fahey, D.W.; Haywood, J.; Lean, J.; Lowe, D.C.; Myhre, G.; et al. *Chapter 2. Changes in Atmospheric Constituents and in Radiative Forcing in Climate Change 2007—The Physical Science Basis. Contribution of Working Group I to the Fourth Assessment Report of the Intergovernmental Panel on Climate Change*; Solomon, S., Qin, D., Manning, M., Marquis, M., Averyt, K., Tignor, M.M.B., Miller, H.L., Eds.; Cambridge University Press: New York, NY, USA, 2007.

42. CBC. *Biofuels and the Bay: Getting it Right to Benefit Farms, Forests, and the Chesapeake*; A Report of the Chesapeake Bay Commission; Chesapeake Bay Commission: Annapolis, MD, USA, 2007; p. 34.

43. Kwiatkowski, J.R.; McAloon, A.J.; Taylor, F.; Johnston, D.B. Modeling the process and costs of fuel ethanol production by the corn dry-grind process. *Ind. Crops Prod.* **2006**, *23*, 288–296. [CrossRef]

44. MacLean, H.L.; Spatari, S. The contribution of enzymes and process chemicals to the life cycle of ethanol. *Environ. Res. Lett.* **2009**, *4*, 014001. [CrossRef]

45. *SimaPro*, version 8.4; PRé Consultants: Amersfoort, The Netherlands, 2018.

46. Thomason, W. *Barley for Grain (Intensive Management)*; Virginia Cooperative Extension: Blacksburg, VA, USA, 2007.

47. Parton, W.J.; Hartman, M.; Ojima, D.S.; Schimel, D.S. DAYCENT and its land surface submodel: Description and testing. *Glob. Planet. Chang.* **1998**, *19*, 35–48. [CrossRef]

48. Wang, M.Q. *GREET1*; Argonne National Laboratory: DuPage County, IL, USA, 2019; p. 49.

49. US Department of Agriculture NASS Quick Statistics Database. Available online: http://www.nass.usda.gov/QuickStats/ (accessed on 17 November 2011).

50. Kim, S.; Dale, B.; Jenkins, R. Life cycle assessment of corn grain and corn stover in the United States. *Int. J. Life Cycle Assess.* **2009**, *14*, 160–174. [CrossRef]

51. Kim, S.; Dale, B.E. Environmental aspects of ethanol derived from no-tilled corn grain: Nonrenewable Energy consumption and greenhouse gas emissions. *Biomass Bioenergy* **2005**, *28*, 475–489. [CrossRef]

52. Wang, M.; Han, J.; Dunn, J.B.; Cai, H.; Elgowainy, A. Well-to-wheels energy use and greenhouse gas emissions of ethanol from corn, sugarcane and cellulosic biomass for US use. *Environ. Res. Lett.* **2012**, *7*, 045905. [CrossRef]

53. IPCC. *2006 IPCC Guidelines for National Greenhouse Gas Inventories, Volume 4: Agriculture, Forestry and Other Land Use*; 4-88788-032-4; Institute for Global Environmental Strategies (IGES), Hayama, Japan on behalf of the IPCC: Hayama, Japan, 2006.

54. Wernet, G.; Bauer, C.; Steubing, B.; Reinhard, J.; Moreno-Ruiz, E.; Weidema, B. The ecoinvent database version 3 (part I): Overview and methodology. *Int. J. Life Cycle Assess.* **2016**, *21*, 1218–1230. [CrossRef]

55. Wakeley, H.L.; Hendrickson, C.T.; Griffin, W.M.; Matthews, H.S. Economic and Environmental Transportation Effects of Large-Scale Ethanol Production and Distribution in the United States. *Environ. Sci. Technol.* **2009**, *43*, 2228–2233. [CrossRef]

56. Pourhashem, G.; Adler, P.R.; McAloon, A.J.; Spatari, S. Cost and greenhouse gas emission tradeoffs of alternative uses of lignin for second generation ethanol. *Environ. Res. Lett.* **2013**, *8*, 025021. [CrossRef]

57. Adams, D.; Alig, R.; McCarl, B.; Murray, B. *FASOMGHG Conceptual Structure, and Specification: Documentation*; Texas A&M University: College Station, TX, USA, 2005; p. 447.

58. Beach, R.H.; Zhang, Y.W.; McCarl, B.A. Modeling Bioenergy, Land Use, and GHG Emissions with FASOMGHG: Model Overview and Analysis of Storage Cost Implications. *Clim. Chang. Econ.* **2012**, *3*, 1250012. [CrossRef]

59. Devadoss, S.W.; Patrick, C.; Helmar, M.D.; Grundmeier, E.; Skold, K.D.; Meyers, W.H.; Johnson Stanley, R. *The FAPRI Modeling System at CARD: A Documentation Summary*; Iowa State University: Ames, IA, USA, 1989.

60. U.S. Office of Management and Budgets (OMB). *Circular A-4, Regulatory Analysis*; Last modified 17 September 2003 ed.; U.S. Office of Management and Budgets (OMB): Washington, DC, USA, 2003.

61. Spatari, S.; MacLean, H.L. Characterizing Model Uncertainties in the Life Cycle of Lignocellulose-Based Ethanol Fuels. *Environ. Sci. Technol.* **2010**, *44*, 8773–8780. [CrossRef]

62. Pourbafrani, M.; McKechnie, J.; Shen, T.; Saville, B.A.; MacLean, H.L. Impacts of pre-treatment technologies and co-products on greenhouse gas emissions and energy use of lignocellulosic ethanol production. *J. Clean. Prod.* **2014**, *78*, 104–111. [CrossRef]

Article

Life Cycle Assessment and Energy Balance of a Novel Polyhydroxyalkanoates Production Process with Mixed Microbial Cultures Fed on Pyrolytic Products of Wastewater Treatment Sludge

Luciano Vogli [1,*], Stefano Macrelli [1], Diego Marazza [2,3], Paola Galletti [3,4], Cristian Torri [3,4], Chiara Samorì [3,4] and Serena Righi [2,3]

[1] CIRSA (Interdepartmental Research Centre for Environmental Sciences), University of Bologna, via Sant'Alberto, 163, 48123 Ravenna, Italy; stefano.macrelli4@unibo.it

[2] Department of Physics and Astronomy, University of Bologna, viale Berti Pichat, 6/2, 40127 Bologna, Italy; diego.marazza@unibo.it (D.M.); serena.righi2@unibo.it (S.R.)

[3] CIRI FRAME (Interdepartmental Centre for Industrial Research in Renewable Resources, Environment, Sea and Energy), University of Bologna, via Sant'Alberto, 163, 48123 Ravenna, Italy; paola.galletti@unibo.it (P.G.); cristian.torri@unibo.it (C.T.); chiara.samori3@unibo.it (C.S.)

[4] Department of Chemistry "G. Ciamician", University of Bologna, via Selmi, 2, 40126 Bologna, Italy

* Correspondence: luciano.vogli@unibo.it or luciano.vogli@gmail.com; Tel.: +39-338-9333561

Received: 8 April 2020; Accepted: 25 May 2020; Published: 28 May 2020

Abstract: A "cradle-to-grave" life cycle assessment is performed to identify the environmental issues of polyhydroxyalkanoates (PHAs) produced through a hybrid thermochemical-biological process using anaerobically digested sewage sludge (ADSS) as feedstock. The assessment includes a measure of the energy performance of the process. The system boundary includes: (i) Sludge pyrolysis followed by volatile fatty acids (VFAs) production; (ii) PHAs-enriched biomass production using a mixed microbial culture (MMC); (iii) PHAs extraction with dimethyl carbonate; and iv) PHAs end-of-life. Three scenarios differing in the use of the syngas produced by both pyrolysis and biochar gasification, and two more scenarios differing only in the external energy sources were evaluated. Results show a trade-off between environmental impacts at global scale, such as climate change and resources depletion, and those having an effect at the local/regional scale, such as acidification, eutrophication, and toxicity. Process configurations based only on the sludge-to-PHAs route require an external energy supply, which determines the highest impacts with respect to climate change, resources depletion, and water depletion. On the contrary, process configurations also integrating the sludge-to-energy route for self-sustainment imply more onsite sludge processing and combustion; this results in the highest values of eutrophication, ecotoxicity, and human toxicity. There is not a categorical winner among the investigated configurations; however, the use of a selected mix of external renewable sources while using sludge to produce PHAs only seems the best compromise. The results are comparable to those of both other PHAs production processes found in the literature and various fossil-based and bio-based polymers, in terms of both non-biogenic GHG emissions and energy demand. Further process advancements and technology improvement in high impact stages are required to make this PHAs production process a competitive candidate for the production of biopolymers on a wide scale.

Keywords: LCA; energy metrics; PHAs; bio-based polymers; biodegradable plastics; pyrolysis; volatile fatty acids

---

1. Introduction

According to the European Commission [1], the transition to a more circular economy is an essential contribution to the efforts to develop a sustainable, low-carbon, resource-efficient, and competitive economy. In a circular economy, closed material cycles should be encouraged where possible [1]. Bio-based materials, i.e., those derived from renewable resources, such as wood, crops, or fibers, have various applications in a large variety of industries (e.g., construction, furniture, packaging, coatings, textiles, cardboard, chemicals, etc.) and energy uses (e.g., biofuels). Their characteristic of being made of organic carbon, which can be recycled and reused many times, in many ways, goes towards the principles of the waste hierarchy and, more generally, results in better overall environmental performances [1]. The bioeconomy, consequently, offers alternatives to fossil-based energy and products, and can make an important contribution to the circular economy. Moreover, bio-based materials can provide benefits connected to their renewability, biodegradability, or compostability. Nevertheless, the use of biological resources requires attention and a careful assessment through their life-cycle environmental impacts. Indeed, their use can also create competition for them and generate pressure on land use [2]. In this context, bioplastics are a very promising research area, usually indicated as a sustainable alternative to conventional fossil-based products. They have the advantage, over traditional plastics, of diminishing the use of non-renewable resources and also decreasing the environmental impact related to fossil resources' consumption [3,4].

Polyhydroxyalkanoates (PHAs) are polymers belonging to a group of polyesters that are generated by some bacteria as carbon and energy intracellular reserve granules. Currently, more than 90 bacterial species that produce PHAs and about 150 diverse monomers of PHAs have been recognized [5]. Their accumulation is usually observed when one nutrient (e.g., nitrogen, phosphorus, and oxygen) is present in the fermentation broth in a limiting concentration, while, at the same time, there is an available excess source of carbon. Most bacterial strains, such as *Cupriavidus necator*, accumulate PHAs as secondary products under nutrient-limiting conditions. However, some bacterial strains, such as recombinant *Escherichia coli* and *Alcaligenes latus*, synthesize PHAs as a primary metabolite during microbial growth [6]. Generally, PHAs are produced by pure microbial cultures grown on renewable feedstocks (i.e., sugar or oils) under sterile conditions, but recently, several authors have investigated the exploitation of residues and waste as growing substrates [7–9]. Commonly, PHAs are considered eco-friendly because they are produced from renewable natural resources instead of petrochemicals, and because they biodegrade without producing harmful or toxic by-products and leaving no worrisome waste [10]. Anyway, in order to conclude if biopolymers are environmentally advantageous over petrol-based plastics, it is necessary to analyze their entire life cycle. Moreover, to make PHAs really competitive, first of all they need to equal their petrochemical counterparts both in terms of quality and economic performances [11]. Sustainable PHAs production is multifaceted and several criticisms need to be tackled in the process in order to make the manufacture of PHAs on an industrial scale convenient both environmentally and economically. For this purpose, cheaper microbial cultures and reutilization of waste streams as growing substrates appear among the main points for large-scale industrial production [12–14].

Among the many residues and wastes tested as growth substrates [9], agro-industrial and municipal effluents and sewage sludge have received considerable attention in recent years. Combinations of food waste and sewage sludge [15,16], industry and municipal primary sludge [17], municipal secondary sludge [18], and sewage sludge after hydrothermal carbonization and acidogenic microbial fermentation [14,19] are examples of the substrates investigated in order to obtain volatile fatty acids (VFAs) suitable as a carbon source for PHAs producing mixed microbial cultures (MMCs). At the same time, agro-industrial wastewater and sludge have also attracted great interest. A plethora of agro-industrial effluents have been tested as substrate for PHAs production: Sugarcane molasses, paper mill effluent, dairy effluent [20], distillery effluents [21], rice winery wastewater [22], and yeast industry wastewater [23]. Authors often agree in concluding that the reutilization of wastewater and sludge as a carbon substrate not only abates the cost of PHAs production [20,21] but also provides for a significant reduction of the sludge disposal cost [19], constituting a sustainable waste management [15,18] and significantly decreasing the environmental impact of PHAs [17].

The European Bioplastics association [3] supports life cycle assessment (LCA) in order to validate the eco-sustainability of bioplastics. LCA encompasses the energy and material flows within the system boundary and calculates the relevant impacts generated by each unit process. The energetic and environmental sustainability of PHAs production has usually been evaluated by means of LCA. The first LCA studies were focused on PHAs production based on carbon sources from dedicated cultures, mainly glucose [24], in particular from corn [25–28] and corn grain integrated by corn stover [29], but also soybean oil [30] and sucrose from sugar cane [31,32] as an alternative to glucose. PHAs from a genetically modified corn have also been analyzed [33]. Later, the focus shifted to evaluation of the eco-performances of PHAs produced from alternatives to dedicated carbon sources, such as fermentable sugars from lignocellulosic biomass [31], industrial wastewater [12,34], biomethane [35], municipal solid waste [36], potato [37], switchgrass [38], etc. In the last few years, the increasing attention given to wastewater and sludge as a PHAs-accumulating bacteria growth substrate has pushed many LCA studies in that direction. Heimersson and co-authors [39], Fernández-Dacosta and co-authors [40], and Morgan-Sagastume and co-authors [41] investigated the environmental performances of wastewater treatment plants (WWTPs) with integrated PHAs production. Dietrich and co-authors [42] analyzed the sustainability of PHAs production in integrated lignocellulose biorefineries. Vega and co-authors [43] examined the eco-sustainability of a biorefinery treating a mixture of cow manure and grape marc. Very recently, a review on the link between sustainability and industrial waste streams as feedstock for the production of PHAs has been published [44]; many different industrial streams have been taken into consideration, including activated sludge and industrial aqueous streams. The authors agree that the consideration of waste stream exploitation as a bacteria growth substrate is an interesting contribution towards a circular bioeconomy [44], economic competitiveness against fossil-based products [34,42], and eco-sustainability [43,45]. However, there is also a general agreement on the need for further investigations in order to improve the PHAs production process [40,44,46,47], and to deeply explore the environmental performances [41], thus avoiding possible burden shifting [43] and widening the analysis to include disregarded impacts like water use, land use, and eutrophication [39]. The reviews of Narodoslawsky and co-authors [48] and Cristóbal and co-authors [49] show a great variability among the LCA results obtained by the different authors due to technological differences but also different choices in the LCA applications. In any case, the power of LCA as tool to address eco-sustainability is reasserted.

This study aims to exploit LCA to assess the energy and environmental burdens of a hybrid thermochemical-biological process [50] that couples pyrolysis and anaerobic-aerobic fermentations to convert industrial sludge into PHAs. Specifically, the system valorizes anaerobically digested sewage sludge (ADSS) coming from wastewater treatment plants of agri-food industries for the production of PHAs via VFAs. Five different scenarios for the use of thermochemical process products were compared.

2. Materials and Methods

2.1. PHAs Production at the Lab Scale

The initial steps of the PHAs production process were tested at the laboratory scale by the Chemistry Lab of CIRI FRAME of the University of Bologna, Ravenna Campus; then, a hypothesis of process upscaling at the industrial scale was carried out.

The process for PHAs production involving MMCs consists of several phases (Figure 1) that can be summarized into three main steps: (1) Biomass feedstock pre-treatment through pyrolysis, followed by anaerobic digestion of organic carbon to produce mixtures of volatile fatty acids (VFAs); (2) PHAs-enriched microbial biomass production; and (3) PHAs extraction using organic solvents. A detailed explanation of each step is provided below.

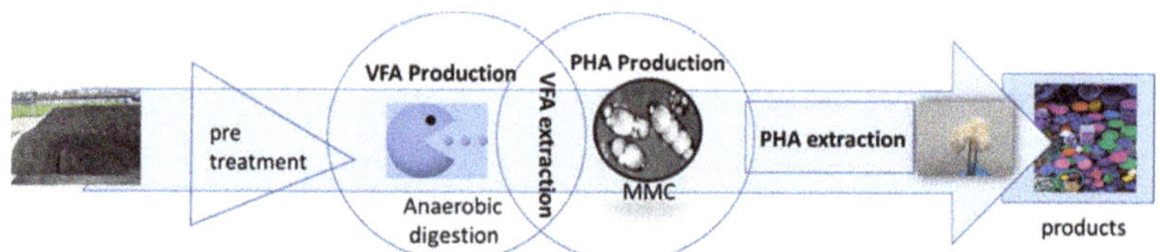

Figure 1. Process for PHAs production involving mixed microbial cultures.

2.1.1. Pyrolysis and Anaerobic Digestion (PyAD)

The pyrolytic pre-treatment was performed in order to improve the ADSS fermentability, which can generally be low due to the high content of recalcitrant compounds and complex poorly degradable inorganic matter [51]. Pyrolysis of biomass is a thermochemical decomposition converting biomass into energy and materials and occurring in the absence of oxygen or under an inert gas flow. Pyrolysis provides three products: A liquid fraction (bio-oil), a solid carbonaceous material (biochar), and a gaseous fraction (syngas) [52]. The bio-oil and the syngas can be used as the substrate to feed the anaerobic fermenter to produce VFAs [51]. Besides, the syngas can also be converted into energy in a combined heat and power (CHP) unit [53]. The biochar fraction can be used either as a carbon storage for long-term sequestration, or as a soil amendment bringing it back to the soil, or to produce energy through combustion [54].

Pyrolysis tests were carried out on ADSS, using a fixed bed tubular quartz reactor placed into a refractory furnace, and were performed at 500 °C for 30 min. The pyrolizer was connected downstream to two cold traps (ice bath) for trapping the condensable vapors (liquid fraction). The bio-oil obtained was a dark-brown biphasic liquid characterized by a low viscous aqueous phase and a tarry dark-brown one (organic phase). The two phases were separated by centrifuge. Bio-oil and biochar were collected, and the yields determined by weight difference (see Table 1).

Subsequently, the anaerobic digestion tests of pyrolysis bio-oil to produce VFAs were carried out following the procedure by Torri and co-authors [55]. Sludge generated by agro-industrial wastewater treatment itself was used as inoculum after sterilization at 120 °C and 2 bar for 60 min; this step should eliminate the methanogenic bacteria while preserving the sporogenic ones. This has the effect of avoiding the production of biogas or, in other terms, maximizing the yield of VFAs. Indeed, the production of VFAs is an anaerobic process involving hydrolysis and acidogenesis, known as acidogenic fermentation. During hydrolysis, the complex organic polymers present in ADSS are subdivided into simpler organic monomers due to the effect of enzymes excreted by the hydrolytic microorganisms [15]. Subsequently, acidogenic bacteria ferment these monomers into VFAs, mainly as acetic, propionic, and butyric acids. Both processes involve a wide range of obligate and optional anaerobes, such as *Bacteriocides*, *Clostridia*, *Bifidobacteria*, *Streptococci*, and *Enterobacteriaceae* [56]. A 100-mL syringe reactor was used, filled with 20 mL of bacterial inoculum and an aliquot of 200 mg of chemical oxygen demand equivalent (COD_{eq}) of bio-oil added to feed the bacterial inoculum. The syringe was closed and stored at 42 °C. The reactor was analyzed daily in terms of the chemical oxygen demand (COD) and VFAs content. Biogas production was assessed on the basis of the COD content in the liquid phase, assuming that its decrease was equal to the biogas produced (namely CH_4 and H_2).

Table 1. Main parameters used for the industrial scale-up modeling of the process and related values. A reference is provided for already published data; otherwise, data are unpublished. HB: hydroxybutyrate; HV: hydroxyvalerate; HH: hydroxyhexanoate.

Process Step	Parameter	Unit	Value	Reference
Pyrolysis	yield of syngas from ADSS	g COD/g COD	0.31	-
	yield of oil from ADSS	g COD/g COD	0.22	-
	yield of biochar from ADSS	g COD/g COD	0.47	-
Biochar gasification	yield of syngas from biochar	g COD/g COD	0.70	-
Anaerobic acidogenic fermenter	yield of VFAs from oil	g COD/g COD	0.33	-
	yield of VFAs from syngas	g COD/g COD	0.80	-
	residence time	days	7	-
	organic load	g COD/L/day	7	-
Pertraction system	VFAs concentration after pertraction	g COD/L	6.9	estimate based on [57]
Sequencing Batch Reactor (SBR)	VFAs to SBR	% of total VFAs	35	-
	residence time	hours	24	-
Accumulation Reactor (AR)	VFAs to AR	% of total VFAs	65	-
	residence time	hours	6	-
	yield of PHAs from VFAs	g COD/g COD	0.5	estimate based on [14]
	PHAs concentration in microbial cells	g/g	0.6	estimate based on [14]
	HB:HV:HH ratio		88:11:1	[14]
PHAs extraction system	biomass to DMC ratio	kg/L	0.025	[47]
	recovered DMC	%	97	[47]
	recovered polymer	%	96	[47]

As for the VFAs extraction, reference was made to the experimental data obtained by Torri and co-authors [57], where the innovative methodology of pertraction making use of new liquid membranes (LMs) based on lipophilic amines and biodiesel was proposed for this purpose. The VFAs flux rate obtained with the best performing liquid membrane based on trioctylamine (TOA) at 10 wt% (TOA10-B) demonstrates the feasibility of a closed loop for a selective conversion of VFAs extracted from anaerobic fermentation systems into PHAs-enriched microbial biomass.

2.1.2. PHAs-Enriched Biomass Production

VFAs were converted into PHAs using two aerobic batch reactors in sequence. The physical separation allows process optimization, as it has been shown that different conditions are required at each stage [58,59]. In the first reactor, mixed cultures are subjected to "feast and famine" conditions: High availability and shortage of the substrate are alternated in order to select microbial populations capable of incorporating the VFAs with greater efficiency. The microbial sludge retention time value was set to ensure that all carbon was consumed for cell growth and maintenance. In the latter, the selected microorganisms were fed exclusively with the VFAs produced in the acidogenic fermentation with the aim of promoting the accumulation of PHAs [14,60].

2.1.3. PHAs Extraction

The applied extraction method is based on the solubilization of PHAs with dimethyl carbonate (DMC). This process can be applied either directly on concentrated microbial sludge or on dry biomass, allowing a very high polymer recovery and an excellent purity. The direct extraction from microbial sludge applied in this case study required a biomass to solvent ratio of 2.5% weight to volume ratio. This concentration was obtained by centrifuging and concentrating the microbial culture after the accumulation phase. The sludge underwent extraction with DMC for 4 h at 90 °C. Subsequently, the DMC phase, containing the extracted PHAs, and the biomass sludge were centrifuged and separated, and then the extracted polymer was recovered after filtering and evaporating the solvent. The polymer recovery was very high, around 96% [47].

2.2. PHAs Production at the Industrial Scale

The above described lab-scale pyrolytic system was scaled-up to treat 2500 t dry matter (DM) sludge/year at 3.8 wt% DM received from an average-sized WWTP. To obtain the mass and energy balance, a mixed approach was adopted considering data from laboratory experiments, assumptions, and vendors' data sheets for the equipment. Rigorous calculations for the mass, enthalpy, and COD balances were performed at the pyrolysis stage to ensure comparability between scenarios, given the composition heterogeneity of pyrolysis (bio-oil, syngas, biochar) and gasification products (syngas, ash). Table 1 summarizes the main parameters determined by the tests carried out at CIRI FRAME and used for the industrial scale-up modeling of the process.

2.2.1. Pyrolysis and Anaerobic Digestion (PyAD)

The initial process step was sewage sludge dewatering, first by centrifugation up to 25% DM then by thermal drying up to 85% DM. The dryer was assumed to recover the latent heat by condensing the evaporated water, which was then recirculated into the process water system. As explained, the pyrolytic pre-treatment of ADSS enhances the soluble, and thus bioavailable, COD of ADSS itself, facilitating the conversion yield of organic matter into VFAs during the anaerobic acidogenic fermentation. The thermal energy required for pyrolysis was assumed to be 1.28 MJ per kg of ADSS at 85% DM. VFAs were produced with different yields from the pyrolysis syngas, bio-oil and water phases, and also from syngas obtained through biochar gasification (see Table 1 and Section 2.3.1). As for this technology, a dataset representing an average indirectly heated atmospheric fixed-bed gasifier followed by a low-temperature wet gas treatment was selected. Energy self-production of the plant was modelled by a syngas-driven CHP plant dataset representing an average combustion situation without further flue gas treatment.

In the anaerobic acidogenic fermenter (AAF), the syngas and bio-oil were converted into VFAs with yields equal to 80% and 33% on the COD basis, respectively, and a hydraulic residence time of 7 days. No methane was produced as a result of the sterilization process applied on the inoculum to inhibit methanogenic bacteria.

The energy required by the AAF consists of the thermal energy necessary to heat up inlet streams and to keep the AAF at 40 °C, and of the electricity needed for pumping the inlet/outlet flows and for the mixing of the AAF volume. The heat necessary for increasing the temperature of the inlet flow from the environment temperature was calculated on the basis of the specific sensible heat of the water phase. The outlet stream from the AAF was separated into a biological sludge stream and a water phase by a centrifuge consuming 7 MJ per kg of sludge at 3% DM. The biological sludge was recirculated at the initial drying stage, while VFAs were concentrated by the aforementioned pertraction system, which required a centrifugal pump power of 344 J per kg of treated liquid.

2.2.2. PHAs-Enriched Biomass Production

The sequencing batch reactor (SBR) is an aerobic reactor where microbial biomass is produced using VFAs as carbon source and PHAs start to accumulate in microbial cells. Biomass is subsequently transferred to another aerobic batch reactor (accumulation reactor, AR) where MMCs convert VFAs into PHAs with a 50% yield on COD basis. PHAs accumulate in microbial cells up to 60 wt%. Air sparging and liquid mixing are needed in both SBR and AR. The oxygen required for microbial biomass growth was modelled assuming a stoichiometric uptake by microorganisms. A first stage of microbial biomass dewatering was carried out by filtration, in order to reach a 13% DM content. A further dewatering stage was performed by a second thermal dryer to increase the DM to 20% before PHAs extraction. Additionally, in this dryer, the evaporated water was condensed to supply the process water system, but no heat recovery was considered in this case.

2.2.3. PHAs Extraction

The model includes the extraction of PHAs from the MMC by means of the solubilization with DMC as explained in Section 2.1.3. The PHAs recovery rate was set to 96%. The extraction processes were composed by a series of equipment units: Centrifuges, batch reaction vessels, heaters and pumps.

2.2.4. PHAs End-of-Life

The "end-of-life" phase was modelled based on data provided by COREPLA [61]. In Italy, in 2016, 2.2 million tons of post-consumer plastics waste ended up in the waste stream. A share of 85% was recovered, both through material recycling (43%) and energy recovery (42%), while the remaining 15% still went to landfill. Since a supply chain for the material recovery of bioplastics has not yet been implemented, neither at the Italian nor at European level, and since the amount of bioplastics recovered with the organic fraction of municipal solid waste for composting is still very small, it was assumed that all the recovered bioplastic undergoes energy recovery.

2.3. Energy and LCA Analysis

An energy analysis of the system was performed by means of the energy metrics described in Section 2.3.4.

An attributional LCA modelling was adopted following the ISO 14040:2006 and ISO 14044:2006 [62,63]. This approach was chosen because it is the most applied and best established.

2.3.1. Scenarios

Three different scenarios were conceived to model the use of syngas produced by both ADSS pyrolysis and biochar gasification: (1) Pyrolysis syngas and biochar syngas to AAF for VFAs production (scenario A—AllSyn2VFA); (2) pyrolysis syngas to AAF for VFAs production, biochar syngas to energy production (scenario B—CharSyn2E); and (3) both pyrolysis syngas and biochar syngas to energy production up to complete fulfilment of the electricity and thermal energy request of the plant, the exceeding part to AAF for VFAs production (scenario C—MostSyn2E).

The electricity and thermal energy external demand were set to null by an optimization algorithm able to adjust the energy fraction produced by the plant CHP and boiler. The algorithm ensures alternatively the electric self-sustainment of the plant (scenario B) and both the electric and thermal self-sustainment of the plant (scenario C) as primary objective.

In order to assess the impact of the external energy sources choice, three scenarios were modelled on the basis of scenario A, which is the one using only external energy sources.

In the first one, named A1—ConvE, electricity is supplied by the Italian grid mix and thermal energy by the combustion of natural gas. In scenario A2—RE/Pv+Bg, electricity is provided by photovoltaic systems and thermal energy by the combustion of biogas from anaerobic digestion of energy crops, such as the so-called waxy maize. In the A3—RE/Mix+SB scenario, electricity is supplied by a mix of renewable sources and thermal energy by the combustion of solid biomass. The renewable electricity mix was determined on the basis of the 2030 outlook contained in the Integrated National Plan for Energy and Climate [64]. The plan predicts that renewable sources will provide for 55% of the electricity consumption in 2030, but in A3—RE/Mix+SB, we envisage that renewable sources will provide for 100% of the consumption mix of PHAs production plant while maintaining the same mutual relationship (Table 2).

Table 2. Renewable electricity mix composition for scenario A3—RE/Mix+SB.

Electric Energy Source	Unit	Value
Photovoltaic	%	39.92
Hydroelectric	%	26.92
Wind	%	22.66
Geothermal	%	3.88
Biogas	%	3.69
Solid biomass	%	1.87
Waste	%	1.07

2.3.2. Functional Unit and System Boundary

The functional unit (FU) was defined as 1 kg of biopolymer ready for the product's manufacturing.

The system boundary was from "cradle-to-grave": It was assumed that the bioplastic item is landfilled or incinerated. In detail, the system boundary includes (Figure 2): Sewage sludge pyrolysis, VFAs production through anaerobic digestion, PHAs-enriched biomass production using a MMC, PHAs extraction with DMC, and bioplastic items end-of-life. The analysis was carried out considering four main stages, called: "Pyrolysis and anaerobic digestion" (PyAD), "PHAs-enriched biomass production", "PHAs extraction", and "PHAs end-of-life" (PHAs EoL). The system boundary does not include product manufacturing and the additives used in polymer resins to achieve desirable material properties since these phases are the same for all scenarios; this choice is consistent with previous comparative LCA studies of biopolymers and conventional polymers [38,65–67]. Moreover, we assumed that the collection and transport of waste to various treatment plants are not relevant [68], and that even these phases are identical in all scenarios; therefore, they were not included in the system boundary. Finally, sewage sludge coming from the agro-industrial sector entering the process was considered as entering the system without any associated impact, according to the 'zero-burden-boundary' hypothesis.

The end-of-life (EoL) phase for plastic items at the European level envisages different types of treatment: Mechanical material recycling, organic carbon recycling (i.e., composting and anaerobic digestion), energy recovery, and landfilling [69,70]. Only energy recovery and landfilling for bioplastics was considered because the separated collection and recycling of bioplastic waste is scarcely implemented at the European level and in Italy, in agreement with information reported by COREPLA, the Italian national consortium for collection, recycling, and recovery of plastic packaging, as mentioned in Section 2.2.4.

Model decisions regarding the treatment of waste, and co- and by-products, are a potentially important contributor to the differences among LCA studies. System expansion, where applicable, is the baseline method for handling co- and by-products, consistent with ISO 14044 [63]. In the PHAs production model, different fates are foreseen for waste and by-products. Biological sludge from the AAF and from the PHAs extraction phase is recycled to the initial drying stage: This waste biomass was assumed to have the same composition as the inlet ADSS because the ash content concentrates in biochar; in other terms, ash exit the system and does not accumulate after every cycle. Process water is internally reused, while wastewater is sent to a WWTP and the sewage sludge produced is treated as other biosolids and used in agriculture. As regards pyrolysis co-products, bio-oil and water phases are sent to the AAF to produce VFAs, while biochar is gasified to produce biochar syngas. The latter, together with pyrolysis syngas, is alternatively sent to the AAF to produce VFAs or to energy production through CHP and a boiler in order to satisfy the energy request of the plant, depending on the scenario.

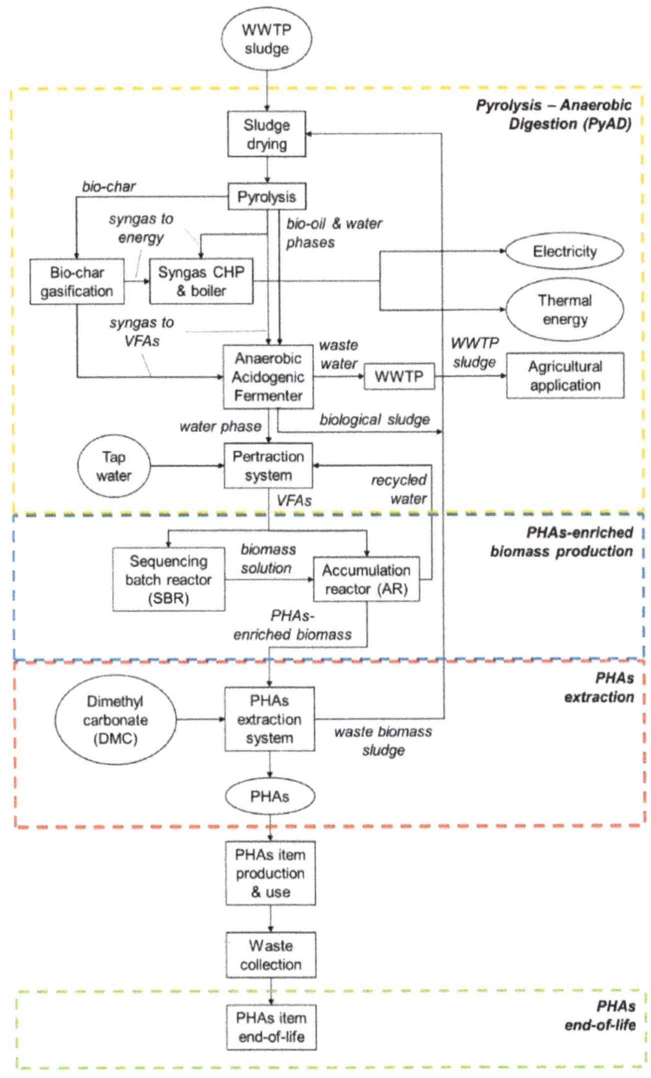

Figure 2. System boundary. The colored dotted lines delimit the four life cycle phases considered in the analysis, with the same color code used in the results graphs. The feedstock and the main product and co-products are indicated in the oval boxes, the processes in the rectangular boxes, and the flows in italics.

2.3.3. Inventory Data

The study was based on laboratory and pilot-scale data collection in accordance with the current level of maturity of the technology: An operational commercial-scale facility has not yet been implemented. Primary data were used for foreground processes, which took place at CIRI FRAME laboratories, and LCA databases were used for background processes, while estimates based on commercial and literature data were used for other processes, such as the biochar gasification, and to scale-up the whole PHAs production system.

LCA was performed using GaBi ts 8.7 software. The LCA databases used for background data were GaBi Professional Database [71] and Ecoinvent Database version 2 [72].

In Table 3, the life cycle inventory data of the main flows referred to the FU for the scenarios are presented.

Table 3. Life cycle inventory data of the main flows, referred to the functional unit (1 kg of PHAs produced). I = input, O = output.

Process Step	Flow	Input/Output	Unit	Scenarios		
				A [1]	B	C
Biomass inlet	ADSS from WWTP	I	kg	203.60	440.89	688.64
Feedstock centrifuge	sludge	I	kg	209.97	448.04	695.22
	wastewater	O	kg	176.36	378.16	588.06
Feedstock dryer	dried sludge (85% DM)	O	kg	10.33	21.16	32.17
Pyrolysis	pyrolysis water	O	kg	2.02	4.15	6.31
	pyrolysis bio-oil	O	kg	0.68	1.40	2.12
	pyrolysis biochar	O	kg	5.06	10.37	15.76
	pyrolysis syngas	O	kg	2.56	5.25	7.98
Biochar gasification	ash	O	kg	3.38	7.28	11.33
	biochar syngas	O	kg	4.88	8.99	12.89
Energy production	pyrolysis syngas to energy production	I	kg	0.00	0.00	3.54
	biochar syngas to energy production	I	kg	0.00	8.99	12.89
	electricity from syngas	O	MJ	0.00	23.71	29.21
	thermal energy from syngas	O	MJ	0.00	31.28	95.69
External energy supply [2]	electricity supply	I	MJ	18.53	0.00	0.00
	thermal energy supply	I	MJ	51.33	42.01	0.00
Anaerobic acidogenic fermenter	syngas to VFAs	I	kg	7.44	5.25	4.44
	VFAs	O	kg	4.49	4.49	4.49
AAF sludge centrifuge	concentrated biological sludge (25% DM) recycled	O	kg	1.52	2.06	2.23
	concentrated biological sludge (25% DM) to agricultural spreading	O	kg	0.42	0.16	0.27
Pertraction system	VFAs solution	O	kg	700.37	700.37	700.37
Water supply	tap water	I	kg	0.00	0.00	0.00
	recycled water	I	kg	695.88	695.88	695.88
Sequencing Batch Reactor (SBR)	VFAs solution to SBR	I	kg	243.86	243.86	243.86
	biomass solution to AR	O	kg	240.53	240.53	240.53
Accumulation Reactor (AR)	VFAs solution to AR	I	kg	456.51	456.51	456.51
	PHAs-enriched biomass solution	O	kg	68.33	68.33	68.33
PHAs extraction system	PHAs-enriched biomass dried	I	kg	8.33	8.33	8.33
	DMC	I	kg	0.19	0.19	0.19
	PHAs	O	kg	1.00	1.00	1.00
	waste biomass sludge recycled	O	kg	7.48	7.48	7.48
PHAs End-of-Life	PHAs to incineration	I	kg	0.85	0.85	0.85
	PHAs to landfill	I	kg	0.15	0.15	0.15
	electricity	O	MJ	5.72	5.72	5.72
	steam	O	MJ	10.10	10.10	10.10

[1] Data for the three A-based scenarios are reported in a single column since they differ only in the external energy source used, and not in flow figures. [2] Various energy sources, depending on scenario.

As it is possible to observe in Table 3, data for the analyzed scenarios differ only in the pyrolysis and anaerobic digestion phase.

As regards the energy recovered through the incineration process in the EoL phase, a dataset for the combustion of bio-based polypropylene (bio-PP) with a net calorific value (NCV) of 43.5 MJ/kg in a waste incineration plant was considered [71]. It was assumed that this is also the PHAs' NCV, on the basis of the physical properties and chemical structure being similar to bio-PP. To be consistent, this value was also assumed for the calculation of energy indicators.

2.3.4. Energy Balance and Energy Performance Metrics

Besides the mass balance, the energy balance was also calculated as it constitutes part of the inventory for impact assessment. Every energy figure, including primary energy (PE), is provided on an NCV basis and electricity is equally provided as NCV equivalent.

In Table 4, the energy balance for the three analyzed scenarios of the PHAs production system referring to the FU is presented. The end-of-life phase is excluded.

Table 4. Energy balance of the PHAs production system, referring to the functional unit (1 kg of PHAs produced).

Energy Demand/Supply	System Stage and Energy Type	Unit	Scenarios		
			A [1]	B	C
	VFA production				
	electricity	MJ	8.49	13.67	19.18
	thermal energy	MJ	36.14	58.11	80.50
	PHAs-enriched biomass production				
	electricity	MJ	6.09	6.08	6.08
Energy demand	thermal energy	MJ	15.20	15.19	15.19
	PHAs extraction				
	electricity	MJ	3.95	3.95	3.95
	thermal energy	MJ	0.00	0.00	0.00
	PHAs production system (total)				
	electricity	MJ	18.53	23.71	29.21
	thermal energy	MJ	51.33	73.30	95.69
	Onsite production				
	electricity from CHP	MJ	0.00	23.71	29.21
	thermal energy from CHP & boiler	MJ	0.00	31.28	95.69
Energy supply	External production [2]				
	electricity	MJ	18.53	0.00	0.00
	thermal energy	MJ	51.33	42.01	0.00

[1] Data for the three A-based scenarios are reported in a single column since they differ only in the external energy source used, and not in figures. [2] Various energy sources, depending on the scenario.

As in Table 3, also in Table 4, data of the three A-based scenarios are reported in a single column since they differ only in the external energy source used, and not in energy figures.

Several energy indicators, such as returns and ratios, are derived from the energy balance and adopted to better investigate the effects of the main factors affecting the results: The amount of ADSS used as feedstock, the energy recovery by end-of-life, and the renewable energy ratio.

Two indicators on energy return (ER) are defined:

- $ER_{PHAs, \, overall}$ (Equation (1)) expresses the energy contained in PHAs as a rate of the total energy invested in their production; the total energy invested is defined as the sum of the energy contained in input sewage sludge and the primary energy of heat and electricity sources; and all factors are measured in terms of their NCV:

$$ER_{PHAs, \, overall} = \frac{energy \ in \ PHAs}{total \ energy \ invested} = \frac{E_{PHAs}}{PE_{fossil} + PE_{renewable \ for \ H+E} + E_{sludge, \ for \ H+E} + E_{sludge, for \ PHAs}}. \tag{1}$$

- $ER_{PHAs, \, EoL}$ (Equation (2)) is a similar indicator to $ER_{PHAs, \, overall}$ but includes a term representing the energy recovered from PHAs EoL; thus, it expresses the energy return based on the net energy required for PHAs production considering also the end-of-life phase:

$$ER_{PHAs, \, EoL} = \frac{energy \ in \ PHAs}{total \ energy \ invested - energy \ recovered \ from \ EoL} = \frac{E_{PHAs}}{PE_{fossil} + PE_{renewable \ for \ H+E} + E_{sludge, \ for \ H+E} + E_{sludge, \ for \ PHAs} - E_{EoL}}. \tag{2}$$

Two other indicators named renewable energy ratio (RER) are adopted to determine the rate of renewable energy used:

- $RER_{for\ H+E}$ (Equation (3)) describes the renewable quota of energy used as heat and electricity only, thus excluding the fraction of sewage sludge transformed into PHAs:

$$RER_{for\ H+E} = \frac{\text{external renewable energy} + \text{energy in sludge for heat and electricity}}{\text{total energy invested} - \text{energy in sludge for PHAs}} = \frac{PE_{\text{renewable for H+E}} + E_{\text{sludge, for H+E}}}{PE_{\text{fossil}} + PE_{\text{renewable for H+E}} + E_{\text{sludge, for H+E}}}. \tag{3}$$

- $RER_{overall}$ (Equation (4)) describes the renewable quota of energy used in the PHAs production process as heat, electricity, and material, thus including the fraction of sewage sludge transformed into PHAs; in fact, the denominator is the total energy invested:

$$RER_{overall} = \frac{\text{external renewable energy} + \text{energy in sludge}}{\text{total energy invested}} = \frac{PE_{\text{renewable for H+E}} + E_{\text{sludge, for H+E}} + E_{\text{sludge, for PHAs}}}{PE_{\text{fossil}} + PE_{\text{renewable for H+E}} + E_{\text{sludge, for H+E}} + E_{\text{sludge, for PHAs}}}. \tag{4}$$

The last energy indicator adopted is the PHAs-to-fossil-energy ratio (Equation (5)), i.e., the energy in PHAs per unit of primary energy from fossil sources invested, as an indicator of the fossil energy use efficiency:

$$FER_{PHAs} = \frac{E_{PHAs}}{PE_{fossil}}. \tag{5}$$

2.3.5. Life Cycle Impact Assessment

In the life cycle impact assessment (LCIA) phase, environmental burdens were assessed using the flows and data referred to the FU modelled in the inventory phase.

The ILCD/PEF Recommendations v1.09 method [73,74] was used; this method considers the environmental impact under 16 impact categories, listed below: Acidification midpoint (AP); climate change midpoint, excluded biogenic carbon (GWPexc.); climate change midpoint, included biogenic carbon (GWPinc.); ecotoxicity freshwater midpoint (FE); eutrophication freshwater midpoint (EuF); eutrophication marine midpoint (EuM); eutrophication terrestrial midpoint (EuT); human toxicity midpoint, cancer effects (HTc); human toxicity midpoint, non-cancer effects (HTnc); ionizing radiation midpoint, human health (IR); land use midpoint (LU); ozone depletion midpoint (OD); particulate matter/respiratory inorganics midpoint (PM); photochemical ozone formation midpoint (POF); resource depletion water midpoint (WD); and resource depletion mineral, fossils, and renewables midpoint (RD).

The impacts due to the energy demand of the process were specifically taken into consideration through the category "primary energy from renewable and non-renewable resources" (PED, R+NR), which is not part of the aforementioned method.

Moreover, the optional phases of normalization and weighting were applied, in order to identify the most relevant impact categories in determining the overall impact of the process. The normalization factors and the weight vector used in this analysis were those proposed by PEFCR guidance [75].

The environmental performances in terms of the primary energy demand and climate change obtained for PHAs within this study were compared with those of two polymers of fossil origin, i.e., polyethylene terephthalate (PET) and polypropylene (PP), and two biobased polymers, i.e., bio-polypropylene (bio-PP) and polylactic acid (PLA). The environmental performances of these four polymers were also obtained by modeling their production and end-of-life phases through GaBi software and the related databases [71].

3. Results and Discussion

3.1. Energy Performance Indicators Results

In Table 5, the energy indicators results for the five analyzed scenarios of PHAs production system referring to the FU are presented.

Table 5. Energy indicators results referring to the functional unit (1 kg of PHAs produced). All energy indicators are measured in terms of net calorific value.

Energy Indicator	Unit	Scenarios				
		A1	A2	A3	B	C
PHAs output (energy)	MJ	43.5	43.5	43.5	43.5	43.5
ADSS input (mass)	kg	203.6	203.6	203.6	440.9	688.6
ADSS input (energy)	MJ	91.6	91.6	91.6	198.4	309.9
Renewable energy in sewage sludge transformed into PHAs	MJ	91.6	91.6	91.6	91.3	103.0
Renewable energy in sewage sludge used for H+E	MJ	0.0	0.0	0.0	107.1	206.9
PE – external supply	MJ	130.9	250.5	154.6	66.4	19.3
PE – external supply, fossil	MJ	106.2	23.5	17.1	64.3	16.5
PE – external supply, renewable	MJ	24.7	227.0	137.5	2.1	2.8
Total energy invested	MJ	222.5	342.1	246.2	264.8	329.2
Total renewable energy supply	MJ	116.3	318.6	229.1	200.5	312.7
Energy recovered by PHAs EoL	MJ	−28.7	−28.7	−28.7	−28.7	−28.7
Rate of energy in PHAs recovered by EoL	%	66.0	66.0	66.0	66.0	66.0
$ER_{PHAs,\ overall}$	%	19.5	12.7	17.6	16.4	13.2
$ER_{PHAs,\ EoL}$	%	22.4	13.9	20.0	18.4	14.5
$RER_{for\ H+E}$	%	18.9	90.6	88.8	62.9	92.7
$RER_{overall}$	%	52.3	93.1	93.0	75.7	95.0
FER_{PHAs}	-	0.41	1.85	2.53	0.68	2.63

Considering the five scenarios from an energy perspective, several factors affect the energy balance. Such factors are (a) the amount of sewage sludge used for onsite energy production besides the PHAs production, (b) the energy recovery from PHAs EoL, and (c) the renewable energy ratio. Energy indicators are used to describe and explain the impact of factors' variation.

As previously mentioned (see Section 2.3.3), it is assumed that the energy contained in PHAs is 43.5 MJ/kg, while the energy contained in ADSS is 0.45 MJ/kg.

A key indicator is the total energy invested, this being the sum of ADSS input (energy) and PE—external supply. The latter includes the contribution of both fossil-based energy and renewable sources, such as biogas and photovoltaic, depending on the scenario. The scenario with the lowest total energy invested to produce 1 kg of PHAs is A1—ConvE, because of the lowest amount of ADSS input required by the process. As an alternative to fossil sources, scenarios A2 and A3 were considered to evaluate the use of renewable energy (heat and electricity) to fuel the process. Despite the same ADSS input value of scenario A1, both in scenarioA2 and in scenario A3, the amount of total energy invested is higher because of the higher value of PE—external supply. Such a rise depends on the renewable component of the indicator (PE—external supply, renewable), whose increase more than compensates for the simultaneous decrease of the non-renewable component (PE—external supply, fossil).

In contrast to the A-based scenarios, where the entire amount of sewage sludge is used to produce PHAs, scenarios B and C are designed to use a fraction of pyrolytic products to produce onsite the energy to fulfil either the internal demand for electricity only (scenario B) or the internal demand for both electricity and heat (scenario C). For this reason, an increasing amount of ADSS is needed to produce 1 kg of PHAs in scenarios B and C: 2.17 and 3.38 times the mass flow of scenarios A, respectively. The total energy invested per kg of PHAs ranges from 222.6 MJ (scenario A1) to 329.2 MJ (scenario C), i.e., a growth of 48% of total energy invested is needed if the energy production is sourced onsite by the ADSS instead of the external energy supply. Such a rise depends on the ADSS input (energy) increase, and in particular on the renewable energy in sewage sludge used for H+E, whose increase more than compensates for the simultaneous decrease of the PE — external supply. This is partly explained by the difference in the efficiency between the CHP and boiler used to produce energy onsite, and the external power plants. Moreover, the sludge-to-energy process comprises several steps, each one characterized by energy costs and losses. In more detail, sewage sludge must be centrifuged and dried to be first pyrolyzed, the resulting biochar is gasified, and then biochar syngas and pyrolysis syngas are combusted together in the CHP and boiler.

To better highlight the differences among scenarios, we did not use the energy return on energy invested (EROEI), which is specifically adopted for energy systems and fuels, but a similar indicator, the energy return (ER) defined in Equations (1) and (2). This was used to evaluate the overall efficiency of the entire PHAs life cycle, including the end-of-life stage (Equation (2)).

Energy return indicators show that PHAs contain about one eighth up to a fifth of all the inputs used for its production, both energy and materials, all of them on an energy basis. If energy recovery from the EoL phase is also taken into account, such figures increase. $ER_{PHAs, \, overall}$ ranges from 12.7% to 19.5% and $ER_{PHAs, \, EoL}$ from 13.8% to 22.2%. Comparing $ER_{PHAs, \, overall}$ and $ER_{PHAs, \, EoL}$, the latter is higher in the range between 1.2 and 2.9 percentage points due to the energy recovery by PHAs EoL. If a life-cycle perspective is assumed, EoL contributes to a reduction of the total energy invested to produce PHAs; it was calculated that 66% of the NCV contained in PHAs is recovered at the PHAs EoL stage by incineration of PHAs and the combustion of biogas from PHAs degradation in landfill. Moreover, the contribution of the EoL energy recovery can completely offset the external supply of PE, such as in scenario C, where the external PE is even lower than the energy recovery by EoL. For both these energy return indicators, it is found that the lower the ADSS amount used for energy, the greater the return (A1 > B > C).

From a process perspective, the amount of energy used per FU is an indicator of the production efficiency, but it is not enough to address the viability and impacts of the process. For this purpose, other factors, such as the renewability of energy sources, must be taken into account. The renewable energy used is expressed as the renewable energy ratio considering the sole energy used for process heat and electricity ($RER_{for \, H+E}$) or including every energy input (i.e., heat, electricity, sewage sludge NCV, PE) in the PHAs production ($RER_{overall}$). The onsite production of energy from ADSS enhances the renewable energy ratio ($RER_{for \, H+E}$) from 19% in scenario A1 up to 63% in scenario B and 93% in scenario C. Results comparable with the latter are achieved in scenarios A2 (91%) and A3 (89%), where the entire heat and electricity supply comes from external and different renewable sources. The overall renewable energy ratio ($RER_{overall}$) takes into account also the NCV of the sewage sludge transformed into PHAs. Compared to the $RER_{for \, H+E}$ results, $RER_{overall}$ shows higher values, especially for scenario A1, due to the sewage sludge NCV, which almost equals the fossil PE. For all the other scenarios, $RER_{overall}$ is just slightly higher than $RER_{for \, H+E}$. Scenario C is the one with the highest RER and with the lowest ER (excluding scenario A2) for the same reason: The use of sewage sludge to produce onsite renewable heat and electricity.

3.2. LCIA Results

The overall LCIA results for the five scenarios are reported in Table 6.

Detailed results for the five scenarios are shown in Figure 3, where the relative contribution of each LCA phase—pyrolysis and anaerobic digestion (PyAD), PHAs-enriched biomass production, PHAs extraction, polymer end-of-life (PHAs EoL)—to each midpoint impact category score are reported and compared.

Table 6. Life cycle impact assessment results for the five analyzed scenarios (A1, A2, A3, B, C).

Impact Category	Abbrev.	Unit	Scenarios				
			A1	A2	A3	B	C
Primary energy demand from ren. and non ren. resources (as NCV)	PED, R+NR	MJ	1.0×10^2	2.2×10^2	1.3×10^2	3.8×10^1	-9.3×10^0
Acidification midpoint	AC	mole of H+ eq.	6.1×10^{-3}	1.9×10^{-2}	6.8×10^{-3}	8.3×10^{-3}	1.4×10^{-2}
Climate change midpoint, excluded biogenic carbon	GWPexc.	kg CO_2 eq.	5.2×10^0	8.7×10^{-1}	-3.0×10^{-1}	2.5×10^0	-3.0×10^{-1}
Climate change midpoint, included biogenic carbon	GWPinc.	kg CO_2 eq.	2.0×10^1	1.6×10^1	1.4×10^1	2.2×10^1	2.4×10^1
Ecotoxicity freshwater midpoint	FE	CTUe	5.7×10^1	5.7×10^1	5.7×10^1	1.2×10^2	1.9×10^2
Eutrophication freshwater midpoint	EuF	kg P eq.	8.0×10^{-4}	1.1×10^{-3}	8.5×10^{-4}	1.7×10^{-3}	2.6×10^{-3}
Eutrophication marine midpoint	EuM	kg N eq.	2.8×10^{-3}	1.0×10^{-2}	4.1×10^{-3}	5.8×10^{-3}	9.8×10^{-3}
Eutrophication terrestrial midpoint	EuT	mole of N eq.	2.4×10^{-2}	6.8×10^{-2}	2.4×10^{-2}	4.8×10^{-2}	8.2×10^{-2}
Human toxicity midpoint, cancer effects	HTc	CTUh	1.1×10^{-6}	1.1×10^{-6}	1.1×10^{-6}	2.3×10^{-6}	3.5×10^{-6}
Human toxicity midpoint, non-cancer effects	HTnc	CTUh	1.5×10^{-6}	1.8×10^{-6}	1.8×10^{-6}	3.2×10^{-6}	5.0×10^{-6}
Ionizing radiation midpoint, human health	IR	kBq U235 eq.	2.4×10^{-2}	-2.3×10^{-1}	-2.6×10^{-1}	-2.5×10^{-1}	-2.2×10^{-1}
Land use midpoint	LU	kg C deficit eq.	2.3×10^0	4.2×10^1	1.7×10^1	1.4×10^0	2.2×10^0
Ozone depletion midpoint	OD	kg CFC-11 eq.	2.3×10^{-8}	2.3×10^{-8}	2.3×10^{-8}	2.3×10^{-8}	2.3×10^{-8}
Particulate matter/Respiratory inorganics midpoint	PM	kg PM2.5 eq.	2.5×10^{-4}	7.7×10^{-4}	3.1×10^{-4}	3.3×10^{-4}	5.7×10^{-4}
Photochemical ozone formation midpoint, human health	POF	kg NMVOC	7.4×10^{-3}	1.3×10^{-2}	7.5×10^{-3}	1.3×10^{-2}	2.1×10^{-2}
Resource depletion water, midpoint	WD	m^3 eq.	5.3×10^{-2}	4.4×10^{-2}	-2.1×10^{-3}	-2.5×10^{-2}	-2.0×10^{-2}
Resource depletion, mineral, fossils and renewables, midpoint	RD	kg Sb eq.	3.2×10^{-6}	3.6×10^{-5}	1.0×10^{-5}	-1.6×10^{-6}	-3.1×10^{-6}

3.2.1. Scenarios' Performance Comparison

Table 6 shows the total impact scores of the 5 analyzed scenarios over the 17 chosen impact categories. A clear pattern emerges from the data when considering the number of impact categories in which each scenario shows better results than the others or, in other words, when all 17 impact categories are considered equivalent. From this point of view, scenario A1—ConvE performs better than the other ones but only slightly better than scenario A3—RE/Mix+SB. In more detail, scenario A1 scores best in six categories (AC, EuF, EuM, POF, PM, HTnc), and in four more categories, it is first on an equal footing (EuT, FE, HTc, and OD). Scenario A3 shows the best results in two categories (GWPinc. and IR), and in five more categories, it is first on an equal footing (EuT, FE, GWPexc., HTc, and OD), four of them with scenario A1, and is the second best in six categories (AC, EuF, EuM, POF, PM, HTnc, the latest on an equal footing). Moreover, A1 outperforms scenario A3 in nine categories, including PED, R+NR, and RD, while scenario A3 outperforms scenario A1 in four categories, including GWPexc. and GWPinc. Scenario B—CharSyn2E has medium performances, showing the best results in two categories (LU and WD); in another one, it is first on an equal footing with all the other scenarios (OD), and it never shows the worst results. Scenario A2—RE/Pv+Bg shows the worst results in six categories (AC, EuM, LU, PM, RD, PED, R+NR) and the best in three categories (FE, HTc, and OD), all of them on an equal footing with other scenarios. Scenario C—MostSyn2E shows the worst performances in seven categories (GWPinc., EuF, EuT, FE, HTc, HTnc, POF) but also the best in the other four, even if in two of them (GWPexc. and OD) on an equal footing. It is worth noting that three of the impact categories where it shows the best results are GWPexc., PED, R+NR, and RD, which is the main impacts—often the only ones—considered when comparing the environmental performances of plastics and/or bioplastics. These categories can be considered on a global scale, since the impact measured by the midpoint indicator, e.g., the global warming potential, has an effect on a global scale. On the other hand, it is possible to state that scenario C generally shows higher values than the A-based scenarios in the impact categories AC, EuF, EuM, EuT, FE, HTc, HTnc, PM, and POF. These categories can be considered on the local and regional scale, since the impact measured by the midpoint indicator, e.g., the acidifying or eutrophicating potentials, has an effect on a local or regional scale. Therefore, the results exclude that scenarios A1 and A3 are superior to scenario 3 when considering only a small number of impact categories, or when their relative importance is also taken into account. In other words, the results cannot establish in a straightforward way whether one scenario prevails over the others, as the ranking depends on the method adopted to establish it. An insight into the reasons and origins of this trade-off is given in the following Section 3.2.2, after analyzing in detail the impacts in relevant categories.

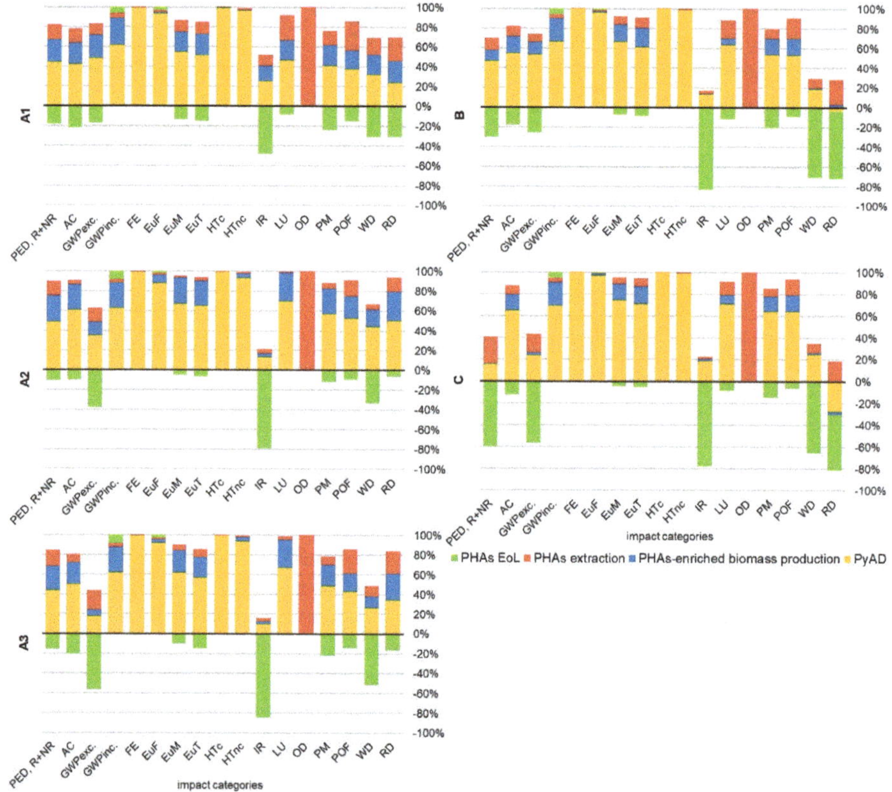

Figure 3. Relative contribution of LCA phases to midpoint impact categories scores for each of the five analyzed scenarios (A1, A2, A3, B, C).

The only category equally impacted in all five scenarios is ozone depletion (2.3×10^{-8} kg CFC-11 eq.). This finding can be easily explained since the impact derives almost completely from the PHAs extraction process that is the same for each scenario. In particular, the impact is attributable to the emission into the atmosphere of organic halogenates, mainly bromochlorodifluoromethane (Halon 1211), deriving from the methanol production process that is the base-chemical of DMC synthesis.

The differences among the results of the five scenarios are high for many impact categories, even one order of magnitude. Generally, differences are higher than ±50%; apart from the OD category, climate change included biogenic carbon is the one showing the lowest percent differences.

3.2.2. Detailed Analysis of Relevant Impact Categories Results

As explained in Section 2.3.5, results were also calculated using the Environmental Footprint 2.0 method in order to determine relevant impact categories, i.e., those cumulatively accounting for 80% of the total impact score. Applying the normalization and weighting factors proposed by PEFCR guidance [75], the identified categories are (in decreasing order): GWP, WD, RD, EuF, POF, and respiratory inorganics.

Considering this, the analysis of the relative contribution of the processes was focused on the following impact categories: GWPexc., EuF, WD, POF, PM (as proxy for "respiratory inorganics"), and RD. The PED, R+NR impact category is also discussed. Figure 3 shows the relative contributions of PyAD, PHAs-enriched biomass production, PHAs extraction, and PHAs EoL phases.

Differences among the five scenarios for the primary energy demand from renewable and non-renewable resources are high. The energy recovered in the EoL phase is the same for every scenario, while energy required in PHAs production varies widely (see Table 4) as a function of both the energy source and the input feedstock, which increases from scenarios A to scenario B, and from scenario B to scenario C. In fact, a greater quantity of processed ADSS is progressively used, until energy independence is reached in scenario C. It must be remembered that in the LCA analysis, the input ADSS is considered as "zero-burden", i.e., without any associated environmental impact, in accordance with what is generally applied in life cycle analyses of waste. This implies that the PED, R+NR value of ADSS is zero. On the contrary, external sources supplying energy in A-based scenarios and in scenario B have positive PED, R+NR values, i.e., they have associated environmental impacts. For this reason, the PED, R+NR decreases as the processed ADSS increases. The results for A-based scenarios clearly show that the PED, R+NR values for external renewable sources are not lower than those of fossil sources; quite obviously, their high value mainly derives from the renewable part of the indicator, while that of the fossil sources mainly derives from the non-renewable part. This is in accordance also with the findings from the analysis of energy performance indicators (see Section 3.1).

In scenarios A1, A2, A3, and B, the life cycle phase causing the greatest impacts in terms of PED, R+NR is PyAD, and in particular the pertraction system, due to its high energy consumption. In scenario C, the phase responsible for the greatest impacts is PHAs extraction, due to the background process of DMC production, an energy-intensive process. Thanks to the energy recovery from the bioplastic products, the EoL phase leads to environmental credits. These credits, together with the energy self-sufficiency of the process, lead to the negative PED, R+NR value for scenario C.

Differences among the five scenarios for the climate change, excluded biogenic carbon category, are high, in some cases even more than one order of magnitude. The contributions to climate change scores are mainly due to PyAD and PHAs-enriched biomass production in scenario A1; to PyAD and PHAs EoL in scenarios A2, B, and C; and to PHAs extraction and PHAs EoL in scenario A3. Relevant flows analysis indicates that the carbon dioxide emissions due to thermal energy and electricity production processes are key contributors in every scenario. Generally, the lower the non-renewable energy use, the better the results for GWPexc. In scenarios A3 and C, i.e., the two best performing scenarios, carbon dioxide savings are due to credits for energy production in the PHAs EoL phase. These credits are subtracted from the emissions due to the other life cycle phases, resulting in an overall environmental benefit for this impact category.

Differences among the five scenarios for eutrophication of freshwater are less than one order of magnitude. EuF is mainly related to PyAD in every scenario. In particular, it is related to the phosphorous emissions to freshwater deriving from WWTP and related sludge agricultural spreading, and to phosphate emissions to freshwater from biochar gasification ash disposal. Coherently, the impact increases as the amount of input ADSS to be processed increases.

As regards water depletion, scenarios are grouped into three clusters: The highest impact values are observed for A1 and A2, and the lowest ones for B and C, while A3 shows intermediate values. The relevant processes related to water depletion are different for the various scenarios. In scenario A1, this impact is roughly equally distributed among the different processes and PHAs EoL leads to a water saving of about 30%. In scenario A2, the contribution of PHAs extraction and PHAs-enriched biomass production decreases, while that of PyAD increases, as well as the percentage saving due to EoL. In scenario A3, the contribution of PyAD and PHAs-enriched biomass production decreases substantially, while that of PHAs extraction increases slightly; however, credits obtained from PHAs EoL lead to an overall environmental benefit. In scenarios B and C, WD is an "avoided impact" too, and for more than 50%, thanks to PHAs EoL. These savings are due to credits for energy production through plastics incineration, while the impacts in the other life cycle stages in every scenario are always due to electricity consumption from the grid mix.

Resources depletion shows a very similar pattern to that of water depletion for scenario A1, where the non-renewable elements and energy resource consumption for energy production have a high impact. The impact grows further in scenarios A3 and A2, in particular for the PyAD and PHAs-enriched biomass production phases, even if in these scenarios are due to the consumption of renewable energy sources. PyAD in scenario B and both PyAD and PHAs-enriched biomass production in scenario C show negative scores. This is due to credits obtained for the use in agriculture of WWTP sludge deriving from process water treatment, which implies an avoided consumption of phosphate ore for chemical fertilizer production. In scenarios B and C, the negative score is also due to the avoided consumption of non-renewable elements and energy resources for energy production. There are other impacts coming from non-renewable elements and resource consumption for methanol production in the PHAs extraction phase in all scenarios.

As mentioned above, two other relevant impact categories are POF and respiratory inorganics. Differences among the five scenarios for photochemical ozone formation are less than one order of magnitude and the results show a pattern similar to the eutrophication of freshwater. Impacts are mainly related to PyAD, in particular to nitrogen oxide emissions from syngas CHP and the boiler used for electricity and thermal energy production in scenarios B and C, and from biogas combustion in scenario A2. Coherently, the impact increases as the amount of syngas and biogas used to produce energy increases. As a proxy for respiratory inorganics, not present among the ILCD-recommended impact categories, it is possible to analyze the impact category of particulate matter. Additionally, for this category, the differences among the five scenarios are less than one order of magnitude, and the impact score is primarily due to PyAD, in particular to PM2.5, i.e. PM with a diameter of 2.5 µm or less, and nitrogen oxides emissions from syngas CHP and the boiler used for electricity and thermal energy production in scenarios B and C, and from thermal energy production through biogas combustion in scenario A2. In this case, biogas combustion causes a greater impact than syngas combustion.

Summarizing the contributions analysis, it is possible to claim that PyAD is generally the most relevant phase for each scenario and each relevant impact category, and that when there are exceptions, they are related to the avoided impacts obtained by PHAs EoL. The environmental impacts of PyAD in all scenarios are essentially due to the energy production processes, both from fossil and renewable sources, and also from the syngas CHP and boiler when present. Among the processes with the highest energy needs is the pertraction system used to transfer VFAs from dilute streams of anaerobic digestion into a VFAs-rich stream for PHAs production. The temperature difference between mesophilic anaerobic digestion with sporogenic bacteria and PHAs production with MMC results in a high thermal energy demand in the pertraction system because of the cooling of large volumes of liquid from AAF, which must be heated again to a mesophilic temperature before being recirculated back to AAF. As demonstrated by comparing the different scenarios, and in particular the A-based ones, the replacement of fossil energy sources with renewable energy sources does not allow ipso facto for a reduction of the environmental impacts due to this process in all the analyzed impact categories, but there is a trade-off between global and local impacts.

Generally, renewable energy sources cause lower fossil carbon emissions and a lower use of fossil primary energy and fossil resources, that is, lower impact values in global-scale categories; however, their total primary energy and their resource use is higher when renewable energies and resources are taken into account, but this is not the case when organic waste, such as ADSS, which is considered "impact free", i.e., without any associated environmental impact, is used as an energy source. On the other hand, renewable energy production and particularly the production and combustion of biomass shows higher scores in regional- and local-scale impact categories; this is mainly due to the fact that biomass production requires vast agricultural areas and intensive cultivation methods, including the use of chemical fertilizers, and that biomass combustion is often less efficient than fossil fuel combustion, generating emissions and waste with high acidifying and eutrophicating potential, and with toxic effects.

However, it is possible to note that the A3—RE/Mix+SB scenario generally has good performances both on global impacts, such as climate change, thanks to the use of renewable resources, and on local ones, such as the water depletion, particulate matter emission, and eutrophication categories. The latter is thanks to the fact that the combustion processes of solid biomass (consisting mainly of by-products) have lower impacts compared to the combustion of biogas produced from ad hoc grown biomass (A2—RE/Pv+Bg), and that the renewable electricity mix has lower impacts than the electricity mix used in the "conventional energy" scenario (A1—ConvE).

The modelled end-of-life phase leads to environmental credits being obtained. In fact, electricity and thermal energy is produced from both PHAs' direct combustion in waste-to-energy plants and the combustion of recovered biogas generated by PHAs' degradation in landfills. Thanks to this energy production, an equivalent amount of energy production from non-renewable resources, with greater environmental impact, is avoided.

An undeniable advantage of PHAs is their biodegradability, which at the moment is not fully considered and evaluated within the LCA methodology. While methodological research in this field shall go on, in the near future, a different end-of-life for these polymers is foreseeable as their diffusion increases, that is, their collection together with organic waste, followed by the recovery of the carbon content through composting or anaerobic digestion. At that stage, it will be appropriate and interesting to assess which end-of-life of the polymer will bring the greatest benefits.

3.2.3. Comparison with PHAs Literature and Other Polymers' Results

In terms of environmental performance, in general, bio-based products tend to compare poorly against their respective conventional counterparts on impact categories, such as eutrophication, an impact typically caused by fertilizers during biomass cultivation [76,77]. Mixed results have been reported for other categories, including acidification and tropospheric ozone formation. Claims of environmental benefits for bioproducts often rely on reduced greenhouse gas (GHG) emissions and non-renewable energy use. In any case, the wide ranges of objectives, scopes, methodological choices, and results of LCA studies on bio-based products makes it difficult not only to infer a general trend on their environmental impacts but also to suggest a widely applicable way to reduce their impacts [76]. However, as regards PHAs production specifically, and as it is shown in the following comparison with the literature, the choice of feedstock and the use of residues and co-products for energy production can reduce the environmental impacts of this process [12,31,33,34,36,40,41].

In Figure 4, the values of non-biogenic GHGs emissions per kilogram of PHAs obtained by various authors are compared; data are in chronological order and the whole range of results found in each paper, including this study, is represented when available. Figure 4 shows also the values for four other polymers with which the PHAs are compared with, two of them are fossil-based (PET and PP) and two bio-based (bio-PP and PLA).

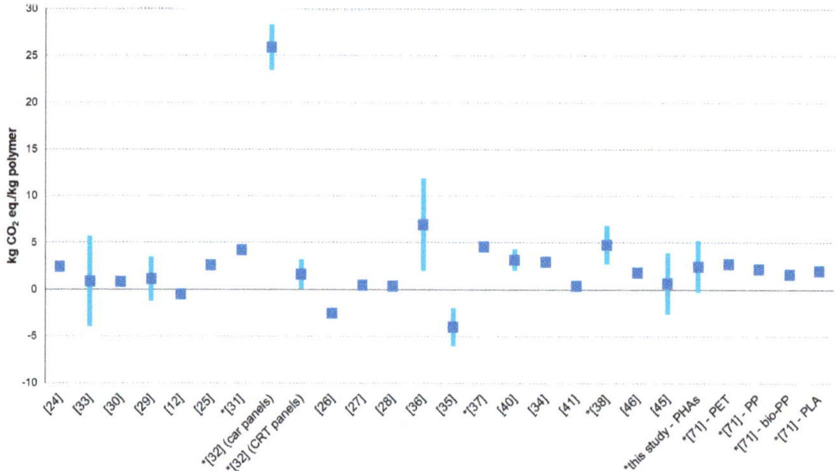

Figure 4. Greenhouse gases emissions reported in previous PHAs production studies compared to those of PHAs production investigated in this study and those calculated for other polymers. The symbol "*" before reference number indicates that the system boundaries are "cradle-to-grave", otherwise they are "cradle-to-gate". The whole range of results of each study are represented when available.

Generally, values range from 0.5 to 5 kg of CO_2-equivalent per kilogram of PHAs. The results of the present study fall within this range. However, great variability exists among the studies. A potential impact up to about 25 kg CO_2-equivalent per kg of PHAs was assessed by Pietrini and coauthors [32] and an avoided impact of up to about 6 kg CO_2-equivalent per kg of PHAs was evaluated by Rostkowski and coauthors [35]. Higher values were reported by Pietrini and coauthors [32], Kendall [36], and Posen and coauthors [38] and they correspond to sugar cane- and corn-derived polyhydroxybutyrate (PHB). Lower values (i.e., below 0.5 kg CO_2-equivalent emitted per kilogram of biopolymer) were reported by studies that account for carbon uptake during biomass growth, thereby considering carbon storage in the bio-based polymer [26–28,30] or when the only carbon source is a waste [12,41]. Negative impacts were reported when the energy source [33] or also when waste streams from PHAs production are used for energy recovery [35].

In Figure 5, the fossil energy requirement for PHAs production reported by various authors is compared; data are in chronological order and the whole range of results found in each paper, including this study, are represented when available. Figure 5 shows also the values for four other polymers with which the PHAs are compared; two of them are fossil-based (PET and PP) and two bio-based (bio-PP and PLA).

Data range from 320 to −12 MJ per kilogram of PHAs, with median values spanning between 20 to 80 MJ per kilogram of biopolymer. The present work shows values ranging from about 80 to −10 MJ per kg of PHAs, depending on the considered scenarios. The highest values were reported by Pietrini and coauthors [32] and Sakamoto [37], where carbon sources are from dedicated crops and the whole life cycle of PHB-based end products is considered. The lowest values were reported in studies where the use of organic residues as fuel to generate electricity and steam is assumed [26,29].

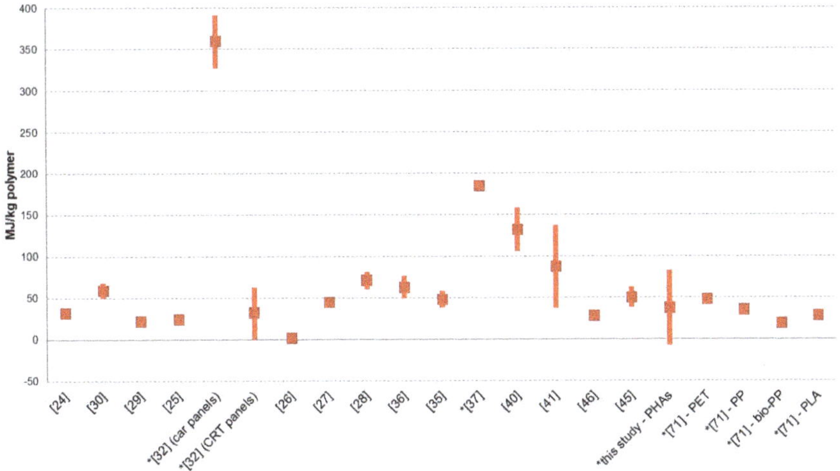

Figure 5. Non-renewable energy demand reported in previous PHAs production studies compared to that of PHAs production investigated in this study and that calculated for other polymers. The symbol "*" before reference number indicates that the system boundaries are "cradle-to-grave", otherwise they are "cradle-to-gate". The whole range of results of each study are represented when available.

As regards the comparison of PHAs from ADSS with other polymers, first of all, it can be noted that the bio-based polymers tend to have both lower non-biogenic GHGs emissions and lower fossil energy demand values than the fossil-based polymers, as expected. For both indicators, the PHAs' range of values overlaps the range of other polymers' values, being wider. PHAs' average value is only slightly higher than that of bio-PP and PLA, and falls within the range of fossil-based PET and PP.

In summary, the performances of PHAs produced from sewage sludge of the agro-industry analyzed in this study are of the same order of magnitude as those of other PHAs production processes found in the literature, in terms of both non-biotic GHG emissions and energy demand. Moreover, their performances also overlap with those of other fossil-based and bio-based polymers. This is a good omen, considering that there is still room for improvement for this process and that no primary data is available at the industrial scale yet.

4. Conclusions

From the analysis of the energy indicators, it was possible to observe that the use of ADSS for onsite energy production (scenario C—MostSyn2E) results in an increase of the overall renewable energy ratio (95%), while decreasing the PE (−85%), the fossil energy used (−84%), and the dependency from the external supply, despite a lower energy return and a higher total energy invested. On the other hand, the use of a mix of external sources of renewable energy, such as in scenario A3—RE/Mix+SB, can allow for a higher energy return and a lower total renewable energy invested, while keeping both the renewable energy ratio and fossil energy efficiency use at high levels. PHAs EoL can further enhance the energy return by allowing for a recovery of up to 66% of PHAs energy.

From the LCA analysis, it was possible to observe that differences among the five scenarios for PED, R+NR, GWPexc., WD, and RD are high. Smaller differences, less than one order of magnitude, were observed for EuF, PM, and POF. The only category equally impacted in all five scenarios was OD. It is possible to conclude that scenario C—MostSyn2E is the best in climate change, primary energy demand from renewable and non-renewable resources, and resources depletion categories, while scenario A1—ConvE prevails on acidification, eutrophication, and generally in categories related to local and regional impacts. The results highlight a trade-off between local/regional impacts and global ones. It cannot be established in a straightforward way whether one scenario prevails over the others, as the ranking depends on the method adopted to establish it. However, also from the LCA perspective, the use of a mix of renewable sources can help find a balance between opposite scenarios, and contribute to significantly lowering the global impacts while keeping local ones at low levels.

PyAD is generally the life cycle phase showing the highest impacts, essentially due to the high energy demand and the related combustion processes; on the other hand, PHAs EoL usually generates avoided impacts. The innovative pertraction system has high thermal energy needs, due to the unavoidable small temperature gap, but multiplied by very large volumes, between VFAs-producing anaerobic acidogenic fermentation and PHAs-producing aerobic reactors with MMC.

Anyway, the PHAs produced from sewage sludge analyzed in this study already show environmental performances comparable to those of both fossil-based and bio-based polymers, in terms of both non-biotic GHG emissions and energy demand. New data at the industrial scale and the improvement of technology, besides the use of renewable energy sources, will make this process a competitive candidate for the production of biopolymers on a wide scale, also considering the vast availability of low-value sustainable feedstock.

Author Contributions: Conceptualization, D.M., P.G. and S.R.; Data curation, L.V., S.M., C.T. and C.S.; Formal analysis, L.V., S.M. and C.T.; Funding acquisition, D.M., P.G. and S.R.; Investigation, L.V., S.M., C.T. and C.S.; Methodology, L.V., S.M., D.M. and S.R.; Project administration, P.G. and S.R.; Resources, L.V., P.G., C.T., C.S. and S.R.; Supervision, S.R.; Validation, D.M., P.G. and S.R.; Visualization, L.V.; Writing—original draft, L.V., S.M. and S.R.; Writing—review & editing, L.V., S.M., D.M., P.G., C.T., C.S. and S.R. All authors have read and agreed to the published version of the manuscript.

Funding: This research was funded by both Regione Emilia-Romagna through European Regional Development Fund, grant number PG/2015/735284 ("VALSOVIT" project), and EIT Climate-KIC through both Pathfinder program ("BioGrapPa" project) and Accelerator program ("B-PLAS" project).

Acknowledgments: The authors wish to thank Alberto Novi of thinkstep Ltd. and Filippo Baioli of CIRSA—University of Bologna for their support and their advice in modelling with GaBi software, and also the company Caviro Extra Ltd. for supplying the sludge samples for the experiments.

Conflicts of Interest: The authors declare no conflict of interest. The funders had no role in the design of the study; in the collection, analyses, or interpretation of data; in the writing of the manuscript, or in the decision to publish the results.

Abbreviations

AAF	Anaerobic acidogenic fermenter
AD	Anaerobic digestion
ADSS	Anaerobically digested sewage sludge
AP	Acidification midpoint
AR	Accumulation reactor
Bg	Biogas
Bio-PP	Bio-polypropylene
CHP	Combined heat and power
COD	Chemical oxygen demand
COD_{eq}	Chemical oxygen demand equivalent
DM	Dry matter
DMC	Dimethyl carbonate
E_{EoL}	Energy recovered after PHAs end-of-life treatment, as net calorific value
E_{PHAs}	Energy contained in PHAs, as net calorific value
E_{sludge}	Energy contained in sludge, as net calorific value
$E_{sludge,\ for\ H+E}$	Energy contained in the sludge fraction used to produce heat and electricity, as net calorific value
$E_{sludge,\ for\ PHAs}$	Energy contained in the sludge fraction transformed into PHAs, as net calorific value

EoL	End-of-life
ER	Energy return
$ER_{PHAs,\ EoL}$	Energy return as the rate of the net energy required (invested minus recovered) for PHAs production contained in PHAs
$ER_{PHAs,\ overall}$	Energy return as the rate of the total energy invested in PHAs production contained in PHAs
EROEI	Energy return on energy invested
EuF	Eutrophication freshwater midpoint
EuM	Eutrophication marine midpoint
EuT	Eutrophication terrestrial midpoint
FE	Ecotoxicity freshwater midpoint
FER_{PHAs}	Fossil energy ratio as the energy contained in PHAs per unit of primary energy from fossil resources invested
FU	Functional unit
GHG	Greenhouse gas
GWPexc.	Climate change midpoint, excluded biogenic carbon
GWPinc.	Climate change midpoint, included biogenic carbon
H+E	Heat and electricity production
HB	Hydroxybutyrate
HH	Hydroxyhexanoate
HTc	Human toxicity midpoint, cancer effects
HTnc	Human toxicity midpoint, non-cancer effects
HV	Hydroxyvalerate
IR	Ionizing radiation midpoint, human health
LCA	Life cycle assessment
LU	Land use midpoint
Mix	Electricity grid mix
MMC	Mixed microbial culture
NCV	Net calorific value
OD	Ozone depletion midpoint
PE	Primary energy
PE_{fossil}	Primary energy from fossil sources (heat, electricity, other inputs)
$PE_{renewable\ for\ H+E}$	Primary energy from renewable sources externally supplied for heat and electricity
PED, R+NR	Primary energy demand from renewable and non-renewable resources
PET	Polyethylene terephthalate
PHAs	Polyhydroxyalkanoates
PHB	Polyhydroxybutyrate
PLA	Polylactic acid
PM	Particulate matter / Respiratory inorganics midpoint
PM2.5	Particulate matter with a diameter of 2.5 μm or less
POF	Photochemical ozone formation midpoint
PP	Polypropylene
Pv	Photovoltaic
PyAD	Pyrolysis followed by anaerobic digestion
RD	Resource depletion mineral, fossils and renewables
RE	Renewable energy
RER	Renewable energy ratio
$RER_{for\ H+E}$	Renewable energy ratio as the renewable quota of energy used only for heat and electricity production
$RER_{overall}$	Renewable energy ratio as the renewable quota of total energy invested in PHAs production
SB	Solid biomass
SBR	Sequencing batch reactor
TOA	Trioctylamine
TOA10-B	Biodiesel-based liquid membrane with 10% in weight of trioctylamine
VFAs	Volatile fatty acids
WD	Resource depletion water midpoint
wt	Weight
WWTP	Wastewater treatment plant

References

1. EC (European Commission). *Communication from the Commission to the European Parliament, the Council, the European Economic and Social Committee and the Committee of the Regions*; Closing the loop—An EU action plan for the Circular Economy. COM (2015) 614 final; EC: Brussels, Belgium, 2015.
2. Almendro-Candel, M.B.; Gómez Lucas, I.; Navarro-Pedreño, J.; Zorpas, A.A. Physical Properties of Soils Affected by the Use of Agricultural Waste. In *Agricultural Waste and Residues*, 1st ed.; Aladjadjiyan, A., Ed.; IntechOpen: London, UK, 2018. Available online: https://www.intechopen.com/books/agricultural-waste-and-residues/physical-properties-of-soils-affected-by-the-use-of-agricultural-waste (accessed on 7 April 2020). [CrossRef]
3. European Bioplastics. Sound LCA as a Basis for Policy Formulation. 2019. Available online: https://docs.european-bioplastics.org/publications/pp/EUBP_PP_LCA_as_a_basis_for_policy_formulation.pdf (accessed on 7 April 2020).
4. World Economic Forum; Ellen MacArthur Foundation and McKinsey & Company. The New Plastics Economy—Rethinking the Future of Plastics. Available online: https://www.newplasticseconomy.org/assets/doc/EllenMacArthurFoundation_TheNewPlasticsEconomy_Pages.pdf (accessed on 7 April 2020).
5. Rodriguez-Perez, S.; Serrano, A.; Pantión, A.A.; Alonso-Fariñas, B. Challenges of scaling-up PHA production from waste streams. A review. *J. Environ. Manag.* **2018**, *205*, 215–230. [CrossRef]
6. Serafim, L.S.; Lemos, P.C.; Albuquerque, M.G.E.; Reis, M.A.M. Strategies for PHA production by mixed cultures and renewable waste materials. *Appl. Microbiol. Biotechnol.* **2008**, *81*, 615–628. [CrossRef]
7. Eshtaya, M.K.; Rahman, N.A.; Hassan, M.A. Bioconversion of restaurant waste into Polyhydroxybutyrate (PHB) by recombinant E. coli through anaerobic digestion. *Int. J. Environ. Waste Manag.* **2013**, *11*, 27–37. [CrossRef]
8. Bugnicourt, E.; Cinelli, P.; Lazzeri, A.; Alvarez, V. Polyhydroxyalkanoate (PHA): Review of synthesis, characteristics, processing and potential applications in packaging. *Express Polym. Lett.* **2014**, *8*, 791–808. [CrossRef]
9. Nielsen, C.; Rahman, A.; Rehman, A.U.; Walsh, M.K.; Mille, C.D. Food waste conversion to microbial polyhydroxyalkanoates. *Microb. Biotechnol.* **2017**, *10*, 1338–1352. [CrossRef] [PubMed]
10. CSU (California State University Chico Research Institute). *Performance Evaluation of Environmentally Degradable Plastic Packaging and Disposable Food Service Ware—Final Report*; The California Integrated Waste Management Board: California, CA, USA, 2007; pp. 1–40.
11. Koller, M. Advances in polyhydroxyalkanoate (PHA) production. *Bioengineering* **2017**, *4*, 88. [CrossRef]
12. Gurieff, N.; Lant, P. Comparative life cycle assessment and financial analysis of mixed culture polyhydroxyalkanoates production. *Bioresour. Technol.* **2007**, *98*, 3393–3403. [CrossRef] [PubMed]
13. Koller, M.; Maršálek, L.; de Miranda Sousa Dias, M.; Braunegg, G. Producing microbial polyhydroxyalkanoate (PHA) biopolyesters in a sustainable manner. *New Biotechnol.* **2017**, *37*, 24–38. [CrossRef]
14. Samorì, C.; Kiwan, A.; Torri, C.; Conti, R.; Galletti, P.; Tagliavini, E. Polyhydroxyalkanoates and Crotonic Acid from Anaerobically Digested Sewage Sludge. *ACS Sustain. Chem. Eng.* **2019**, *7*, 10266–10273. [CrossRef]
15. Valentino, F.; Moretto, G.; Lorini, L.; Bolzonella, D.; Pavan, P.; Majone, M. Pilot-Scale Polyhydroxyalkanoate Production from Combined Treatment of Organic Fraction of Municipal Solid Waste and Sewage Sludge. *Ind. Eng. Chem. Res.* **2019**, *58*, 12149–12158. [CrossRef]
16. Battista, F.; Frison, N.; Pavan, P.; Cavinato, C.; Gottardo, M.; Fatone, F.; Eusebi, A.L.; Majone, M.; Zeppilli, M.; Valentino, F. Food wastes and sewage sludge as feedstock for an urban biorefinery producing biofuels and added-value bioproducts. *Chem. Technol. Biotechnol.* **2020**, *95*, 328–338. [CrossRef]
17. Werker, A.; Bengtsson, S.; Korving, L.; Hjort, M.; Anterrieu, S. Consistent production of high quality PHA using activated sludge harvested from full scale municipal wastewater treatment—PHARIO. *Water Sci. Technol.* **2018**, *78*, 2256–2269. [CrossRef] [PubMed]
18. Kumar, M.; Ghosh, P.; Khosla, K.; Thakur, I.S. Recovery of polyhydroxyalkanoates from municipal secondary wastewater sludge. *Bioresour. Technol.* **2018**, *255*, 111–115. [CrossRef]
19. Torri, C.; Weme, T.D.O.; Samorì, C.; Kiwan, A.; Brilman, D.W.F. Renewable alkenes from the hydrothermal treatment of polyhydroxyalkanoates-containing sludge. *Environ. Sci. Technol.* **2017**, *51*, 12683–12691. [CrossRef]

20. Rathika, R.; Janaki, V.; Shanthi, K.; Kamala-Kannan, S. Bioconversion of agro-industrial effluents for polyhydroxyalkanoates production using Bacillus subtilis RS1. *Int. J. Environ. Sci. Technol.* **2019**, *16*, 5725–5734. [CrossRef]

21. Anjali, M.; Sukumar, C.; Kanakalakshmi, A.; Shanthi, K. Enhancement of growth and production of polyhydroxyalkanoates by Bacillus subtilis from agro-industrial waste as carbon substrates. *Compos. Interfaces* **2014**, *21*, 111–119. [CrossRef]

22. Fang, F.; Xu, R.Z.; Huang, Y.Q.; Wang, S.N.; Zhang, L.L.; Dong, J.Y.; Xie, W.M.; Chen, X.; Cao, J.S. Production of polyhydroxyalkanoates and enrichment of associated microbes in bioreactors fed with rice winery wastewater at various organic loading rates. *Bioresour. Technol.* **2019**, *292*, 121978. [CrossRef]

23. Bhalerao, A.; Banerjee, R.; Nogueira, R. Continuous cultivation strategy for yeast industrial wastewater-based polyhydroxyalkanoate production. *J. Biosci. Bioeng.* **2019**, *129*, 595–602. [CrossRef]

24. Gerngross, T.U. Can biotechnology move us toward a sustainable society? *Nat. Biotechnol.* **1999**, *17*, 541–544. [CrossRef]

25. Harding, K.G.; Dennis, J.S.; von Blottnitz, H.; Harrison, S.T.L. Environmental analysis of plastic production processes: Comparing petroleum-based polypropylene and polyethylene with biologically-based poly-b-hydroxybutyric acid using life cycle analysis. *J. Biotechnol.* **2007**, *130*, 57–66. [CrossRef] [PubMed]

26. Kim, S.; Dale, B.E. Energy and greenhouse gas profiles of polyhydroxybutyrates derived from corn grain: A life cycle perspective. *Environ. Sci. Technol.* **2008**, *42*, 7690–7695. [CrossRef]

27. Yu, J.; Chen, L.X.L. The greenhouse gas emissions and fossil energy requirement of bioplastics from cradle to gate of a biomass refinery. *Environ. Sci. Technol.* **2008**, *42*, 6961–6966. [CrossRef]

28. Khoo, H.H.; Tan, R.B.H.; Chang, K.W.L. Environmental impacts of conventional plastic and bio-based carrier bags. *Int. J. Life Cycle Assess.* **2010**, *15*, 284–293. [CrossRef]

29. Kim, S.; Dale, B.E. Life cycle assessment study of biopolymers (polyhydroxyalkanoates)-derived from no-tilled corn. *Int. J. Life Cycle Assess.* **2005**, *10*, 200–210. [CrossRef]

30. Akiyama, M.; Tsuge, T.; Doi, Y. Environmental life cycle comparison of polyhydroxyalkanoates produced from renewable carbon resources by bacterial fermentation. *Polym. Degrad. Stabil.* **2003**, *80*, 183–194. [CrossRef]

31. Hermann, B.G.; Blok, K.; Patel, M.K. Producing bio-based bulk chemicals using industrial biotechnology saves energy and combats climate change. *Environ. Sci. Technol.* **2007**, *41*, 7915–7921. [CrossRef]

32. Pietrini, M.; Roes, L.; Patel, M.K.; Chiellini, E. Comparative life cycle studies on poly(3-hydroxybutyrate)-based composites as potential replacement for conventional petrochemical plastics. *Biomacromolecules* **2007**, *8*, 2210–2218. [CrossRef]

33. Kurdikar, D.; Paster, M.; Gruys, K.J.; Fournet, L.; Gerngross, T.U.; Slater, S.C.; Coulon, R. Greenhouse gas profile of a plastic derived from a genetically modified plant. *J. Ind. Ecol.* **2001**, *4*, 107–122. [CrossRef]

34. Fernández-Dacosta, C.; Posada, J.A.; Ramirez, A. Techno-economic and carbon footprint assessment of methyl crotonate and methyl acrylate production from wastewater-based polyhydroxybutyrate (PHB). *J. Clean. Prod.* **2016**, *137*, 942–952. [CrossRef]

35. Rostkowski, K.H.; Criddle, C.S.; Lepech, M.D. Cradle-to-gate life cycle assessment for a cradle-to-cradle cycle: Biogas-to-bioplastic (and back). *Environ. Sci. Technol.* **2012**, *46*, 9822–9829. [CrossRef]

36. Kendall, A. A life cycle assessment of biopolymer production from material recovery facility residuals. *Resour. Conserv. Recy.* **2012**, *61*, 69–74. [CrossRef]

37. Sakamoto, Y. Life Cycle Assessment of Biodegradable Plastics. *J. Shanghai Jiaotong Univ.* **2012**, *17*, 327–329. [CrossRef]

38. Posen, I.D.; Jaramillo, P.; Griffin, W.M. Uncertainty in the Life Cycle Greenhouse Gas Emissions from U.S. Production of Three Biobased Polymer Families. *Environ. Sci. Technol.* **2016**, *50*, 2846–2858. [CrossRef]

39. Heimersson, S.; Morgan-Sagastume, F.; Peters, G.M.; Werker, A.; Svanström, M. Methodological issues in life cycle assessment of mixed-culture polyhydroxyalkanoate production utilising waste as feedstock. *New Biotechnol.* **2014**, *31*, 383–393. [CrossRef] [PubMed]

40. Fernández-Dacosta, C.; Posada, J.A.; Kleerebezem, R.; Cuellar, M.C.; Ramirez, A. Microbial community-based polyhydroxyalkanoates (PHAs) production from wastewater: Techno-economic analysis and ex-ante environmental assessment. *Bioresour. Technol.* **2015**, *185*, 368–377. [CrossRef]

41. Morgan-Sagastume, F.; Heimersson, S.; Laera, G.; Werker, A.; Svanström, M. Techno-environmental assessment of integrating polyhydroxyalkanoate (PHA) production with services of municipal wastewater treatment. *J. Clean. Prod.* **2016**, *137*, 1368–1381. [CrossRef]

42. Dietrich, K.; Dumont, M.J.; Del Rio, L.F.; Orsat, V. Producing PHAs in the bioeconomy—Towards a sustainable bioplastic. *Sustain. Prod. Consum.* **2017**, *9*, 58–70. [CrossRef]

43. Vega, G.C.; Sohn, J.; Bruun, S.; Olsen, S.I.; Birkved, M. Maximizing environmental impact savings potential through innovative biorefinery alternatives: An application of the TM-LCA framework for regional scale impact assessment. *Sustainability* **2019**, *11*, 3836. [CrossRef]

44. Yadav, B.; Pandey, A.; Kumar, L.R.; Tyagi, R.D. Bioconversion of waste (water)/residues to bioplastics- A circular bioeconomy approach. *Bioresour. Technol.* **2020**, *298*, 122584. [CrossRef]

45. Kookos, I.K.; Koutinas, A.; Vlysidis, A. Life cycle assessment of bioprocessing schemes for poly(3-hydroxybutyrate) production using soybean oil and sucrose as carbon sources. *Resour. Conserv. Recy.* **2019**, *141*, 317–328. [CrossRef]

46. Lopez-Arenas, T.; González-Contreras, M.; Anaya-Reza, O.; Sales-Cruz, M. Analysis of the fermentation strategy and its impact on the economicsof the production process of PHB (polyhydroxybutyrate). *Comput. Chem. Eng.* **2017**, *107*, 140–150. [CrossRef]

47. Righi, S.; Baioli, F.; Samorì, C.; Galletti, P.; Tagliavini, E.; Stramigioli, C.; Tugnoli, A.; Fantke, P. A life cycle assessment of poly-hydroxybutyrate extraction from microbial biomass using dimethyl carbonate. *J. Clean. Prod.* **2017**, *168*, 692–707. [CrossRef]

48. Narodoslawsky, M.; Shazad, K.; Kollmann, R.; Schnitzer, H. LCA of PHA Production—Identifying the Ecological Potential of Bio-plastic. *Chem. Biochem. Eng. Q.* **2015**, *29*, 299–305. [CrossRef]

49. Cristóbal, J.; Matos, C.T.; Aurambout, J.-P.; Manfredi, S.; Kavalov, B. Environmental sustainability assessment of bioeconomy value chains. *Biomass Bioenergy* **2016**, *89*, 159–171. [CrossRef]

50. Moita, R.; Lemos, P.C. Biopolymers production from mixed cultures and pyrolysis by-products. *J. Biotechnol.* **2012**, *157*, 578–583. [CrossRef] [PubMed]

51. Fabbri, D.; Torri, C. Linking pyrolysis and anaerobic digestion (Py-AD) for the conversion of lignocellulosic biomass. *Curr. Opin. Biotech.* **2016**, *38*, 167–173. [CrossRef]

52. Bridgwater, A.V. Review of fast pyrolysis of biomass and product upgrading. *Biomass Bioenergy* **2012**, *38*, 68–94. [CrossRef]

53. Hagos, F.Y.; Aziz, A.R.A.; Sulaiman, S.A. Trends of Syngas as a Fuel in Internal Combustion Engines. *Adv. Mech. Eng.* **2014**, *6*, 1–10. [CrossRef]

54. Lehmann, J.; Joseph, S. Biochar for environmental management: An introduction. In *Biochar for Environmental Management*; Lehmann, J., Joseph, S., Eds.; Earthscan: Sterling, VA, USA, 2015; pp. 33–46.

55. Torri, C.; Cordiani, H.; Samorì, C.; Favaro, L.; Fabbri, D. Fast procedure for the analysis of polyhydroxyalkanoates in bacterial cells by off-line pyrolysis/gas-chromatography with flame ionization detector. *J. Chromatogr.* **2014**, *1359*, 230–236. [CrossRef]

56. Weiland, P. Biogas production: Current state and perspectives. *Appl. Microbiol. Biotechnol.* **2010**, *85*, 849–860. [CrossRef] [PubMed]

57. Torri, C.; Samorì, C.; Ajao, V.; Baraldi, S.; Galletti, P.; Tagliavini, E. Pertraction of volatile fatty acids through biodiesel-based liquid membranes. *Chem. Eng. J.* **2019**, *366*, 254–263. [CrossRef]

58. Dionisi, D.; Majone, M.; Tandoi, V.; Beccari, M. Sequencing batch reactor: Influence of periodic operation on performance of activated sludges in biological wastewater treatment. *Ind. Eng. Chem. Res.* **2001**, *40*, 5110–5119. [CrossRef]

59. Dionisi, D.; Majone, M.; Vallini, G.; Di Gregorio, S.; Beccari, M. Effect of the applied organic load rate on biodegradable polymer production by mixed microbial cultures in a sequencing batch reactor. *Biotechnol. Bioeng.* **2006**, *93*, 76–88. [CrossRef] [PubMed]

60. Bengtsson, S.; Werker, A.; Christensson, M.; Welander, T. Production of polyhydroxyalkanoates by activated sludge treating a paper mill wastewater. *Bioresour. Technol.* **2008**, *99*, 509–516. [CrossRef] [PubMed]

61. COREPLA. Relazione Sulla Gestione. 2016. Available online: http://www.corepla.it/documenti/621248cb-892c-4351-bef3-4e55f4559919/03+Relazione+sulla+gestione+2016.pdf (accessed on 7 April 2020).

62. ISO (International Organization for Standardization). *ISO 14040: Environmental Management—Life Cycle Assessment—Principles and Framework*; ISO: Geneve, Switzerland, 2006.

63. ISO (International Organization for Standardization). *ISO 14044: Environmental Management—Life Cycle Assessment—Requirements and Guidelines*; ISO: Geneve, Switzerland, 2006.

64. Ministero dello Sviluppo Economico (MISE), Ministero dell'Ambiente e della Tutela del Territorio e del Mare (MATTM), Ministero delle Infrastrutture e dei Trasporti (MIT). Integrated National Energy and Climate Plan. Available online: https://www.mise.gov.it/images/stories/documenti/it_final_necp_main_en.pdf (accessed on 7 April 2020).

65. Franklin Associates. *Life Cycle Inventory of Five Products Produced from Polylactide (PLA) and Petroleum-Based Resins*; Athena Institute International: Kansas, KS, USA, 2006.

66. Álvarez-Chávez, C.R.; Edwards, S.; Moure-Eraso, R.; Geiser, K. Sustainability of bio-based plastics: General comparative analysis and recommendations for improvement. *J. Clean. Prod.* **2012**, *23*, 47–56. [CrossRef]

67. Hottle, T.A.; Bilec, M.M.; Landis, A.E. Sustainability assessments of bio-based polymers. *Polym. Degrad. Stab.* **2013**, *98*, 1898–1907. [CrossRef]

68. Salhofer, S.; Schneider, F.; Obersteiner, G. The ecological relevance of transport in waste disposal systems in Western Europe. *Waste Manag.* **2007**, *27*, S47–S57. [CrossRef]

69. PlasticsEurope. Plastics—the Facts 2015. An Analysis of European Plastics Production, Demand and Waste Data. Available online: http://www.plasticseurope.org/documents/document/20151216062602-plastics_the_facts_2015_final_30pages_14122015.pdf (accessed on 7 April 2020).

70. European Bioplastics. Bioplastics—Furthering Efficient Waste Management. 2017. Available online: https://docs.european-bioplastics.org/publications/fs/EUBP_FS_End-of-life.pdf (accessed on 7 April 2020).

71. Thinkstep. GaBi Professional Database. 2017. Available online: http://www.gabi-software.com/support/gabi/gabi-database-2020-lci-documentation/professional-database-2020 (accessed on 7 April 2020).

72. Frischknecht, R.; Jungbluth, N.; Althaus, H.-J.; Doka, G.; Heck, T.; Hellweg, S.; Hischier, R.; Nemecek, T.; Rebitzer, G.; Spielmann, M.; et al. *Overview and Methodology. Data v2.0. Ecoinvent Report No. 1*; Swiss Centre for Life Cycle Inventories: Dübendorf, Switzerland, 2007.

73. EC-JRC (European Commission Joint Research Centre). *International Reference Life Cycle Data System (ILCD) Handbook. Recommendations for Life Cycle Impact Assessment in the European Context*, 1st ed.; European Commission, Joint Research Centre: Ispra, Italy, 2011.

74. EC-JRC (European Commission Joint Research Centre). *Characterisation Factors of the ILCD Recommended Life Cycle Impact Assessment Methods. Database and Supporting Information*, 1st ed.; European Commission, Joint Research Centre, Institute for Environment and Sustainability: Ispra, Italy, 2012.

75. EC-JRC (European Commission Joint Research Centre). *PEFCR Guidance document—Guidance for the development of Product Environmental Footprint Category Rules (PEFCRs)*, version 6.3; EC-JRC: Ispra, Italy, December 2017.

76. Weiss, M.; Haufe, J.; Carus, M.; Brandao, M.; Bringezu, S.; Hermann, B.; Patel, M.K. A Review of the Environmental Impacts of Biobased Materials. *J. Ind. Ecol.* **2012**, *16*, S169–S181. [CrossRef]

77. Righi, S.; Bandini, V.; Marazza, D.; Baioli, F.; Torri, C.; Contin, A. Life Cycle Assessment of high ligno-cellulosic biomass pyrolysis coupled with anaerobic digestion. *Bioresour. Technol.* **2016**, *212*, 245–253. [CrossRef]

Article

Environmental Performance of Innovative Ground-Source Heat Pumps with PCM Energy Storage

Emanuele Bonamente [1,2,*] and Andrea Aquino [3]

1   Department of Engineering, University of Perugia, 06125 Perugia, Italy
2   CIRIAF—Interuniversity Research Center on Pollution and Environment, University of Perugia, 06125 Perugia, Italy
3   Department of Mechanical and Industrial Engineering, University of Brescia, 25123 Brescia, Italy; andrea.aquino@unibs.it
*   Correspondence: emanuele.bonamente@unipg.it

Received: 19 November 2019; Accepted: 23 December 2019; Published: 25 December 2019

Abstract: Space conditioning is responsible for the majority of carbon dioxide emission and fossil fuel consumption during a building's life cycle. The exploitation of renewable energy sources, together with efficiency enhancement, is the most promising solution. An innovative layout for ground-source heat pumps, featuring upstream thermal energy storage (uTES), was already proposed and proved to be as effective as conventional systems while requiring lower impact geothermal installations thanks to its ability to decouple ground and heat-pump energy fluxes. This work presents further improvements to the layout, obtained using more compact and efficient thermal energy storage containing phase-change materials (PCMs). The switch from sensible- to latent-heat storage has the twofold benefit of dramatically reducing the volume of storage (by a factor of approximately 10) and increasing the coefficient of performance of the heat pump. During the daily cycle, the PCMs are continuously melted/solidified, however, the average storage temperature remains approximately constant, allowing the heat pump to operate closer to its maximum efficiency. A life cycle assessment (LCA) was performed to study the environmental benefits of introducing PCM-uTES during the entire life cycle of the system in a comparative approach.

Keywords: ground-source heat pumps; space conditioning; environmental sustainability; life cycle assessment (LCA); phase-change material (PCM)

---

1. Introduction

Buildings dramatically contribute to worldwide energy use and greenhouse gas (GHG) emissions. According to last official European Union reports [1,2], in 2010 the building sector accounted for about 32% of worldwide final energy consumption and 34% of global GHG emissions. More recently, the Intergovernmental Panel on Climate Change (IPCC) updates confirmed the energy share (31%, [3]) and reduced the estimates of greenhouse gas emissions (23%, [4]). However, both values are expected to increase, resulting in further energy consumption (+2 billion tons of oil equivalent [5]) for the construction of new living spaces worldwide (+230 billion square meters), causing an estimated +30% increase in GHG emissions in the next 40 years. In this context, government and international organizations are pursuing new and more severe policies specifically aimed at reducing the energy consumption of the building sector and mitigating associated emissions. As an example, the European Union aims to reduce by 90% the equivalent carbon dioxide (CO_2e) emissions of buildings by the year 2050, estimating that about the 17% of the primary energy saving is achievable by retrofitting existing buildings. The potential energy savings of an existing building are strictly related to its heating and cooling systems. During the entire building's life cycle (from construction to demolition), a major share of the required energy (80–90%) is associated with operating energy [6], and up to 60% of the total energy consumption is due to air-conditioning [7]. The enhancement of air-conditioning energy efficiency, coupled with an integration of renewable-energy technologies, is considered an effective solution to mitigate the environmental impacts of buildings [8–12].

Among the renewable energy systems for space conditioning [13], ground-source heat pumps (GSHPs) have a wide diffusion rate because of their low operative temperature. Low-temperature heat sources (soil or underground water) between 5 and 30 °C are found easily worldwide as they are available at reasonable depths [14,15]. GSHPs exchange heat with the first layer of the ground via properly designed borehole heat exchangers (BHEs) and, due to the low operative temperatures of the working cycle, their feasibility does not depend on the nature of the geothermal resources available at the building site [16]. Like conventional heat pump (HP) systems, GSHPs need electricity as the driving energy in order to transfer heat from a cold source to a hot source. During the heating season, the heat is extracted from the ground and given to the indoor space. Meanwhile, during the cooling season the cycle is reversed. The coefficient of performance (COP), defined as the ratio of total output energy (i.e., heat transfer to/from the building) to the total power input, measures the overall system efficiency. This is the only indicator recognized by technical regulations [17] to evaluate the environmental performance of an HP system, which is classified as a renewable energy system only if the measured COP exceeds a minimum value (3.2 to 3.5 for typical geothermal applications [18]). However, the exploitation of geothermal resources is affected by several environmental implications and a comprehensive environmental impact assessment of a GSHP cannot be limited to its energy performance. Additional negative environmental impacts should be considered, such as the risks of aquifer contamination or the contribution to soil compaction and habitat loss or disturbance. An important source of such impacts is the excavation of BHEs, which can also affect initial costs by up to 50% of the total investment [19].

Thermal energy storage (TES) is a key component of the optimization of the performance of a building energy system and the reduction of its environmental impact [20,21]. Thermal energy can be stored in three main forms: sensible heat, latent heat, and thermo-chemical energy. Sensible storage devices commonly use water as a storage material because of its well-known properties, availability, and negligible cost. However, these systems require important storage volumes with relatively high costs. Latent heat TES makes use of the enthalpy change of a material during its phase transition in order to provide higher energy density than sensible systems. As a consequence, latent heat systems need smaller storage volumes than sensible ones. Conventional materials (e.g., water, diathermic oil, etc.) change phase at temperatures that are not optimal for air-conditioning applications. On the other hand, phase change materials (PCMs) make the latent heat storage available at a large range of operative temperatures, including those favorable for indoor air-conditioning. These materials are easily suited to several heating and cooling applications due to the wide market offer in terms of thermal properties and packaging: PCMs are commonly available both in packaged volumes, as spheres, panels, etc. (i.e., encapsulated PCMs), or without packaging (i.e., free-form PCMs) [22,23].

Previous studies on GSHP systems [24,25] have shown that final COP depends on system configuration and local site conditions (e.g., the BHEs setup, the climate, the ground properties, and the HP nominal power), as well as the transient operative conditions of the system's duty-cycle: the ground's undisturbed temperature, the temperature difference between the surface and the underground, and the temperature of the water leaving the heat pump and the BHEs. These temperature fluctuations lower the system COP. The performance of GSHP gradually decays with operating time because of the continuous injection of heat into the ground (cooling cycle), which increases its undisturbed temperature [26]. The heating cycle produces a similar effect that presents as an inverted temperature trend [27]. A general explanation of this issue is that gradual COP reduction is due to an unbalanced (over the yearly season) heat exchange between the system and the ground [28]. The effects of this thermal imbalance by a two-year on-site measurement campaign were a decrease in system COP by 42% and 26% in the winter and summer season, respectively [29].

The role of a TES is essentially to decouple the thermal energy exchange of the ground from the building demand by accumulating thermal energy (heat or cold) when it is more readily available and using it upon request. This mismatch could attenuate the effects derived from temperature fluctuations in system efficiency and, therefore, several works aim to include a TES system in the standard GSHP layout. The two approaches can be identified as:

1. Enhancement of the thermal capacity of the system by upgrading a specific component (BHEs or air distribution system);
2. Direct installation of a TES.

The first approach is illustrated in [30]. A laboratory-scale BHE with PCM included in its backfill was tested. The system successfully delayed the ground temperature variation and reduced the thermal interference radius (i.e., less area was required for installation). Furthermore, the constant temperature of phase change improved the BHE extraction rate. Another work ([31]) presented the exergy analysis of a hybrid GSHP system, coupled with solar panels, and indoor air conditioning was possible because of radiant panels with encapsulated PCM. The results showed that PCM was beneficial in term of first and second law efficiency; it increased with density and melting temperature.

The work presented in this paper follows the second approach. We propose the inclusion of an upstream (i.e., between the BHE and the HP) thermal energy storage in the standard GSHP layout. A similar approach is given in [32], where five different control methods for an air-to-air HP were implemented and coupled to a PCM-based TES to optimize its charge/discharge step with electricity tariffs and user demand. Solar-assisted GSHP were analyzed in [33], where the PCMs within the hot water puffer increased the amount of stored solar energy and generally improved the global COP. The importance of thermal storage in these hybrid systems was also confirmed by the experimental campaign in [34] and the numerical analysis in [35].

Finally, the works specifically focused on geothermal systems are [36,37], where PCM-based and sensible, respectively, downstream thermal storage (i.e., between the HP and the air distribution system) were coupled to a GSHP. On the one hand, in a comparison of the annual thermal energy use, at variable partial load ratio, of a boiler-based air conditioning system vs. conventional GSHP vs. TES-upgraded GSHP, the latter showed the best performance in terms of energy-savings. On the other hand, a similar configuration was seen for the cooling of an office building using different cooling storage ratios of PCM storage (i.e., the cooling capacity of PCM on the total building cooling capacity). The optimal cooling capacity was estimated to be 40%, which produced a decrease of 34% of the annual energy cost if compared to a GSHP system without TES.

In this work two different prototypal layouts were assessed and compared: a sensible-heat TES (SH-TES), where water was used as thermal storage material, and a latent-heat TES based on PCMs (PCM-TES). A first energy analysis using computational fluid dynamics of both scenarios was available in a previous work [38]. Based on these results, the work presented here is an in-depth evaluation of the respective energy performance and environmental impacts, measured and compared throughout whole life cycles, with the final aim of defining the most efficient and sustainable system configuration.

Section 2 gives a short description of the methodology. The case study is presented in Section 3. The results are shown in Section 4. Conclusions are discussed in Section 5. Extended results are given in the Supplementary Material.

2. Materials and Methods

All the potential environmental impacts of the proposed energy systems were quantified by the life cycle assessment (LCA) approach. This technique can be successfully used as a comparative analysis of different products or technologies providing the same service, in order to recognize the most efficient and environmentally sustainable option [39,40]. A life cycle assessment is a standardized and widespread approach allowing the evaluation of environmental impacts associated with any process, product, or service throughout its entire life, usually starting from raw material acquisition until the end-of-life stage. Thanks to this approach, the life cycle of the studied product could be split into several parts corresponding to relevant components and/or phases (i.e., steps of the production and usage). The relative environmental weight of each part was explicitly assessed in order to spot and, possibly, optimize, the most impactful ones.

Our study, in accordance with standard LCA guidelines [41,42], was undertaken using the following steps:

1. Compilation of the life cycle inventory (LCI)—an input database based on product specifications (e.g., functional unit, system time and spatial boundaries, and quality criteria) that details all of the mass and energy flows, relative to each stage of the considered life cycle. The LCI is a collection of data, usually in a tabular form, used as input of the LCA model;
2. Quantification of the environmental impacts by relevant mid- and end-point indicators—the former evaluates the environmental burden as a function of single physical quantities. The latter takes into account higher aggregation levels, such as human health or biodiversity, for which information from multiple mid-point indicators is necessary. They are quantified by arbitrary units (Pt) instead of physical quantities. Mid-point indicators were taken from the comprehensive suite CML-IA baseline V3.05 [43]. End-point indicators were computed by the ReCiPe 20106 Endpoint (H) V1.03 method [44].

The aim of this work was to measure the environmental impact of two innovative GSHP layouts. The comparative approach was as follows: Starting from a reference layout used as a benchmark (i.e., baseline scenario), the enhancement of its energy performance and environmental impacts gained by the two proposed TES systems was measured in terms of mid- and end-point indicators. Different operative conditions were also evaluated for each system. All results were validated against experimental data measured using an existent prototype with a sensible-heat thermal storage and a small-scale laboratory unit with a latent-heat thermal storage.

The life cycle was modeled using the SimaPro v9.0.0 software [45] and the ecoinvent v3.5 database [46]. The impacts were computed according to the product category rule (PCR) document relative to electricity, steam and hot/cold water generation and distribution version 3.1 [47]. In this study, the following environmental indicators were considered

1. Mid-point indicators:

    -   global warming in equivalent $kgCO_2$;
    -   photochemical oxidation in equivalent kgC_2H_4;
    -   acidification in equivalent $kgSO_2e$;
    -   eutrophication in equivalent $kgPO_4{}^{3-}$.

2. End-point indicators:

    -   human health;
    -   ecosystems;
    -   resources.

3. The Case-Study

The system under investigation is a conventional ground-source heat pump system, used air-condition commercial buildings, characterized by a peak power of 17 kW, a yearly demand of 19,965 kWht of heating energy, and 4918 kWht of cooling energy, for a total of 24,883 kWht/year provided by the heat pump [48]. The baseline configuration (BAS) of this system consists of:

-   a 17-kW reversible ground-source heat pump;

- and a total of 3 double-loop boreholes, each one 120 m deep with an average extraction capacity of 46.5 W/m measured by the ground response test.

This setup was upgraded by including upstream thermal storage (uTES) between the borehole heat exchangers and the heat pump [49]. The uTES was sized based on the maximum daily energy need instead of the maximum power. This new layout was designed to exchange heat with the ground even when it was not needed by the building, and using the thermal storage as a peak-shaving component. While average power exchanged with the ground on the one side and with the heat pump on the other side was the same, the maximum power required by the ground was less than the maximum power required by the heat pump as the former works with a higher duty cycle. On-site monitoring of a real-building prototype [50] confirmed that this layout was able to successfully decouple the energy exchange between the ground and the HP, resulting in a substantial reduction of peak geothermal power and, therefore, reduced borehole size (i.e., shorter/fewer boreholes to be drilled). Figure 1 shows the upgraded layouts with respect to the BAS scenario. They both use the same 17 kW heat pump while using just 1 borehole in combination with an uTES properly configured for each setup as follows:

- The sensible heating thermal energy storage (SH-TES) scenario used a concrete made tank of 12 m^3 and the thermal storage material was water;
- The PCM thermal storage (PCM-TES) scenario was obtained using a PCM-based component instead of water. The storage was made of two PCM stacks of 2057 kg (RT-6 PCM) and 1233 kg (RT-27 PCMs) for nominal latent heat of 80 kWh and 50 kWh, respectively, and a total storage volume of about 1 m^3.

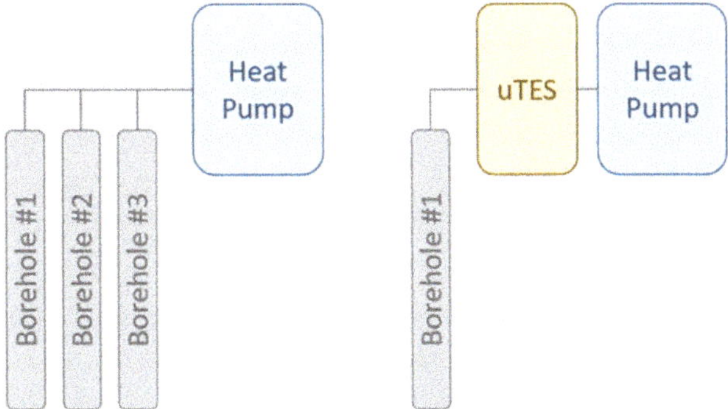

Figure 1. Layouts of the baseline setup (**left**) and the upgraded setup using an upstream thermal storage (uTES) (**right**).

As shown in a previous LCA study [48], the sensible-heat uTES was already able to reduce the environmental impact of the conventional layout. However, this preliminary analysis was limited to the evaluation of the environmental impact of the SH-uTES scenario, proposed first as an upgraded conventional BAS system. The inclusion of a latent-heat TES was already mentioned as a potential future development to reduce the storage volume, without any reference to its environmental issues. Similarly, the preliminary CFD energy analysis [38] detailed in the following paragraph exclusively calculated in terms of global COP, the performance of the PCM-uTES scenario. A more detailed assessment of the environmental impact of this system is presented in the current LCA analysis, where the environmental impacts of the PCM-TES system were measured during the life cycle of the plant (20 years), considering the energy consumption of a reference year. The obtained results were compared with a baseline scenario (BAS) and a sensible-heat thermal storage (SH-TES) scenario. For all three scenarios, two hypotheses were proposed: (a) electric energy is taken from the grid, and (b) electric energy is produced entirely from photovoltaic panels.

3.1. PCM Modeling and Validation

The energy performance of the PCM-TES scenario was already calculated by a comparative CFD analysis in [38], between the sensible and the latent uTES. Both systems abide by the same daily energy demand schedule, implemented on the basis of yearly energy consumption (see Section 3) and building characteristics [50]. The PCM-TES scenario simulated by CFD analysis provided two closed loops connected to the heat pump and the borehole, respectively. The heat transfer fluid within each loop entered a serpentine tube that crossed two different PCM stacks. The heat transfer between the borehole and PCM during the winter season was modeled via the following steps:

1. The heat transfer fluid (water) from the borehole flowed into the PCM storage;
2. This flow crosses the serpentine and, meanwhile, released heat into the PCMs in direct contact with tube shell;
3. Finally, it left the storage and entered the BHE again;
4. While the previous steps were running, the heat transfer fluid from the heat pump flowed to the PCM storage and withdrew the heat stored in latent form. The cycle was inverted in the summer season.

The results of this preliminary analysis clearly showed the twofold benefits of the PCM-TES scenario. On the one hand, the daily needs were fully covered by 10 times less thermal storage and, on the other hand, the reduction of storage temperature oscillations, as an effect of phase transition, increased the overall COP from 3.4, for both heating and cooling, to 4.13 and 5.89 for the heating and cooling modes, respectively.

An experimental campaign for the CFD model validation using data from a 1-kWh scale prototype is currently ongoing [51]. Figure 2 shows the experimental apparatus. It consists of two open loops connected to a boiler and a chiller, respectively. Each loop has its circulation pump. The PCM-storage connects both loops and collects the flows; these get mixed within the storage and exchange heat with the PCMs. (They present the same properties used in the CFD analysis.) The heating cycle took place in the following way:

1. The boiler that reproduced the BHE heated the water;
2. The heat was released to the PCM and stored in latent form;
3. The cooling water, produced by the chiller representing the heat pump, flowed within the storage and withdrew the stored heat.

Data were measured by two calorie flow meters placed at the outlet sections of the storage. The PCM phase transition was modeled as function of their temperature (T_{PCM}) and the thermo-physical properties (Table 1). When T_{PCM} achieved the melting/solidification point, the phase change started and the heat was stored in latent form along the transition region (±0.25 °C with respect to the nominal transition temperature). Preliminary experimental data collected from the laboratory unit are shown in Figure 3. Good agreement between the simulated and measured outlet temperature (T_{out}) was observed using nominal PCM properties before any model calibration, strengthening the above estimates on performance improvement. CFD simulations were performed using a constant power equal to the average power measured during the experimental campaign. More detailed analyses will be presented in future works.

Table 1. PCM thermal properties.

PCM	Melting Range (°C)	Heat of Fusion (kJ/kg)	Sensible Heat (kJ/kg)		Density (kg/m^3)	
			Solid	Liquid	Solid	Liquid
RT6	8.0–8.5	140	1.8	2.4	860	770
RT27	25.0–25.5	146	1.8	2.4	870	750

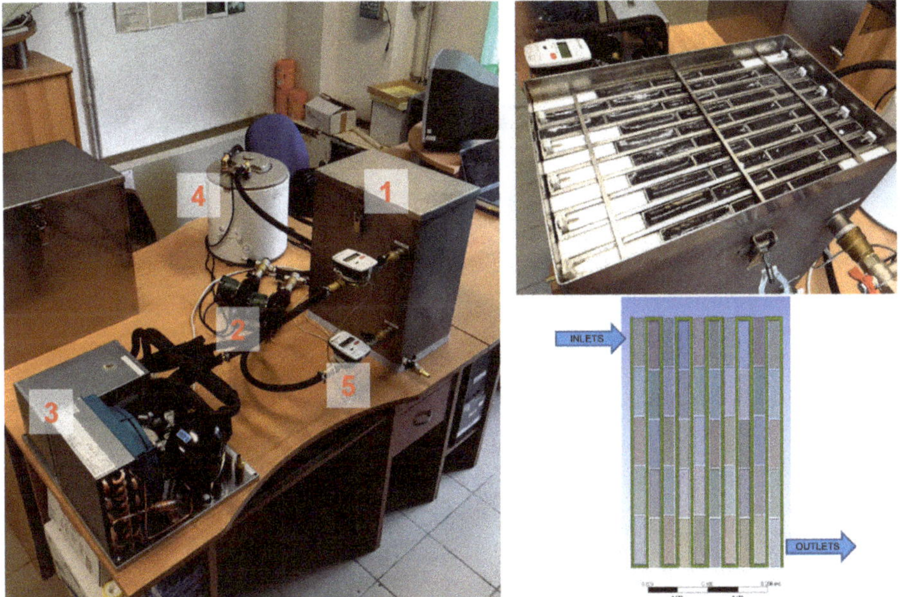

Figure 2. (**left**) Overview of the experimental layout. 1: phase-change-material (PCM) thermal storage, 2: circulating pumps, 3: chiller, 4: heater, 5: temperature/flow meters. (**right**) Close view of the PCM storage and top view of its geometry model.

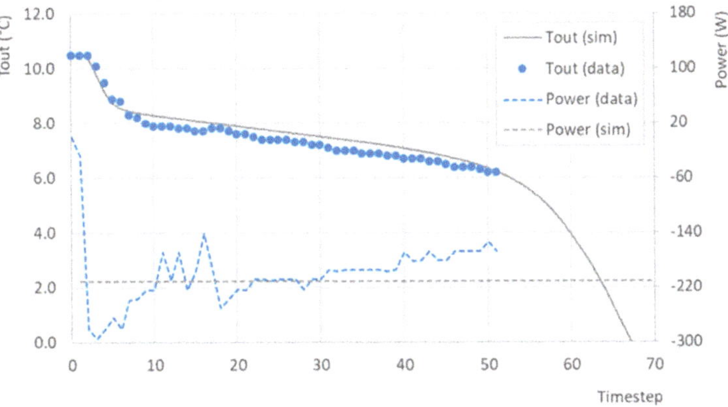

Figure 3. RT-6 PCM transition region: simulated vs. experimental data.

3.2. The Life Cycle Inventory

The three scenarios differ by borehole number, storage type, and electric energy consumption for heat production (the incidence of raw materials is reported in Table 2); the final electric energy consumption of each scenario is given by the respective COP values (Table 3). These values can be considered a conservative estimate as no contribution from cooling (heating) material is considered for the size of the heating (cooling) material (i.e., the effect of sensible heat of the cooling-optimized PCM is not considered on heating and vice versa).

Table 2. Material inventory of each scenario.

Scenario	BAS		SH-TES			PCM-TES		
Material/Process	Boreholes	Heat Pump	Boreholes	SH-uTES	Heat Pump	Boreholes	PCM-uTES	Heat Pump
Polyethylene (kg)	428	-	140	7.57	-	140	7.57	-
Sand (kg)	40,526	-	10,120	10.189	-	10,120	-	-
Gravel (kg)	53,606	-	13,398	13,440	-	13,398	-	-
Excavation (m^3)	55.9	-	14.0	76.4	-	14	-	-
Water (kg)	583	-	191	17,510	-	191	3510	-
Glycol (kg)	258	-	84.4	4.56	-	84.4	4.56	-
Concrete (m^3)	3.63	-	1.21	1.58	-	1.21	1299	81.0
Steel (kg)	105	81.0	43.0	428	81.0	43.0	-	-
Brass (kg)	-	6.00	-	8.00	-	-	8.00	-
Insulation (kg)	-	21.8	-	-	27.80	-	23.9	27.8
PCM (kg)	-	-	-	-	-	-	3290	-
Heat Pump (kW)	-	17.0	-	-	17.00	-	-	17.0
Al tape (m^2)	-	10.0	-	-	10.00	-	-	10.0

Table 3. Coefficient of performance (COP) and yearly electricity consumption of each scenario.

Scenario	COP		Electric Energy (kWh/year)	
	Heating	Cooling	Heating	Cooling
BAS	3.50	4.00	5704	1230
SH-TES	3.39	3.41	5889	1442
PCM-TES	4.13	5.89	4834	835

3.3. Functional Unit and System Boundaries

The functional unit in this study was 1 kWh of thermal energy (kWht) provided by the heat pump to the thermal energy distribution system. The impacts associated with the functional unit were obtained considering the energy inputs during the reference year and allocating the results based on the total production of heating and cooling energy. The system boundaries included the production of input materials for the implementation of the system from the boreholes to the heat pump, excavation activities for the boreholes and the sensible-heat storage, electric energy consumption, and end-of-life of the thermal storage and heat pump.

4. Results

4.1. Mid-Point Indicators

Results for mid-point indicators are shown in Tables 4 and 5, considering electric energy from the grid and from photovoltaic panels, respectively. The percentage variation was relative to the corresponding baseline scenario. We observed that, in terms of overall GHG emissions, the PCM-TES scenario was the most systematically competitive thanks to its higher overall COP that reduced primary energy consumption. This particularly impacted the case where electric energy was taken from the grid (−16% and −17% with respect to BAS and SH-TES scenarios, respectively). It can be noted that the SH-TES scenario had slightly higher impact as a direct consequence of COP deterioration (see Table 3). If electric energy was produced entirely by photovoltaic panels, the average reduction in impact indicators for the PCM-TES scenario would be approximately 69% (76% for CO$_2$e). Even in the PV hypothesis, the PCM-TES scenario performed best (−11% and −3.5% with respect to BAS and SH-TES scenarios, respectively). In all cases, the most impactful process was the consumption of electric energy. For the baseline scenarios, it accounted for an average of 85% using the grid hypothesis (Table 6) and 64% using the PV hypothesis (Table 7). Such values dropped to 64% and 41% for the PCM-TES scenarios, respectively. This result was produced by the fact that emission factors for PV production (multi-Si, 3 kWp) were, on average, 75% lower than emission factors for grid electricity (Italy, low voltage).

Table 4. Mid-point indicators for the three scenarios (electricity from grid).

Indicator	Units	Scenario				
		BAS, Grid	SH-TES, Grid		PCM-TES, Grid	
Global warming	$kgCO_2e/kWht$	0.130	0.132	+1.6%	0.108	−16.6%
Photochemical oxidation	$gC_2H_4e/kWht$	0.0307	0.0305	−0.8%	0.0246	−19.9%
Acidification	$gSO_2e/kWht$	0.728	0.743	+2.0%	0.620	−14.8%
Eutrophication	$gPO_4^{3-}e/kWht$	0.237	0.239	+1.2%	0.209	−11.8%
Average variation (with respect to BAS)				+1.0%		−15.8%

Table 5. Mid-point indicators for the three scenarios (electricity from photovoltaic sources).

Indicator	Units	Scenario				
		BAS, PV	SH-TES, PV		PCM-TES, PV	
Global warming	$kgCO_2e/kWht$	0.0356	0.0323	−9.3%	0.0312	−12.2%
Photochemical oxidation	$gC_2H_4e/kWht$	0.0122	0.0109	−10.8%	0.00948	−22.5%
Acidification	$gSO_2e/kWht$	0.209	0.194	−7.2%	0.195	−6.4%
Eutrophication	$gPO_4^{3-}e/kWht$	0.110	0.105	−4.1%	0.105	−4.4%
Average variation (with respect to BAS)				−7.8%		−11.4%

Table 6. Incidence of highest impact processes (electricity from grid).

BAS, Grid				
Process	Global Warming	Photochemical Oxidation	Acidification	Eutrophication
Electricity, grid	87.85%	82.45%	88.32%	82.85%
Heat pump	4.12%	5.80%	4.88%	9.84%
Propylene glycol	1.86%	3.79%	1.39%	2.43%
Gravel	1.50%	2.06%	1.58%	1.37%
Polyethylene	1.38%	1.81%	0.83%	0.24%
Concrete	1.26%	0.69%	0.64%	0.56%

SH_TES, Grid				
Process	Global Warming	Photochemical Oxidation	Acidification	Eutrophication
Electricity, grid	91.41%	87.92%	91.52%	86.57%
Heat pump	4.05%	5.85%	4.78%	9.72%
Propylene glycol	0.63%	1.32%	0.47%	0.83%
Gravel	0.74%	1.04%	0.78%	0.68%

PCM-TES, Grid				
Process	Global Warming	Photochemical Oxidation	Acidification	Eutrophication
Electricity, grid	86.10%	84.18%	84.79%	76.83%
Heat pump	4.94%	7.24%	5.73%	11.16%
Paraffin	4.56%	5.33%	5.57%	3.16%
Steel, low-alloyed	4.36%	8.91%	3.14%	7.22%
Steel and iron (waste treatment)	−4.11%	−12.07%	−2.93%	−2.65%

Table 7. Incidence of highest impact processes (electricity from photovoltaic).

BAS, PV				
Process	**Global Warming**	**Photochemical Oxidation**	**Acidification**	**Eutrophication**
Electricity, PV	55.65%	55.88%	59.24%	63.01%
Heat pump	15.03%	14.58%	17.02%	21.22%
Propylene glycol	6.78%	9.54%	4.85%	5.23%
Gravel	5.46%	5.18%	5.52%	2.95%
Polyethylene	5.05%	4.55%	2.90%	0.53%
Concrete	4.60%	1.73%	2.24%	1.20%
SH_TES, PV				
Process	**Global Warming**	**Photochemical Oxidation**	**Acidification**	**Eutrophication**
Electricity, PV	64.89%	66.24%	67.47%	69.45%
Heat pump	16.57%	16.34%	18.34%	22.12%
Propylene glycol	2.58%	3.69%	1.80%	1.88%
Gravel	3.02%	2.91%	2.98%	1.54%
Concrete	3.90%	1.49%	1.85%	0.96%
PCM-TES, PV				
Process	**Global Warming**	**Photochemical Oxidation**	**Acidification**	**Eutrophication**
Electricity, PV	51.82%	58.93%	51.72%	53.91%
Heat pump	17.11%	18.81%	18.18%	22.20%
Paraffin	15.82%	13.83%	17.67%	6.29%
Steel, low-alloyed	15.13%	23.13%	9.97%	14.36%
Propylene glycol	2.66%	4.24%	1.78%	1.89%
Steel and iron (waste treatment)	−14.25%	−31.33%	−9.31%	−5.28%

Tables 6 and 7 clearly show that processes related to the PCM-uTES introduced non-negligible impacts. From a life cycle perspective, it is, therefore, worth investigating if possible optimization of such processes might be viable and rewarding. Paraffins and steel, with similar shares, account for approximately 9% using the grid hypothesis, and 31% using the PV hypothesis. Considering these values, an optimal scenario might be foreseen (as discussed in Section 4.2) including the selection of bio-based PCMs and the minimization of steel in the storage structure with respect to the non-optimized prototypal layout.

4.2. Global Warming

The system life cycle was divided into six parts according to physical and usage boundaries. In addition, a detailed analysis on the global warming indicator was performed using this partition. The tree view of the PCM-uTES scenario considering electricity from the grid is given in Figure 4. The LCA indicators were:

1. BOREHOLES—containing borehole materials (tubes, concrete, etc.) and excavation;
2. STORAGE—containing different inputs according to different scenarios, including all raw materials, manufacturing/building processes, and piping;
3. HEAT PUMP—containing the HP machine, piping and insulation;
4. HEATING—containing electric energy consumption for heating (grid/PV hypotheses);
5. COOLING—containing electric energy consumption for cooling (grid/PV hypotheses);
6. END-OF-LIFE—containing recycling processes for separable materials (e.g., steel) and specific disposal for the reminder.

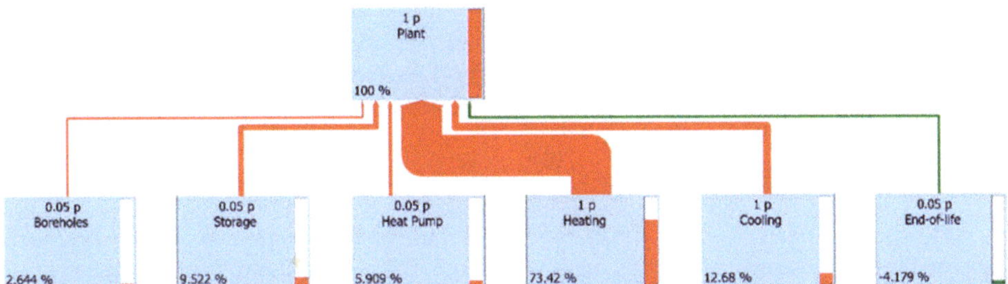

Figure 4. Typical tree view of the system life cycle (shown as global warming for the PCM-uTES scenario with energy from the grid).

The relative incidence of each part of the total GHG emissions is shown in Table 8. The phase-change material thermal energy storage was characterized by an impact higher than the water storage, mainly because of the steel and PCMs. However, a large part of its impact (44%) was compensated by the end-of-life indicator, during which steel can be recycled.

Table 8. Incidence of different life-cycle assessment (LCA) parts on global warming.

Part Scenario	Boreholes	Storage	Heat Pump	Heating	Cooling	End-of-life
BAS, grid	7.2%	0.0%	4.9%	72.3%	15.6%	0.0%
SH-TES, grid	2.2%	1.5%	4.8%	73.4%	18.0%	0.1%
PCM-TES, grid	2.6%	9.5%	5.9%	73.4%	12.7%	−4.2%

Part Scenario	Boreholes	Storage	Heat Pump	Heating	Cooling	End-of-life
BAS, PV	26.4%	0.0%	18.0%	45.8%	9.9%	0.0%
SH-TES, PV	8.9%	6.0%	19.8%	52.1%	12.8%	0.5%
PCM-TES, PV	9.2%	33.0%	20.5%	44.2%	7.6%	−14.5%

Details of the PCM-uTES are given in Table 9. PCMs and steel accounted for the highest impacts (42% and 40%, respectively). The amount of steel was computed using a direct proportion based on the 1-kWh lab prototype of storage. It was, therefore, straightforward to hypothesize lower incidence of steel for an optimized production-ready component. An optimal scenario was finally foreseen using bio-based PCMs and a reduced amount of steel. In the case of bio-based alternatives, the emission factor for PCMs could be lowered from 0.747 kgCO$_2$e/kg (ecoinvent, generic paraffin) down to 0.278 kgCO$_2$e/kg [52]. Considering the optimization of the PCM storage design, a reduction of 0.5 mm in the wall and tube thickness could be hypothesized for the production step without compromising the structure strength, resulting in a savings of approximately 40% of raw materials with respect to the non-optimized prototypal design.

Table 9. Details of the PCM-uTES.

Process	Global Warming		Photochemical Oxidation		Acidification		Eutrophication	
	kgCO$_2$e	%	gC$_2$H$_4$e	%	gSO$_2$e	%	gPO$_4^{3-}$e	%
Polyethylene	15.7	0.31%	4.87	0.26%	53.1	0.18%	5.05	0.04%
Water	1.98	0.04%	0.50	0.03%	9.10	0.03%	4.18	0.03%
Propylene glycol	21.2	0.41%	10.3	0.55%	89.0	0.30%	50.5	0.41%
Brass	50.5	0.99%	81.0	4.32%	2028	6.89%	1314	10.7%
Paraffin (RT-6)	1537	30.0%	408	21.7%	10,743	36.5%	2051	16.8%
Paraffin (RT-27)	921	18.0%	244	13.0%	6440	21.9%	1230	10.1%
Steel	2351	45.8%	1091	58.2%	9690	32.9%	7494	61.3%
Polystyrene	231	4.50%	34.9	1.86%	386	1.31%	85.2	0.70%
Total	5130	100%	1875	100%	29,439	100%	12,233	100%

The optimal scenario (PCM-TESopt, PV), realized using the above criteria and tested using the PV hypothesis, resulted in a 51% decrease in the global warming impact associated with the PCM-uTES, and an additional reduction of 10% with respect to the PCM-TES, PV scenario. The overall emission factor for the optimal scenario would be 0.028 kgCO$_2$e/kWht, −78% and −21% with respect to BAS, grid and BAS, PV scenarios, respectively.

4.3. End-Point Indicators

The results of the end-point indicators are shown in Figure 5. Impacts on human health are responsible for 93% of average overall impacts. As part of the electricity-from-grid hypothesis, the PCM-TES scenario showed a decrease of 10% with respect to the baseline. On the other hand, an increase of 1.3% was observed using the photovoltaic hypothesis. In this case the PCM-uTES introduced a positive impact (+25.6%) that was slightly higher than the overall reduction due to shorter boreholes (−13.5%) and lower energy demand for heating (−7.4%) and cooling (−3.3%).

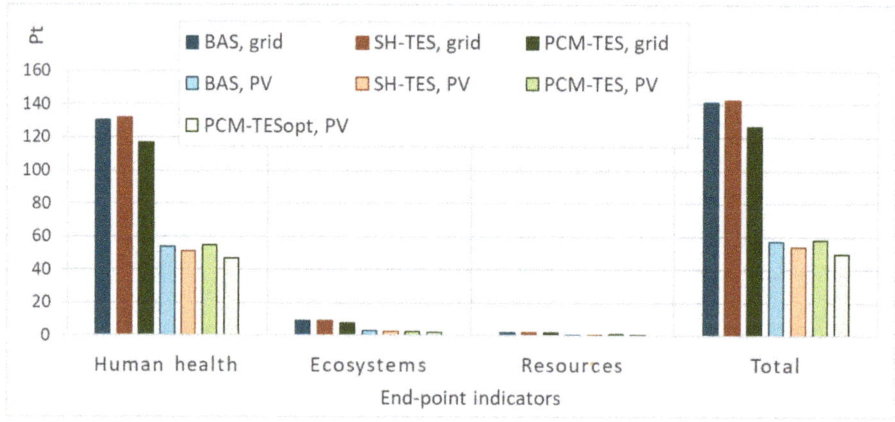

Figure 5. End-point indicator results.

The optimal scenario would likely lower the impacts of storage, resulting in the overall best performing layout, also in terms of end-point indicators (−64% and −13% with respect to BAS, grid and BAS, PV, respectively).

5. Conclusions

The LCA analysis was applied to a customized ground-source heat pump system. The innovative layout featured upstream thermal energy storage that was able to decouple the thermal fluxes from/to the ground and from/to the heat pump. Following the results of the experimental campaign, an upgrade TES component was evaluated using phase change materials instead of water. We showed that PCMs could reduce the volume of storage by a factor of 10 and enhance the system COP because of temperature stabilization during the charge/discharge cycles. The environmental performance of the system was evaluated in terms of mid- and end-point indicators per unit of thermal energy provided to the building considering the following two hypotheses: (1) energy is taken from the grid, and (2) energy is supplied by photovoltaic panels.

We found that the exploitation of phase-change materials allowed for sensible reduction of electric energy consumption (−18%) thanks to an improved COP both in terms of heating (from 3.50 to 4.13) and cooling mode (from 4.00 to 5.89). This resulted in an overall variation of mid-point indicators of −16% and −11% with respect to hypotheses (1) and (2), respectively. End-point indicator variation was −10% and +1.3%, respectively. The increase in the end-point indicators with respect to the PV hypothesis was due to the high relative impact of the PCM storage itself. Considering further, realistic, optimization of the prototypal storage design using the PV hypothesis, mid- and end-point impacts could be as low as −78% and −64%, respectively, with respect to the baseline scenario (grid hypothesis), making it the most sustainable option.

In terms of global warming, the baseline scenario was characterized by 1.30 $kgCO_2e/kWht$ (grid) and 0.0356 $kgCO_2e/kWht$ (PV). The grid-hypothesis result was consistent with a previous study (0.156 $kgCO_2e/kWht$ [50]), the difference being the higher value of the emission factor of electricity in the database used (ecoinvent v 3.2 [53]). The proposed PCM-TES layout could lower impacts by as much as 0.108 $kgCO_2e/kWht$ and 0.0312 $kgCO_2e/kWht$ for the grid and PV hypothesis, respectively. This would make such a solution very attractive to sensibly reduce both energy consumption (−21%) and the environmental impacts associated with space conditioning.

Supplementary Materials: The following are available online at http://www.mdpi.com/1996-1073/13/1/117/s1, Table S1: Mid-point indicators (CML-IA baseline V3.05) for the three scenarios (electricity from grid), Table S2: Mid-point indicators (CML-IA baseline V3.05) for the three scenarios (electricity from PV), Table S3: Incidence of most-impacting processes (BAS, grid), Table S4: Incidence of most-impacting processes (SH-TES, grid), Table S5: Incidence of most-impacting processes (PCM-TES, grid), Table S6: Incidence of most-impacting processes (BAS, PV), Table S7: Incidence of most-impacting processes (SH-TES, PV), Table S8: Incidence of most-impacting processes (PCM-TES, PV).

Author Contributions: Conceptualization, E.B. and A.A.; methodology, E.B.; software, E.B.; data curation, E.B. and A.A.; writing—original draft preparation, E.B.; writing—review and editing, E.B. and A.A.; supervision, E.B.; project administration, E.B.; funding acquisition, E.B. All authors have read and agreed to the published version of the manuscript.

Funding: This research was funded by Fondazione Cassa di Risparmio di Perugia, grant number 2017.0237.021.

Conflicts of Interest: The authors declare no conflict of interest.

References

1. European Commission. *A Roadmap for Moving to a Competitive Low Carbon Economy in 2050*; European Commission: Brussels, Belgium, 2011.

2. European Commission. *Horizon 2020 Work Programme for Secure, Clean, and efficient Energy*; European Commission: Brussels, Belgium, 2019.

3. Schwarz, M.; Nakhle, C.; Knoeri, C. Innovative designs of building energy codesfor building decarbonization and their implementation challenges. *J. Clean. Prod.* **2019**, 119260. [CrossRef]

4. Intergovernmental Panel on Climate Change. *Global Warming of 1.5 °C—Chapter 2: Mitigation Pathways Compatible with 1.5 °C in the Context of Sustainable Development*; Intergovernmental Panel on Climate Change: Geneva, Switzerland, 2018.

5. Atanasiu, B.; Despret, C.; Economidou, M.; Maio, J.; Nolte, I.; Rapf, O.; Laustsen, J.; Ruyssevelt, P.; Staniaszek, D.; Strong, D.; et al. *Europe's Buildings under the Microscope: A Country-by-Country Review of the Energy Performance of Buildings*; Buildings Performance Institute Europe: Brussels, Belgium, 2011.

6. Cabeza, L.F.; Rincón, L.; Vilariño, V.; Pérez, G.; Castell, A. Life cycle assessment (LCA) and life cycle energy analysis (LCEA) of buildings and the building sector: A review. *Renew. Sustain. Energy Rev.* **2014**, *29*, 394–416. [CrossRef]

7. Cappello, M. *Efficienza Energetica Degli Edifici*; Grafill Editoria Tecnica: Palermo, Italy, 2008.

8. Bonamente, E.; Pelliccia, L.; Merico, M.; Rinaldi, S.; Petrozzi, A. The multifunctional environmental energy tower: Carbon footprint and land use analysis of an integrated renewable Energy plant. *Sustainability* **2015**, *7*, 13564–13584. [CrossRef]

9. Bonamente, E.; Cotana, F. Carbon and energy footprints of prefabricated industrial buildings: A systematic life cycle assessment analysis. *Energies* **2015**, *8*, 12685–12701. [CrossRef]

10. Wang, Z.; Zhao, J.; Li, M. Analysis and optimization of carbon trading mechanism for renewable energy application in buildings. *Renew. Sustain. Energy Rev.* **2017**, *73*, 435–451. [CrossRef]

11. Vares, S.; Häkkinen, T.; Ketomäki, J.; Shemeikka, J.; Jung, N. Impact of renewable energy technologies on the embodied and operational GHG emissions of a nearly zero energy building. *J. Build. Eng.* **2019**, *22*, 439–450. [CrossRef]

12. Asaleye, D.A.; Breen, M.; Murphy, M.D. A decision support tool for building integrated renewable energy microgrids connected to a smart grid. *Energies* **2017**, *10*, 1765. [CrossRef]

13. Sartori, I.; Hestnes, A.G. Energy use in the life cycle of conventional and low-energy buildings: A review article. *Energy Build.* **2007**, *39*, 249–257. [CrossRef]

14. Self, S.J.; Reddy, B.V.; Rosen, M.A. Geothermal heat pump systems: Status review and comparison with other heating options. *Appl. Energy* **2013**, *101*, 341–348. [CrossRef]

15. Curtis, R.; Lund, J.; Sanner, B.; Rybach, L.; Hellström, G. Ground source heat pumps–geothermal energy for anyone, anywhere: Current worldwide activity. *Proc. World Geotherm. Congr.* **2005**, *1437*, 1–9.

16. Ghasemi-Fare, O.; Basu, P. Predictive assessment of heat exchange performance of geothermal piles. *Renew. Energy* **2016**, *86*, 1178–1196. [CrossRef]

17. EU. Directive 2009/28/EC of the European Parliament and of the Council of 23 April 2009 on the promotion of the use of energy from renewable sources. *Off. J. Eur. Union* **2009**, *5*, 2009.

18. Braungardt, S.; Bürger, V.; Zieger, J.; Bosselaar, L. How to include cooling in the EU Renewable Energy Directive? Strategies and policy implications. *Energy Policy* **2019**, *129*, 260–267. [CrossRef]

19. Bu, X.; Ma, W.; Li, H. Geothermal energy production utilizing abandoned oil and gas wells. *Renew. Energy* **2012**, *41*, 80–85. [CrossRef]

20. Heier, J.; Bales, C.; Martin, V. Combining thermal energy storage with buildings—A review. *Renew. Sustain. Energy Rev.* **2015**, *42*, 1305–1325. [CrossRef]

21. Dincer, I. On thermal energy storage systems and applications in buildings. *Energy Build.* **2002**, *34*, 377–388. [CrossRef]

22. Cabeza, L.F.; Castell, A.; Barreneche, C.D.; De Gracia, A.; Fernández, A. Materials used as PCM in thermal energy storage in buildings: A review. *Renew. Sustain. Energy Rev.* **2011**, *15*, 1675–1695. [CrossRef]

23. Kenisarin, M.; Mahkamov, K. Passive thermal control in residential buildings using phase change materials. *Renew. Sustain. Energy Rev.* **2016**, *55*, 371–398. [CrossRef]

24. Cui, Y.; Zhu, J.; Twaha, S.; Chu, J.; Bai, H.; Huang, K.; Chen, X.; Zoras, S.; Soleimani, Z. Techno-economic assessment of the horizontal geothermal heat pump systems: A comprehensive review. *Energy Convers. Manag.* **2019**, *191*, 208–236. [CrossRef]

25. Piga, B.; Casasso, A.; Pace, F.; Godio, A.; Sethi, R. Thermal impact assessment of groundwater heat pumps (GWHPs): Rigorous vs. simplified models. *Energies* **2017**, *10*, 1385. [CrossRef]

26. Cui, P.; Yang, H.; Fang, Z. Numerical analysis and experimental validation of heat transfer in ground heat exchangers in alternative operation modes. *Energy Build.* **2008**, *40*, 1060–1066. [CrossRef]

27. Balbay, A.; Esen, M. Experimental investigation of using ground source heat pump system for snow melting on pavements and bridge decks. *Sci. Res. Essays* **2010**, *5*, 3955–3966.

28. Wang, G.; Wang, W.; Luo, J.; Zhang, Y. Assessment of three types of shallow geothermal resources and ground-source heat-pump applications in provincial capitals in the Yangtze River Basin, China. *Renew. Sustain. Energy Rev.* **2019**, *111*, 392–421. [CrossRef]

29. Zhang, S.; Zhang, L.; Wei, H.; Jing, J.; Zhou, X.; Zhang, X. Field testing and performance analyses of ground source heat pump systems for residential applications in Hot Summer and Cold Winter area in China. *Energy Build.* **2016**, *133*, 615–627. [CrossRef]

30. Yang, W.; Xu, R.; Yang, B.; Yang, J. Experimental and numerical investigations on the thermal performance of a borehole ground heat exchanger with PCM backfill. *Energy* **2019**, *174*, 216–235. [CrossRef]

31. Oruc, O.; Dincer, I.; Javani, N. Application of a ground source heat pump system with PCM-embedded radiant wall heating for buildings. *Int. J. Energy Res.* **2019**, *43*, 6542–6550. [CrossRef]

32. Jung, W.; Kim, D.; Kang, B.; Chang, Y. Investigation of heat pump operation strategies with thermal storage in heating conditions. *Energies* **2017**, *10*, 2020. [CrossRef]

33. Han, Z.; Zheng, M.; Kong, F.; Wang, F.; Li, Z.; Bai, T. Numerical simulation of solar assisted ground-source heat pump heating system with latent heat energy storage in severely cold area. *Appl. Therm. Eng.* **2008**, *28*, 1427–1436. [CrossRef]

34. Kaygusuz, K. Investigation of a combined solar–heat pump system for residential heating. Part 1: Experimental results. *Int. J. Energy Res.* **1999**, *23*, 1213–1223. [CrossRef]

35. Kaygusuz, K. Investigation of a combined solar–heat pump system for residential heating. Part 2: Simulation results. *Int. J. Energy Res* **1999**, *23*, 1225–1237. [CrossRef]

36. Seo, B.M.; Hong, S.H.; Choi, J.M.; Lee, K.H. Part load ratio characteristics and energy saving performance of standing column well geothermal heat pump system assisted with storage tank in an apartment. *Energy* **2019**, *174*, 1060–1078.
37. Zhu, N.; Hu, P.; Lei, Y.; Jiang, Z.; Lei, F. Numerical study on ground source heat pump integrated with phase change material cooling storage system in office building. *Appl. Therm. Eng.* **2015**, *87*, 615–623. [CrossRef]
38. Bonamente, E.; Aquino, A.; Cotana, F. A PCMThermal Storage for Ground-source Heat Pumps: Simulating the System Performance via CFD Approach. *Energy Procedia* **2016**, *101*, 1079–1086. [CrossRef]
39. Vollaro, R.D.L.; Calvesi, M.; Battista, G.; Evangelisti, L.; Botta, F. Calculation model for optimization design of low impact energy systems for buildings. *Energy Procedia* **2014**, *48*, 1459–1467. [CrossRef]
40. Finnveden, G.; Hauschild, M.Z.; Ekvall, T.; Guinée, J.; Heijungs, R.; Hellweg, S.; Koehler, A.; Pennington, D.; Suh, S. Recent developments in life cycle assessment. *J. Environ. Manag.* **2009**, *91*, 1–21. [CrossRef]
41. International Organization for Standardization. *ISO 14044: Environmental Management, Life Cycle Assessment, Requirements and Guidelines*; International Organization for Standardization: Geneva, Switzerland, 2006.
42. International Organization for Standardization. *Environmental Management: Life Cycle Assessment; Principles and Framework*; Number 2006; International Organization for Standardization: Geneva, Switzerland, 2006.
43. CML-IA Characterization Factors. Available online: https://www.universiteitleiden.nl/en/research/research-output/science/cml-ia-characterisation-factors (accessed on 13 November 2019).
44. Goedkoop, M.; Heijungs, R.; Huijbregts, M.; De Schryver, A.; Struijs, J.; van Zelm, R. *ReCiPe 2008. A Life Cycle Impact Assessment Method Which Comprises Harmonised Category Indicators at the Midpoint and the Endpoint Level*; The Ministry of Housing, Spatial Planning and the Environment: Hague, The Netherlands, 2018.
45. SimaPro 9.0. Available online: https://simapro.com/2019/whats-new-in-simapro-9-0/ (accessed on 13 November 2019).
46. Ecoinvent 3.5. Available online: https://www.ecoinvent.org/database/older-versions/ecoinvent-35/ecoinvent-35.html (accessed on 13 November 2019).
47. The International EPD System. *Product Category Rules: Electricity, Steam and Hot/Cold Water Generation and Distribution*; The International EPD System: Stockholm, Sweden, 2015.
48. Bonamente, E.; Aquino, A. Life-Cycle Assessment of an Innovative Ground-Source Heat Pump System with Upstream Thermal Storage. *Energies* **2017**, *10*, 1854. [CrossRef]
49. Moretti, E.; Bonamente, E.; Buratti, C.; Cotana, F. Development of innovative heating and cooling systems using renewable energy sources for non-residential buildings. *Energies* **2013**, *6*, 5114–5129. [CrossRef]
50. Bonamente, E.; Moretti, E.; Buratti, C.; Cotana, F. Design and monitoring of an innovative geothermal system including an underground heat-storage tank. *Int. J. Green Energy* **2016**, *13*, 822–830. [CrossRef]
51. ESCC 2016. Available online: http://escc.uth.gr/escc-2019/ (accessed on 13 November 2019).
52. Heidari, M.D.; Mathis, D.; Blanchet, P.; Amor, B. Streamlined Life Cycle Assessment of an Innovative Bio-Based Material in Construction: A Case Study of a Phase Change Material Panel. *Forests* **2019**, *10*, 160. [CrossRef]
53. Ecoinvent 3.2. Available online: https://www.ecoinvent.org/support/documents-and-files/information-on-ecoinvent-3/information-on-ecoinvent-3.html#1980 (accessed on 13 November 2019).

Article

The Environmental Potential of Phase Change Materials in Building Applications. A Multiple Case Investigation Based on Life Cycle Assessment and Building Simulation

Roberta Di Bari [1,*], Rafael Horn [1], Björn Nienborg [2], Felix Klinker [3], Esther Kieseritzky [4] and Felix Pawelz [4]

[1] Institute for Acoustics and Building Physics, University of Stuttgart, 70569 Stuttgart, Germany; rafael.horn@iabp.uni-stuttgart.de
[2] Fraunhofer Institute for Solar Energy Systems (ISE), 79119 Freiburg, Germany; bjoern.nienborg@ise.fraunhofer.de
[3] ZAE Bayern, 97074 Würzburg, Germany; felix.klinker@zae-bayern.de
[4] Rubitherm GmbH, 12307 Berlin, Germany; esther.kieseritzky@rubitherm.com (E.K.); felix.pawelz@rubitherm.com (F.P.)
* Correspondence: roberta.di-bari@iabp.uni-stuttgart.de

Received: 15 April 2020; Accepted: 10 June 2020; Published: 12 June 2020

Abstract: New materials and technologies have become the main drivers for reducing energy demand in the building sector in recent years. Energy efficiency can be reached by utilization of materials with thermal storage potential; among them, phase change materials (PCMs) seem to be promising. If they are used in combination with solar collectors in heating applications or with water chillers or in chilled ceilings in cooling applications, PCMs can provide ecological benefits through energy savings during the building's operational phase. However, their environmental value should be analyzed by taking into account their whole lifecycle. The purpose of this paper is the assessment of PCMs at the material level as well as at higher levels, namely the component and building levels. Life cycle assessment analyses are based on information from PCM manufacturers and building energy simulations. With the newly developed software "Storage LCA Tool" (Version 1.0, University of Stuttgart, IABP, Stuttgart, Germany), PCM storage systems can be compared with traditional systems that do not entail energy storage. Their benefits can be evaluated in order to support decision-making on energy concepts for buildings. The collection of several case studies shows that PCM energy concepts are not always advantageous. However, with conclusive concepts, suitable storage dimensioning and ecologically favorable PCMs, systems can be realized that have a lower environmental impact over the entire life cycle compared to traditional systems.

Keywords: phase change materials; PCM; thermal energy storage; life cycle assessment (LCA); Storage LCA Tool; Speicher LCA

---

1. Introduction

In a context calling for more affordable, sustainable and modern energy [1,2], the building sector is under particular attention as one of the main drivers of energy consumption. While most of the primary energy supply is delivered for energy production [3], heating and hot water in households alone account for 79% of the total final energy use. In addition, despite efforts made to improve energy efficiency, the final energy use for space conditioning grew from 118 million TJ in 2010 to around 128 million TJ in 2018 [4].

A strategy for energy saving is the combination of a source of renewable energy with thermal energy storage, which can be realized for heating and cooling systems not only through sensible heat storage materials but also through phase change materials (PCMs) and thermochemical materials (TCMs) [5,6]. Processes that enable thermal storage are reversible adsorption–desorption reactions, exothermic in adsorption and endothermic in regeneration [7]. Typically, water vapor is used in combination with thermally stable and inexpensive nanoporous materials belonging, for example, to the class of zeolites (Zeolite 13X) or composite sorbents [8,9]. Composite materials, such as multiwalled carbon nanotubes/lithium chloride (MWCN-LiCl) especially, have proven to be advantageous due to their heat storage density with both water and methanol as working fluid [10]. For PCMs, latent heat storage can be achieved through state of matter changes (solid → liquid and liquid → solid). While PCMs also allow for sensible heat storage (i.e., they store heat by raising the temperature of a liquid or solid and they release it with the decrease of temperature if required), they can absorb large amounts of heat at their melting point with constant temperature until all the material is melted [11]. Solidification of PCM occurs when the ambient temperature around the liquid material falls below the crystallization temperature, which leads to the release of the stored latent heat [12]. PCMs, which are mainly hydrated salts or paraffins, are available in any required typology; they can be organic or inorganic and suit any temperature in a range from −50 to 100 °C [13,14].

The advantages of innovative storage systems that incorporate PCMs in the building sector have already been investigated: in comparison with a traditional storage concept, containment volumes are reduced, allowing more storage capacity. The energy input/output occurs longer with almost constant temperatures. As a consequence, insulation of latent storage systems may be less sophisticated and expensive [15]. Together with energy storage systems, they can be directly attached to building components, e.g., walls or chilled ceilings, or included into cooling devices. In such cases, a reduction of direct greenhouse gas emissions during a building's operational stage can be achieved through energy savings [12,16,17]. However, these savings may be overcompensated by additional impacts caused by storage material production or other activities not necessary in conventional systems. As a consequence, innovative storage systems can contribute to tackling climate change if the overall environmental impact during the life cycle can be reduced. This requirement is also indirectly requested by 7th sustainable development goals (SDG7), which calls a reduction in environmental impacts that drive climate change [18].

The environmental impacts can be evaluated through life cycle assessment (LCA), a technique comprised of a set of procedures for compiling and examining the inputs and outputs of materials and energy and the associated environmental impacts directly attributable to the functioning of a product or service system throughout its life cycle. International Standard ISO 14040 establishes LCA principles and framework [19], while ISO 14044 defines requirements and guidelines [20].

With regard to PCMs, only a few LCA studies and investigations are available so far. However, existing studies are often very limited in detail or focus only on a specific application [21]. Existing studies consider only cost aspects or energetic evaluation in operation [22] without a link to the environmental impact evaluation over the entire life cycle of the storage material. When available, the environmental evaluation is oftentimes only carried out at the laboratory scale (1 m × 1 m × 1 m cube) without reference to real application scenarios. Other works carry out analyses on a large scale and neglect a life cycle perspective [23].

In an effort to bridge the gap between the environmental promise of innovative storage materials and their actual environmental impact while considering system complexity and modularity, the "Storage LCA" (German "Speicher LCA") project was conducted, with funding from the German Federal Ministry of Economics and Technology (BMWi) and in collaboration with Fraunhofer ISE, ZAE Bayern and University Stuttgart. Some of the project results are presented in this paper [24]. The research focuses on a wide range of organic paraffins and salt hydrates with melting points between 10–15 °C for cooling systems and 58–62 °C for heating systems. Central heating systems combine the advantage of renewable energy supply through solar collectors with PCM thermal storage. With respect to centralized cooling systems, cooling devices are accompanied by PCM cold storage that allows for a cooling load shift from day to night, increasing the efficiency of cold production. Furthermore, room-integrated chilled PCM ceilings or PCM ventilation systems are considered. The consideration of sensible, latent and thermochemical storage concepts, as well as their energetic performances evaluated through energy simulation, enables comprehensive environmental assessments of the materials and systems used for thermal energy storage in buildings. Furthermore, the data collection for LCA is based on up-to-date information coming from manufacturers and material developers [13,14]. Previous works related to the project provided results at the material and component levels and already demonstrated the benefits of PCM configurations with high storage density in comparison with, for example, water storage [15].

In the present study, the created "Storage LCA Tool" is presented [25]. Based on the obtained LCA and energy simulation data platform, both practitioners and experts in energy and storage can carry out analyses on different levels. With the help of graphs and numerical results, users are able to decide whether innovative storage systems are advantageous in comparison with conventional systems. This paper analyses the overall environmental assessment of PCMs applied in buildings and provides insight into the general effectiveness and environmental benefits of PCM storage systems.

2. Materials and Methods

For the analysis, a three-level approach was established. Terms used in this work and their respective definitions are given below [6].

- Storage material: In this case, the pure PCM storage material.
- Storage component: All auxiliary equipment such as containment, insulation and heat exchangers required to unlock the storage capacity of the material.
- Storage system concept: All components required for the building supply and for the provision of heat and/or cold. Storage materials, storage components and conventional building service systems for energy provision are included. The main function of the storage system concept is to supply a defined building (building type, energetic standard and climate location) over a defined heating or cooling period. Examples of the possible benefits of PCM systems are the increased use of environmental heat and cold or the reduction of heating or cooling load peaks, which enables lower installed heating or cooling capacities.

Environmental database and energy simulation results were combined in a common data platform, on which the final building LCA was provided consistently with the selected analyzed case study. The environmental database contained results coming from LCA analyses carried out in Gabi ts software (Version 8.0, Thinkstep, a sphera company, Stuttgart, Germany) [26] (updated in 2019), according to the ISO 14040 and 14044 standards [19,20] and following specifications and suggestions for buildings [27–32]. Building energy simulations were carried out in TRNSYS (Version 18, Thermal Energy System Specialists LLC, Madison, WI, USA) [33].

2.1. LCA Specifications

In this work, LCAs were carried out at the material, component and energy concept levels (see Table 1). For the cradle-to-grave analyses, the considered life cycle modules were production stage (A1–A3, including raw material supply, transport and manufacturing) and end-of life (C3, C4 and D modules, including waste processing, disposal and eventual benefits due to recycling) [27]. For storage concept analyses, the operational building energy use was included (B6 module according to EN 15804 [27]). At the building level, a 20-year lifespan was considered, with reference to a nominal service life of the installation system without component replacement (module B4 according to EN 15804 [27]). Functional units are related to each level and specified in the following subsections.

Table 1. Life cycle assessment (LCA) specification according to [24,25,28].

Goal and Scope	Analyses of PCMs, Storage Components and Energy Storage Systems
System Boundaries	Cradle to grave analyses. Lifecycle modules [22]: Production (A1–A3) Use phase, including operational building energy demand (B6)—storage concept system only End-of-Life (C + D), including waste + credits due to recycling
Functional Unit (f.u.)	Defined for each level
Lifespan	20 years
Impact Categories	Global warming potential (GWP) (kg CO_2 eq./f.u.) Primary energy demand total (PEtot) (MJ/f.u.) Primary energy renewable total (PERT) (MJ/f.u.) Primary energy nonrenewable total (PENRT) (MJ/f.u.)

The investigated impact categories were selected in the project due to the goal and scope defined therein, including expectations of energy storage experts in regard to the tool content. For the evaluation of overall energy savings, the primary energy demand total (PEtot) indicator was chosen; this can be divided into primary energy renewable total (PERT) and primary energy nonrenewable total (PENRT). The global warming potential (GWP) expresses the emission of greenhouse gases in kg CO_2 equivalent. The life cycle impact assessment (LCIA) characterization method suggested for buildings in EN15084-Annex 2 [27] was used.

2.2. Analysis at the Material Level

For the PCM analysis, a wide range of commercially available materials were investigated with the support of Rubitherm Technologies GmbH, a PCM producing company. The investigated materials were mainly organic paraffins (RTXX) [13] and inorganic salt hydrates (SPXX) [14] with different melting points in both encapsulated and non-encapsulated variants (Table 2).

Table 2. Analyzed phase change materials (PCMs), listed as organic materials and salt hydrates.

Paraffins	Salt Hydrates
RT10HC	SP15
RT11HC	SP21
RT18HC	SP58
RT24	
RT62HC	

For storage material analyses, impacts were given by the functional unit (impact/kWh material storage capacity). These were obtained through the unit transformation as shown by Equations (1) and (2):

$$\text{GWP}\left(\frac{\text{kg } CO_2 \text{ eq.}}{\text{kWh}}\right) = \text{GWP}\left(\frac{\text{kg } CO_2 \text{ eq.}}{\text{kg}}\right) \cdot \frac{1}{E_d\left(\frac{\text{kWh}}{\text{kg}}\right)}, \tag{1}$$

$$\text{PEtot}\left(\frac{\text{MJ}}{\text{kWh}}\right) = \text{PEtot}\left(\frac{\text{MJ}}{\text{kg}}\right) \cdot \frac{1}{E_d\left(\frac{\text{kWh}}{\text{kg}}\right)}, \tag{2}$$

where E_d is the PCM energy density, calculated by taking into account the PCM's physical properties and the working temperatures of the distribution system. The material properties were based on our own experimental testing supported by technical datasheets of PCM producers as well as the collection of information about manufacturing.

The material analysis was supported by the theoretical cycle-based payback time, namely the ratio between impacts due to PCM lifecycle and conventional reference systems (Equation (3)).

$$\text{Payback}_{GWP}\left(\frac{\text{kg CO}_2 \text{ eq.}}{\text{kWh}}\right) = \text{GWP}_{PCM}\left(\frac{\text{kg CO}_2 \text{ eq.}}{\text{kWh}}\right) \cdot \frac{1}{\text{GWP}_{ref}\left(\frac{\text{kg CO}_2 \text{ eq.}}{\text{kWh}}\right)}, \tag{3}$$

The energetic payback-cycles indicate the theoretical minimum number of full charge and discharge cycles the material must endure to have environmental benefits compared to conventional systems (without storage). At the material level, this number does not consider any capacity losses or other use-phase-related impacts. It thus provides a theoretical minimum value that systems have to achieve from an environmental perspective to provide an advantage compared to the chosen reference. The selected reference systems were the following:

- Gas heating with 0.24 kg CO_2 eq./kWh and PENRT = 3.85 MJ/kWh [13];
- Split cooling (SEER = 3.5) with 0.18 kg CO_2 eq./kWh and PENRT = 2.21 MJ/kWh [13];
- Water chiller (SEER = 4.7) with 0.13 kg CO_2 eq./kWh and PENRT = 1.65 MJ/kWh [13].

2.3. Analysis at the Component Level

At the component level, functional units were established singularly. By considering, for instance, a storage containment, the storage volume (m^3) could be recommended as functional unit. In order to expand the data basis for LCA, several materials were selected at the component level based on implemented systems or recommendations from experts and users. Together, the components formed the storage system (see Table 3), which, in addition to the actual storage, also included other building installations such as the piping. For most components, an LCA on constructive aspects was sufficient, and production (A1–A3) and end-of-life (C + D) were considered. Since the focus of this work is on PCM storage systems, it does not show water-based energy storage systems and their respective components.

Table 3. Storage component list with their respective available materials.

Storage Component	Available Materials
Storage material	Water [4] or PCM (Table 2)
Storage containment	HDPE [1], steel [4], stainless steel [1,4]
Insulation storage	Mineral foam, EPDM [2] foam [4]; XPS [3]
PCM containment/capsules	Aluminum, steel, HDPE [1]

[1] High-density polyethylene. [2] Ethylene propylene diene monomer rubber. [3] Extruded polystyrene. [4] Only for sensible water storage.

2.4. Analysis at the System Level

Analysis at the whole system level considered the building components and the operational stage. Environmental impacts were calculated by multiplying material impacts by their quantities. These were scaled by the established functional unit, namely the yearly impact per net surface unit (kg CO_2 eq./m^2 net surface year).

In order to determine the energy demand of all investigated systems, they were examined in several building types in a building simulation using TRNSYS 17 or TRNSYS 18 (Table 4) [32]. Each building type was considered in different climate zones (Athens, Strasbourg and Helsinki) and with different insulation levels. The insulation levels refer to insulation standards in the three different climate zones around the year 1990, where "no insulation", "little" and "moderate" refer the insulation standards of Greece, Central Europe and Northern Europe, respectively. Buildings of this age are particularly interesting for system-level analyses, as the existing heating and cooling systems of these buildings are at the end of their life cycle and therefore need to be replaced. Simulations of energy-efficient buildings are included as well. Together with the system provided with thermal storage, the reference system without thermal storage is simulated for comparison (Table 4).

Table 4. Energy concept systems list.

Category	Energy Supply and Storage	Distribution System
Central heating	(1) Gas boiler [1] (2) Hot water storage + solar collector (3) PCM storage + solar collector	Radiators Underfloor heating
Central cooling	(1) Water chiller [1] (2) Split device [1] (3) Water chiller + cold water storage (4) Water chiller + PCM storage	Surface cooling Fan coil
Room-integrated	(1) Air cooling [1] (2) PCM surface cooling + Water chiller (3) PCM ventilation system	Natural ventilation [1] Surface cooling Air cooling

[1] Reference system.

2.5. Storage LCA Tool

"Storage LCA Tool" is available online for free [22]. The tool supports storage experts and practitioners by providing scientific support in the selection of environmentally suitable thermal storage materials and concepts for building applications. Comprehensive environmental assessments based on the LCA software Gabi ts [26] allow environmental assessments and comparisons at the material, component and system levels. In addition to the environmental database, simulation results have been included under the previously mentioned different boundary conditions.

All relevant components have been identified through expert surveys among the participating institutions and participants of the IEA Task 58—"Materials & components for thermal energy storage of the solar Heating and cooling" program of the International Energy Agency [33]. Based on the surveys, standard component and system setups have been derived and their parts have been modeled for LCA [15,22]. The models were based on ÖKOBAUDAT datasets [34] and our own models and were complemented by a material database to create and assess specific components that were not natively considered.

The tool is a useful instrument for all expert and nonexpert users. For this reason, two different tool usage modes are available (Figure 1).

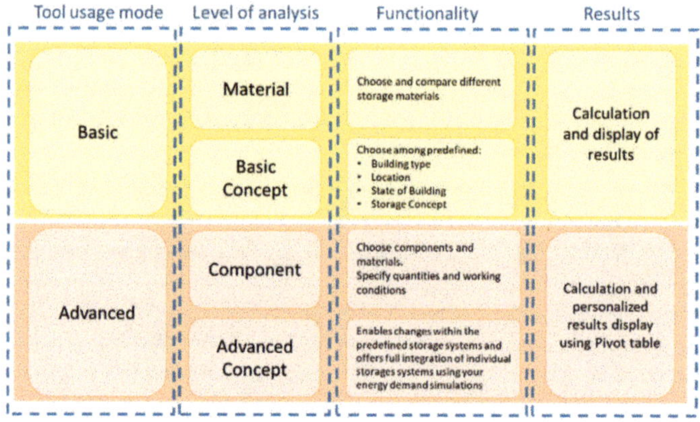

Figure 1. Storage LCA Tool functionality: advanced and basic modes.

The Basic Mode enables the user to perform comparisons of storage materials at predefined working temperatures. Minimum and maximum temperatures are defined according to the melting temperature range of the selected PCM and operating temperature of the selected distribution system. At the system level, a simplified analysis of innovative storage system layouts at the building level can be conducted and compared with a reference system. User enters information about (1) building type, (2) location, (3) insulation level, (4) energy storage and supply concept. Depending on such specifications, a drop-down list of available storage system layouts is generated and the chosen one analyzed. The final analysis is based on default storage components/combinations as well as default energetic TRNSYS simulations. Analogously, reference systems layouts are automatically generated and analyzed based on LCIA and energy simulation results. Results can be visualized in numerical or graphical simplified form easy to understand.

Within the Advanced Mode, default storage components and systems can be customized individually and analyzed. Moreover, individual energy demand simulations may be integrated and a more detailed analysis performed. To ensure consistent results, advanced mode analyses are suggested for expert users and only if comprehensive information about storage system and building installations are available. Results visualization can be personalized as well through Pivot tables and diagrams according to the scope of the analysis.

3. Results

In this section, with the support of "Storage LCA Tool", significant innovative storage materials as well as storage system concepts are selected and compared to their respective reference systems as demonstrative examples.

The chosen examples of storage system concepts for heating and cooling using innovative centralized PCM storage all use salt hydrates as storage materials. In the selected innovative cooling system, the relevance of the location for the selection of suitable and effective storage systems was emphasized. A heating concept has been selected in order to prove the great advantages of coupling PCM storage systems and solar collectors.

Generalized results for all energy concepts are presented in Section 4.

3.1. Material Level

To provide a broad spectrum of materials, the LCA makes use of a generic modeling approach, assuming similar production processes and routes wherever possible. This not only allows for comparison of materials with different technology readiness levels (TRLs), but also focuses on the impacts of the underlying raw materials. Detailed information at the material level is provided in the tool documentation [25]. As an example, a comparison between paraffin RT10HC, the salt hydrate SP15 and water is presented (Tables 5 and 6). Physical properties are derived from producers' technical documentation (Table 5), while the environmental assessment is carried out by Storage LCA Tool (Table 6), which is able to compare up to three materials [25]. Both PCMs are assumed to be operated in the range of 10–16 °C. As already noticed by previous works, paraffins show higher GWP impacts and PENRT [6,15]. The required raw materials strongly influence the material's global warming potential. For instance, the paraffin model only considers the petroleum-based route as a refinery by-product. In comparison with salt hydrates, paraffins consequently have higher environmental impacts (GWP of RT10HC is almost 50% higher in comparison with SP15, and the PENRT is 5 times higher) [35]. The selected system conditions result in long payback times for RT10HC (Figure 2). Therefore, from the environmental perspective, the salt hydrate is preferable in this case. Results analysis for all PCMs are available in Appendix A.

Table 5. Material properties of RT10HC (paraffin) and SP15 (salt hydrate) in comparison with water. Reproduced from [13,14], Rubitherm: 2019.

Name	Melting Enthalpy (Wh/kg)	Melting Range (°C)	Specific Heat Capacity (kJ/kg K)	Melting Point (°C)	Density (kg/m^3)
RT10HC	57.22	10–12	1.7	10	770
SP15	52.22	15–17	2.0	15	1350
Water	92.64		4.2	0	1000

Table 6. LCA analyses for RT10HC (paraffin) and SP15 (salt hydrate) in comparison with water. Reproduced from [25], IABP: 2019.

Name	GWP (kg CO$_2$ eq./kWh [1])	PENRT (MJ/kWh [1])	Payback Cycles GWP	Payback Cycles PENRT
RT10HC	18.30	880.85	82.7	226
SP15	12.02	180.95	49.5	47
Water	0.0	0.02	0.0	0.012

[1] kWh material storage capacity.

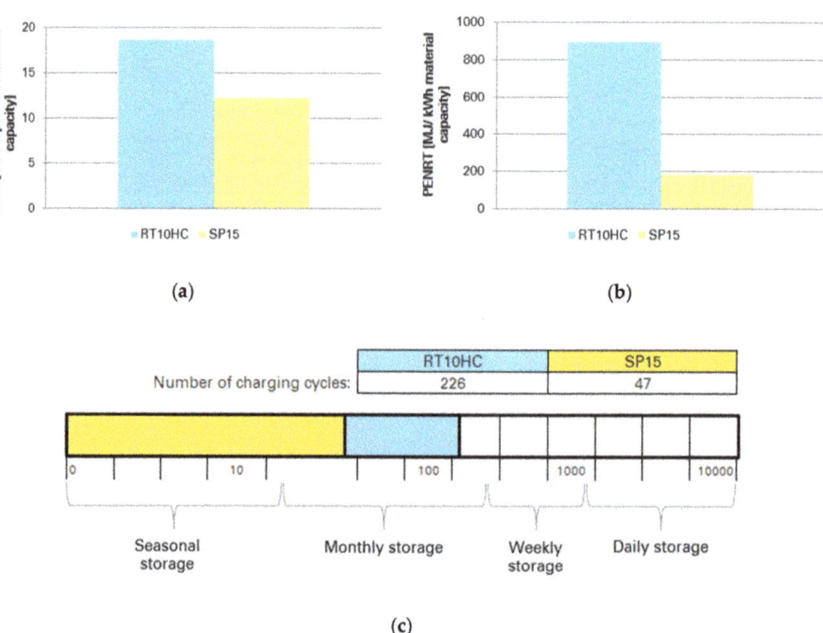

Figure 2. Environmental assessment through "Storage LCA Tool" of RT10HC and SP15: (**a**) global warming potential (GWP); (**b**) primary energy nonrenewable total (PENRT); (**c**) PENRT payback cycles. Reproduced from [25], IABP: 2019.

3.2. Component and System Levels

3.2.1. Helsinki—North European Insulation Standard

A cooling system with PCM storage in an office block located in Helsinki with moderate insulation level (North European insulation level) serves as an example. Together with the storage components (see Table 4), further elements constitute the overall storage energy concept, as shown in Figure 3a. The PCM storage is connected through valves and pipework to the building where the cold is distributed via chilled ceilings. In Figure 3b, the associated reference system for the comparison is represented. Unlike the innovative one, it does not entail any thermal storage. The water chiller and the distribution system have the same features [24]. For the selected location, with a mean temperature of 6.05 °C, the annual cooling demand is 4.08 kWh/m^2 a (see Appendix B).

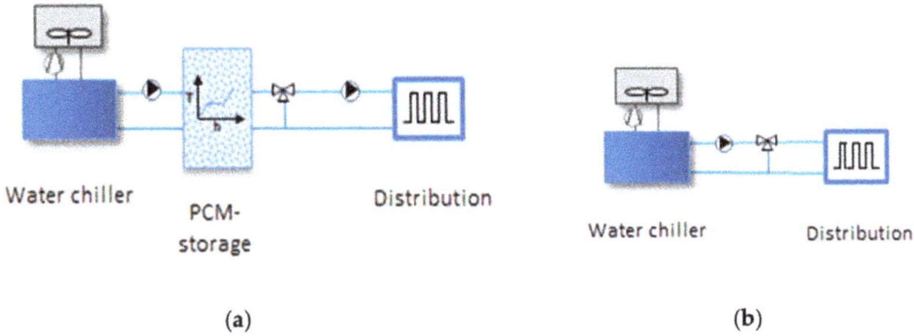

Figure 3. System layouts of (**a**) PCM storage for a cooling system and (**b**) a reference system. Reproduced from [24], Fraunhofer ISE 2019.

Table 7 reports comprehensive system information concerning materials and quantities. On this basis, LCIA is carried out, with the results reported in Table 8. Here, impacts are grouped according to their respective system (storage, cooling, distribution) and shown in both absolute and relative form.

Table 7. System information, including storage components. Reproduced from [25], IABP: 2019.

Storage component	Material	Amount	Unit	
Storage material	SP15	1755	kg	20% recycling rate
Containment	HDPE	85	kg	
Insulation (storage)	XPS	11.3	kg	
Heat exchanger	PP capillary tube	35	kg	
System component	**Material**	**Amount**	**Unit**	
Pipework	Steel	1755	kg	
Pipework insulation	XPS	2.97	kg	
Heat exchanger	PP capillary tube	45.5	kg	
Valves	Stainless steel	32	kg	
Circulation pump	Standard	250–1000	W	
Heat transfer fluid	Propylene glycol/water	30.85	kg	
Water chiller		11	kW	
Cooling surface	Copper (200 mm distance)	516	m^2	
Use Phase	**Process**	**Amount**	**Unit**	
Electricity	Electricity mix DE	616.0	kWh/a	

Table 8. Environmental assessment of system, including storage components Reproduced from [25], IABP: 2019.

Storage Component	A1–A3			C+D		
	GWP	**PENRT**	**PERT**	**GWP**	**PENRT**	**PERT**
Storage material	1125.5	169,937	1042.7	26.9	358.6	27.0
Containment	165.9	6224	286.6	96.5	−1902.8	−289.7
Insulation (storage)	959.7	29,376.7	538.4	18.5	−278.5	−54.8
Heat exchanger	98	2845	304.2	44.8	−160.4	−160.4
System component	**GWP**	**PENRT**	**PERT**	**GWP**	**PENRT**	**PERT**
Pipework	97	893.1	58.1	−54.6	−490.0	32.6
Pipework insulation	251.3	7683	141.0	4.8	−72.9	−14.4
Heat exchanger	127.4	3699.2	395.4	58.3	−1056.8	−208.5
Valves	33	394	61.6	−14.4	−129.8	8.2
Circulation pump	117.8	1588	254.4	−17.8	−254.9	−26.0
Water chiller	421.3	5731	987.8	−213.2	−2991.6	−336.5
Cooling surface	5627	98,806	12,733.5	−567.4	−26,682	−4409.2
Use Phase	**B6 yearly**			**B6 Total (20 years)**		
	GWP	**PENRT**	**PERT**	**GWP**	**PENRT**	**PERT**
Electricity mix	376.42	4935.8	2062.1	7528.30	95,695.05	41,241.86

LCIA results (Table 8) show that the storage system is the main factor responsible for the evaluated impacts. Among storage components, the PCM (SP15) and its insulation (XPS) present the highest global warming potential (GWP) and primary energy nonrenewable demand (PENRT), respectively.

Finally, the system is compared with its respective reference system. As demonstrated by the results (see Figure 4), compared with the reference system, the inclusion of a PCM storage system entails greater impact due to production. Energy savings during the operational stage due to the increased efficiency in cold production enabled by the inclusion of thermal energy storage do not compensate for this initial impact over a 20-year analysis. As a result, the total GWP and PEtot are slightly higher for the innovative system in the considered setup and climate zone.

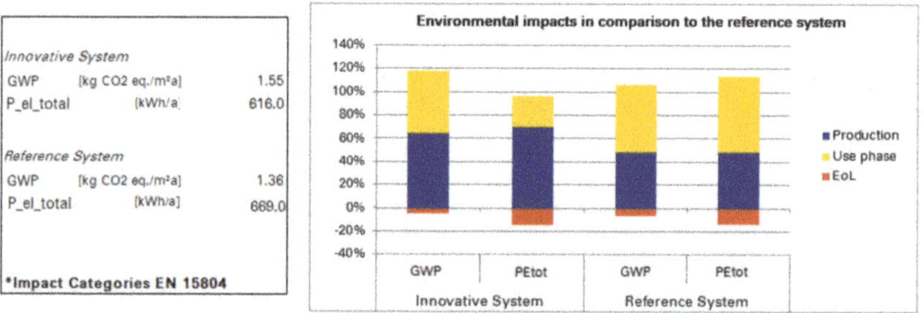

Figure 4. PCM storage for a cooling system in comparison with the reference system. Reproduced from [25], IABP: 2019.

The high environmental impact may be due to the selected boundary conditions. The energy simulation results included in the tool show an electricity demand of 616 kWh per year (kWh/a) for cold production and distribution. The reference system in turn consumes 669 kWh/a. The electricity savings for the innovative system amount to only 8%. Hence, in this case, the investigated innovative cooling system in a building with standard insulation level may not be the most effective solution due to the selected rather cool Helsinki climate zone.

3.2.2. Athens—North European Insulation Standard

A further analysis can be carried out for the same cooling system, water chiller + SP15 storage, in an office building with moderate insulation located in Athens. The new selected location has a mean temperature of 16.54 °C, and the annual cooling demand is 48.01 kWh/m^{2}a (see Appendix B). The total primary energy demand of the innovative system is reduced compared to the reference, and there are slight reductions of the total GWP. According to the simulation results included in the tool, this system has an electrical energy demand of 7445 kWh/a due to water chiller and cooling distribution. The reference system in turn has a total demand of 9039 kWh/a. On one hand, the cooling demand is higher due to the selected location, but on the other hand, an innovative system can provide greater benefits, with an 18% reduction in electricity demand for the cooling system. As a result, despite the high GWP due to SP15 production, the impacts can be more than compensated for. For the whole lifecycle, the selected innovative system records a GWP reduction of almost 10% (Figure 5).

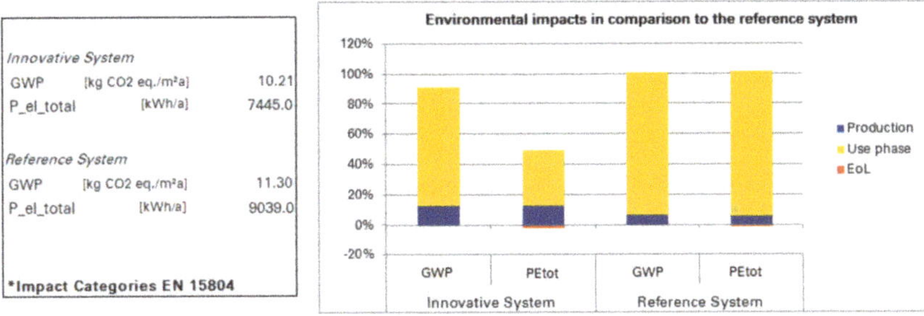

Figure 5. PCM storage for a cooling system in comparison with the reference system. Reproduced from [25], IABP: 2019.

3.2.3. Centralized Heating System with PCM Storage: Office Block in Helsinki

In contrast to the previous case study, the same building with an equal energy standard is now provided with a centralized heating system (see Figure 6a). A storage system using the salt hydrate SP58 is considered. The PCM storage with a volume of 16.49 m^3 is combined with solar collectors (with an area of 149.40 m^2). The reference system consists of a gas boiler with domestic hot water storage and underfloor heating (see Figure 6b) [24]. For the selected location with a mean temperature of 6.05 °C, the annual heating demand is 137.39 kWh/m^{2}a (see Appendix B).

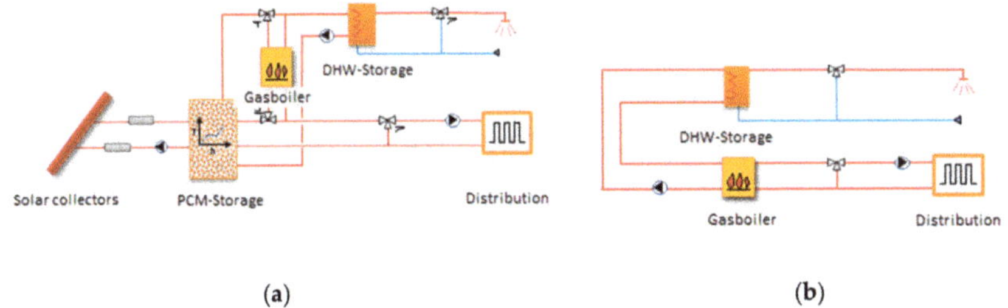

(a)   (b)

Figure 6. System layouts of (a) PCM storage for a heating system and (b) a reference system. Reproduced from [24], Fraunhofer ISE 2019.

The results of the analysis demonstrate the effectiveness of this choice. In terms of both GWP and PEtot, high savings are recorded. This is in the first place due to the solar collectors, which increase the total energy savings. The simulation results included in the tool show an electricity demand related to pumps and the collector circuit of 881.8 kWh/a and an energy demand of 52.74 kWh/a for heating provided by the gas boiler. Cooling demand is not included. For the reference system, the lack of solar collectors reduces electricity demand to 71.4 kWh but considerably increases the gas demand, which reaches 79.38 kWh/a (+44% in comparison with the innovative system). The gas demand strongly affects the final environmental assessment. The innovative system shows environmental advantages with a 46% reduction of the total GWP (Figure 7).

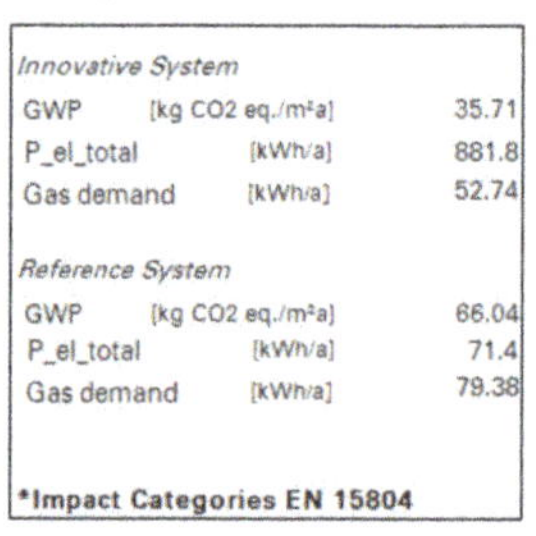

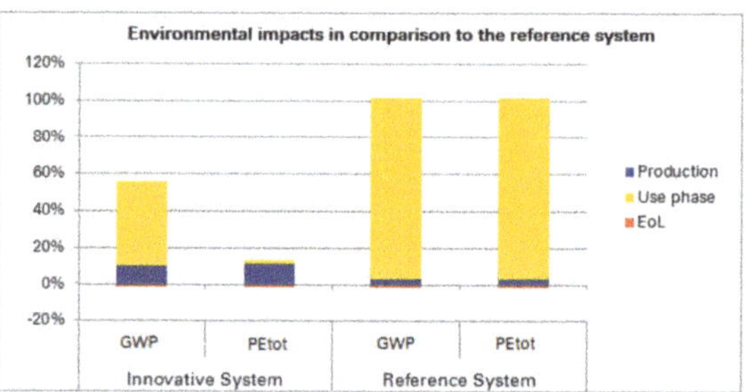

Figure 7. PCM storage for a heating system in comparison with the reference system. Reproduced from [25], IABP: 2019.

As the previous example, the storage material (SP15) is the main factor responsible for evaluated impacts. Storage thermal insulation (XPS) has only a minor influence on the overall evaluation (Table 9).

Table 9. Environmental assessment of the system, including storage components. Reproduced from [25], IABP: 2019.

Storage Component	A1–A3			C + D		
	GWP	**PENRT**	**PERT**	**GWP**	**PENRT**	**PERT**
Storage material	39,112.6	578,430	26,345.5	340.9	4548.3	342.5
Containment	121.0	4540	209.1	70.4	−1387.9	−211.3
Insulation (storage)	6011.1	183,994	3372.2	115.8	−1744.4	−343.2
Heat exchanger	98	394	304.2	44.8	−160.4	−160.4
System component	**GWP**	**PENRT**	**PERT**	**GWP**	**PENRT**	**PERT**
Pipework	543.3	4918.8	320.0	−300.8	−2698.9	179.3
Pipework insulation	35.1	1073.4	19.7	0.7	−10.2	−2.0
Heat exchanger	1616	46,922.3	5015.4	739.2	−13,404.8	−2644.4
Valves	33	1588.2	254.4	−17.8	−254.9	−26.0
Circulation pump	117.8	172,718.2	54,986.6	−8014.0	−111,185.4	−32,834.6
Gas boiler	526.6	6364.3	983.4	−137.0	−1622.8	−79.6
Underfloor heating	5627	98,805.6	12,733.5	−567.4	−26,681.8	−4409.2
Use Phase	**B6 yearly**			**B6 Total (20y)**		
	GWP	**PENRT**	**PERT**	**GWP**	**PENRT**	**PERT**
Electricity mix	538.8	6848.9	2951.7	10,776	136,979	59,034
Gas low temperature boiler	14,944	240,366	4576.5	298,881	4,807,327	91,530

4. Discussion

Since results typically vary from case to case, evaluations of the utility of a PCM application cannot be based on the results of a single case, such as those considered in the previous section. In order to derive generalizable conclusions and to identify eventual common characteristics from the combined energetic–environmental investigations, all results coming from "Storage LCA Tool" have been assessed in a meta-analysis, i.e., data from multiple case studies were combined.

Environmental impacts have been gathered and sorted by storage system, PCM storage material, building type, insulation level and location. For each case, impacts (GWP$_{\text{inn sys}}$ and PENRT$_{\text{inn sys}}$) of the innovative system have been divided by the impacts of the associated reference system ($\text{GWP}_{\text{ref sys}}$ and $\text{PENRT}_{\text{ref sys}}$), according to Equations (4) and (5). Two ratios are thus calculated: (1) a GWP ratio over the whole lifecycle, which describes the environmental performance, and (2) a primary nonrenewable energy demand ratio over the building use phase (B6 module), which is used to analyze the efficiency of the storage system.

$$\text{GWP ratio} = \frac{\text{GWP}_{\text{inn sys}}\left(\text{kg CO}_2 \text{ eq.}\right)}{\text{GWP}_{\text{ref sys}}\left(\text{kg CO}_2 \text{ eq.}\right)}, \tag{4}$$

$$\text{PENRT ratio}_{B6} = \frac{\text{PENRT}_{\text{inn sys,B6}}\left(\text{MJ}\right)}{\text{PENRT}_{\text{ref sys,B6}}\left(\text{MJ}\right)} \tag{5}$$

Results are visualized in an x–y diagram, where the GWP ratio is plotted on the y-axis and energy efficiency ratio is plotted on the x-axis (see Figure 8). If both ratios are less than 1, the system is deemed advantageous from both environmental and energy perspectives. If the PENRT ratio is less than 1, savings due to energy storage are recorded and the storage system can be deemed efficient. In the worst-case scenario, in which both ratios are greater than 1, the storage system is found to have very energy-intensive production processes and high nonrenewable energy demand. In such cases, storage systems are not advantageous.

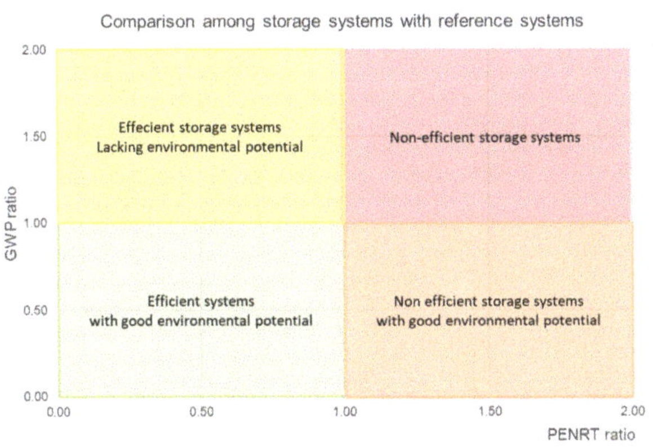

Figure 8. Interpretation of results for following figures (scheme).

4.1. Centralized Heating Systems

In Figure 9a centralized heating systems with PCM storage are considered and compared to the corresponding reference systems and are differentiated by PCM type. In the Figure 9b–d, results are filtered by location. The evaluated innovative systems use two different storage materials, namely RT62HC (organic) and SP58 (salt hydrate), and two different distribution systems, namely radiators and underfloor heating. The following parameter variations were implemented:

- 6 collector sizes (1.03 to 228.8 m^2);
- 12 storage sizes (daily to seasonal storage: 0.05 to 1181 m^3);
- 3 building types (single-family house, multifamily house, office building);
- 4 building insulation standards (none, little, moderate, efficient);
- 3 locations (Athens, Strasburg, Helsinki);
- 2 PCMs (RT62HC, SP58).

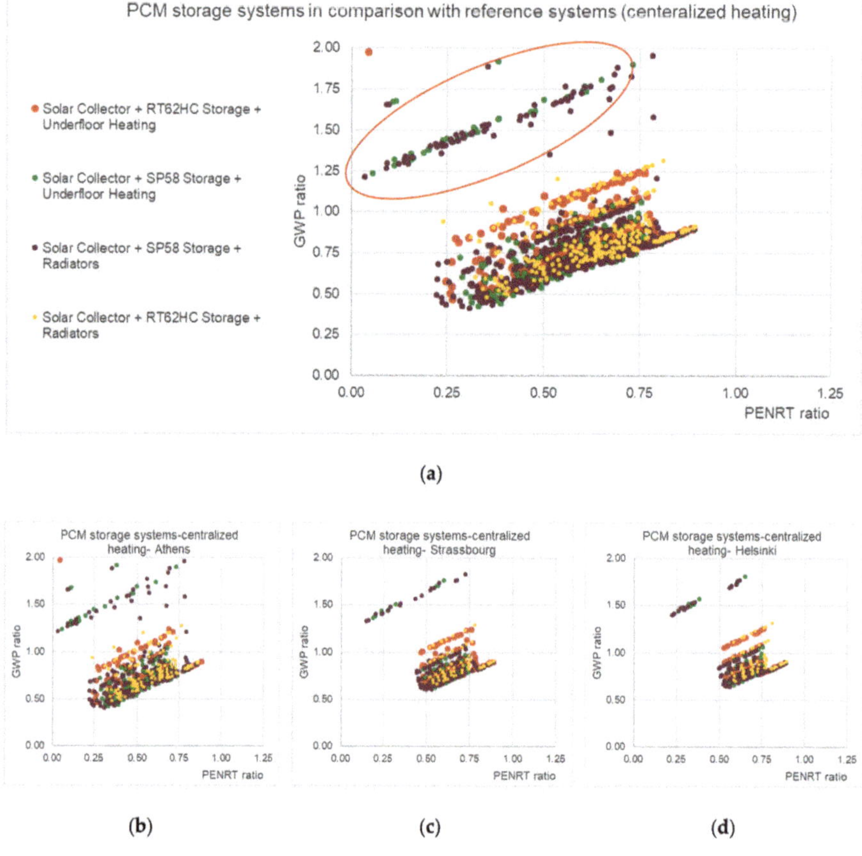

(a)

(b)   (c)   (d)

Figure 9. Environmental assessment through "Storage LCA Tool" of centralized heating systems, showing (**a**) comparison with reference systems and (**b**–**d**) results filtered by location: (**b**) Athens; (**c**) Strasbourg; (**d**) Helsinki.

The combination of PCM storage tanks with solar collectors seems to be largely advantageous and enables energy savings and emission reductions in most cases, as indicated by the accumulation of points in the lower left quadrant in Figure 9a. Different distribution systems seem to be not relevant to the whole lifecycle. There are also results in the upper left quadrant (red circle in Figure 9a) which show high GWP ratios. These cases mainly belong to systems with SP58 and seasonal storage, which enable energy savings but lead to high GWP ratios above 1.3 due to the high amount of PCM and its infrequent use. They mostly occur in Athens (Figure 9b) and decrease in North European locations such as Helsinki (Figure 9d).

The combination of solar collector + RT62HC (with both distribution systems) presents less variation in terms of environmental potential and energy storage. The best savings are recorded by solar collector + SP58 storage + radiators for a single-family house located in Athens with little insulation (GWP ratio = 0.44; energy ratio = 0.32). The worst performance (GWP ratio 2.13; energy ratio = 0.77) is recorded for a solar collector + SP58 storage + underfloor heating system in an energy-efficient office block in Strasbourg with high storage volume (121.82 m^3) and small collector surface (5.65 m^2). A more detailed visualization of results can be found in Appendix C (Figure A1).

4.2. Centralized Cooling Systems

The following parameter variations were implemented:

- Three water chillers (2 to 31 kW capacity);
- Nine storage sizes (daily to seasonal storage: 0.25 to 19 m^3);
- Three building types (single-family house, multifamily house, office building);
- Four building insulation standards (none, little, moderate, efficient);
- Three locations (Athens, Strasburg, Helsinki);
- Three different PCMs (RT10HC, RT11HC, SP15).

As already mentioned in Section 3.2.3, the achievable energy savings in central cooling systems are lower in comparison with those recorded for heating systems (Figure 10a). As in the example above, the results in Figure 10b–d are filtered by location. Compared with the reference systems, only a few innovative PCM storage concepts achieve positive environmental balances. In these cases, the energy savings due to the efficiency improvements in cold generation, which are achieved by shifting the cooling load into the night by integrating a PCM cold storage, more than compensate for the higher environmental impact due to the additional storage components.

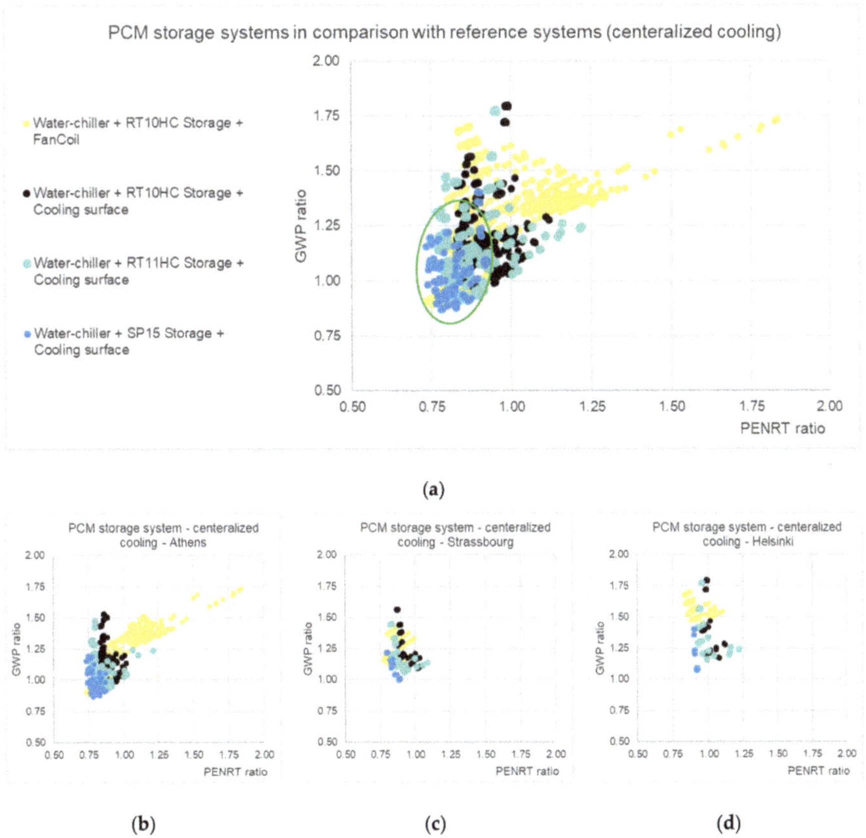

Figure 10. Environmental assessment through "Storage LCA Tool" of centralized cooling systems, showing (**a**) comparison with reference systems and (**b–d**) results filtered by location: (**b**) Athens; (**c**) Strasbourg; (**d**) Helsinki.

Unlike heating systems, better performances are reached in Mediterranean area (Athens, see Figure 10b), while PCM systems in Helsinki lack good environmental performances overall (Figure 10d).

Water chiller + RT10HC + fan coil systems show high performance variability, especially in Athens, while the SP15 storage + cooling surface combination offers a better performance in more cases with lower variability (green circle in Figure 10a). An office block located in Athens (no insulation) provided with RT10HC storage + fan coil and a storage volume of 7.11 m^3 has the best performance (GWP ratio = 0.89; energy ratio = 0.75). The worst performance (GWP ratio = 1.73; energy ratio = 1.84) is recorded by the same energy concept used in a single-family house in Athens, with low storage volume (0.34 m^3) and little insulation [22]. More results details are available in Appendix C (Figure A2).

4.3. Decenteralized Systems

Finally, decentralized PCM ventilation systems were analyzed. The two different systems are a water chiller with a chilled PCM ceiling (Figure 11a) and a water chiller with a PCM ventilation system (Figure 11b).

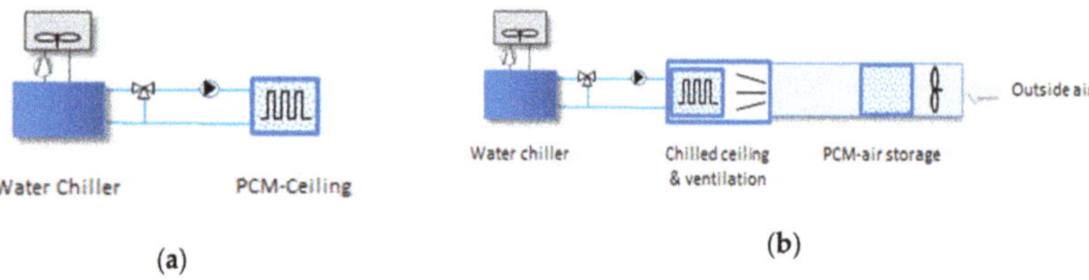

Figure 11. PCM storage for decentralized systems. System layouts of (**a**) water chiller + PCM cooling surface and (**b**) water chiller + PCM ventilation systems. Reproduced from [24], Fraunhofer ISE 2019.

The following parameter variations were implemented:

- Three water chiller powers (18 to 36 kW);
- Three water chiller temperatures (6, 10 or 14 °C);
- Three PCM mass distribution for chilled PCM ceilings (11, 16 or 22 kg/m^2);
- Three storage volumes for PCM ventilation systems (1, 2 or 3 m^3);
- Three volume flow rates for PCM ventilation systems (500, 1000 or 2000 m^3/s);
- One building type (office building);
- Four building insulation standards for chilled PCM ceiling simulations (none, little, moderate, efficient);
- One building insulation standard for PCM ventilation systems (efficient);
- Three locations (Athens, Strasburg, Helsinki);
- Four different PCMs (RT18HC, SP21EK, RT24, SP24).

Results are shown in Figure 12a. Here, a linear relationship between environmental impacts and energy demand is found.

Among the simulated examples (only office buildings), most cases located in Strasbourg (Figure 12c) offer advantageous energy performances, and all are environmentally advantageous. In comparison, the applications located in Helsinki offer even greater environmental savings while showing higher variability in energetic performance (Figure 12d). The systems simulated for Athens did not provide relevant energetic or environmental advantages (Figure 12c).

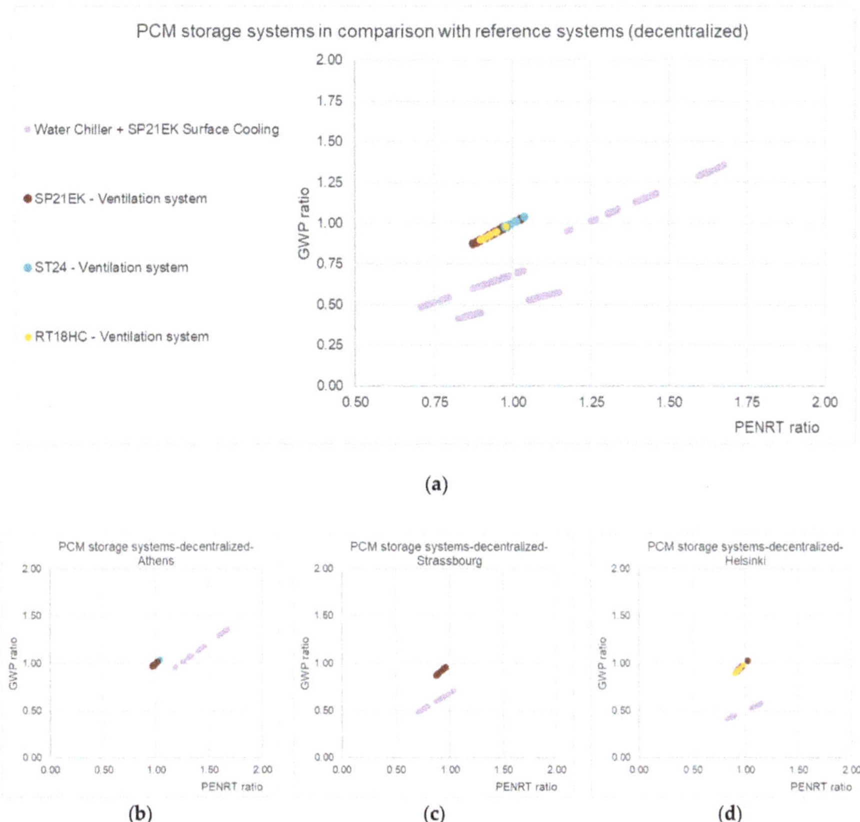

Figure 12. Environmental assessment through "Storage LCA Tool" of decentralized systems, showing (a) comparison with reference systems and (**b–d**) results filtered by location: (**b**) Athens; (**c**) Strasbourg; (**d**) Helsinki.

Water chiller + SP21EK surface cooling systems show a wide range of performance variability depending on location (Figure 12b–d). The best performance (GWP ratio = 0.42; energy ratio = 0.82) is recorded by this energy concept, located in a highly insulated office block in Helsinki with a PCM mass distribution for surface cooling of 22.4 kg/m^2. The worst result (GWP ratio = 1.02; energy ratio = 1.26) is recorded in an energy-efficient office block located in Athens (PCM mass distribution for surface cooling of 22.4 kg/m^2) [22]. A more detailed visualization is available in Appendix C (Figure A3).

5. Conclusions

Within this work, further LCA data were generated and provided by "Storage LCA Tool" for the support of decision-making in field of innovative storage materials, which are available or being researched for application in building services engineering. Through "Storage LCA Tool", analyses were carried out at different levels.

At the pure PCM level, the integration of updated information coming from PCM producers proved to be relevant for the assessment of the environmental potential of storage systems. A wide range of capacity-specific environmental effects depending on materials and their applications in heating, cooling and ventilation systems was demonstrated. Outcomes of analyses at the material level demonstrate the advantages of paraffins in terms of thermal storage but also indicate their higher environmental impacts. Conversely, the consideration of these environmental impacts during material production (e.g., when determining the synthesis route) offers the potential to minimize environmental impacts throughout the life cycle. This calls for more research on organic paraffins, which is actually ongoing, in order to minimize the primary energy (nonrenewable) demand for PCM production by replacing the raw materials by renewable sources and furthermore increase their recycling potential [6]. The environmental optimization of salt hydrate raw materials offers optimization potential as well.

At a higher level, two significant cooling and heating systems have been analyzed. Together with the amount of PCM in the thermal storage, the composition of the required auxiliary components (storage insulation, containment, heat exchanger, etc.) shows a great influence on the overall LCA. For this reason, it is necessary to evaluate each quantity carefully, in order to avoid excessively adverse environmental impacts and, at the same time, guarantee enough storage capacity.

By coupling such results with building energy simulation results, innovative PCM storage concepts seemed to be advantageous, especially if associated with a source of renewable energy such as solar collectors. In this case, PCM storage systems can represent an advantageous alternative to a traditional gas boiler in a heating system. Other factors that affect the benefits of PCM thermal energy storage are boundary conditions, i.e., building type, location and insulation level. These factors influence the yearly energy demand and the efficiency of the considered storage systems. Not surprisingly, by ensuring enough thermal insulation on the building envelope, advantageous PCM applications for cooling systems are recorded in warmer locations.

PCM thermal storage systems are, in some cases, advantageous alternatives to traditional water storage. To prove that, analog analyses have been carried out by assuming hot water tank storage (HWT) within heating systems and cold water tank storage (CWT) for cooling systems. In Figure 13, each water storage system is compared with all of the above-reported PCM storage systems. Systems have been distinguished as heating or cooling systems and divided by distribution system. With regard to heating systems, hot water storage (HWT) systems show greater advantages.

It is well known that solar thermal heating systems can save significant amounts of energy compared to conventional gas heating if they are correctly dimensioned. This is confirmed by this study. The impact is quantified and lies in the PENRT ratio range of 0.05 to 0.98, depending on the load and system size. In terms of GWP, the results are more ambiguous. Large PCM storage systems for long-term storage have a high impact due to production. Due to the low cycle number they are not used effectively, so the savings during use phase cannot always compensate for the footprint of production. Water storage systems, on the other hand, have a much lower impact during production and therefore yield better results here. The rather wide temperature range available for heat storage additionally results in a comparable volumetric energy density of water- and paraffin-based thermal storage. All case studies are located in the diagram area belonging to efficient systems with environmental potential.

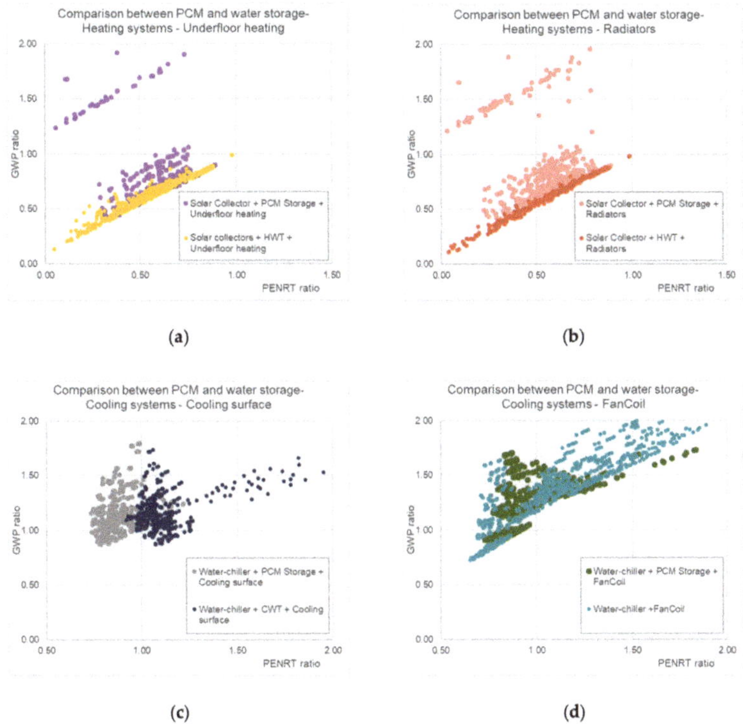

Figure 13. Environmental assessment through "Storage LCA Tool" of water storage systems for (**a**) centralized heating with underfloor heating; (**b**) centralized heating with radiators; (**c**) cooling systems with surface cooling; and (**d**) cooling systems with fan coil.

In contrast, PCM cooling systems are more likely to provide more efficient energy storage and have lower environmental impact compared to a cold water storage reference system. In cooling applications, the temperature interval usable for cold storage is significantly smaller than in heating applications. A PCM storage unit can fully exploit its advantages here due to its high heat storage capacity in a small temperature range. A PCM storage unit allows a significantly smaller storage volume and higher storage temperatures than a comparable water storage unit. On the one hand, this reduces the environmental impact caused by the production of the storage insulation material and the storage cylinder (especially since the PCM storage unit can be made of plastic, whereas the water storage unit is made of steel or stainless steel); on the other hand, the cold can be generated more efficiently due to a higher storage temperature, so that the environmental impact during the utilization phase is reduced.

This strongly affects the final environmental assessment of the innovative cooling systems (Figure 13c,d). Hence, in cooling systems, a PCM storage may be preferable to water tanks.

Despite the large number of energy concepts simulated, the data platform still needs to be improved and enriched. In this study, no degradation of the PCM's thermal properties over thousands of cycles is considered, which would affect the LCA results. At the storage system concept level, in this work, PCM applications for decentralized energy systems showed benefits, although the results are restricted to only one building type. In general, further innovative PCM storage materials; energy concepts; or boundary conditions, including locations, building types and insulation standards, can be included and analyzed through the "Storage LCA Tool". So far, this tool has been assessed (including validation of results) through beta tests, both internal (with help of project partners) and external, with feedback coming from LCA experts and potential tool users. The tool is freely available on the website https://www.iabp.uni-stuttgart.de/new_downloadgallery/GaBi_Downloads/SpeicherLCA.zip for further testing.

Author Contributions: Conceptualization, R.D.B. and R.H.; methodology, R.H., B.N. and F.K.; software, R.D.B.; validation, B.N., F.K. and R.H.; investigation, R.D.B.; data curation, R.H., B.N., F.K., E.K. and F.P.; writing—original draft preparation, R.D.B.; writing—review and editing, R.H., B.N., F.K. and E.K.; visualization, R.D.B.; supervision, R.H.; project administration, B.N. All authors have read and agreed to the published version of the manuscript.

Funding: Research and APC were funded by the German Federal Ministry for Economic Affairs and Energy for the Project "Speicher-LCA" (Grant number: 03ET1333) by a decision of the German Bundestag.

Acknowledgments: The authors thank the company Rubitherm GmbH for providing specific data on the production and encapsulation of various PCMs.

Conflicts of Interest: The authors declare no conflict of interest.

Appendix A

In Table A1, material properties of the analyzed PCMs are reported according to [13,14]. In Table A2, results of LCA analysis of PCMs are presented. All PCMs are applied for use in a cooling surface with operating temperature range of 10–16 °C.

Table A1. Material properties of PCMs and water. Reproduced from [13,14], Rubitherm: 2019.

Name	Melting Enthalpy (Wh/kg)	Melting Range (°C)	Heat Capacity (kJ/kg K)	Melting Point (°C)	Density (kg/m^3)
RT10HC	0.055	10–12	2.0	10	770
RT10HC (Enc [1])	0.041	10–12	1.7	10	1270
RT11HC	0.055	10–12	2.0	11	770
RT11HC (Enc [1])	0.041	10–12	1.7	11	1270
RT18HC	0.072	16–20	2.0	18	825
RT18HC (Enc [1])	0.053	16–20	1.7	18	1311
RT21	0.043	18–23	2.0	21	825
RT21 (Enc [1])	0.032	18–23	1.7	21	1311
RT24	0.043	21–24	2.0	24	825
RT24 (Enc [1])	0.032	21–24	1.7	21	1311
RT62HC	0.043	60–63	2.0	62	825
SP15	0.050	15–17	2.0	15	1350
SP15 (Enc [1])	0.037	15–17	1.7	15	1700
SP21EK	0.047	21–23	2.0	21	1350
SP21EK (Enc [1])	0.035	21–23	1.7	21	1551
SP58	0.069	56–59	2.0	58	1350
SP58 (Enc [1])	0.051	56–59	1.7	58	1551
Water	2.000	0	4.2	0	1000

[1] Macroencapsulated.

Table A2. LCA analyses of PCMs and water applied on a cooling surface with operating temperatures 16–20 °C. Reproduced from [25], IABP: 2019.

Name	GWP (kg CO$_2$ eq./kWh [1])	PENRT (MJ/kWh [1])	Payback Cycles GWP	Payback Cycles PENRT
RT10HC	18.3	880.9	82.8	226.0
RT10HC (Enc)	91.9	1877.9	372.8	476.4
RT11HC	18.1	877.01	81.9	225.0
RT11HC (Enc)	91.6	1873.8	371.9	475.3
RT18HC	316.4	15,348.6	1309.5	3981.9
RT18HC (Enc)	1398.9	28,602.2	5751.4	7414.2
RT21	314.6	15,325.5	1302.1	3975.9
RT21 (Enc)	1397.3	28,582.2	5745.0	7409.0
RT24	314.2	15,318.6	1300.4	3974.1
RT24 (Enc)	1396.9	28,576.1	5743.5	7407.4
RT62HC	314.2	15,318.6	1300.4	3974.1
SP15	12.0	180.95	49.6	47.0
SP15 (Enc)	88.9	1266.1	360.9	324.6
SP21EK	127.1	1794.2	523.0	465.8
SP21EK (Enc)	777.7	10,673.4	3195.7	2766.8
SP58	527.1	7795.0	2169.2	2023.7
SP58 (Enc)	1173.5	16,616.6	4824.2	4309.6
Water	0.0	0.01	0.009	0.0

[1] kWh material storage capacity.

Appendix B

In this Appendix, relevant information for building energy simulations is reported. Table A3 presents mean, maximal and minimal temperatures of each location. Table A4 gives an overview of the established insulation level. Lastly, in Table A5, specific heating and cooling demands are listed for simulation of the chosen office building [24].

Table A3. Overview of the climate data used to calculate the soil surface temperature. Reproduced from [24], Fraunhofer ISE 2019.

Location	Helsinki	Strasbourg	Athens
$\theta_{\text{Mean, year}}$ (°C)	6.05	11.22	16.54
$\theta_{\text{Monthly\_min}}$ (°C)	−4.97	2.09	6.63
$\theta_{\text{Monthly\_max}}$ (°C)	18.36	20.07	25.87
Amplitude (°C)	11.67	8.99	9.62

Table A4. U-values were used for energy simulation of office buildings at the different locations. The building efficiency is examined at all locations. Reproduced from [24], Fraunhofer ISE 2019.

	Office Building			
Insulation Level	**None**	**Little**	**Moderate**	**Efficient**
	(Mediterranean)	(Central Europe)	(North Europe)	
U exterior wall $W/(m^2K)$	2.10	0.93	0.38	0.35
U roof $W/(m^2K)$	2.77	0.56	0.20	0.15
U floor $W/(m^2K)$	2.83	0.95	0.30	0.20
U window $W/(m^2K)$	2. 83	2.83	1.40	0.70

Table A5. Specific heating and cooling energy demands for simulations of office buildings. Reproduced from [24], Fraunhofer ISE 2019.

Building Type	Insulation Level	Location	Heating Demand kWh/m²/y	Cooling Demand kWh/m²/y
	None	Athens	196.79	53.79
	Little	Athens	75.53	47.84
	Moderate	Athens	29.56	48.01
	Efficient	Athens	17.93	42.74
Office Building	Little	Strasbourg	160.20	8.02
	Moderate	Strasbourg	72.76	13.88
	Efficient	Strasbourg	48.53	13.87
	Moderate	Helsinki	137.39	4.08
	Efficient	Helsinki	93.32	4.95

Appendix C

In this Appendix, the overall environmental performances are reported for each single energy concept for better understanding. In Figure A1, results for centralized heating systems are presented.

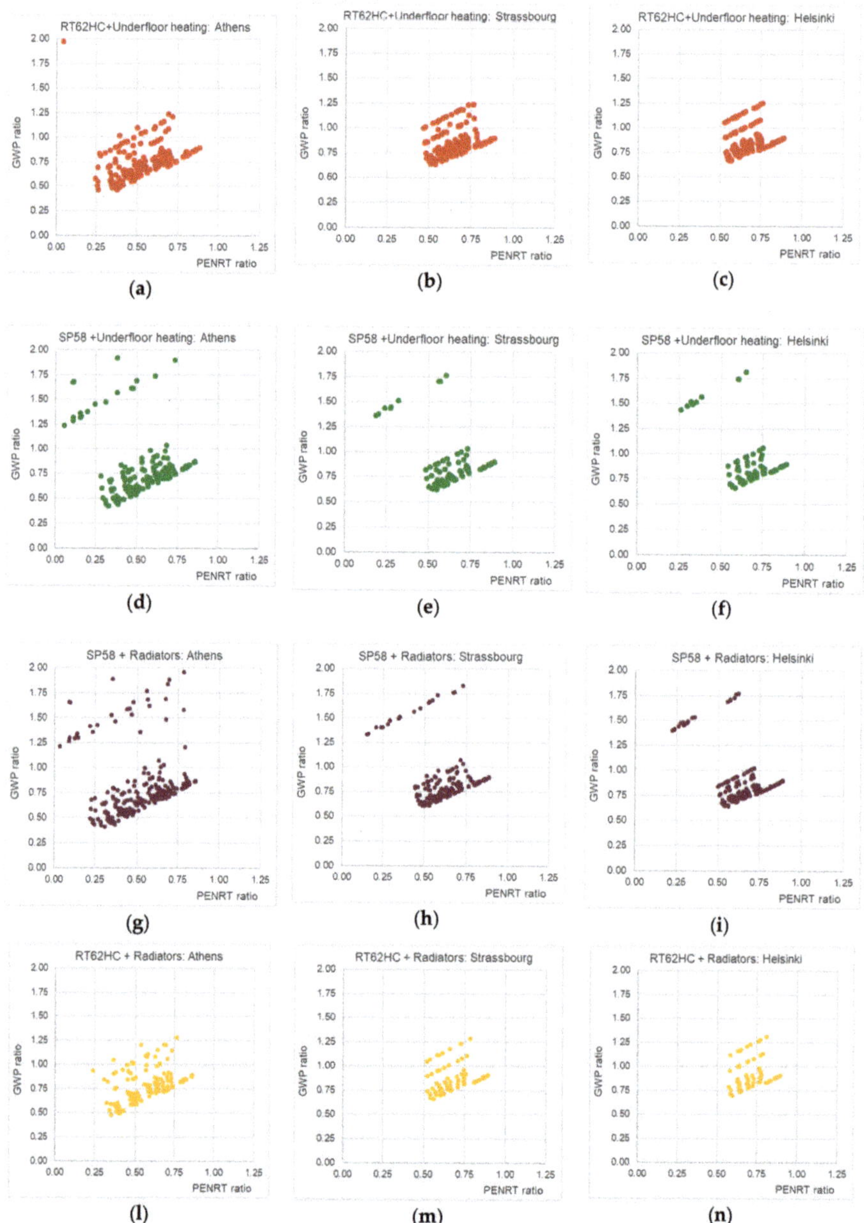

Figure A1. Environmental assessment through "Storage LCA Tool" of PCM storage systems applied for use in centralized heating systems, divided by location and energy concept.

For a better understanding, results for centralized heating systems are presented in Figure A2.

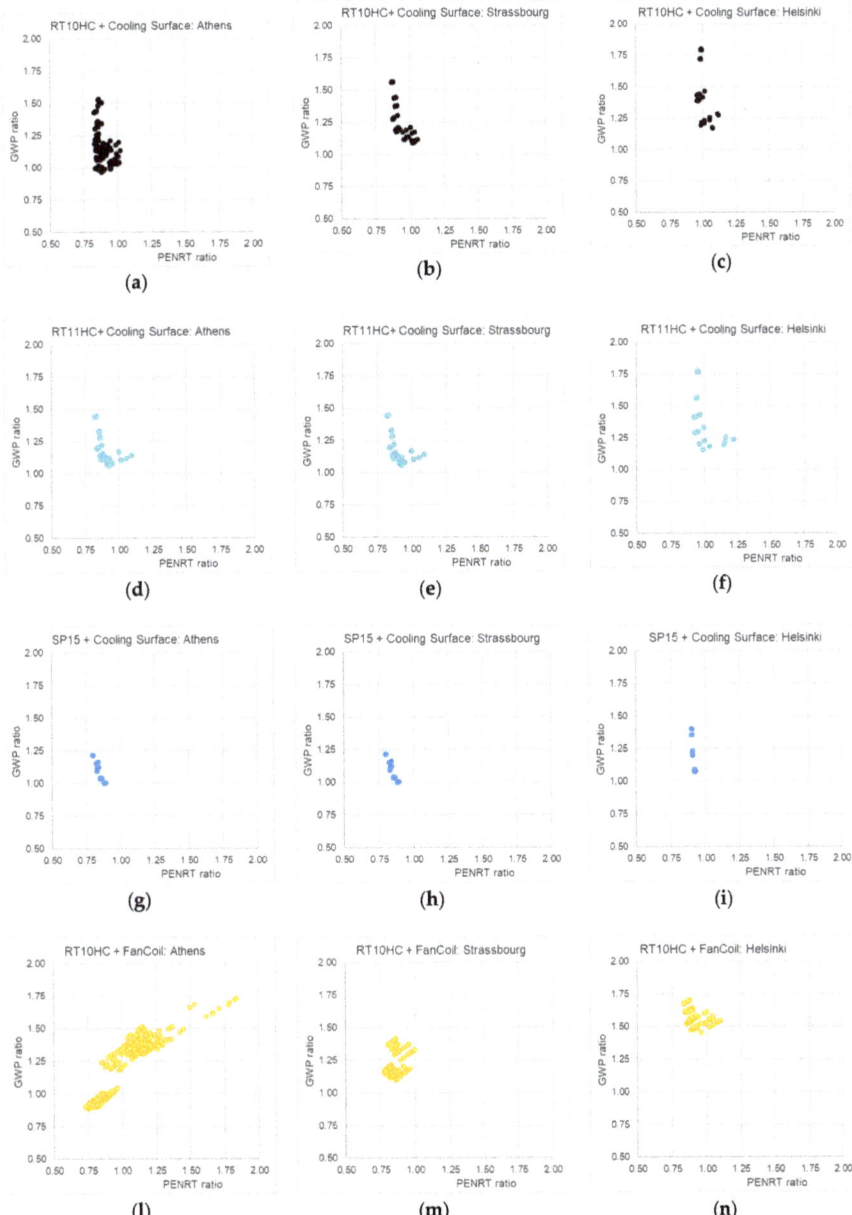

Figure A2. Environmental assessment through "Storage LCA Tool" of PCM storage systems applied for use in centralized cooling systems, divided by location and energy concept.

In Figure A3, results for decentralized systems are presented.

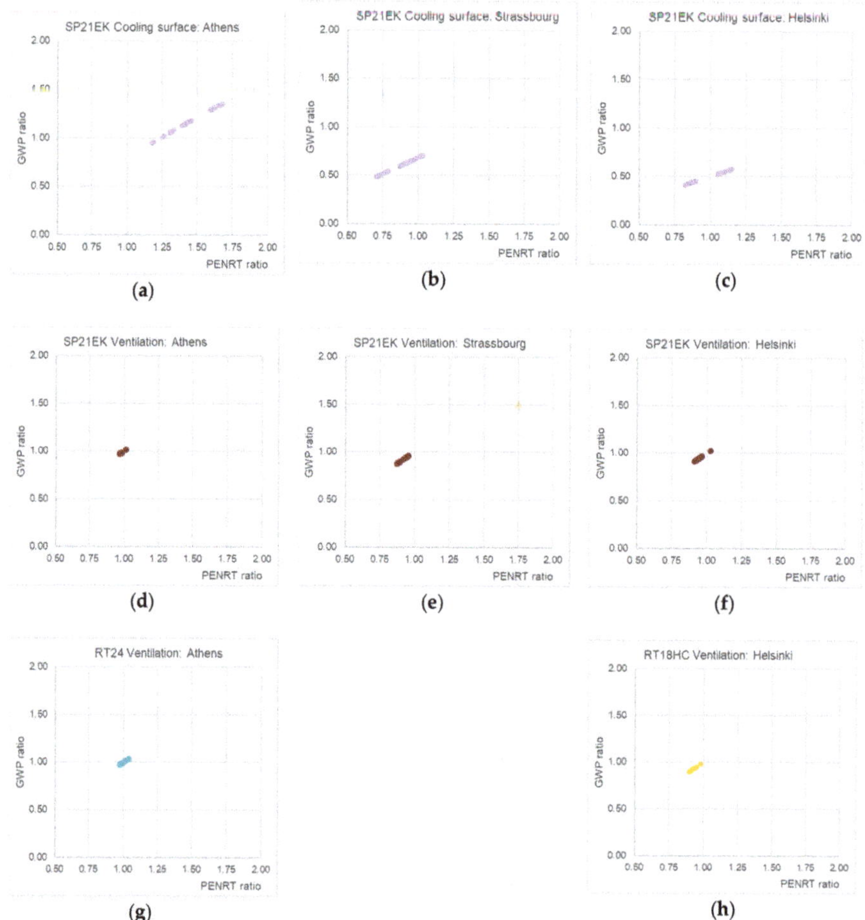

Figure A3. Environmental assessment through "Storage LCA Tool" of PCM storage systems applied for use in decentralized systems, divided by location and energy concept.

References

1. United Nations. Transforming Our World: The 2030 Agenda for Sustainable Development. A/RES/70/1. Available online: https://sustainabledevelopment.un.org/ (accessed on 31 March 2020).
2. United Nations. Analysis of the Voluntary National Reviews Relating to Sustainable Development Goal 7: Ensuring Access to Affordable, Reliable, Sustainable and Modern Energy for All. Available online: https://sustainabledevelopment.un.org/sdg7 (accessed on 31 March 2020).
3. Energy European Commission. Heating and Cooling. Available online: https://ec.europa.eu/energy/topics/energy-efficiency/heating-and-cooling_en (accessed on 11 May 2020).
4. Dulac, J.; Abergel, T.; Delmastro, C. Buildings: Tracking Clean Energy Progress. Available online: https://www.iea.org/tcep/buildings/ (accessed on 18 March 2020).
5. Arce, P.; Medrano, M.; Gil, A.; Oró, E.; Cabeza, L.F. Overview of thermal energy storage (TES) potential energy savings and climate change mitigation in Spain and Europe. *Appl. Energy* **2011**, *88*, 2764–2774. [CrossRef]

6. Horn, R.; Burr, M.; Fröhlich, D.; Gschwander, S.; Held, M.; Lindner, J.P.; Munz, G.; Nienborg, B.; Schossig, P. Life Cycle Assessment of Innovative Materials for Thermal Energy Storage in Buildings. *Procedia CIRP* **2018**, *69*, 206–211. [CrossRef]

7. Rindt, C.; Lan, S.; Gaeini, M.; Zhang, H.; Nedea, S.; Smeulders, D.M.J. Phase Change Materials and Thermochemical Materials for Large-Scale Energy Storage. In *Continuous Media with Microstructure*, 2nd ed.; Albers, B., Kuczma, M., Eds.; Springer: Cham, Switzerland, 2016; pp. 187–197. ISBN 978-3-319-28241-1.

8. Shigeishi, R.A.; Langford, C.H.; Hollebone, B.R. Solar energy storage using chemical potential changes associated with drying of zeolites. *Sol. Energy* **1979**, *23*, 489–495. [CrossRef]

9. Aprea, P.; de Gennaro, B.; Gargiulo, N.; Peluso, A.; Liguori, B.; Iucolano, F.; Caputo, D. Sr-, Zn- and Cd-exchanged zeolitic materials as water vapor adsorbents for thermal energy storage applications. *Appl. Therm. Eng.* **2016**, *106*, 1217–1224. [CrossRef]

10. Frazzica, A.; Freni, A. Adsorbent working pairs for solar thermal energy storage in buildings. *Renew. Energy* **2017**, *110*, 87–94. [CrossRef]

11. Kenisarin, M.; Mahkamov, K. Solar energy storage using phase change materials. *Renew. Sustain. Energy Rev.* **2007**, *11*, 1913–1965. [CrossRef]

12. Pomianowski, M.; Heiselberg, P.; Zhang, Y. Review of thermal energy storage technologies based on PCM application in buildings. *Energy Build.* **2013**, *67*, 56–69. [CrossRef]

13. Rubitherm. PCM RT-Serie. Available online: https://www.rubitherm.eu/index.php/produktkategorie/organische-pcm-rt (accessed on 18 March 2020).

14. Rubitherm. PCM SP-Serie. Available online: https://www.rubitherm.eu/index.php/produktkategorie/anorganische-pcm-sp (accessed on 18 March 2020).

15. Nienborg, B.; Gschwander, S.; Munz, G.; Fröhlich, D.; Helling, T.; Horn, R.; Weinläder, H.; Klinker, F.; Schossig, P. Life Cycle Assessment of thermal energy storage materials and components. *Energy Procedia* **2018**, *155*, 111–120. [CrossRef]

16. Ostermann, E.; Butala, V.; Stritih, U. PCM thermal storage system for 'free' heating and cooling of buildings. *Energy Build.* **2015**, *106*, 125–133. [CrossRef]

17. Souayfane, F.; Fardoun, F.; Biwole, P.-H. Phase change materials (PCM) for cooling applications in buildings: A review. *Energy Build.* **2016**, *129*, 396–431. [CrossRef]

18. Life Cycle Initiative. Paris Agreement, Sustainable Development Goals and Life Cycle Thinking. Available online: https://www.lifecycleinitiative.org/paris-agreement-sustainable-development-goals-life-cycle-thinking/ (accessed on 8 May 2020).

19. International Organization for Standardization. *Environmental Management—Life Cycle Assessment—Principles and Framework*; ISO: Vernier/Geneva, Switzerland, 2009; DIN EN ISO 14040:2006. Available online: https://www.iso.org/standard/37456.html (accessed on 18 March 2020).

20. International Organization for Standardization. *Environmental Management—Life Cycle Assessment—Requirements and Guidelines*; ISO: Vernier/Geneva, Switzerland, 2009; DIN EN ISO 14044:2006. Available online: https://www.iso.org/standard/38498.html (accessed on 18 March 2020).

21. Kylili, A.; Fokaides, P.A. Life Cycle Assessment (LCA) of Phase Change Materials (PCMs) for building applications: A review. *J. Build. Eng.* **2016**, *6*, 133–143. [CrossRef]

22. Llorach-Massana, P.; Peña, J.; Rieradevall, J.; Montero, J.I. Analysis of the technical, environmental and economic potential of phase change materials (PCM) for root zone heating in Mediterranean greenhouses. *Renew. Energy* **2017**, *103*, 570–581. [CrossRef]

23. Menoufi, K.; Castell, A.; Farid, M.M.; Boer, D.; Cabeza, L.F. Life Cycle Assessment of experimental cubicles including PCM manufactured from natural resources (esters): A theoretical study. *Renew. Energy* **2013**, *51*, 398–403. [CrossRef]

24. Nienborg, B.; Henninger, S.; Gschwander, S.; Horn, R.; Di Bari, R.; Klinker, F.; Weinläder, H. *Ökologische Bewertung Ausgewählter Konzepte und Materialien zur Wärme-und Kältespeicherung (Speicher-LCA)*; Fraunhofer ISE: Freiburg, Germany, 2019.

25. TIK Universität Stuttgart. Speicher-LCA Tool: Ökologische Bewertung Ausgewählter Konzepte und Materialien zur Wärme-und Kältespeicherung. Available online: https://www.iabp.uni-stuttgart.de/new_downloadgallery/GaBi_Downloads/SpeicherLCA.zip (accessed on 11 June 2020).

26. GaBi ts; thinkstep: A sphera company: Leinfelden-Echterdingen, Germany. 2019. Available online: http://www.gabi-software.com/international/support/gabi-version-history/gabi-ts-version-history/zip (accessed on 11 June 2020).

27. European Commitee for Standardization. *Sustainability of Construction Works—Environmental Product Declarations—Core Rules for the Product Category of Construction Products*; CEN: Brussel, Belgium, 2018; EN 15804:2018. Available online: https://standards.cen.eu/dyn/www/f?p=204:110:0::::FSP_PROJECT,FSP_ORG_ID:40703,481830&cs=1B0F862919A7304F13AE6688330BBA2FF (accessed on 18 March 2020).

28. European Commitee for Standardization. *Sustainability of Construction Works—Assessment of Buildings. Part. 2: Framework for the Assessment of Environmental Performance*; CEN: Brussel, Belgium, 2010; EN 15643-2. Available online: https://standards.cen.eu/dyn/www/f?p=CENWEB:110:0::::FSP_PROJECT:31551&cs=3C10098881B7F32BADFB51EC51B33CFB9 (accessed on 18 March 2020).

29. European Commitee for Standardization. *Sustainability of Construction Works—Assessment of Environmental Performance of Buildings—Calculation Method*; CEN: Brussel, Belgium, 2012; EN 15978:2012. Available online: https://standards.cen.eu/dyn/www/f?p=CENWEB:110:0::::FSP_PROJECT:31325&cs=31465126A6847927D74F7803C93FBA4E7 (accessed on 18 March 2020).

30. Gantner, J.; Wittstock, B.e.a. EeBGuide Part B: BUILDINGS: Operational Guidance for Life Cycle Assessment Studies of the Energy-Efficient Buildings Initiative. Available online: https://www.eebguide.eu/eebblog/ (accessed on 4 December 2019).

31. Joint Research Centre. ILCD Handbook—General Guide on LCA—Detailed Guidance. Available online: https://eplca.jrc.ec.europa.eu/ilcdHandbook.html (accessed on 4 March 2020).

32. Thermal Energy System Specialists, LLC. *TRNSYS*; Thermal Energy System Specialists, LLC: Madison, WI, USA, 2018.

33. IEA—International Energy Agency. Solar Heating & Cooling Programm Task 58/ECES Annex 33: Material and Component Development for Thermal Energy Storage. Available online: https://task58.iea-shc.org/ (accessed on 8 May 2020).

34. Bundesinstitut für Bau-, Stadt- und Raumforschung. ÖKOBAUDAT. Available online: https://oekobaudat.de/ (accessed on 3 April 2020).

35. Krendlinger, E.; Wolfmeier, U.; Schmidt, H.; Heinrichs, F.-L.; Michalczyk, G.; Payer, W.; Dietsche, W.; Boehlke, K.; Hohner, G.; Wildgruber, J. Waxes. In *Ullmann's Encyclopedia of Industrial Chemistry*; Wiley: Chichester, UK, 2010; pp. 1–63. ISBN 9783527306732.

Article

A Building Life-Cycle Embodied Performance Index—The Relationship between Embodied Energy, Embodied Carbon and Environmental Impact

Ming Hu

School of Architecture, Planning and Preservation, University of Maryland, College Park, MD 20742, USA; mhu2008@umd.edu

Received: 28 March 2020; Accepted: 11 April 2020; Published: 13 April 2020

Abstract: Knowledge and research tying the environmental impact and embodied energy together is a largely unexplored area in the building industry. The aim of this study is to investigate the practicality of using the ratio between embodied energy and embodied carbon to measure the building's impact. This study is based on life-cycle assessment and proposes a new measure: life-cycle embodied performance (LCEP), in order to evaluate building performance. In this project, eight buildings located in the same climate zone with similar construction types are studied to test the proposed method. For each case, the embodied energy intensities and embodied carbon coefficients are calculated, and four environmental impact categories are quantified. The following observations can be drawn from the findings: (a) the ozone depletion potential could be used as an indicator to predict the value of LCEP; (b) the use of embodied energy and embodied carbon independently from each other could lead to incomplete assessments; and (c) the exterior wall system is a common significant factor influencing embodied energy and embodied carbon. The results lead to several conclusions: firstly, the proposed LCEP ratio, between embodied energy and embodied carbon, can serve as a genuine indicator of embodied performance. Secondly, environmental impact categories are not dependent on embodied energy, nor embodied carbon. Rather, they are proportional to LCEP. Lastly, among the different building materials studied, metal and concrete express the highest contribution towards embodied energy and embodied carbon.

Keywords: embodied energy; embodied carbon; environmental impact; life-cycle embodied performance

---

1. Introduction

The environmental impact from operating energy use is a well-established area of research and practice, with well-defined metrics, methodologies, building codes and regulations. There has been substantial progress made by practitioners to improve building operating energy efficiencies, which in turn leads to carbon emission reductions. As the building codes become more stringent, operating energy and related emissions will decrease dramatically, thus decreasing the role of operating energy in a building's life cycle.

Numerous studies have demonstrated the increasingly important role that embodied energy plays in the building life cycle. For a conventional single-family house, the percentage of embodied energy could account for up to 40%–50% of the total life-cycle primary energy use [1]. For low-energy buildings (energy-efficient buildings) and net zero energy buildings, the percentage embodied energy accounts for could be as high as 74%–100% [2]. Regardless, the commonly accepted guidelines and methods of assessment and measurement for embodied energy have not been established. Previous studies demonstrate considerable variation in reported embodied energy values due to the high number of variables [3,4], including building materials [5] and building construction types [2]: there is inadequate published information on whole building life-cycle embodied energy reports [6]. Aside from a lack of consensus on measurement and procedures, embodied energy emissions and related carbon emissions are being largely ignored [7] as the focus is solely on operating energy.

Embodied energy is the energy consumed during a building's whole life cycle. This excludes the operating energy, but includes raw material extraction, product production, manufacturing, installation, on-site construction, maintenance, repair and replacement, and finally the demolition and disposal of a building [8]. Embodied carbon is used to measure the building's contribution to climate change, which is closely related to, but not equal to, embodied energy [8]. There are three principal differences between embodied energy and embodied carbon: (1) the same amount of embodied energy could be converted to a different amount of embodied carbon, depending on the energy mix of the regional energy resources [9] and other factors. For example, if coal comprises a higher percentage of the energy source than wind, there will be a higher conversion rate from embodied energy to embodied carbon. (2) Carbon can be emitted due to chemical processes and reactions that do not involve energy consumption; the carbon emitted during cement production is one example [9]. (3) Carbon can also be sequestered, as is the case with wood during its growth phase [8]. Hence, the material can consume energy and reduce emissions at the same time. For these reasons, the ratio between embodied energy and embodied carbon could be a more meaningful tool to assess the life-cycle embodied performance (LCEP) of a building.

The main purpose of this study is to investigate the utility of the ratio between embodied energy and embodied carbon. This ratio has the potential to measure the building's embodied and environmental impact. There are three essential investigative questions that will aid in this pursuit: (1) Is there correlation between life-cycle embodied energy (LCEE) and life-cycle embodied carbon (LCEC)? (2) Can a building's environmental impact be predicted by the life-cycle embodied performance (LCEP)? (3) Which building components and materials contribute most to the overall embodied energy, embodied carbon and environmental impact?

2. Background and Terminology

2.1. Embodied Carbon

The term embodied carbon (EC) has been used in a variety of ways, which can be confusing. In this paper, life-cycle embodied carbon (LCEC) is not the carbon encapsulated in building materials. Instead, it refers to all the life-cycle greenhouse gas emissions related to the building outside of building operations. It includes carbon associated with energy consumed throughout the entire life-cycle, and excludes carbon saved through recycle and reuse at the end of building service life. The building life-cycle stages are defined in BS EN 15798 norms (BSI [10]). Embodied carbon is also known as "value chain emissions", and it includes upstream and downstream emissions. Figure 1 illustrates the embodied carbon through the building life-cycle, including upstream and downstream emission.

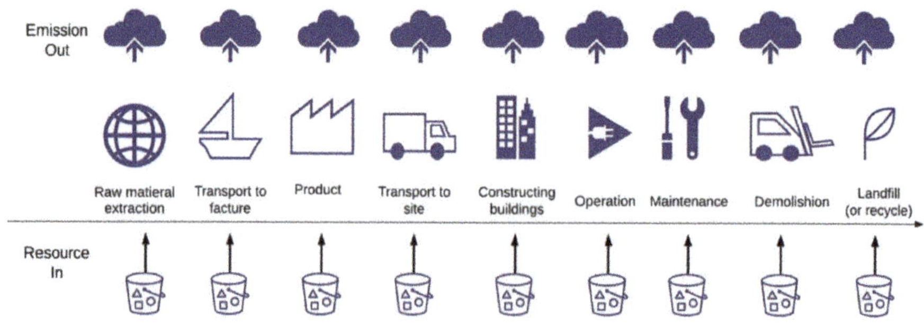

Figure 1. Embodied energy and embodied emission (by authors).

Hu (2019) defines four primary categories that comprise life-cycle embodied carbon (LCEC): Initial embodied carbon (IEC), recurring embodied carbon (REC), demolition embodied carbon (DEC) and recycled embodied carbon (REYC); refer to Equation (1) [11]. In this study, life-cycle embodied carbon (LCEC) is represented by the most commonly used indicator: global warming potential (GWP) [12], measured in $kgCO_2eq$.

This project uses two variables to measure the LCEC: (a) life-cycle embodied carbon coefficient (LCECC), demonstrated in Equation (2), and (b) life-cycle embodied carbon intensity (LCECI), demonstrated in Equation (3). These variables are investigated and compared for their reliability to evaluate building performance. In 1996, Alcorn proposed the term "Embodied Energy Coefficients" (EEC), which they used to measure the change of embodied energy for a variety of building materials used in New Zealand Housing between 1983 and 1996. The results showed 32% to 56% percentage for different materials reflecting the changes in construction and manufacturing methods and processes [13]. EEC was then later used by Dias and Polliyadda (2004) as "embodied carbon coefficients" [14] to measure the embodied performance of buildings. "Life-cycle embodied energy intensity" is the new unit proposed in this project; it is most determined by building materials and assemblies, and is measured in kgCO2e /m2/yr.

$$LCEC = \sum_{c=end}^{c=1} (IEC_c + REC_c + DEC_c) - REYC_e \tag{1}$$

$$LCECC_{building} = \frac{LCEC}{W \times L} \tag{2}$$

where LCECC is life-cycle embodied carbon coefficient, measured by $kgCO_2eq/kg/yr$. LCEC is the life-cycle embodied carbon of the building, measured by kgCO2e. W is the total weight of the building, calculated by kg. L is the total building life span, in years.

$$LCECI_{building} = \frac{LCEC}{A \times L} \tag{3}$$

where LCECI represents life-cycle embodied carbon intensity, measured in $kgCO_2eq/m^2/yr.$, LCEC is the life-cycle embodied carbon of the building, measured in $kgCO_2eq$. A represents the total floor area of the building (conditioned and unconditioned), measured in square meters (m^2). L is the total building life span, in years.

2.2. Embodied Energy

Life-cycle embodied energy (LCEE) comprises all energy consumed during the entire building's life span, except the operating energy. In this project, LCEE is measured by the life-cycle embodied energy intensity (LCEEI), measured in $MJ/m^2/yr$ from Equation (4). The life-cycle embodied energy coefficient (LCEEC), measured in MJ/kg/yr, refers to Equation (5). These measurements allow buildings with different sizes, life spans and construction types to be compared, which will provide a more accurate assessment of how energy intensive the buildings are:

$$LCEEC_{building} = \frac{LCEE}{W \times L} \tag{4}$$

where LCECC is the life-cycle embodied carbon coefficient, measured in $kgCO_2eq/ m^2/yr$. LCEE is the total life-cycle embodied carbon of the building, measured in $kgCO_2eq$. W is the total weight of building, calculated in kg. L is the total building life span, in years:

$$LCEEI_{building} = \frac{LCEE}{A \times L} \tag{5}$$

where LCECI represents life-cycle embodied carbon intensity, measured in $kgCO_2eq/m^2/yr$. LCEC is the total life-cycle embodied carbon of the building, measured in $kgCO_2eq$. A represents the total floor area of the building (conditioned and unconditioned), measured in square meters(m^2). L is the total building life span, in years.

2.3. Embodied Environmental Impact

To better understand a building's embodied environmental impact, four midpoint impact categories are included in the study: acidification potential (AP), measured in $kgSO_2eq$; ozone depletion potential (OD), measured in CFCeq; smog formation potential (SF), measured in kgO3eq; and eutrophication potential (EP), measured in kgNeq. Each category is related to LCEC and LCEE, respectively. Carbon emissions can have acidifying effects [15], therefore the use of building materials with lower embodied carbon can reduce this acidification effect. The primary causes of ozone depletion are electricity generation and motor vehicles [16,17]. Using less-harmful materials and optimizing construction methods could help to reduce the ozone depletion potential. Smog is specific type of air pollution whose formation is caused by NOx, SOx, CO and VOC in the presence of sunlight [18,19], which can be intensified in dense urban areas. Reducing energy use, using less building material, and improving construction methods can all contribute to the reduction of smog formation. Lastly, eutrophication is excessive nutrient enrichment, that can cause undesirable shifts in aquatic ecosystem. Production of building materials, such as concrete is one cause of eutrophication [20,21].

2.4. Current Status of Research

Table 1 summaries a broad range of data from published studies, categorized in terms of the building type, carbon emission, and energy attributed to LCECI and LCEEI (the functional unit is floor area). The results show that previous studies on embodied energy of buildings varies. The different methods of analysis produce results that range from a maximum of 18,000–33,000 MJ/m^2 /yr with a hybrid analysis method to a minimum of 1000–12,000 $MJ/m^2/yr$ with a process analysis method for nearly net zero residential buildings [2]. Table 1 illustrates some of the most current case studies with embodied energy and embodied carbon assessments. The sample consists of more than 100 case studies from around the world, with a time span between 1994 to 2018. The assessed embodied energy range between 20.2 MJ/m^2 /yr to 660 $MJ/m^2/yr$ using different assessment methods for buildings of different construction types, and from different geographic locations. Such large variations make it difficult to compare these case studies and gain a deeper understanding of the primary impact factors driving the embodied energy value. In the same table, embodied carbon information is also provided, and compared to the embodied energy. However, there are a lack of sufficient/published embodied carbon values for buildings. Out of more than 100 case studies, only 29 included embodied carbon in their research process. Even then, only half of those studies used assumed data instead of actual project construction documents. Based on this data, the embodied carbon coefficient ranges from 1.44 to 3200 $(kgCO_2eq/m^2/yr)$, which is a wider proportional variance than the embodied energy. The general ratio between LCECI and LCEEI is 0.25–3.94 $kgCO_2eq/MJ$, represented by LCEP and Equation (6).

Table 1. Literature review: state of art studies on embodied energy and embodied carbon.

	Study	Building Type	Country	Yr	LCEEI (MJ/m^2/yr)	Construction Types	LCECI (kgCO$_2$e/m^2/yr)	Life Span of Building	LCEP (kgCO$_2$eq)/MJ
1	Buchananand Honey [22]	COR	New Zealand	1994	76–1300	Wood	100–1000	25	0.7–1.32
2	Debnath et al [23]	COR	India	1995	82–100	Load bearing – Reinforced Concrete	N/A	*50 **	N/A
		COR			74–84				
		COR			62–86				
3	Suzuki et al [24]	COR (multi-family)	Japan	1995	216–270	Steel Reinforced Concrete	850	*37 ***	3.15–3.94
		COR (single-family)			100	Wood	250	*30	2.5
		COR (single-family)			122	Lightweight Steel	400	*37	3.28
4	Winther and Hestnes [25]	COR	Norway	1999	36–40.1	Timber	N/A	50	N/A
		LER			49.4			50	
		Super insulated			88.48			50	
5	Keoleian et al (2000) [26]	COR	USA	2000	126	N/A	32	50	0.25
		LER			145		89	50	0.61
6	Mithraratene and Vale [27]	LER	New Zealand	2004	44.25	Light timber	N/A	100	N/A
		Super insulated			50.41			100	
7	Horne et al. [28]	LER	Australia	2006	41–57	N/A	N/A	50	N/A
8	Casals [29]	COR	Spain	2006		N/A	N/A	30	N/A
		LER						30	
9	Thormark [30]	LER	Sweden	2006	60.3–96.2	Timber	N/A	50	N/A
10	Szalay [31]	COR	Hungary	2007	71	N/A	N/A	50	N/A
		LER			227–243			50	
11	Citherlet and Defaux [32]	COR	Switzerland	2007	108		N/A	50	
		LER			105–113			50	

LER: Low energy residential. COR: Conventional residential. NZER: Net zero-energy residential. PH: Passive house. *50: assumed life span of building based on national average data. **. Based on national building code (NCB) by Indian government, life span for concrete structure is 75–100 years, the average life span of apartment building is 50–60 years and house is 40 years. https://medium.com/@marsmount.com/what-is-the-lifespan-of-an-apartment-in-india-c2db01928a82. ***. According to the Ministry of Land, Infrastructure, Transport and Tourism (MLIT), the average life span for wooden house is 27–30 years, for reinforced-concrete building is 37 years. https://japanpropertycentral.com/2014/02/understanding-the-lifespan-of-a-japanese-home-or-apartment/.

3. Method and Materials

This study is based on life-cycle assessment; a variety of approaches were used, including a cost-optimality approach that originated in industry [33,34] and an energy-savings approach [35,36]. The study is organized into the following steps: (a) collecting data from case buildings; (b) defining systems and boundaries; (c) building 3D models and creating a bill of materials; (d) conducting embodied energy and embodied carbon analysis; (e) conducting environmental analysis; (f) comparing embodied energy, embodied carbon and their correlation to environmental impact.

3.1. Life-Cycle Embodied Performance (LCEP)

This project proposes a new measure: life-cycle embodied performance (LCEP). It is the ratio between embodied carbon intensity and embodied energy-use intensity. The ratio is measured in $kgCO_2eq/MJ$. The smaller the LCEP value, the less carbon emitted is from the equal amount embodied energy used, whereas the higher LCEP value indicate higher embodied carbon emission with same amount of energy consumption. Therefore, the lower LCEP value, the better the life-cycle embodied performance of the building.

$$LCEP = \frac{LCECI}{LCEEI} \qquad (6)$$

3.2. Case Project Specification

The building types include in this study are academic (educational) buildings (A1, A2), residential buildings (R1, R2) and office buildings (O1, O2). Building floor plans and 3D models are presented in Figure 2. Floor area, building height, year of construction, and year of renovation are listed in Table 2. The total floor area of buildings ranges between 982 m^2 to 7015 m^2. Floor heights range between 2 stories to 4 stories. The buildings are all over 45 years old, and three buildings have had major renovations since initial construction, while the other three have not.

Figure 2. Case buildings.

Table 2. Case project information.

Building #	Building Function	Floor Area (Sq.m)	Floor #	Yr of Construction	Yr of Renovations
A1 (A)	Academic	7,015	2	1972	-
A2 (C)	Academic	2,256	4	1957	2011
O1 (L)	Officea	4,218	4	1969	-
O2 (H)	Office	5,585	4	1964	2004
R1 (W)	Residential	982	4	1948	1984
R2 0	Residential	2768	4	1955	2010

3.3. Data Collection

Due to intellectual property concerns, the data from buildings used for this study are represented in an anonymous format, as opposed to individual projects. Data was collected on each buildings' physical properties from the following two sources:

- Original construction documents and renovation documents (if any), provided by facility managers.
- Field measurements by the research team.

Life-cycle inventory (LCI) data was collected from the following three sources:

- Existing data on material quantities and embodied carbon dioxide are extracted and gathered from literature.
- New LCI data on new materials is obtained through leading practitioners, from major AEC companies, working with authors.
- The open-source Inventory of Carbon and Energy database from the University of Bath was published in 2008. (It is currently the most frequently used embodied carbon database in this industry, due to its comprehensive summary of the best available embodied carbon data [37]).

3.4. System, Boundaries, Building Models and Software

For the purposes of this inquiry, only primary systems are included in the analysis for each building. The primary system includes the structure system, foundation, building roof, building façade, and building interior partition/ceiling. The prescribed building life span for this assessment is 50 years. The energy mix for initial construction, repair, replacement and maintenance is assumed to remain the same over the entire building life span. Autodesk's Revit is used to construct 3D models based on construction documents and other collected data. Information on the building's materials and components is manually inputted in the 3D models. Next, a bill of materials (BOM) is created for each studied case building. Then BOMs are exported into an Excel-format file; the Excel files are cleaned, organized and simplified in order to edit out the non-essential information. Finally, a clean spreadsheet is imported to a software called Athena IE4B for life-cycle analysis. The essential building components included in the analysis are shown in Table 3.

Table 3. Building assemblies and materials included in the studies.

Building Components	A1	A2
Foundation	Concrete spread footing	Concrete spread footing
Exterior wall	10 cm brick masonry supported by 20 cm concrete masonry block	10 cm brick masonry supported by 20 cm concrete masonry block
Exterior window	Steel window frames with single paned glass, no coating. PVC window frame double-glazed no coating air	Wood window frames with single paned glass, no coating.
Exterior door	Aluminum frame, single pane, sliding and swing door	Wood frame, single panel, swing door
Interior wall	Concrete masonry block and brick masonry wall	10 cm concrete masonry block
Partition wall	3 5/8″ metal stud wall with 1 layer of 5/8″ gypsum board on either exterior side	10 cm mtl stud wall with with 1 layer of 5/8″ gypsum board on either exterior side, 8cm batt insulation inside
Floor	Concrete	Asphalt tile on concrete floor
Columns	Concrete	Concrete
Roof	Concrete roof support, built up roofing and reflecting aggregate surface with 1″ rigid insulation	Flat seam metal roof on 3″ gypsum roof tile Slate roof on 3″ gypsum roof tile
Beams	Concrete	Concrete

Table 3. *Cont.*

Building Components	A1	A2
	O1 (L)	**O1 (H)**
Foundation	Concrete spread footing @ 2500 psi supporting a 12 cm (5 inch) concrete slab	Concrete spread footing
Exterior wall	10 cm (4 inch) Brick masonry, supported by 20 cm (8 inch) concrete masonry block, with 5cm (2 inch) rigid insulation	10 cm (4 inch) brick masonry supported by 25 cm (10 inch) concrete masonry block, noinsulation
Exterior window	Wood window frames with single paned glass, no coating	Wood window frames with double-glazed glass
Interior wall	10 cm (4 inch) Brick masonry wall with 1 cm (1/2 inch) gypsum wall board	10 cm (4 inch) metal stud with gypsum board on both sides
Partition wall	10 cm (4 inch) concrete masonry block, painted	20 cm (8 inch) concrete masonry block, with gypsum board on one side; 10 cm (4 inch) Wood stud with plywood boards on both side
Floor	15 cm (6 inch) concrete one-way joists floor with #4 continuous steel rebar, pan width of 36 cm (14 inch)	15 cm (6 inch) concrete one-way joists floor over reinforced concrete slab
Columns	20 cm (8 inch) concrete column @ 3000 psi	20 cm (8 inch) concrete column @ 3000 psi
Roof	(2 × 4 inch) wood rafters supported by (1/4 inch) slate tile with (3/4 inch) plywood underneath, 10 cm (4 inch) batt insulation inbetween rafters	5 × 10 cm (2 × 4 inch) wood rafters supported by (1/4 inch) slate tile with (3/4 inch) plywood underneath, 10 cm (4 inch) batt insulation inbetween rafters
Beams	Concrete joist beam @ 3000 psi	concrete joist beam @ 3000 psi
	R1	**R1**
Foundation	30 × 76 cm Concrete spread footing	Concrete spread footing
Exterior wall	10 cm (4 inch) brick masonry supported by 20 cm (8 inch) concrete masonry block, no insulation	10 cm (4 inch) Brick masonry supported by 10 cm (4 inch) concrete masonry block, without air gap, without insulation
Exterior window	Metal window frames with single paned glass, no coating	
Interior wall	10 cm (4 inch) metal stud with gypsum board on both sides	10 cm (4 inch) concrete block
Partition wall	20 cm (8 inch) concrete masonry block, with Gypsum board one side	10 cm (4 inch) Wood stud with plywood board on both side
Floor	-	-
Columns	20 × 20 cm (8 inch) concrete columns @ 3000 psi	35 × 35 cm (14 inch) concrete columns @ 3000 psi
Roof	(2 × 4 inch) Wood rafters supported (1/4 inch) slate tile with (3/4 inch) plywood underneath, 10 cm (4 inch) batt insulation in-between rafters	Built-up light weight concrete roof over reinforced concrete slab
Beams	Concrete @ 3000 psi	Concrete @ 3000 psi

LCEE and LCEI are calculated from input information in Athena. Four environmental impact categories are assessed as well. LCEEI, LCEEC, LCECI, LCECC, LCEP are calculated using Equations (2)–(6), for each of the studied buildings (shown in Table 4).

3.5. Statistical Analyses

Four single variable regression models are used to determine the dependency between the environment impact categories (AP, OD, SF, EP) and LCEP. A 95% confidence interval for each outcome measure and a Pearson's value of 0.05 were used determine statistical significance:

$$Y_i = \beta_0 + \beta_1 X_i + \epsilon_I \tag{7}$$

where Y_i is the life-cycle embodied performance (LCEP), X_i is the environmental impact category, β_0 is the intercept, and ϵ_I is standard deviation. Tables 4 and 5 represent the variables included in the four models.

4. Analysis Findings

4.1. Correlation between Embodied Energy and Embodied Carbon

Two findings illustrated in Figure 3. First, the measured intensity has different results compared to the coefficient. For example, building O2 and R1 have similar coefficient (LCECC and LCEEC) value, although R1 has > 97% higher intensity (LCECI and LCEEI) value than those of O2. Also, O1 and O2 have comparable intensities, whereas, the coefficient of O2 is twice that of O1. Secondly, findings reveal that the buildings' function (type), size and height do not have a direct influence on life-cycle-embodied carbon and life cycle embodied energy. For instance, building O2 area is almost 6 times over the R1, however, those two buildings have similar embodied energy coefficient. A2 and R2 have similar building area, but very different embodied energy coefficients.

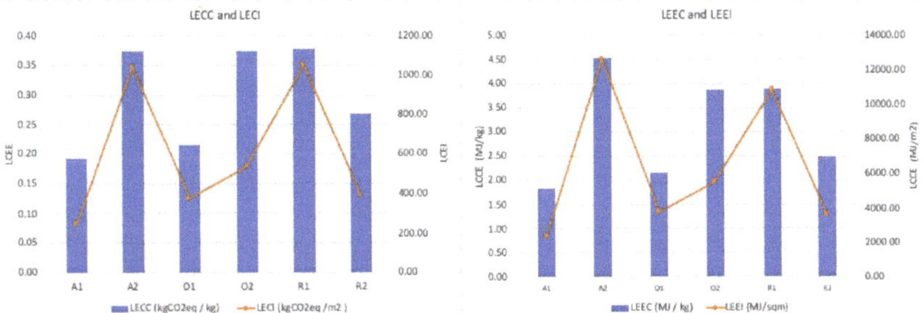

Figure 3. Embodied energy intensity and embodied carbon coefficient.

Table 4 summarizes values for the six case buildings for life-cycle embodied energy coefficient (LCEEC), life-cycle embodied carbon coefficient (LCECC), life-cycle embodied carbon intensity (LCECI), life-cycle embodied energy intensity (LCEEI), and life-cycle embodied performance (LCEP).

Table 4. Embodied energy and embodied carbon analysis of case projects.

Building Number	LCEEC (MJ/kg/yr)	LCEEI (kgCO$_2$eq/kg/yr)	LCECC (MJ/m^2/yr)	LCECI (kgCO$_2$eq/m^2/yr)	LCEP (/kgCO$_2$eq/MJ/yr)
A1	1.84	2324.87	0.19	245.07	0.105
A2	4.53	12,614.20	0.37	1042.97	0.083
O1	2.15	3746.88	0.22	374.39	0.100
O2	3.87	5518.85	0.38	535.42	0.097
R1	3.90	10,876.95	0.38	1053.81	0.097
R2	2.49	3619.27	0.27	392.08	0.108

A1: academic building 1; O1: office building; R1: residential building 1.

Table 4 and Figure 4 demonstrate that the ratio between energy and carbon is a better measurement for building performance compared to coefficient or intensity alone. When we look at the embodied energy and embodied carbon independently from each other, A2 has highest LCEEI, 12,614.20 kgCO$_2$eq/kg (illustrated in blue), and the second highest LCECC, 0.37 MJ/m^2 (illustrated in orange). Based on these scores, A2 can be rated with the lowest performance, which opposes the results when using the ratio between embodied energy and embodied carbon. As explained previously, a lower LCEP value implies a better embodied performance on the part of the building. Among the six buildings studied, A2 has the lowest LCEP score, 0.083, which means A2 emits the least amount of carbon while consuming the same amount of energy. Therefore, using the proposed model, building A2 has the best life-cycle embodied performance within the sample size.

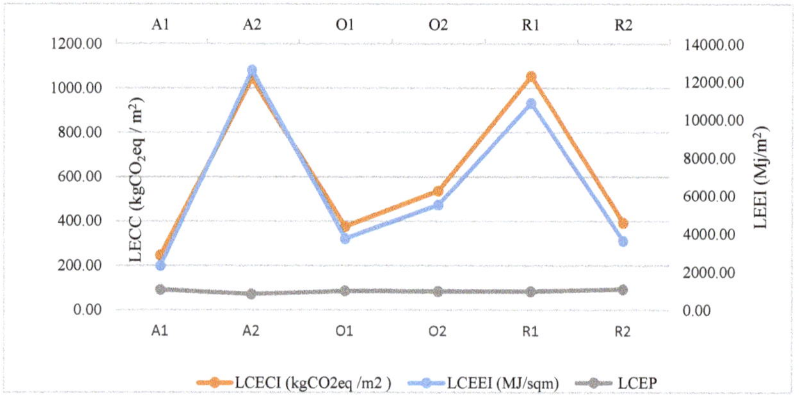

Figure 4. Life-cycle embodied energy intensity and life-cycle embodied carbon coefficient and life-cycle embodied performance.

4.2. Correlation between Environmental Impact and LCEP

The results in Section 4.2 are derived from the four single- regression models (Equation (7)) using input from Equation (2)–(6). There are large variations in all four environmental categories; AP intensities range between 0.94 to 4.37 $kgSO_2eq/m^2$, EP Intensities range between 0.05 to 0.26 $kgNeq/m^2$, SP intensities range between 14.85 to 58.87 $kgO_3eq/m^2/yr$, and OD intensities range between 7.51×10^{-7} to 3.27×10^{-5} CFC-11eq/ m^2/yr.

It is difficult to compare buildings' embodied performance based on the total environmental impact through their entire life, so impact intensity (measured in floor area per year) was used. Figure 5 demonstrates the clear correlation between AP intensity and SF intensity: high AP couples with high SF, which indicates acidification potential as a causal factor for smog formation potential. Figure 5 also demonstrates that higher AP and SF do not always result in a higher EP. For example, office building 2 (O2) has higher AP and SF compared to office building 1 (O1), but lower EP than that of O1. This result indicates that causal factors differ between embodied EP, and AP and SF.

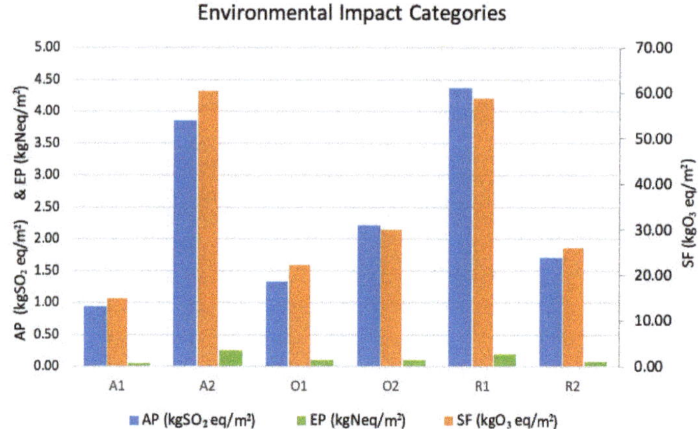

Figure 5. Environmental impact categories: AP, EP and SF.

Ozone depletion (OD) potential intensity was looked separately from the other three categories due to its different unit of measurement. Figure 6 shows that building A2 has a substantially higher impact intensity than the other buildings. To investigate this further, we looked into the buildings' materials and components, which contribute to high OD ratings.

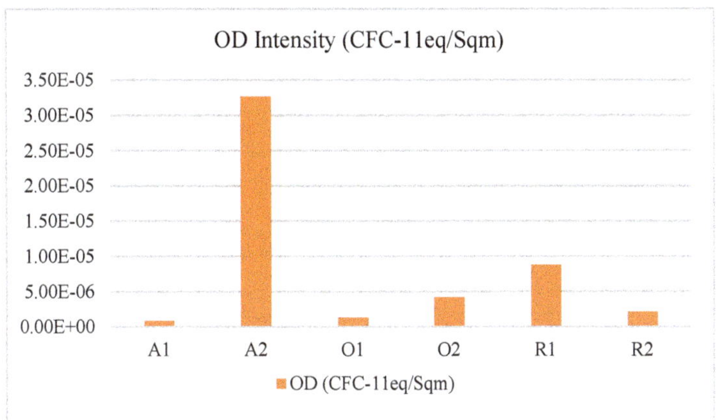

Figure 6. Environmental impact categories: ozone depletion (OD).

Figure 7 shows that metals (steel and alumina) contribute the most to the ozone depletion potential across all of the buildings. A2 building has a higher metal use compared to other buildings; it uses metals primarily in the building roof, and a portion of the exterior wall is made of alumina panel. However, if we look closely at A2 building materials use (refer to Figure 8), calculated by weight (kg), metal only acounts for 19% of the total material use, but masonry counts for 56% and constitutes a much smaller percentage of the overall ozone depletion potential. Previous research has proven that the most significant factors responsible for ozone depletion are industrial production and disposal of halogenated hydrocarbons [38]. In A2, metal manufacturing and production processes are the hotspots that contribute to a high impact on ozone depletion.

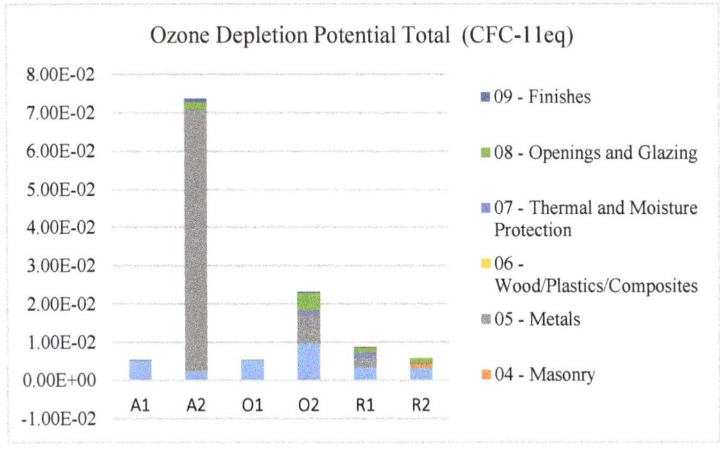

Figure 7. Building materials contribution to ozone depletion.

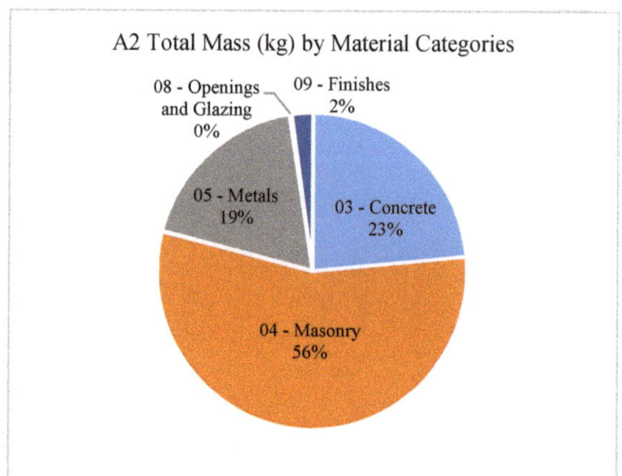

Figure 8. A2 building total mass (kg) by material categories.

The general finding from this study is demonstrate in Table 5, which shows the results from four linear regression models. Among the four environment categories, it shows statistical significance in Ozone Depletion potential, with a p-value of 0.006 (less than 0.05). The R-squared value of OD is 0.877, which means 87.7% LCEP value in the data set could be predicted or interpreted by the value of OD value. This result means Ozone Depletion potential could be used as an indicator to predict the value of LCEP, or, the building life cycle embodied performance is correlated with OD.

Table 5. Linear regression model of correlation between environment impact and LCEP.

Category	R-Squared	Adjusted R-Squared	Significance F	p-Value	Significance
AP	0.161	−0.049	0.431	0.431	No
OD	0.877	0.846	0.006	0.006	Yes
SF	0.214	0.017	0.356	0.356	No
EP	0.387	0.234	0.187	0.187	No

4.3. Building Components and Materials' Contribution

The results in Section 4.3 are from environmental impact analysis conducted in Athena (refer to Section 3.4), using the input from data listed in Table 3. In order to gain a better understanding of what building components or materials contribute the most to embodied carbon, embodied energy, and environmental impact, detailed analyses are conducted for each building. For the embodied carbon, illustrated in Figure 9a, concrete accounts for 51%, and metal accounts for 31% in the O2 building. In the A2 building, the concrete contribution is 17%, and metal is 51%. In the A1 building, concrete is responsible for 51% of LCEC. Figure 9b reveals the top 3 buildings with highest LCEE are O2, A2 and A1 again. Among all the material categories, concrete and metal are the primary contributors to embodied energy. In the A2 building, concrete contributes to 17% of LCEE and metal accounts for 52% of LCEE; and in the O2 building, concrete contributes 51% of LCEE and metal accounts for 31% of LCEE. Overall, the two residential buildings have lower LCEC and LCEE than the other buildings. This is not because of the smaller building's footprint, it is mostly determined by the building materials used.

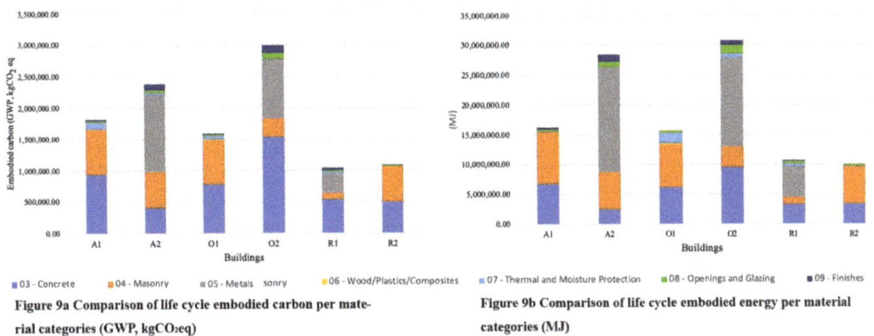

Figure 9a Comparison of life cycle embodied carbon per material categories (GWP, kgCO₂eq)

Figure 9b Comparison of life cycle embodied energy per material categories (MJ)

Figure 9. (**a**) Comparison of life cycle embodied carbon per material categories (GWP, kgCO$_2$eq); (**b**) Comparison of life cycle embodied energy per material categories (MJ).

As far as building assembly groups, shown in Figure 10a, the building floors contribute the most to embodied carbon in A1, A2 and O1. In O2 and R1, walls, including exterior walls and interior walls, are the largest contributor. In R2, windows contribute the most to embodied carbon. For embodied energy, walls are the largest contributor in A1, A2, O1, O2 and R1, shown in Figure 10b. R2 is the exception, where windows accounts for more than 50% of embodied energy. When examining the embodied energy and embodied carbon together, building walls, especially exterior walls, are a common significant factor. For future building renovations, replacing or upgrading existing exterior walls with low embodied energy and carbon components can effectively reduce the overall embodied energy and carbon.

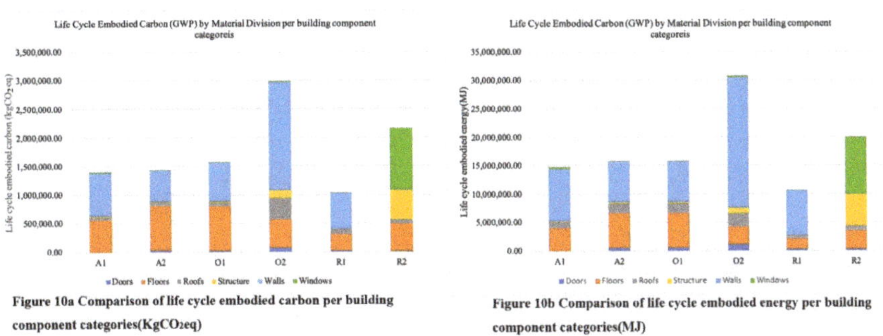

Figure 10a Comparison of life cycle embodied carbon per building component categories(KgCO₂eq)

Figure 10b Comparison of life cycle embodied energy per building component categories(MJ)

Figure 10. (**a**) comparison of life cycle embodied carbon per building component categories (KgCO$_2$ eq); (**b**) comparison of life cycle embodied energy per building component categories (MJ).

For environmental impact, Figure 11 illustrates how residential buildings perform better in all four environmental categories, and commercial buildings and academic buildings' performance varies quite a lot depending on the impact categories.

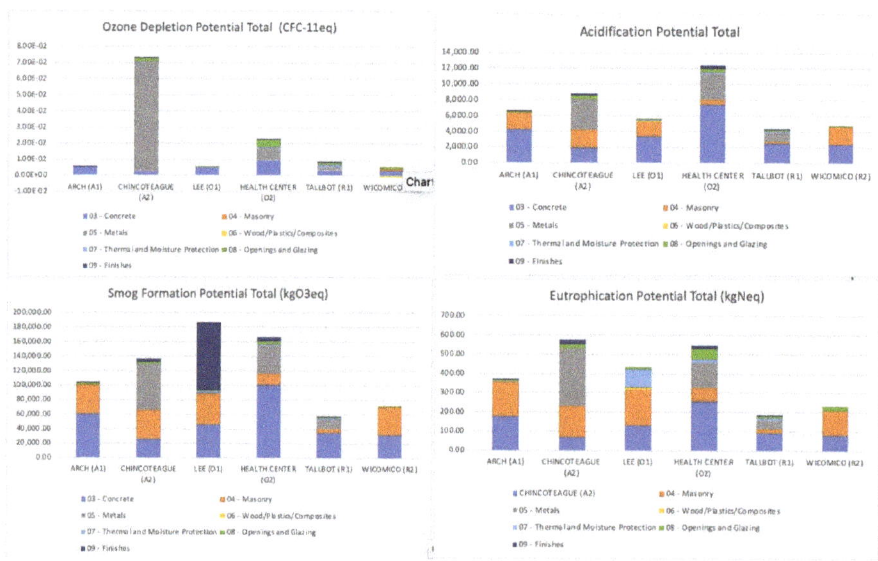

Figure 11. Contribution to total environmental impact per building.

5. Discussion and Conclusions

From the results of this analysis, several conclusions can be drawn. First, the results illustrate a clear difference between embodied energy and embodied carbon. There are two different units of measurement, which do not always correlate to each other. In addition, building function, size, and construction year vary considerably. In order to get a better sense of the embodied performance of a building with a long life span, a more manageable and comparable measurement unit is needed. When embodied carbon and embodied energy are used separately, the results will not provide a comprehensive understanding of the building's embodied performance. Instead, the ratio between embodied energy and embodied carbon can serve as a genuine indicator of embodied performance. This ratio appears to be correlated to ozone depletion potential, and not any of the other environmental impact categories measured in this study.

Second, environmental impact categories are not dependent on embodied energy, nor embodied carbon. Rather, they are proportional to LCEP. A2 and R1 have the highest environmental impact intensities in all four categories (AP, OD, EP, SF), however, LCEP indicates that A2 and R1 perform better (with lowest LCEP score) in terms of reducing their embodied emissions. The LCEP is proportionally inverse to environment impact potentials. Potentially, with more data, a statistical model could be created to predict the potential environmental impact in all four categories, using LCEP as an indicator when designing new buildings. This could reduce the complexity of current environmental impact assessments and could, therefore, help designers overcome the challenges of including environmental impact potentials as design criteria. Also, the results reveal hotspots that contributing to ozone depletion: metal manufacturing and production processes, which provide a direction for mitigation strategies.

Third, among the different building materials studies, metal and concrete express the highest contribution to embodied energy and embodied carbon. For building components, building exterior wall systems are the biggest embodied energy consumers and polluters, which indicates that building façade and wall systems could play significant roles in reducing embodied carbon and energy. This, in turn, would improve buildings' embodied performance.

Three primary observations can be extrapolated from this study:

1. Ozone depletion potential may be usable as an indicator to predict the value of LCEP

2. Using LCEE and LCEC independently from each other can lead to incomplete assessments
3. Regardless of the large variation in the performance of different building types, building exterior assemblies, particularly exterior walls, are a common significant factor influencing embodied energy and embodied carbon.

The significance of this study can be explained in three areas. Firstly, the actual building data is recorded and analyzed: original construction documents and historical records are collected and used to perform embodied energy, embodied carbon, and environment impact analysis. Secondly, this study investigates the case buildings at a detailed level to identify the contribution from each building assemblies' categories towards energy, carbon and environmental impact. Lastly, four environmental impact categories are assessed to gain a broad understanding of building's impact in addition to its contribution to global warming.

This study also has limitations that must be taken into account. First, the limited number of case buildings is an important limiting factor; more buildings need to be included in these studies. Second, the results of the analysis are dependent on the reliability and accuracy of the data provided by facility management offices and manufacturers. In order to make a more accurate assessment, detailed data is required from actual buildings. There are multiple barriers to acquiring this actual data, especially for existing buildings. Most older existing buildings do not have archives with complete, original construction documents. Often these buildings have also undergone multiple renovations, which can make collecting real data very challenging. There is potential that an algorithm could overcome such uncertainty, and a sensitivity analysis could be used to verify the robustness of the analysis results The third limitation is related to the scalability of the proposed method. It is possible to generalize construction types and methods for buildings built around the same time period, in a similar climate zone and in a geographic location. We can then use one or two buildings as a prototype to represent a portfolio of similar buildings, and then apply the proposed method on a much larger scale, such as an entire campus [11,39], neighborhood [40], city [41], or industry [42]. However, overgeneralizing could distort the findings and undermine the reliability of the analysis results as well. In order to prevent this overgeneralization, it is critical in the next steps to look into climate, geographic location and construction types as key influencing factors on the results.

Author Contributions: The author has read and agreed to the published version of the manuscript.

Funding: This project was funded by the Office of Sustainability, University of Maryland.

Acknowledgments: I would like to acknowledge the support given from research assistant David Milner.

Conflicts of Interest: The authors declare no conflict of interest.

Nomenclature

LCEC	life-cycle embodied carbon
LCECC	life-cycle embodied carbon coefficient
LCECI	life-cycle embodied carbon intensity
LCEE	life-cycle embodied energy
LCEEI	life-cycle embodied energy intensity
LCEEC	life-cycle embodied energy coefficient
EC	embodied carbon
IEC	initial embodied carbon
REC	recurring embodied carbon
DEC	demolition embodied carbon
REC	recycled embodied carbon
EEC	embodied energy coefficients
AP	acidification potential
OD	ozone depletion potential
SF	smog formation potential
EP	eutrophication potential

References

1. Gustavsson, L.; Joelsson, A.; Sathre, R. Life cycle primary energy use and carbon emission of an eight-storey wood-framed apartment building. *Energy Build.* **2010**, *42*, 230–242. [CrossRef]
2. Chastas, P.; Theodosiou, T.; Bikas, D. Embodied energy in residential buildings-towards the nearly zero energy building: A literature review. *Build. Environ.* **2016**, *105*, 267–282. [CrossRef]
3. Hu, M. The Embodied Impact of Existing Building Stock. In *Toward Sustainability Through Digital Technologies and Practices in the Eurasian Region*; IGI Global: Hershey, PA, USA, 2020; pp. 1–31.
4. Dixit, M.; Fernandez-Solis, J.L.; Lavy, S.; Culp, C.H. Need for an embodied energy measurement protocol for buildings: A review paper. *Renew. Sustain. Energy Rev.* **2012**, *16*, 3730–3743. [CrossRef]
5. Cabeza, L.F.; Barreneche, C.; Miró, L.; Morera, J.M.; Bartolí, E.; Fernandez, A.I. Low carbon and low embodied energy materials in buildings: A review. *Renew. Sustain. Energy Rev.* **2013**, *23*, 536–542. [CrossRef]
6. Optis, M.; Wild, P. Inadequate documentation in published life cycle energy reports on buildings. *Int. J. Life Cycle Assess.* **2010**, *15*, 644–651. [CrossRef]
7. Kneifel, J.; O'Rear, E.; Webb, D.; O'Fallon, C. An Exploration of the Relationship between Improvements in Energy Efficiency and Life-Cycle Energy and Carbon Emissions using the BIRDS Low-Energy Residential Database. *Energy Build.* **2017**, *160*, 19–33. [CrossRef]
8. De Wolf, C.; Pomponi, F.; Moncaster, A.M. Measuring embodied carbon dioxide equivalent of buildings: A review and critique of current industry practice. *Energy Build.* **2017**, *140*, 68–80. [CrossRef]
9. Hendriks, C.; Worrell, E.; Price, L.; Martin, N.; Meida, L.O.; De Jager, D.; Riemer, P. Emission reduction of greenhouse gases from the cement industry. In *Greenhouse Gas Control Technologies 4*; Elsevier: Amsterdam, The Netherlands, 1999; pp. 939–944.
10. British Standard Institutions (BSI). *BS EN 15978 Sustainability of Construction Works Assessment of Environmental Perforamnce of Buildings—Calculation Method*; British Standards Institution: London, UK, 2011.
11. Hu, M. Life-cycle environmental assessment of energy-retrofit strategies on a campus scale. *Build. Res. Inf.* **2019**, *4*, 1–22. [CrossRef]
12. Athena Sustainable Materials Institute. About Whole-Building LCA and Embodied Carbon. Available online: http://www.athenasmi.org/resources/publications/ (accessed on 20 January 2020).
13. Andrew, A. *Embodied Energy Coefficients of Building Materials*; Centre for Building Performance Research, Victoria University of Wellington: Kelburn, New Zealand, 1996.
14. Dias, W.; Pooliyadda, S. Quality based energy contents and carbon coefficients for building materials: A systems approach. *Energy* **2004**, *29*, 561–580. [CrossRef]
15. Zeebe, R.E.; Zachos, J.C.; Caldeira, K.; Tyrrell, T. OCEANS: Carbon Emissions and Acidification. *Science* **2008**, *321*, 51–52. [CrossRef]
16. Azari, R.; Garshasbi, S.; Amini, P.; Rashed-Ali, H.; Mohammadi, Y.; Najafabadi, R.A. Multi-objective optimization of building envelope design for life cycle environmental performance. *Energy Build.* **2016**, *126*, 524–534. [CrossRef]
17. Scheuer, C.; A Keoleian, G.; Reppe, P. Life cycle energy and environmental performance of a new university building: Modeling challenges and design implications. *Energy Build.* **2003**, *35*, 1049–1064. [CrossRef]
18. Bina, R.; Upma, S.; Chuhan, A.K.; Diwakar, S.; Raaz, M. Photochemical Smog Pollution and Its Mitigation Measures. *J. Adv. Sci. Res.* **2011**, *2*, 28–33.
19. Energy Education. Smog. Available online: https://energyeducation.ca/encyclopedia/Smog (accessed on 6 April 2020).
20. Kim, T.-H.; Chae, C.U. Environmental Impact Analysis of Acidification and Eutrophication Due to Emissions from the Production of Concrete. *Sustainability* **2016**, *8*, 578. [CrossRef]
21. Hill, C.A.S.; Dibdiaková, J. The environmental impact of wood compared to other building materials. *Int. Wood Prod. J.* **2016**, *7*, 215–219. [CrossRef]
22. Buchanan, A.H.; Honey, B.G. Energy and carbon dioxide implications of building construction. *Energy Build.* **1994**, *20*, 205–217. [CrossRef]
23. Debnath, A.; Singh, S.; Singh, Y. Comparative assessment of energy requirements for different types of residential buildings in India. *Energy Build.* **1995**, *23*, 141–146. [CrossRef]
24. Suzuki, M.; Oka, T.; Okada, K. The estimation of energy consumption and CO_2 emission due to housing construction in Japan. *Energy Build.* **1995**, *22*, 165–169. [CrossRef]

25. Winther, B.; Hestnes, A. Solar Versus Green: The Analysis of a Norwegian Row House. *Sol. Energy* **1999**, *66*, 387–393. [CrossRef]

26. Keoleian, G.A.; Blanchard, S.; Reppe, P. Life-Cycle Energy, Costs, and Strategies for Improving a Single-Family House. *J. Ind. Ecol.* **2000**, *4*, 135–156. [CrossRef]

27. Mithraratne, N.; Vale, B. Life cycle analysis model for New Zealand houses. *Build. Environ.* **2004**, *39*, 483–492. [CrossRef]

28. Ralph, H.; Opray, L.; Grant, T. Integrating Life Cycle Assessment into housing environmental performance assessment. In Proceedings of the 5th Australian Conference on Life Cycle Assessment: Achieving Business Benefits from Managing Life Cycle Impacts, Melbourne, Australia, 22–24 November 2006.

29. Casals, X.G. Analysis of building energy regulation and certification in Europe: Their role, limitations and differences. *Energy Build.* **2006**, *38*, 381–392. [CrossRef]

30. Thormark, C. The effect of material choice on the total energy need and recycling potential of a building. *Build. Environ.* **2006**, *41*, 1019–1026. [CrossRef]

31. Szalay, Z. What is missing from the concept of the new European Building Directive? *Build. Environ.* **2007**, *42*, 1761–1769. [CrossRef]

32. Citherlet, S.; Defaux, T. Energy and environmental comparison of three variants of a family house during its whole life span. *Build. Environ.* **2007**, *42*, 591–598. [CrossRef]

33. Lucchi, E.; Tabak, M.; Troi, A. The "Cost Optimality" Approach for the Internal Insulation of Historic Buildings. *Energy Procedia* **2017**, *133*, 412–423. [CrossRef]

34. Bogdan, A.; Kouloumpi, I.; Thomsen, K.E.; Aggerholm, S.; Enseling, A.; Loga, T.; Witczak, K. Implementing the cost-optimal methodology in EU countries: Lessons learned from three case studies. *Energy Procedia* **2015**, *78*, 2022–2027.

35. Andra, B.; Kašs, K.; Edīte, K.; Gatis, Ž.; Agris, K.; Dagnija, B.; Armands, G.; Reinis, P.; Marika, R.; Lelde, T.; et al. Robust Internal Thermal Insulation of Historic Buildings-State of the art on historic building insulation materials and retrofit strategies. *Lirias* **2019**, *21*, 1–87.

36. Akkurt, G.; Aste, N.; Borderon, J.; Buda, A.; Calzolari, M.; Chung, D.; Costanzo, V.; Del Pero, C.; Evola, G.; Huerto-Cardenas, H.; et al. Dynamic thermal and hygrometric simulation of historical buildings: Critical factors and possible solutions. *Renew. Sustain. Energy Rev.* **2020**, *118*, 109509. [CrossRef]

37. Pomponi, F.; Moncaster, A.M. Scrutinising embodied carbon in buildings: The next performance gap made manifest. *Renew. Sustain. Energy Rev.* **2018**, *81*, 2431–2442. [CrossRef]

38. Pielke, R.A., Sr. *Climate Vulnerability: Understanding and Addressing Threats to Essential Resources*; Elsevier: Amsterdam, The Netherlands, 2013.

39. Stephan, A.; Muñoz, S.; Healey, G.; Alcorn, J. Analysing material and embodied environmental flows of an Australian university—Towards a more circular economy. *Resour. Conserv. Recycl.* **2020**, *155*, 104632. [CrossRef]

40. Stephan, A.; Crawford, R.H.; De Myttenaere, K. Multi-scale life cycle energy analysis of a low-density suburban neighbourhood in Melbourne, Australia. *Build. Environ.* **2013**, *68*, 35–49. [CrossRef]

41. Ripa, M.; Fiorentino, G.; Vacca, V.; Ulgiati, S. The relevance of site-specific data in Life Cycle Assessment (LCA). The case of the municipal solid waste management in the metropolitan city of Naples (Italy). *J. Clean. Prod.* **2017**, *142*, 445–460. [CrossRef]

42. Crawford, R.H.; Stephan, A.; Prideaux, F. A comprehensive database of environmental flow coefficients for construction materials: Closing the loop in environmental design. In Proceedings of the 53rd International Conference of the Architectural Science Association 2019: Revisiting the Role of Architecture for 'Surviving' Development, Roorkee, India, 28–30 November 2019.

Article

Bridging Tools to Better Understand Environmental Performances and Raw Materials Supply of Traction Batteries in the Future EU Fleet

Silvia Bobba [1,*], Isabella Bianco [2], Umberto Eynard [1,2,3], Samuel Carrara [4], Fabrice Mathieux [1] and Gian Andrea Blengini [1,2]

[1]   European Commission, Joint Research Centre (JRC), 21027 Ispra, Italy; umberto.eynard@ext.ec.europa.eu (U.E.); fabrice.mathieux@ec.europa.eu (F.M.); gianandrea.blengini@ec.europa.eu (G.A.B.)

[2]   Department of Environment, Land and Infrastructure Engineering, Politecnico di Torino, Corso Duca degli Abruzzi 24, 10129 Torino, Italy; isabella.bianco@polito.it

[3]   SEIDOR SBS Services, c/Pujades 350, 08019 Barcelona, Spain

[4]   European Commission, Joint Research Centre (JRC), 1755 LE Petten, The Netherlands; samuel.carrara@ec.europa.eu

*   Correspondence: silvia.bobba@ec.europa.eu

Received: 10 April 2020; Accepted: 12 May 2020; Published: 15 May 2020

Abstract: Sustainable and smart mobility and associated energy systems are key to decarbonise the EU and develop a clean, resource efficient, circular and carbon-neutral future. To achieve the 2030 and 2050 targets, technological and societal changes are needed. This transition will inevitably change the composition of the future EU fleet, with an increasing share of electric vehicles (xEVs). To assess the potential contribution of lithium-ion traction batteries (LIBs) in decreasing the environmental burdens of EU mobility, several aspects should be included. Even though environmental assessments of batteries along their life-cycle have been already conducted using life-cycle assessment, a single tool does not likely provide a complete overview of such a complex system. Complementary information is provided by material flow analysis and criticality assessment, with emphasis on supply risk. Bridging complementary aspects can better support decision-making, especially when different strategies are simultaneously tackled. The results point out that the future life-cycle GWP of traction LIBs will likely improve, mainly due to more environmental-friendly energy mix and improved recycling. Even though second-use will postpone available materials for recycling, both these end-of-life strategies allow keeping the values of materials in the circular economy, with recycling also contributing to mitigate the supply risk of Lithium and Nickel.

Keywords: Life Cycle Assessment (LCA); Material Flow Analysis (MFA); Criticality; traction batteries; forecast; supply

---

1. Introduction

Sustainable and smart mobility, when articulated with appropriate energy systems, is a key asset to decarbonise the EU and develop a clean, resource efficient and carbon-neutral future. This is confirmed by several policy initiatives, among others: the European Green Deal [1] and the EC COM(2019) 22 [2]. In addition, the transition towards a low-carbon mobility contributes to the United Nations Sustainable Development Goals (UN SGDs), for instance Goal 7—ensure access to affordable, reliable, sustainable and modern energy for all [1].

To achieve the 2030 and 2050 targets, technological and societal changes concerning mobility are needed. To improve circular economy and resource efficiency in the automotive sector, the technological aspects should cover the whole value chain of vehicles, from their design to their End-of-Life (EoL), e.g., new vehicles, light-weighting materials, easier reusability and recyclability of materials, recycled content in vehicles and transport infrastructure [2,3]. In addition, the behaviour of consumers should change toward choices oriented to more environmental-friendly practices, e.g., increase of the occupancy rate of vehicles and the adoption of public transport, sharing of vehicles, teleworking options [3,4]. Automation of vehicles is already proceeding but further efforts are required in the research field to reduce emission and avoid rebound effects (e.g., higher demand for mobility) [4].

This transition will inevitably change the composition of the EU fleet in the future: the increasing share of electric vehicles (xEVs) is already occurring and this trend is expected to accelerate in the next decade [3,5,6]. Meanwhile, new technologies can have an important role in the next future, even though nowadays they are at a very early stage, e.g., fuel cells electric vehicles [7]. Several scenarios are already available in the literature, considering new technologies appearing in the EU market and their evolution according to various parameters (environmental targets, consumers' lifestyles and behaviour, electric driving ranges, economic factors, etc.) (e.g., [6,8]). Such forecasts are quite complex to compare due to the adoption of different assumptions behind the models. Although there is no consensus in the scientific literature on a specific model to adopt in forecasting the uptake of xEVs, the trend is quite clear: xEVs will rapidly increase and will have a very important share on the EU fleet up to 2050, with a consequent increase of traction batteries required for such vehicles.

Among the traction batteries, the most common and most promising chemistry currently used in the EU market are the Li-ion batteries (LIB). The characteristics of LIBs are suitable for their use in different type of xEVs, especially battery electric vehicles (BEVs) and plug-in electric vehicles (PHEVs) [5]. Note that According to the scope of the analysis, since HEVs need relatively small batteries and will have a lower impact on the LIBs market, HEVs are not included in the following analysis.

When traction LIBs reach about 70–80% of their capacity, they are usually extracted from xEV and they have to be properly collected and recycled according to the in-force Directives (2000/53/EC and 2006/66/EC) [9]. However, pilots and research projects are demonstrating that the residual capacity of extracted batteries could be even used in less energy-demanding applications, e.g., residential buildings [9–11]. Even though the second-use of LIB is an industrial practice, there is potential for the creation of a business case beyond 2030 [5]. The second-use of batteries, before their recycling, is an example of circular economy practice where the value of batteries (and therefore embedded materials) is retained within the economy and the resource efficiency is maximised. On the other hand, the flow of batteries available for recycling is delayed in time, which means that also the availability of Secondary Raw Materials (SRMs) from waste batteries is postponed in time due to their higher lifetime [12]. Therefore, an in-depth knowledge of the processes along the value chain of complex products as LIBs is fundamental to assess the effects and the trade-offs of different strategies, e.g., the EoL options.

The New Circular Economy Action Plan [13] and the New Industrial Strategy for Europe (EC, 2020), two of the main building blocks of the European Green Deal [1], stress the role of Critical Raw Materials (CRMs) to achieve a climate-neutral, circular and competitive economy. Future sustainability requirements for batteries could in the future consider, for instance, the carbon footprint of battery manufacturing, ethical sourcing of raw materials and facilitating reuse, repurposing and recycling. Secure and sustainable supply of both primary and secondary raw materials for key technologies such as e-mobility is hence a prerequisite to achieve climate neutrality. With the transition of Europe's industry to climate-neutrality, the reliance on available fossil fuels could be replaced with reliance on non-energy raw materials, many of which are sourced from abroad and for which global competition is becoming more intense [14]. According to OECD forecasts, global demand for raw materials will more than double by 2060, making diversified sourcing essential to increase Europe's security of supply. CRMs are also crucial for markets such as e-mobility, batteries, renewable energies, pharmaceuticals, aerospace, defence and digital applications [15].

The performance of LIBs is related to various aspects, in particular to which materials are used in their cathodes and anodes [5]. The combination of the increasing demand for xEVs and the dynamics related to battery chemistries translate into a growing and diversified demand for raw materials; among the key materials for the manufacturing of high-performant LIBs, some have been classified as CRMs in the EU and/or worldwide [16], e.g., cobalt, natural graphite and lithium. The EU heavily relies on imports for many materials used in batteries. According to the EC Raw Materials Information System (RMIS—https://rmis.jrc.ec.europa.eu/), as the future demand of materials for strategic sectors is concerned, the EU will continue to be almost entirely dependent on third countries, in particular for traction batteries. In fact, the EU produces only 1% of all battery raw materials overall. Materials needed for the batteries manufacturing are mainly extracted in China (32%), Africa and Latin America (both 21%), but also the manufacturing process is mainly occurring in Asian countries. In this framework, some EU initiatives are focusing on improving the EU capacity in manufacturing batteries and improving the EU value chains (e.g., European Battery Alliance and Strategic Action Plan on Batteries).

Due to high cost of specific cathodes, e.g., Li-Co-based such as NMC (nickel-manganese-cobalt) or NCA (nickel-cobalt-aluminium), chemistries with lower cobalt content are already available [17,18]. For instance, the NMC111 is already replaced by cathodes with lower Co content, e.g., NMC811 or NMC9.5.5, which means a strong reduction of the cobalt and manganese content and an increase of nickel. Increased collection of end-of-life vehicles (ELVs) and boosting recycling could increase the amount of recovered materials that can be potentially recirculated in the economic system, i.e., their value is retained within the EU and the EU dependency can reduce [2,14]. The recovery of such materials is particularly relevant for the EU since they are materials with a high supply risk and high economic importance [19]. Note that nickel needed for the batteries manufacturing should belong to CLASS I nickel, i.e., 99% pure nickel [20,21], which should be considered when assessing the recirculation of recovered nickel from recycling processes.

Recycling of the battery at its EoL can be advantageous from both resource conservation and environmental perspectives [22]. Recycling can provide SRMs that according to their quality can be used by the battery sector (e.g., cobalt) or by other industries [23], avoid the extraction of virgin raw materials and generally have lower environmental impacts.

Directive 2006/66/EC14 defines the minimum recycling rates of batteries: 45% of LIBs at their EoL have to be collected and at least 50% of the average weight of LIBs should be recycled, excluding energy recovery. Nevertheless, as underlined by Ellingsen et al. (2018) [22], this Directive often incentivises batteries industries to recover base metals, which are massively used and relatively abundant in nature and available in commodity markets (such as iron and copper). On the other hand, it is known that the market value of metals contained in batteries is an important economic driver for battery recycling. In particular, the higher prices of cobalt and nickel relatively can explain why recycling processes are currently focusing on these metals [22].

To assess the potential contribution of batteries in decreasing the environmental burdens of the EU mobility in the future, several aspects should be included in the assessment. Focusing on the environmental impacts of batteries along their life-cycle, in the literature Life-Cycle Assessment (LCA) studies are already available (e.g., [24–26]). However, *"guidelines or harmonized approaches do not yet exist"* [27] and some issues emerging from the available literature still need to be addressed [28], which make it even more challenging to capture the environmental performances of different type of traction batteries in xEVs, especially in relation to the adoption of energy mix and taking into account different EoL options. Again, a single assessment tool does not likely provide a complete overview of such a complex system; in fact, some aspects are not captured through LCA, e.g., resource efficiency of some EoL options, hence LCA should be integrated with other assessment tools [29]. To obtain a more complete understanding of products' status [30], different assessment tools should be combined [29,31,32]. There are already some studies demonstrating the added value of combining LCA and (dynamic) Material Flow Analysis (MFA), which identify and quantify the stocks and flows of products/materials along their whole value chain. Bridging tools and the complementary use of their results support a more prospective decision-making, especially when different strategies are assessed, e.g., waste strategies [31]. Finally, the relevance of specific materials in terms of availability and vulnerability of a system, can be captured by the criticality assessments, providing relevant information of supply disruption and mitigation measures, taking into account flows of both primary and secondary materials [16].

Well established tools can be adapted and combined according to the product/system' characteristics to provide a wider understanding of the environmental performances in a life-cycle perspective. Hence, it is possible to capture different aspects of the assessed system/product and provide information according to the interests of the specific stakeholders. In addition, a flexible model and a common structure of the data collection (e.g., identification of best available sources, common data when possible) eases the update of the assessment according to the availability of data/information. Finally, consistency of input for different analyses is improved and the comparability of results is strengthened.

For a sustainable management of traction batteries, in an exponentially growing market and with a life-cycle perspective, some key questions need to be answered:

1. To what extent will the environmental performance of future mobility systems improve due to the uptake of EVs and therefore batteries? Is this improvement in line with expectations (e.g., the EU Green Deal and the SGDs)? How relevant is the relative contribution of traction LIBs life-cycle impacts in terms of environmental performance?
2. To cover the forecasted demand of traction LIBS for the EU fleet in the future, will the CRMs used for their manufacturing be available in adequate quantity and quality?
3. What can the role of recycling in terms of improving the environmental performances of LIBs be? To what extent can it contribute until the production of traction batteries peak and stabilise? At what stage of the LIB value chain are the CRMs to be recycled in the future (i.e., SRMs)?
4. In which way are CRMs key to the change of mobility patterns? In which way will the change in mobility patterns affect criticality, e.g., in terms of growing demand for S(C)RMs?
5. How much can circular economy strategies help speed up the change and improve the overall environmental performance of mobility systems? How can trade-offs between different EoL strategies be quantitatively considered?

Aim and Structure of the Paper

This paper builds on a toolbox of existing assessment methodologies and bridges them to assess the environmental performances of complex systems. The adoption of different methodologies providing different type of information of the potential effects of the rapid evolution of a key sector for the EU is the core of the paper. For this reason, well established assessment tools are integrated: (1) to consider specific characteristics of the assessed LIBs; (2) to provide a more holistic and comprehensive understanding of traction LIBs; and (3) to build structured and comprehensive responses to relevant questions made by different stakeholders of the whole value chain of traction LIBs (e.g., manufacturers, recyclers, consumers and policy makers).

The authors believe that bridging tools and a more structured use of their results, as well as a mutual informing among the tools, is an added value in improving the knowledge of complex system and can support decision-making. The main focus of the performed assessment is the environmental performances of traction batteries (mainly LIBs) in decreasing the environmental burdens of the EU fleet up to 2050. In particular, LCA, MFA and criticality assessments, with emphasis on supply risk, are the three tools to bridge and use in a complementary manner. As mentioned in the Introduction, for this paper, the criticality of materials is used as filter to prioritise materials. The criticality of materials is used as filter to prioritise materials. In particular, among the key materials for the future development of batteries in the EU, as defined by the SWD (2018) 45, criticality of materials is used to identify those to be firstly studied. LCA is used to understand the environmental performance of traction batteries in the current and future EU fleet. MFA is used to trace flows in anthropogenic flow cycles, to understand the current and future demand of (C)RMs, including where they are stocked, when they will be available for recycling, etc. In a mutual fashion, criticality is used to prioritise the materials that undergo LCA and MFA, while LCA and MFA provide information to assess potential effects of variation of primary/secondary materials in terms of supply risk in the future (e.g., potential decrease of supply risk due to higher recycling).

The study also aimed at identifying synergies in gathering data and information to be used in the three methodologies and applied them to a specific case study.

A concise literature review about the main aspects affecting the environmental performances of traction batteries (in particular LIBs) is provided in Section 2. A short description of the methodologies is reported in Section 3 and the application to traction LIBs in the current and future EU fleet is described in Section 4. Section 5 summarises the main outcomes of the analysis, highlighting also the main limitations of the study. The main conclusions and further research needs are illustrated in Section 6.

2. Literature Review: Main Aspects Affecting the Environmental Assessment of LIBs in the Future EU Fleet

The research on xEV batteries is currently significantly active and rapidly evolving, as proved by the annual number of publications focusing on EV increased since the 1990s [33,34] and the increasing number of patents worldwide focusing on EV [35]. This section reports the main outcomes of the performed literature review, mostly referred to scientific publications between 2016 and 2019, with some exceptions for particularly relevant data published during 2010-2015. The critical review was carried according to the main research questions presented in the Introduction.

Traction batteries are recognised as key components for the future uptake of xEV and for the decrease of the environmental impacts of the future EU mobility system [13,17,36]. Forecasts and market trends of LIBs are available in the literature from both policy and research studies and manufacturers declarations. Most of the consulted studies are characterised by an exponential increase of both BEVs and PHEVs in EU, although other studies suggest that the uptake of new technologies can be described through an S-curve [6,37]. The modelling of such a curve entails the definition of the saturation level, which depends on technological/economic/social aspects, e.g., increase of occupancy rate of cars, price of new vehicles, concept of mobility and social acceptance of new technologies (see Alonso Raposo et al. [3]). Due to the different scopes, but also modelling, boundaries and assumptions of the explored studies, the comparability of the future EU fleet is quite complex.

Many recent studies have focused on the life-cycle impacts of LIBs, recognising in LIBs the main element differentiating xEVs from ICEVs [38,39]. Even though LCA is a standardised (ISO 14040-44) [40] and mature methodology, more methodological efforts to quantify the life-cycle impact of LIBs are needed [9]. In 2018, the Product Environmental Footprint Category Rules (PEFCR) provided a "detailed and comprehensive technical guidance on how to conduct a PEF study" [41]; among the application fields covered by the document, traction LIBs are included. Despite the guidance, available results from different LCA studies on LIBs end up being very heterogeneous and it is still difficult to clearly define the environmental battery performances [42,43]. These discrepancies are due to different factors: lack, in many cases, of primary data; necessary simplifications and assumptions of the LCA model; and different chemistries of LIBs and therefore different performances [42]. In this context, Peters and Weil [44] started a deep work of review, selection of data and unification of the Life Cycle Inventories (LCIs) of LCA studies that were, until that moment, published on the manufacturing stage of batteries. Peters and Weil [44] analysed 79 studies developed between 2010 and 2016, but just five of these studies used exclusively own primary inventories and clearly disclosed the data [24–26,38]. Other LCA studies provide detailed LCI of LMO-NCM battery based on primary data [45], of NCA cell from its dismantling [46], of Li-Sulphur batteries integrating lab experimentations and theoretical modelling [47], and of lithium manganese batteries (LMO) and lithium iron phosphate batteries (LFP) collecting data from a manufacturer [48]. Among the above-mentioned LIBs, NMC could become the most used Li-ion battery chemistry in 2030, followed by LFP and NCA with a 40% combined market share [5].

Because of the growth of LIBs, the demand of raw materials for manufacturing will increase according to their content in different LIBs chemistries. An increasing number of studies aimed at quantifying the future demand of materials for LIBs, especially if such materials are critical (e.g., cobalt). MFA is often used to quantify the stocks and flows of materials in and between processes along the value chain of LIBs. Among the consulted studies, the majority adopt a global approach, while few focus on the EU value chain; in addition, the analyses often focus on the demand of primary raw materials without including the contribution deriving from SRMs. In addition, more circular options than recycling are arising in the EU. This is the case of extending the lifetime of LIBs, e.g., through their second-use in less energy-demanding applications [9,49]. Thus, a proper management of EoL can have benefits in terms of environmental impacts and supply of SRMs, as well as effects and trade-offs between different EoL options should be further explored to provide information to be used in properly manage waste batteries.

Focusing on the environmental impacts of reuse and second-use of batteries, relevant aspects to be considered in assessing the impacts of LIBs were identified. As the reuse is concerned, key aspects to be considered are electricity mix [50–52], efficiency losses of batteries [50,52] and the characteristics of both the battery and the second-use application [9,45,53–55]. Due to the novelty of the topic and the scarce availability of data, input data are often based on warranties of LIBs and assumptions [9].

As far as concern recycling, guidelines for the impact calculation of LIBs recycling are provided in the PEFCR on batteries [41]. Bobba et al. [12], using information from industries [56], assessed the impacts of different EoL options, mainly focusing of the materials assessment, i.e., through a dynamic MFA. R&D projects and industrial companies are currently investing some efforts to improve the recovery of materials embedded in batteries to increase the sustainability of LIBs and tackle with economic barriers, which in some cases are important obstacles to the development of recycling at industrial scale, e.g., in the case of lithium recycling. For that reason, the amount of SRMs available in the future is expected to increase and contribute to partially cover the demand of raw materials for LIBs.

Available studies assessing criticality of raw materials were critically reviewed by Schrijvers et al. [16]. In this study, it is highlighted that different methods have been developed to identify criticality assessment factors and indicators at different levels (global, country or region, company, technology or specific products) [16]. In addition, data availability is recognised as a key factor that limits the evaluation of criticality. Proxies are needed to overcome this lack of data. Furthermore, data quality, including both data uncertainty and data representativeness, is rarely addressed in the interpretation and communication of results [16]. Focusing on the EU, the list of CRMs and the criticality methodology are a key instrument in the context of the EU raw materials policy, a precise commitment of the Raw Material Initiative [57]. Since the publication of the first list in 2011 and subsequent updates in 2014 and 2017, the EC criticality methodology responded to the needs of governments and industry to better monitor raw materials and inform decision makers on how security of supply can be achieved through diversification of supply, resource efficiency, recycling and substitution. To prioritise needs and actions at the EU level, the list of CRMs supports in negotiating trade agreements, challenging trade distortions and in programming the research and innovation funding. The EC methodology [58] defines CRMs as the combination of high economic importance (EI) for the EU and high risk of supply disruption (SR, supply risk). The assessment in essentially based on past data, e.g., the 2017 list was based on the five-year average (i.e., 2010-2014). Demand growth is often considered by technology-oriented methods, but not always considered by studies focusing on a national economy. This makes this exercise suitable to describe current economic situation, disregarding the future development of the economy [16].

The performed literature review highlights the complexity of the topic and the fact that several aspects should be taken into account to provide valuable and complete information on the environmental performances of traction LIBs in the EU in the future. In this framework, the integration of LCA and MFA to better understanding complex systems and environmental impacts is an added value [29,30,32]. Studies combining LCA and MFA of products are already available in the literature [31,59] and synergies between LCA and criticality were already proposed by Mancini et al. [60]. Specific consideration of future availability and demand of primary/secondary (critical) raw materials for traction LIBs in the EU were explored by Golroudbary et al. [61] and Pillot [62]. Song et al. [63] developed a detailed study on dynamic MFA of the CRMs for the Chinese LIB industry combining both the MFA and a CRMs evaluation model based on Blengini et al. [58] considering future scenarios up to 2025. Studies on the criticality of raw materials embedded in LIBs were investigated by Olivetti et al. [18] and Helbig et al. [64]. The results of criticality assessment and LCA were combined by Gemechu et al. [65]. However, the results do not provide specific information related to the potential variation of the supply risk due to the potential improvement of specific circular strategies and related environmental impacts. Finally, synergies among LCA, MFA and potential supply risk should be further improved at inventory level since some input data can serve all (e.g., processes efficiency and materials content).

The literature review confirms that, to the authors' knowledge, there are no studies specifically addressing traction batteries in the EU along their whole value chain integrating information provided by all LCA, MFA and supply risk considerations together. In addition, a future-oriented approach requires more scientific efforts in terms of methodology, to develop models taking into account key aspects of foreseeable future, e.g., physical scarcity or the future development of the economy [16].

3. Methodology: Modelling Flows and Impacts of LIBs in the EU Fleet

The integration of assessment tools able to analyse different but complementary aspects is a key feature to improve an in-depth and comprehensive knowledge of the EU fleet. In the following paragraphs, the main features of performed assessment of the environmental impact of traction LIBs in the EU fleet are illustrated.

A Life-Cycle Thinking (LCT) approach is adopted to include in the assessment all the relevant aspects along the whole value chain of products, taking also into account external aspects affecting environmental performances of products, e.g., socioeconomic aspects.

3.1. LCA of Traction LIBs

The developed LCA follows the (ISO 14040-44) and the PEFCR for batteries [41]. The LCA tool provides the necessary background information on environmental impacts of products/services under analysis along their whole value chain. Considering the potential development of technology and the complexity of some products, the development of modular LCAs and the adoption of parameters make the LCA model flexible: (1) to update according to available input data; (2) to speed-up the LCAs of different products; and (3) to enlarge the analysis (e.g., new materials and/or components).

According to PEFCR for batteries, the functional unit (F.U.) for rechargeable batteries can be defined as *1 kWh of the total energy provided over the service life by the battery system*. Nevertheless, this functional unit requires referring to the expectancy life of the battery, which is often hard to estimate because it is affected by many different parameters [10,42]. Moreover, since the majority of available LCA studies on batteries show impact results for 1 kg of battery or for 1 Wh of storage capacity, the developed LCA tool provides results for both 1 kWh of energy provided and 1 kg of battery pack. The LCA tool provides information on the specific LCI datasets used for the analysis, with reference to Environmental Footprint (EF) and Ecoinvent databases (see Supplementary Materials for details). Datasets are connected to the related impacts, evaluated with the EF method; this enables the user to easily assess the batteries for all the impact categories available within this method. In this manuscript, attention is however focused on the Global Warming Potential (GWP), as one of the most robust categories and of high societal and policy interest [66]. Results for the other EF impact categories are reported in the Supplementary Materials.

The unified database of Peters and Weil [44] is used to assess the environmental impacts of different LIBs chemistries according to available inventories in the literature. In addition, this analysis updates and extends the unified database with recent data on the manufacturing of LIBs data [45,46] and on other stages of batteries life-cycle (use and EoL stages). Table 1 lists the main characteristics of the analysed batteries and the selected source of the LCI data.

Table 1. Main characteristics of the analysed batteries and the selected source of the LCI data.

Battery Chemistry	Type of Vehicle	Weight [kg]	Capacity [kWh]	LCI Data Source for Manufacturing Stage	LCI Data Source for Use Stage	LCI Data Source for EoL Stage
NMC 111	EV	253	26.6	[24,44]	[41]	[41,45,67]
NMC 424	EV	n.a.	n.a.	[26,44]	[41]	[41,45,67]
NCA	EV	142	18.9	[44,68]	[41]	[41,45,67]
NCA	EV	154	20	[46]	[41]	[41,45,67]
LMO/NMC	PHEV	175	11.4	[45]	[41]	[41,45,67]

The use stage is defined by the energy losses due to the battery and charger efficiency [41]. The model was built in a way that the total energy used by a xEV before replacing the traction LIB is obtained by multiplying the distance covered by vehicles and the average fuel economy (energy necessary to cover the distance of 1 km). The change of the energy mix along time is considering through an increasing share of renewables in the energy mix, based on EC projections [69,70].

The EoL stage includes dismantling of components, the conversion into recycled material, other operations and credits connected to the re-availability of material after the recycling process [41]. The LCI of the EoL stage was unified with reference to data provided by the PEFCR on batteries, with exception of input/output data of lithium, nickel, manganese, cobalt, graphite, copper and aluminium. For these latter, quantities of recycled material (and consequent credits) were calculated considering the amount of materials available in each battery and the recovery of the same material after the recycling process.

To make the LCA replicable and updatable, the LCI of each LIB chemistry is provided in a spreadsheet (included in the Supplementary Materials), where each input/output flow of material, energy, waste and emission is related to 1 kg of battery pack. Each flow is connected to the related impacts, enabling the automatic calculation of the impact of the battery life-cycle. The changing of input/output quantities or parameter values allows the evaluation of different scenarios, as shown in Section 5. Moreover, the modularity of the model allows quick and consistent comparisons between environmental performances of different batteries. Additionally, due to the fast development of the technology, the modularity of the LCA model allows enlarging it, adding e.g., new materials and/or components.

3.2. MFA of traction LIBs in the EU

MFA is used to better understand the value chain of products through the representation of the main processes along the value chain but also to quantify the stocks and flow of products/materials over time [71].

According to Bobba et al. [12], the adoption of parameters in the MFA model makes it customisable and flexible to assess different scenarios and identifying e.g., circular economy aspects and/or effects of EoL options along the whole value chain of LIBs. In addition, the modularity of the model allows easily adding/updating modules within the MFA model in case of new/more data would be available. In case some modules are not of interest of the assessment (e.g., second-use of specific LIBs' chemistries or in addressing some future EoL scenarios), parameters allow simply not considering these modules for the quantification of stocks and flows. In addition, different aspects of LIBs can be assessed, i.e., stocks and flows of both batteries, materials embedded in LIBs and storage energy capacity.

Figure 1 shows the MFA model created to estimate the stocks and flow of the EU fleet in the future. Differently from Bobba et al. [12], the model was enlarged to also include the recirculation of SRMs in the system and incineration/landfilling of LIBs. Due to the difficulty in collecting all the needed data, some flows were estimated through the adoption of a parameter which was made varying in-time. In line with the goal of the study, the model is used to quantify the stocks and flows of traction LIBs, lithium, nickel and the energy storage capacity for various LIBs chemistries and applications (i.e., PHEV and BEVs).

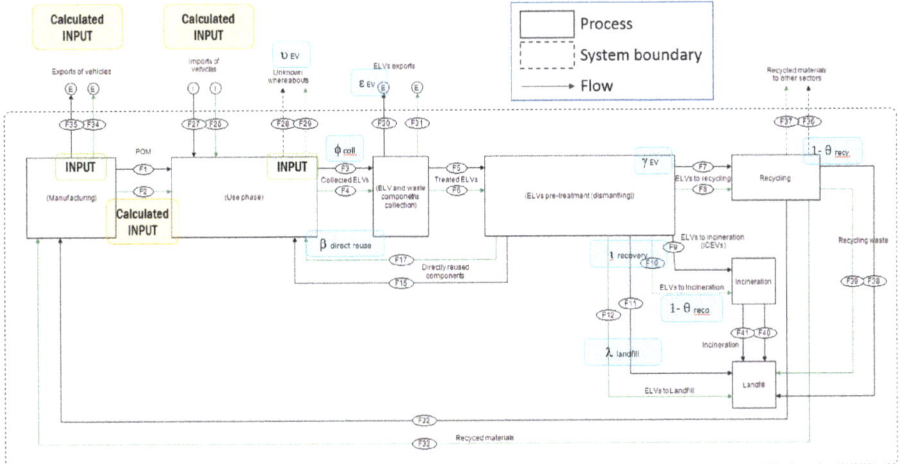

Figure 1. Modelling of stocks and flows of LIBs in the EU (adapted from Bobba et al. [12]).

Where:

$\upsilon_{ICEV/EV, y}$	= Unknown whereabouts
$\phi_{coll, y}$	= Collected ELVs
$\varepsilon_{EV, y}$	= ELVs exports
γ_{EV}	= ELVs to recycling
$\iota_{recovery}$	= ELVs to recovery
$\beta_{direct\ reuse}$	= ELVs to direct reuse
$\lambda_{recovery}$	= ELVs to landfill
θ_{recy}	= Recycling efficiency
θ_{reco}	= Recovery efficiency

3.3. Criticality and Supply Risk of LIBs Raw Materials

According to Schrijvers et al. [16], "the raw material criticality is the field of study that evaluates the economic and technical dependency on a certain material, as well as the probability of supply disruptions, for a defined stakeholder group within a certain time frame". In line with the system boundary of the study, the analysis on materials supply risk is applied at EU level, focusing on battery's raw materials. As mentioned in Section 2, the EC methodology [58] defines CRMs as the combination of high economic importance (EI) for the EU and high risk of supply disruption (SR, supply risk).

In this paper, we focus on the contribution of recycling as a supply risk mitigation factor, according to the EC methodology [58]. Recycling of end-of-life products is in fact considered a more secure source of raw materials, in comparison to primary production [58].

The battery materials initially prioritised in this study include cobalt, lithium, natural graphite and nickel; manganese is considered too, although less often identified as CRMs [16,72] due to the lower supply concentration and subsequent lower supply risk. Among the batteries' materials, technologies for recycling Li are currently available in the EU, even though not yet at industrial scale [73,74].

Concerning nickel, it is to be considered that nickel used for manufacturing traction LIBs needs to be of high quality; in fact, Nickel Class I has a Ni content higher than 99% [20]. Despite some examples of high-quality nickel recovered from batteries (e.g., [75]), most of the nickel recycled nowadays is not suitable for the manufacturing of new LIBs [76]. In the EU, stainless steel is the biggest user of primary and scrap nickel, which already uses nickel scrap recycling for stainless steel production and will increase by 2025.

Despite limitations, data provided by MFA, in particular amount of materials recovered from traction LIBs recycling, are used to estimate the potential reduction of the SR for both Li and Ni in relation to the expected evolution of the EU fleet.

3.4. Common Inventory and Data Gaps

Availability of data is a key point for all the above illustrated assessment methodologies. In some cases, the same data can serve two or three components, which is an added value for an already difficult data collection. Moreover, the adoption of the same data and/or information improve the consistency of the obtained results and ease the replicability of the assessment in case new data will be available in the future. Finally, a common inventory facilitates the clear definition of the needed assumptions for the three components.

Table 2 provides an overview of the data and assumptions included in the study that are common to at least two out of the three components. However, it is highlighted that the adoption of the same data for multiple components requires more efforts in terms of data quality and geographical/temporal representativeness.

Table 2. Overview of possible synergies between LCA, MFA and Criticality (supply risk) in terms of inventory data.

	LCA	MFA	Criticality (Supply Risk)	Unit	Notes
Weight of the battery	X	X		[kg]	
Lifetime	X	X		[year] or [provided kWh]	
Materials content	X	X	X	[kg/kg battery] of [kg/kWh]	
Process efficiency (i.e., losses)	X	X		[kg/kg battery]	
Import/export	for transport	for flows and stocks	X	[tonne]	- impacts of transport - outbound/inbound flows - import reliance
Collection rate	X	X		[-]	- indirectly availability of SRMs
Battery reuse	X	X	X	[%]	- lower impact of the battery life-cycle (longer lifetime) - stocks increase, creation of new stocks - indirectly availability of SRMs
Battery dismantling efficiency	X	X		[%]	
Recycling efficiency	X	X	X	[%]	- impacts of recycling process / avoided materials - available SRMs to be recirculated in the system
Quality of recycled materials	X	X	X	[-]	- closed/open loop - available SRMs for specific sectors

Table 2. *Cont.*

	LCA	MFA	Criticality (Supply Risk)	Unit	Notes
Materials substitutability	X	X	X	[-]	- increase/decrease of materials content - LCA of different chemistries
Future technological change	X	X	X	[-]	- different chemistries, materials, components - potential improve of recycling technologies
Geographical considerations	X	X	X (import reliance and production)	[-]	- evaluate transports and import/export flows - EU dependency on third Countries - import reliance
WGI	Social (not assessed in this study)		X	[-]	
New energy sources	X		X	[-]	
Trade agreements and restrictions		X	X	[-]	

4. Case-Study: Traction Batteries in the Future EU Fleet

The selected methodologies were applied to traction LIBs in the EU fleet between 2015 and 2030 to assess the environmental contribution of traction LIBs to the potential decrease of the impacts of the EU fleet in the next decades, but also the role of the key materials embedded in batteries. In this section, a brief description of the case-study is provided; to assess the potential added value of the coordinated approach, different scenarios were considered.

4.1. Description of the Case-study and the Assessed Scenarios

The environmental assessment focuses on traction batteries used in both PHEVs and BEVs, especially on LIBs embedding Ni and Li. Note that, even though some LIBs are already used for HEVs, this technology was excluded from the analysis since the main power source is a combustion engine. A Base-Case Scenario was created to capture most of the aspects mentioned in the research questions (Section 1); in addition, the effects of some key aspects (i.e., EoL management and change of energy mix) are considered through the creation of ad-hoc scenarios hereinafter described (see Table 3).

4.1.1. "Base-Case Scenario"

The evolution of BEVs and PHEVs in EU between 2015 and 2030 is based on the EU Long-Term Strategy (LTS 1.5°C Technical), which assumes a reduction of the EU greenhouse gas emissions for 2030 of about 50% and zero emissions in 2050 [4].

The case-study is mainly focused on traction NMC and NCA chemistries, which are expected to dominate the traction LIBs EU market in the future (Section 2); LMO/NMC chemistry is also included in the analyses of current scenarios, while it is excluded in the calculations of future scenarios because of the scarce availability of data on possible trends for these chemistries. For the use stage of LIBs, the default amount for energy density is 9.6 kWh/kg (aligned with the PEFCR for batteries) and fixed for all the LIBs. Default values for the battery use are: 100,000 km for the driven distance (in line with the warranty generally given for batteries by BEV producers) and a fuel economy of 0.2 kWh/km (according to Fuel Economy data provided by EPA). For the EoL, no remanufacturing and second-use are assumed to develop at industrial scale in the EU. The recuperation of materials is calculated as the product of the collection rate, the dismantling rate and recycling rate. A collection rate of 95% is assumed, according to the PEFCR [41], and the recycling rate is based on Lebedeva et al. [74] and Cusenza et al. [45]. These rates are parameters which can be easily modified for assessing future scenarios.

The specific energy mix considered in the study is based on literature data [69,70]. The estimation of the stocks and flows of LIBs/energy storage capacity/embedded materials is adapted from [12], with additional modules and flows; the value chain of batteries in the EU (Figure 2) is assumed to mainly maintain the current characteristics. This means that the manufacturing of batteries is assumed in the EU, according to Peters and Weil [44].

Concerning the assessment of Li and Ni, as representative of key materials in the battery sector for the EU, their content in LIBs is assumed to vary in time according to available roadmaps (e.g., [5,72,74,77]). Currently, the recirculation of Li and Ni in a closed loop (i.e., to manufacture new LIBs) is almost null. The recovery of Li is not yet developed at industrial scale in the EU, mainly due to economic reasons [73,74]; therefore, it is assumed a current recycling efficiency equal to 1% in 2018 [41], linearly increasing in the future thanks to the ongoing research activities (up to 3% in 2030). On the other hand, the recycling rate of Ni is already quite high, i.e., 96% [41]; however, to be used for manufacturing new cathodes, the purity of Ni has to be very high, and, according to the authors' knowledge, there are few examples of companies that are using secondary Ni to manufacture new cathode (e.g., [78]).

4.1.2. Scenario A: Extension of the LIBs Lifetime Through Their Second-use

Second-use of batteries in less energy demanding applications is an EoL option that can reduce the environmental impacts and boost resource efficiency [79]. Despite not yet occurring at large scale in the EU, the ongoing pilots and research activities have demonstrated that this option is valid and can increase in the next decades, if also supported by an adequate regulatory framework. Therefore, based on the Base-Case Scenario, Scenario A was created to assess the environmental effects of extending the lifetime of batteries. In particular, the life-cycle impacts include both the first- and second-use, the energy storage capacity of LIBs is further exploited before their recycling and the embedded materials are locked in the in-use stock for longer compared to the Base-Case Scenario.

The total amount of energy that is provided by the battery during its first and eventually second life is quite complex, since it depends on different connected variables, e.g., depth of discharge (DoD), charging-rate and operation temperature [10,42], and the environmental benefits depends on both the LIB's and the system's characteristics [9]. For this study, Scenario A considers a default value of 20,000 kWh provided by the xEV battery first life and a second life providing 5143 kWh, according to Bobba et al. [9,12]. A linear increase of second-use of batteries from 0% (current situation) to 10% in 2030 and 30% in 2050 is assumed.

4.1.3. Scenario B: Improved EoL Extension of the LIBs Lifetime Through Their Second-use and Improvement of Recycling

Recovery of key materials is essential to decrease the dependency of the EU from third countries (Section 3.3). Then, the effects of an improvement in the management of EoL practices in terms of both second-use of LIBs and improved recycling efficiency is assessed in Scenario B. The trend of LIBs in second-use applications is the same as in Scenario A. In addition, it is assumed that the recycling technology allows higher recycling efficiency of Li (15% in 2030 and 2050) and higher level of purity for covered Ni, which can be used again to manufacture LIBs (linear increase of the closed loop of Ni, from 0% to 25% between 2018 and 2030).

4.1.4. Scenario C: Renewable Energy for the Manufacturing Stage

The amount and the source of energy used in the life cycle of LIBs can highly influence its environmental performance [24,80–82]. Currently, there are examples of producers of LIBs' cells that are using high share of renewables in the production process, e.g., Tesla and Northvolt Ett. In particular, Tesla claims to design its Gigafactory 1 in Nevada to be completely powered by solar array installed on its roof and wind turbine installed nearby, while Northvolt Ett declares that its plant in Sweden will rely on clean electricity from wind and hydroelectric power [83].

Hence, to observe the effects of a more environmental-friendly energy mix in the life-cycle of LIBs, Scenario C assumes that electricity for battery manufacturing is equally provided by photovoltaic panels, wind turbines and hydroelectric plants. All other input/output flows and related variables follow the Base-Case Scenario.

Table 3. Summary of the main differences between assessed scenarios.

Variables	Scenarios			
	Base-Case Scenario	Scenario A	Scenario B	Scenario C
Change in European energy mix (current/2030/2050)	X	X	X	X
Change in battery material contents (current/2030/2050)	X	X	X	X
Batteries are reused in a second life (10% in 2030; 30% in 2050)		X	X	
The recycling rate of lithium and nickel is enhanced (2030/2050)			X	
Energy for manufacturing is completely provided by renewable sources (current/2030/2050)				X

5. Results and Discussion

Section 5 reports the main outcomes of the assessment. In particular, for the LCA analysis, results are reported on GWP impacts provided by the LCA tool for 1 kWh of the total energy provided by the battery during its entire life-cycle. Note that, as previously discussed, this latter F.U. is influenced by the battery lifetime, which is by default set to 20,000 kWh for all the analysed batteries. In addition, results reported in Section 5 mainly refer to the NMC and NCA chemistries as most of the consulted sources provide information of future trends of these two chemistries and almost no data about the uptake of LMO/NMC batteries are available.

5.1. Results

The Life-Cycle Impact Assessment (LCIA) of the assessed LIBs shows that the current life-cycle GWP is on average 0.16 kg CO_2 eq/kWh. The impact of NMC batteries results slightly higher than the NCA one. This difference could be partly due to the different material composition and partly to the higher mass of NCM batteries. The contribution analysis confirmed that the manufacturing stage highly contributes to the life-cycle GWP. In addition, the EoL recycling can reduce the impact of the battery by 22% on average. Among the LIBs' chemistries, the NMC 111 battery manufacturing has the highest impacts per kWh of provided energy. At the same time, NMC shows the highest benefit from its recycling, mainly related to the credits due to the availability of secondary raw materials after the battery recycling, mainly copper, nickel sulphate and cobalt sulphate. In the Base-Case Scenario, it is highlighted that the change of materials in different LIBs (e.g., from NMC111 to NMC811) and the increase of renewable energy share lead to a decrease of GWP in time (Figure 2). The manufacturing impacts of NMC batteries will decrease by 22% and 31% in 2030 and 2050 compared to the NMC nowadays in the market (2010-2018), while the reduction is lower (9% and 15%) for NCA batteries manufacturing. High reductions are observed for the use stage: about 42% in 2030 and 56% in 2050 for both chemistries.

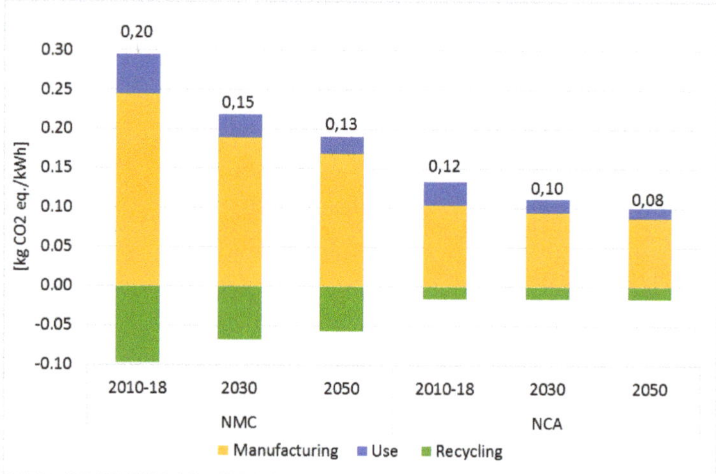

Figure 2. Life-cycle Global Warming Potential (GWP) of NMC and NCA chemistries in the Base-Case Scenario for different years. Labels at the top of each column indicate the respective values of the life-cycle GWP.

Focusing on the MFA, the results of the Base-Case Scenario show that the increasing demand of LIBs in the EU will not significantly affect the energy capacity storage and materials flows in the EU until 2030; then, waste flows start to be relevant in terms of quantities, especially recycling flows. Figure 3 shows an example of stocks and flows of energy capacity and materials in the studied system (for interpretation of the references to colour in this figure legend, the reader is referred to the web version of this article). Combining the above-illustrated GWP and energy storage capacity of LIBs placed on the EU market, it is estimated that the GWP of traction LIBs entering in EU fleet in 2015 is about 12 kt of CO_2 eq., and it will increase up to 35 kt in 2030 and 95 kt in 2050.

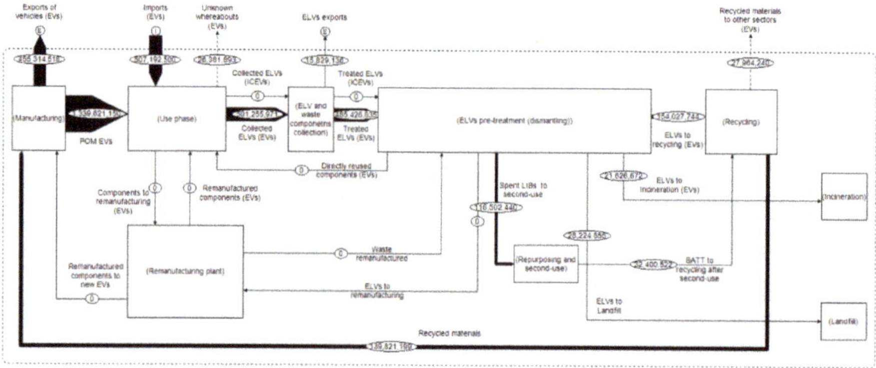

Figure 3. Flows of energy capacity storage in the EU in 2050.

Focusing on materials embedded in batteries, is observed that the amount of Li required for traction LIBs demand in 2030 and 2050 is, respectively, five and seven times higher than the 2020 demand (Figure 4). For Ni, these values increase to 7 and 14 times. Once extracted from EVs, the amount of Li/Ni entering in the recycling process can potentially provide 7% of the Li/Ni demand in 2030 and 26% in 2050. However, the recovered Li in 2030 is lower than 40 tonnes in 2030 and about 400 tonnes in 2050, mainly due to the lack of recycling processes at industrial scale; similarly, the recovered Ni in

2030 is about 900 tonnes and 11,000 tonnes in 2050. Note that almost all the recovered Ni is recycled in an open-loop, hence not used for manufacturing new LIBs. Secondary Li is lower than 0.5% of the Li demand in 2030, while secondary Ni is 0.4% of Ni demand for LIBs; these values slightly increase up to 1% for Li and to 2.5% for Ni in 2050.

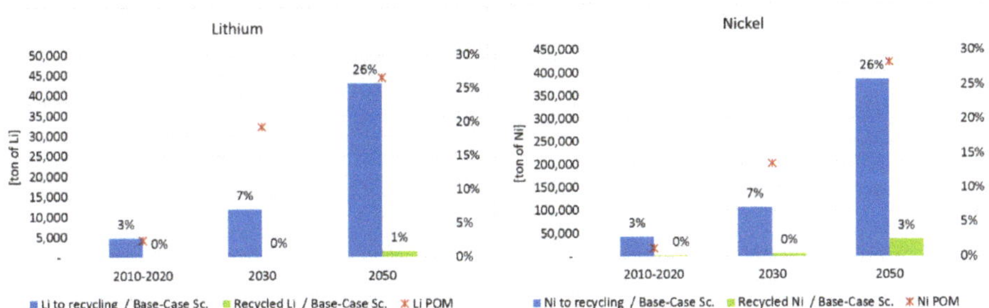

Figure 4. Li and Ni embedded in traction LIBs placed on the EU market (POM), available for recycling and recovered in different years.

If second-use will develop in the EU (Scenario A), LIBs and embedded materials will last longer within the system, decreasing the life-cycle impacts and postponing the amount of materials available for recycling. Assuming a second-life providing additional 5143 kWh, the GWP impact per kWh of provided energy necessarily decreases. With the assumption that the percentage of reused batteries is of 10% in 2030 and 30% in 2050, the LCIA results show an average GWP reduction, respectively, of 3% and 8% (Figure 5). If, in addition to the reuse of batteries, the recycling efficiency of LIBs will increase in time (Scenario B), reductions of 11% and 17% are observed for NMC in 2030 and 2050, respectively. The reduction is lower for the NCA chemistry: 4% and 9% in 2030 and 2050, respectively (Figure 5).

The adoption of a completely renewable energy mix for the NMC and NCA manufacturing (Scenario C) decreases the GWP of the manufacturing stage for both the current and future scenarios (see also Supplementary Materials). Benefits of manufacturing plants powered by renewables become progressively less evident over time, because of the enhanced sustainability of the European energy mix in 2030 and 2050. Compared to the Base-Case Scenario, Scenario C shows a GWP reduction of 23% in the current situation, 15% in 2030 and 12% in 2050.

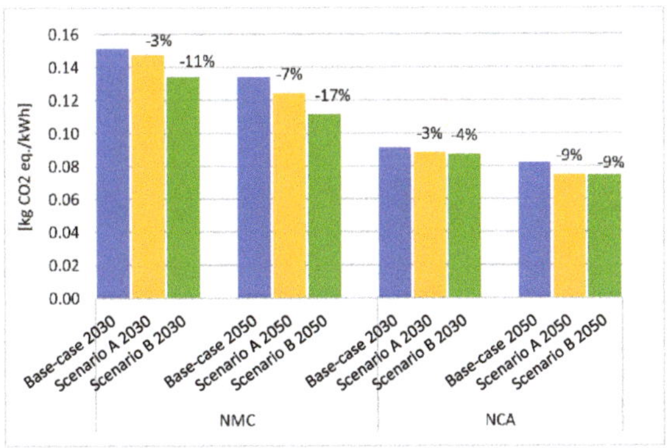

Figure 5. Life-cycle Global Warming Potential (GWP) of NMC and NCA batteries for Base-Case Scenario, Scenario A and Scenario B in different years. Percentages indicates the reduction of GWP impacts with reference to the Base-Case Scenario for the same year.

The development of second-use of LIBs (Scenario A) confirmed a delay in availability of Li and Ni for recycling and the consequent decrease of secondary Li and Ni in the EU: in 2050 about 400 tonne of secondary Li and 9,500 tonne of Ni will be available (about 0.9% and 2% of the Li and Ni demand in 2050) (yellow bars in Figure 6). Difference of secondary Li and Ni compared to the Base-Case Scenario are almost null since the stock of LIBs in second-use application will become more relevant around 2040. Finally, from the analysis of stocks and flows of materials, it emerged that the increase of recycling efficiency will result in a significant flow of SRMs (green bars in Figure 6). In this case, the secondary Li and Ni in 2050 will be respectively 7% and 9% of the Li and Ni demand in 2050, i.e., 3000 and 38,000 tonnes.

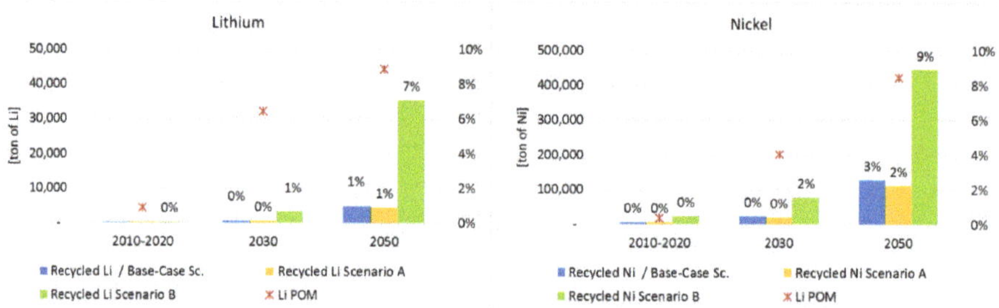

Figure 6. Li and Ni embedded in traction LIBs placed on the EU market (POM) and recovered from recycling processes for the Base-Case Scenario, Scenario A and Scenario B in different years.

Finally, Figure 7 reports the EOL-RIR and the EOL-RR [58] calculated for all the assessed scenarios. The increasing recovery of Li as SRMs (Scenario B, green bars) significantly increases both indices; a similar result is observed if the technological development of recycling Ni will allow reaching a high level of purity, i.e., >99%, in order to use Ni as SRMs for manufacturing new LIBs' cathodes. Note that the amount of materials stocked in second-use application is not importantly affecting the indices since second-use is still a limited EoL option (Scenario A, yellow bars).

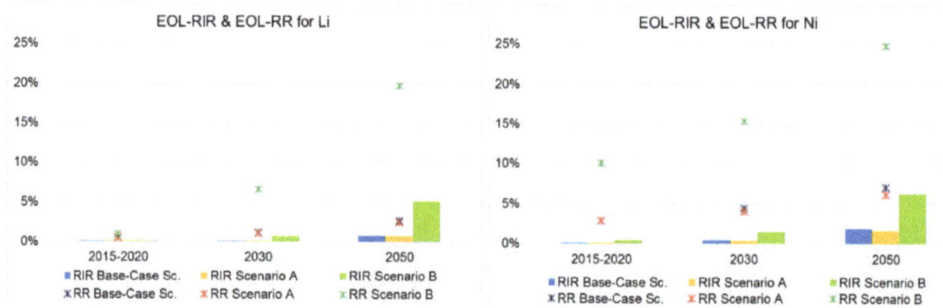

Figure 7. EOL-RIR and EOL RR of Li and Ni according to the Base-Case Scenario, Scenario A and Scenario B in different years.

The EOL-RIR of Li and Ni is used to better understand the role of recycling in mitigating the SR in 2050 according to the expected evolution of the EU fleet. The current and future EoL-RIR is estimated by the MFA tool (average 2015–2020) and in the future (2050), based on the assessed scenarios.

The supply risk reduction of Ni is < 1% in the Base-Case Scenario (2015-2020) and 6.3% in Scenario B (2050). Ni used to manufacture LIBs requires a higher grade (Class I) than Ni used in, e.g., steel applications, which is its main use according to the Ni Institute [21] and the overall EoL-RIR is estimated to be about 34% [84]. Even though the recycling rate (RR) of Ni from LIBs is quite high and its trend is rising, the required grade is higher to feed the input for batteries with SRM. Thus, the EoL-RIR can slightly reduce the supply risk.

The figures are quite similar for Li, with a supply risk reduction <1% in the Base-Case Scenario (2015-2020) and 5% in Scenario B (2050).

Moreover, it is well known that the supply risk for Ni is much lower than that of Li according to several criticality assessments run internationally [16], thus it appears that recycling as a risk mitigation factor seems even more important for Li than for Ni. Unfortunately, at present and in the future, recycling, alone, does not seem a relevant mitigation factor for the supply risk of neither Li nor Ni (all other factors being equal, and taking into account the limits of the model).

5.2. Discussion

The complexity of batteries and the assessment of their (environmental) performances is certainly a major challenge in this and in other studies. A single assessment tool cannot provide a complete overview of the impacts of LIBs and can unlikely capture all the effects of different EoL options [29,85]. This is particularly relevant for those new options that have shown potential to develop in the next future (e.g., second-use of batteries), as well as in case of raw materials with high supply risk (e.g., Li).

LCA and MFA are applied to traction LIBs in the current and future EU fleet, to assess the life-cycle GWP of specific LIBs chemistries currently available in the market and LIBs entering in the EU fleet up to 2050 (Section 3.1); in addition, stocks and flows of LIBs/storage capacity/embedded materials were quantified along the EU value chain of LIBs for different scenarios (Section 3.2). The criticality of materials in this study was used to filter the raw materials to firstly focus on and to assess the supply risk of such materials using data provided by the MFA results. Among the key materials for the future development of batteries [72], those selected for the study are Li and Ni embedded (Section 3.3). Due to the uncertainty of input data related to the fast evolution of the technology, the complexity of the LIBs' components, the globalised market and different assumptions behind available LCA and MFA studies, assessment models are built using modules and parameters. This makes the model flexible and updatable according to available data (e.g., in case of new components/processes/stocks) and addresses uncertainty through the creation of different scenarios (e.g., change in energy mix and increase/decrease of processes efficiency). Obtained results confirm the added value of the adoption of different assessment tools to improve the knowledge of complex systems, as traction LIBs [29,30,32].

Since the vehicle electrification certainly plays an important role in the decarbonisation of mobility, the investigation on the environmental performance of traction batteries is of crucial importance, being the main element that differentiates ICEVs and EVs. The lack of reliable data highlighted by several authors (e.g., Peters et al. [42] and Zackrisson et al. [25]), and therefore the difficulty in consistent comparison between LCIA results is addressed in the study through the adoption of the unified database proposed by Peters and Weil [44], used to assess the GWP of NMC, NCA and LMO/NMC chemistries. To bypass the barriers for information sharing and reuse often caused by the only partial interoperability between the main LCA software and database currently available [86], to ease the replicability of results and to further enlarge the proposed analysis, the LCA tool is provided in the Supplementary Materials as a spreadsheet file. LCIA results obtained for 1 kg of battery pack (provided in the Supplementary Material) are aligned with the available literature [24,26,44]. The results expressed for the F.U. of 1 kWh provided by the battery are not easily comparable with previous literature since most authors did not provide results with this F.U. In addition, it is necessary to underline that LCIA results for 1 kWh of provided energy directly depend on the lifetime of the battery. This latter is fixed in this paper to 20,000 total kWh provided, but can vary among different batteries. This parameter can however be easily modified in the LCA tool provided in the Supplementary Material and according to available data, its variation is recommended to understand its relevance in line to the goal of the analysis. The change of the energy mix plays a key role in decreasing both the manufacturing impacts and the life-cycle impacts of LIBs in 2030 and 2050. In fact, manufacturing LIBs with renewable energy sources (Scenario C) reduces the impacts of LIBs placed on market in 2020 of almost 4000 tonnes of

CO_2 compared to the impacts of LIBs placed on market the same year and manufactured with the current EU energy mix. In addition, the increased share of renewables (Base-Case Scenario) is the most important factor in reducing the life-cycle GWP of NMC and NCA chemistries: respectively, by 24% and 22% in 2030 and by 32% and 30% in 2050. To better understand the contribution of LIBs in decreasing the impacts of the EU mobility system in the future, a wider analysis should be performed to include the life-cycle impacts of different types of vehicles, even though this means a further level of uncertainty related to needed assumptions and simplifications, e.g., losing the detail on different chemistries and performances due to the high amount of information to be processed. In addition, electricity mix importantly affects the life-cycle impacts of EV vehicles, thus is a key aspect requiring more in-depth analysis, especially in the case of future energy mix [28]. The developed spreadsheet file can be used to model the contribution of LIBs taking into account key parameters and without losing details about performances of batteries (see Supplementary Information).

In 2030, 32 kt of Li and 202 kt of Ni will be required according to the targets established by the EU Long-Term Strategy [4]. These values increase up to 44 kt and 423 kt in 2050. The assessment of different trend of xEVs uptake, as discussed in Section 2, is recommended. In fact, input data concerning materials content in LIBs are currently poor and should be updated with new available data, and it is expected that new technologies will appear in the market, like fuel cell EVs [3,7].

The results of the analysis confirm that the growth of LIBs in the future EU fleet corresponds to an increasing flow of energy storage capacity, which is better exploited through second-use of LIBs [12]. As a consequence of the extension of LIBs' lifetime, Li and Ni are locked in the second-use stock, and therefore they cannot be available for recycling and potentially be recovered as SRMs.

Even though LIBs are not second-used, current recycling of Li is quite poor [73,74] and the secondary Ni is not pure enough to be used again for LIBs manufacturing [20,21]. In fact, without any improvements in recycling processes (Base-Case Scenario), the contribution of secondary Li and Ni to the EU materials demand is and will remain quite limited (1% of the Li demand and 2.5% of the Ni demand in 2050). Even though the availability of Li and Ni is delayed in time according to the extended lifetime of LIBs in second-use applications (Scenario A), the potential increase of recycling (Scenario B) of Li and Ni entails environmental benefits in terms of GWP (17% and 9% of the life-cycle GWP for, respectively, NMC and NCA chemistries in 2050, compared to the GWP of currently LIBs in the market). Moreover, this improvement also means an increasing amount of SRMs, hence a slight mitigation of the supply risk in the future (2050). In fact, although supply risk of Ni is relatively low at present and the recycling could lead to a more sustainable production, the overall risk would not be affected significantly. The supply risk of Li is very high due to the high concentration of EU sourcing in Chile and completely reliant on import from third countries. Despite a rising trend of Li in the future, EoL RIR will not be able to mitigate this important supply risk. It can be concluded that, with the aim of pursuing strategies of mitigation, fostering European domestic production and diversifying the EU suppliers seems the best option to reduce the supply risk, rather than the way of recycling.

Batteries require specific grade of materials for their chemistry and a sectorial level assessment would draw a more precise picture for their risk of supply disruption [63]. In addition, the EU consumption of raw materials used in batteries is extremely low compared to other sectors, e.g., Li is mostly consumed by the glass and ceramics industries, and only 1% feeds the battery industry. Ni is mainly used to produce different stainless and alloy steels, which is the biggest user of primary and scrap Ni [21]. The Ni consumption in the EU battery industry is negligible compared to the other sectors. Within this context, the competition among sectors which use the same raw materials for different purpose could play an important role to define the supply risk and/or economic vulnerability of battery raw materials. According to Nassar et al. [87], the industry's relative vulnerability can be quantified by calculating "the ratio of an industry's expenditures for a given commodity relative to that industry's profitability". This approach seems to fit well to a sectorial level such as battery manufacturing, although it requires a well-structured breakdown of economic sectors and data availability.

It is worth noting that assumptions about, e.g., the share of materials recovered in a closed/open-loop, evolution of recycling efficiency, extended lifetime of LIBs, materials content, variation of LIBs capacity, supply data, etc., are based on available literature and often refer to past and current situation. Modularity and parameters in the developed models are used: (1) to make the application of selected tools flexible and updatable according to available data; and (2) to address uncertainty through the creation of different scenarios. Relevant parameters identified in the performed analysis are energy mix, recycling efficiency and supply risk. More efforts are needed to further explore the effects of, e.g., improved lifetime, one of the key parameters identified in the literature [10,42] and future supply risk.

Finally, the proposed approach could in the future be used to enlarge the performed analysis covering more aspects, e.g., different deployment scenarios, improved EoL practices and new materials. An interesting example on new materials expected to be used in manufacturing future traction LIBs is Niobium [88]. Niobium, which belongs to the CRMs list [19], is mainly produced in Brazil and Canada, even though exploration activities are taking place in several countries [89]. Currently, niobium is not massively used for manufacturing batteries, but its application has some potential, e.g., for anodes [88,90,91], which means a potential increase of demand. Very few studies on niobium in batteries are available and environmental impacts are almost unknown [92]. According to the authors' knowledge, the GWP of ferro-niobium were provided by Dolganova [93], who used primary data provided by CBMM (Companhia Brasileira de Metalurgia e Mineração). Such gap of data requires more efforts in assessing the environmental impacts of niobium for batteries, but also in investigating various aspects affecting both MFA and its supply, e.g., niobium content in anodes, which materials it will substitute (see Table 2).

Overall, the combination of information provided by different tools and experts in different fields, as well as the assessment of different scenarios, is recommended to identify key aspects that can contribute to decrease the environmental impacts of a strategic sector for the EU, such as mobility.

6. Conclusions

LCA, MFA and criticality (supply risk) considerations are contrasted and discussed in a mutually interactive manner, with focus on the environmental impacts of traction LIBs and some of their constituting materials in the current and future EU fleet from different perspectives. Modules and parameters used in both the LCA and MFA models allow identifying relevant aspects in terms of impacts and assessing different scenarios, including different possible EoL options. Criticality mainly contributes to prioritise materials to be studied and improve knowledge on the potential contribution of SRMs in the supply risk for specific materials.

The Supplementary Materials support the replicability of the assessment, as well as future studies. Users can easily access to relevant information on both the LCI and LCIA results, being able to: (1) directly compare input/output flows of the life-cycle of different LIBs; (2) further enlarge the assessment; and (3) modify and/or add input/output flows. The dynamic MFA tool describes the value chain of traction LIBs in the EU including all EoL options, and it allows quantifying the stocks and flows of LIBs/energy storage capacity/embedded materials in the EU for different scenarios and under different assumptions. Criticality assessments run worldwide suggested to initially focus on Li and Ni as key raw materials embedded in LIBs. The paper provided information of the role of recycling as mitigation factor of the supply risk of both Li and Ni used in LIBs.

Results point out that the life-cycle GWP of traction LIBs will likely improve in the future, mainly due to more environmental-friendly energy mix and improved LIBs' recycling. Even though second-use will postpone the availability of materials for recycling, recycling improvement can importantly increase the flows of SRMs, boosting resource efficiency and keeping the values of materials in the EU. In addition, recycling can further contribute to reducing the supply risk of both Li and Ni in 2050. Such enhancements are related to both the development of recycling technologies at large scale (in case of Li) and the higher grade of recovered materials (in case of Ni).

Lack of knowledge about the LIBs value chain, lack of robust data as input for the assessments and adoption of data based on past and current trends are importantly reflected in the obtained results, thus suggesting to intensify research efforts. Obviously, further methodological work, concerning, e.g., scenario setting, uptake and impacts of future technologies, adoption of consequential LCA, substitution of materials, effects of stocks and flows, etc. would also be advisable to appropriately support more prospective decision-making.

Both the literature review and the performed analysis confirmed the added-value of performing a multi-criteria analysis, especially when addressing complex systems. Involvement of different expertise and running scenarios varying relevant parameters are keys in updating both the modelling and the input data in order to provide reliable information to identify circular economy aspects that support decision making to properly manage the whole value chain of a strategic sector for the EU, such as batteries for e-mobility.

Supplementary Materials: The following tables in excel file that provides the main assumptions of the assessment reported in this paper are available online at http://www.mdpi.com/1996-1073/13/10/2513/s1. Moreover, to make the LCA replicable and updatable, the LCI of each LIB chemistry is provided in the spreadsheets of this excel file, where each input/output flow of material, energy, waste and emission is related to 1 kg of battery pack.Each flow is connected to the related impacts, enabling the automatic calculation of the impact of the battery life-cycle. The LCIA for all the impact categories included in the assessment are reported in the last spreadsheet ("Impact_calculation").

Author Contributions: Conceptualization, S.B., G.A.B. and I.B.; methodology, S.B., G.A.B., I.B. and U.E.; validation, G.A.B. and F.M.; investigation, S.B., G.A.B., I.B. and U.E.; resources, F.M.; data curation, S.B., I.B., U.E. and S.C.; writing—original draft preparation, S.B., I.B. and U.E.; writing—review and editing, G.A.B, F.M. and S.C.; visualization, S.B., I.B. and U.E.; supervision, G.A.B. and F.M.; project administration, G.A.B. and F.M. All authors have read and agreed to the published version of the manuscript.

Funding: This research received no external funding.

Conflicts of Interest: The authors declare no conflict of interest.

Disclaimer: The views expressed in the article are personal and do not necessarily reflect an official position of the European Commission.

References

1. EC. *Communication from the Commission to the European Parliament, the European Council, the Council, the European economic and social Committee and the Committee of the Regions. The European Green Deal; COM(2019) 640;* Publications Office of the EU: Luxemburg, 2019.
2. EC. *Reflection Paper—Towards a Sustainable Europe By 2030; COM(2019) 22 final;* Publications Office of the EU: Luxemburg, 2019.
3. Alonso Raposo, M.; Ciuffo, B.; Ardente, F.; Aurambout, J.P.; Baldini, G.; Braun, R.; Christidis, P.; Christodoulou, A.; Duboz, A.; Felici, S.; et al. *The Future of Road Transport—Implications of Automated, Connected, Low-Carbon and Shared Mobility;* Publications Office of the EU: Luxemburg, 2019; ISBN 978-92-76-03409-4.
4. EC. *In-Depth Analysis in Support of the Commission Communication COM(2018) 773—A Clean Planet for all A European Long-Term Strategic Vision for a Prosperous, Modern, Competitive and Climate Neutral Economy;* Publications Office of the EU: Luxemburg, 2018.
5. Zubi, G.; Dufo-López, R.; Carvalho, M.; Pasaoglu, G. The lithium-ion battery: State of the art and future perspectives. *Renew. Sustain. Energy Rev.* **2018**, *89*, 292–308. [CrossRef]
6. Witkamp, B.; van Gijlswijk, R.; Bolech, M.; Coosemans, T.; Hooftman, N. The transition to a Zero Emission Vehicles fleet for cars in the EU by 2050. In *Pathways and Impacts: An Evaluation of Forecasts and Backcasting the COP21 Commitments;* European Alternative Fuel Observatory (EAFO): Brussels, Belgium, 2017.
7. Cano, Z.P.; Banham, D.; Ye, S.; Hintennach, A.; Lu, J.; Fowler, M.; Chen, Z. Batteries and fuel cells for emerging electric vehicle markets. *Nat. Energy* **2018**, *3*, 279–289. [CrossRef]
8. Thiel, C.; Nijs, W.; Simoes, S.; Schmidt, J.; van Zyl, A.; Schmid, E. The impact of the EU car CO2 regulation on the energy system and the role of electro-mobility to achieve transport decarbonisation. *Energy Policy* **2016**, *96*, 153–166. [CrossRef]

9. Bobba, S.; Mathieux, F.; Ardente, F.; Blengini, G.A.; Cusenza, M.A.; Podias, A.; Pfrang, A. Life Cycle Assessment of repurposed electric vehicle batteries: An adapted method based on modelling energy flows. *J. Energy Storage* **2018**, *19*, 213–225. [CrossRef]

10. Podias, A.; Pfrang, A.; Di Persio, F.; Kriston, A.; Bobba, S.; Mathieux, F.; Messagie, M.; Boon-Brett, L. Sustainability Assessment of Second Use Applications of Automotive Batteries: Ageing of Li-Ion Battery Cells in Automotive and Grid-Scale Applications. *World Electr. Veh. J.* **2018**, *9*, 24. [CrossRef]

11. EUROBAT. E-Mobility Battery R&D Roadmap 2030—Battery Technology for Vehicle Applications; Association of European Automotive and Industrial Battery Manufacturers (EUROBAT). Available online: https://eurobat.org/sites/default/files/e-mobility_roadmap_presentation_olivier_amiel_final.pdf (accessed on 2015).

12. Bobba, S.; Mathieux, F.; Blengini, G.A. How will second-use of batteries affect stocks and flows in the EU? A model for traction Li-ion batteries. *Resour. Conserv. Recycl.* **2019**, *145*, 279–291. [CrossRef] [PubMed]

13. EC. *Communication from the Commission to the European Parliament, the European Council, the Council, the European economic and social Committee and the Committee of the Regions; A new Circular Economy Action Plan; For a cleaner and more competitive Europe; COM: 2020/98 final*; Publications Office of the EU: Luxemburg, 2020.

14. Winslow, K.M.; Laux, S.J.; Townsend, T.G. A review on the growing concern and potential management strategies of waste lithium-ion batteries. *Resour. Conserv. Recycl.* **2018**, *129*, 263–277. [CrossRef]

15. EC. *Communication from the Commission to the European Parliament, the European Council, the Council, the European economic and social Committee and the Committee of the Regions; A New Industrial Strategy for Europe; COM: 2020; 102 final*; Publications Office of the EU: Luxemburg, 2020.

16. Schrijvers, D.; Hool, A.; Blengini, G.A.; Chen, W.-Q.; Dewulf, J.; Eggert, R.; van Ellen, L.; Gauss, R.; Goddin, J.; Habib, K.; et al. A review of methods and data to determine raw material criticality. *Resour. Conserv. Recycl.* **2020**, *155*, 104617. [CrossRef]

17. IEA. *Global EV Outlook 2018: Towards Cross-Modal Electrification*; International Energy Agency (IEA): Paris, France, 2018; ISBN 9789264302365.

18. Olivetti, E.A.; Ceder, G.; Gaustad, G.G.; Fu, X. Lithium-Ion Battery Supply Chain Considerations: Analysis of Potential Bottlenecks in Critical Metals. *Joule* **2017**, *1*, 229–243. [CrossRef]

19. EC. *Communication from the Commission to the European Parliament, the Council, the European Economic and Social Committee and the Committee of the Regions on the 2017 list of Critical Raw Materials for the EU; COM: 2017; 490 final*; Publications Office of the EU: Luxemburg, 2017.

20. Mistry, M.; Gediga, J.; Boonzaier, S. Life cycle assessment of nickel products. *Int. J. Life Cycle Assess.* **2016**, *21*, 1559–1572. [CrossRef]

21. INSG. International Nickel Study Group—The World Nickel Factbook 2018. 2018. Available online: https://insg.org/wp-content/uploads/2019/03/publist_The-World-Nickel-Factbook-2018.pdf (accessed on 15 May 2020).

22. Ellingsen, L.A.-W.; Hung, C.R. Research for TRAN Committee—Battery-Powered Electric Vehicles: Market Development and Lifecycle Emissions. European Union, 2018. Available online: https://www.europarl.europa.eu/RegData/etudes/STUD/2018/617457/IPOL_STU(2018)617457_EN.pdf (accessed on 15 May 2020).

23. Mathieux, F.; Ardente, F.; Bobba, S.; Nuss, P.; Blengini, G.A.; Alves Dias, P.; Blagoeva, D.; Torres De Matos, C.; Wittmer, D.; Pavel, C.; et al. *Critical Raw Materials and the Circular Economy—Background Report; JRC-EC (Joint Research Centre—European Commission) Science-for-Policy Report, EUR 28832 EN*; Publications Office of the European Union: Luxembourg, 2017; ISBN 9789279742828.

24. Ellingsen, L.A.-W.; Majeau-Bettez, G.; Singh, B.; Srivastava, A.K.; Valøen, L.O.; Strømman, A.H. Life Cycle Assessment of a Lithium-Ion Battery Vehicle Pack: LCA of a Li-Ion Battery Vehicle Pack. *J. Ind. Ecol.* **2014**, *18*, 113–124. [CrossRef]

25. Zackrisson, M.; Avellán, L.; Orlenius, J. Life cycle assessment of lithium-ion batteries for plug-in hybrid electric vehicles – Critical issues. *J. Clean. Prod.* **2010**, *18*, 1519–1529. [CrossRef]

26. Majeau-Bettez, G.; Hawkins, T.R.; Strømman, A.H. Life Cycle Environmental Assessment of Lithium-Ion and Nickel Metal Hydride Batteries for Plug-In Hybrid and Battery Electric Vehicles. *Environ. Sci. Technol.* **2011**, *45*, 4548–4554. [CrossRef]

27. Ruiz, V.; Boon-Brett, L.; Steen, M.; Berghe, L. Van den Putting Science into Standards: Workshop—Summary & Outcomes. In Proceedings of the Driving Towards Decarbonisation of Transport: Safety, Performance, Second life and Recycling of Automotive Batteries for e-Vehicles, Workshop Organised by JRC and CEN-CENELEC JRC, Petten, The Netherlands, 22–23 September 2016.

28. Marmiroli, B.; Messagie, M.; Dotelli, G.; Van Mierlo, J. Electricity Generation in LCA of Electric Vehicles: A Review. *Appl. Sci.* **2018**, *8*, 1384. [CrossRef]

29. Ardente, F.; Tecchio, P.; Bobba, S.; Mathieux, F. Assessment of resource efficiency in a life cycle perspective: The case of reuse. In Proceedings of the Atti del XI Convegno della Rete Italiana LCA—Resource Efficiency e Sustainable Development Goals: Il ruolo del Life Cycle Thinking, ENEA, Siena, Italy, 22–23 June 2017; ISBN 978-88-8286-352-4.

30. Nuss, P.; Blengini, G.A. Towards better monitoring of technology critical elements in Europe: Coupling of natural and anthropogenic cycles. *Sci. Total Environ.* **2018**, *613*, 569–578. [CrossRef] [PubMed]

31. De Meester, S.; Nachtergaele, P.; Debaveye, S.; Vos, P.; Dewulf, J. Using material flow analysis and life cycle assessment in decision support: A case study on WEEE valorization in Belgium. *Resour. Conserv. Recycl.* **2019**, *142*, 1–9. [CrossRef]

32. Mancini, L.; Benini, L.; Sala, S. Resource footprint of Europe: Complementarity of material flow analysis and life cycle assessment for policy support. *Environ. Sci. Policy* **2015**, *54*, 367–376. [CrossRef]

33. Ramirez, D.A.B.; Ochoa, G.E.V.; Peña, A.R.; Escorcia, Y.C. Bibliometric analysis of nearly a decade of research in electric vehicles: A dynamic approach. *ARPN J. Eng. Appl. Sci.* **2018**, *13*, 4730–4736.

34. Zhao, X.; Wang, S.; Wang, X. Characteristics and Trends of Research on New Energy Vehicle Reliability Based on the Web of Science. *Sustainability* **2018**, *10*, 3560. [CrossRef]

35. Schmitt, G.; Scott, J.; Davis, A.; Utz, T. Patents and progress; intellectual property showing the future of electric vehicles. In Proceedings of the EVS29 Symposium, Montréal, QC, Canada, 19–22 June 2016; Volume 8, pp. 635–645.

36. Messagie, M.; Boureima, F.S.; Coosemans, T.; Macharis, C.; Mierlo, J. Van A range-based vehicle life cycle assessment incorporating variability in the environmental assessment of different vehicle technologies and fuels. *Energies* **2014**, *7*, 1467–1482. [CrossRef]

37. Harvey, L.D.D. Resource implications of alternative strategies for achieving zero greenhouse gas emissions from light-duty vehicles by 2060. *Appl. Energy* **2018**, *212*, 663–679. [CrossRef]

38. Notter, D.A.; Gauch, M.; Widmer, R.; Wäger, P.; Stamp, A.; Zah, R.; Althaus, H.-J. Contribution of Li-Ion Batteries to the Environmental Impact of Electric Vehicles. *Environ. Sci. Technol.* **2010**, *44*, 6550–6556. [CrossRef] [PubMed]

39. Girardi, P.; Gargiulo, A.; Brambilla, P.C. A comparative LCA of an electric vehicle and an internal combustion engine vehicle using the appropriate power mix: The Italian case study. *Int. J. Life Cycle Assess.* **2015**, *20*, 1127–1142. [CrossRef]

40. ISO 14040:2006. *Environmental Management—Life Cycle Assessment—Requirements and Guidelines*; UNI EN ISO (International Standard Organisation), 2006.

41. Recharge Association. PEFCR—Product Environmental Footprint Category Rules for High Specific Energy Rechargeable Batteries for Mobile Applications. 2018. Available online: https://ec.europa.eu/environment/eussd/smgp/pdf/PEFCR_Batteries.pdf (accessed on 15 May 2020).

42. Peters, J.F.; Baumann, M.; Zimmermann, B.; Braun, J.; Weil, M. The environmental impact of Li-Ion batteries and the role of key parameters—A review. *Renew. Sustain. Energy Rev.* **2017**, *67*, 491–506. [CrossRef]

43. Ellingsen, L.A.-W.; Hung, C.R.; Strømman, A.H. Identifying key assumptions and differences in life cycle assessment studies of lithium-ion traction batteries with focus on greenhouse gas emissions. *Transp. Res. Part D Transp. Environ.* **2017**, *55*, 82–90. [CrossRef]

44. Peters, J.F.; Weil, M. Providing a common base for life cycle assessments of Li-Ion batteries. *J. Clean. Prod.* **2018**, *171*, 704–713. [CrossRef]

45. Cusenza, M.A.; Bobba, S.; Ardente, F.; Cellura, M.; Di Persio, F. Energy and environmental assessment of a traction lithium-ion battery pack for plug-in hybrid electric vehicles. *J. Clean. Prod.* **2019**, *215*, 634–649. [CrossRef]

46. Philippot, M.; Álvarez, G.; Ayerbe, E.; Van Mierlo, J.; Messagie, M. Eco-Efficiency of a Lithium-Ion Battery for Electric Vehicles: Influence of Manufacturing Country and Commodity Prices on GHG Emissions and Costs. *Batteries* **2019**, *5*, 23. [CrossRef]

47. Deng, Y.; Li, J.; Li, T.; Gao, X.; Yuan, C. Life cycle assessment of lithium sulfur battery for electric vehicles. *J. Power Sources* **2017**, *343*, 284–295. [CrossRef]

48. Wang, Q.; Liu, W.; Yuan, X.; Tang, H.; Tang, Y.; Wang, M.; Zuo, J.; Song, Z.; Sun, J. Environmental impact analysis and process optimization of batteries based on life cycle assessment. *J. Clean. Prod.* **2018**, *174*, 1262–1273. [CrossRef]

49. EC. *Commission Staff working document Accompanying the document Communication from the Commission to the European Parliament, the Council, the European Economic and Social Committee and the Committee of the Regions EUROPE ON THE MOVE An Agenda for a Socially; SWD: 2017; 177 final*; Publications Office of the EU: Luxemburg, 2017.

50. Faria, R.; Marques, P.; Garcia, R.; Moura, P.; Freire, F.; Delgado, J.; de Almeida, A.T. Primary and secondary use of electric mobility batteries from a life cycle perspective. *J. Power Sources* **2014**, *262*, 169–177. [CrossRef]

51. Erkisi-Arici, S.; Egede, P.; Cerdas, F.; Kaluza, A.; Herrmann, C. Life Cycle Assessment of Electric Vehicles—The Influence of Regional Aspects and Future Renewable Energy Targets. In Proceedings of the EVS30 Symposium, Stuttgart, Germany, 9–11 October 2017.

52. Ahmadi, L.; Yip, A.; Fowler, M.; Young, S.B.; Fraser, R.A. Environmental feasibility of re-use of electric vehicle batteries. *Sustain. Energy Technol. Assess.* **2014**, *6*, 64–74. [CrossRef]

53. Casals, L.C.; Amante García, B.; Canal, C. Second life batteries lifespan: Rest of useful life and environmental analysis. *J. Environ. Manag.* **2019**, *232*, 354–363. [CrossRef]

54. Koch-Ciobotaru, C.; Saez-de-Ibarra, A.; Martinez-Laserna, E.; Stroe, D.-I.; Swierczynski, M.; Rodriguez, P. Second life battery energy storage system for enhancing renewable energy grid integration. In Proceedings of the 2015 IEEE Energy Conversion Congress and Exposition (ECCE), Montreal, QC, Canada, 20–24 September 2015; pp. 78–84. [CrossRef]

55. Weniger, J.; Tjaden, T.; Quaschning, V. Sizing of residential PV battery systems. *Energy Procedia* **2014**, *46*, 78–87. [CrossRef]

56. UMICORE. Battery Recycling—Our Recycling Process. Available online: http://csm.umicore.com/en/recycling/battery-recycling/our-recycling-process/ (accessed on 15 February 2020).

57. EC. *The Raw Materials Initiative—Meeting our Critical Needs for Growth and Jobs in Europe; COM: 2008; 699 final*; Publications Office of the EU: Luxemburg, 2008.

58. Blengini, G.A.; Blagoeva, D.; Dewulf, J.; Torres de Matos, C.; Baranzelli, C.; Ciupagea, C.; Dias, P.; Kayam, Y.; Latunussa, C.E.L.; Mancini, L.; et al. Methodology for establishing the EU list of Critical Raw Materials. *Publ. Off. Eur. Union* **2017**.

59. Turner, D.A.; Williams, I.D.; Kemp, S. Combined material flow analysis and life cycle assessment as a support tool for solid waste management decision making. *J. Clean. Prod.* **2016**, *129*, 234–248. [CrossRef]

60. Mancini, L.; Sala, S.; Recchioni, M.; Benini, L.; Goralczyk, M.; Pennington, D. Potential of life cycle assessment for supporting the management of critical raw materials. *Int. J. Life Cycle Assess.* **2015**, *20*, 100–116. [CrossRef]

61. Golroudbary, S.R.; Calisaya-Azpilcueta, D.; Kraslawski, A. The life cycle of energy consumption and greenhouse gas emissions from critical minerals recycling: Case of lithium-ion batteries. *Procedia CIRP* **2019**, *80*, 316–321. [CrossRef]

62. Pillot, C. Lithium ion Battery Raw Material Supply and Demand 2016–2025. In Proceedings of the 7th International AABC Advanced Automotive battery Conference, Mainz, Germany, 12–16 January 2017.

63. Song, J.; Yan, W.; Cao, H.; Song, Q.; Ding, H.; Lv, Z.; Zhang, Y.; Sun, Z. Material flow analysis on critical raw materials of lithium-ion batteries in China. *J. Clean. Prod.* **2019**, *215*, 570–581. [CrossRef]

64. Helbig, C.; Bradshaw, A.M.; Wietschel, L.; Thorenz, A.; Tuma, A. Supply risks associated with lithium-ion battery materials. *J. Clean. Prod.* **2018**, *172*, 274–286. [CrossRef]

65. Gemechu, E.D.; Sonnemann, G.; Young, S.B. Geopolitical-related supply risk assessment as a complement to environmental impact assessment: The case of electric vehicles. *Int. J. Life Cycle Assess.* **2017**, *22*, 31–39. [CrossRef]

66. Eynard, U.; Bobba, S.; Cusenza, M.A.; Blengini, G.A. Lithium-ion batteries for electric vehicles: Combining Environmental and Social Life Cycle Assessments. In Proceedings of the Life Cycle Thinking in Decision-Making for Sustainability: From Public Policies to Private Businesses, Messina, Italy, 11–12 June 2018.

67. Gaines, L. Lithium-ion battery recycling processes: Research towards a sustainable course. *Sustain. Mater. Technol.* **2018**, *17*, e00068. [CrossRef]

68. Bauer, C. *Ökobilanz von Lithium-Ionen Batterien*; Paul Scherrer Institut, Labor für Energiesystem-Analysen (LEA): Villigen, Switzerland, 2010.

69. Banja, M.; Jégard, M. Renewable technologies in the EU electricity sector: trends and projections. In *Analysis in the Framework of the EU 2030 Climate and Energy Strategy*; EUR 28897 EN; Publications Office of the European Union: Luxemburg, 2017.

70. Decker, M.; Vasakova, L. Energy Roadmap 2050. Impact assessment and scenario analysis. In *Eur. Comm. Energy Unit A1 Energy Policy Anal Commission Staff Working Paper, SEC(2011) 1565 final*; Publications Office of the European Union: Luxemburg, 2011.

71. Nakamura, S.; Kondo, Y.; Kagawa, S.; Matsubae, K.; Nakajima, K.; Nagasaka, T. MaTrace: Tracing the fate of materials over time and across products in open-loop recycling. *Environ. Sci. Technol.* **2014**, *48*, 7207–7214. [CrossRef] [PubMed]

72. EC. *Report on Raw Materials for Battery Applications*; Commission Staff Working Document SWD(2018) 245; Publications Office of the European Union: Brussels, Belgium, 2018.

73. Dahllöf, L.; Romare, M.; Wu, A. *Mapping Lithium-Ion Batteries for Vehicles: A study of their Fate in the Nordic Countries*; Lithium-Ion Batter; Nordic Council of Ministers: Copenhagen, Demark, 2019.

74. Lebedeva, N.; Di Persio, F.; Boon-Brett, L. *Lithium Ion Battery Value Chain and Related Opportunities for Europe*; EUR 28534 EN; Publications Office of the European Union: Luxembourg, 2017; ISBN 9789279669484.

75. Schitech Europa 99% Metal Purity Achieved from Recycled Batteries. Available online: https://www.scitecheuropa.eu/99-metal-purity-achieved-from-recycled-batteries/98373/ (accessed on 14 February 2020).

76. Campagnol, N.; Hoffman, K.; Lala, A.; Ramsbottom, O. The Future of Nickel: A Class Act. 2017. Available online: https://www.mckinsey.com/industries/metals-and-mining/our-insights/the-future-of-nickel-a-class-act (accessed on 15 May 2020).

77. Berckmans, G.; Messagie, M.; Smekens, J.; Omar, N.; Vanhaverbeke, L.; Van Mierlo, J. Cost projection of state of the art lithium-ion batteries for electric vehicles up to 2030. *Energies* **2017**, *10*, 1314. [CrossRef]

78. Green Car Congress Audi and Umicore Start Closed Loop for Cobalt and Nickel, More than 90% of Co and Ni in e-tron Batteries can be Recovered. Available online: https://www.greencarcongress.com/2019/12/20191218-audiumicore.html (accessed on 15 February 2020).

79. Bobba, S.; Andreas, P.; Persio, F.D.; Maarten, M.; Paolo, T.; Cusenza, M.; Umberto, E.; Fabrice, M. *Sustainability Assessment of Second Life Applications of Automotive Batteries (SASLAB)*; Publications Office of the European Union: Luxembourg, 2018; ISBN 978-92-79-92835-2.

80. Arvidsson, R.; Janssen, M.; Svanström, M.; Johansson, P.; Sandén, B.A. Energy use and climate change improvements of Li/S batteries based on life cycle assessment. *J. Power Sources* **2018**, *383*, 87–92. [CrossRef]

81. Kim, H.C.; Wallington, T.J.; Arsenault, R.; Bae, C.; Ahn, S.; Lee, J. Cradle-to-Gate Emissions from a Commercial Electric Vehicle Li-Ion Battery: A Comparative Analysis. *Environ. Sci. Technol.* **2016**, *50*, 7715–7722. [CrossRef]

82. Dai, Q.; Kelly, J.C.; Gaines, L.; Wang, M. Life cycle analysis of lithium-ion batteries for automotive applications. *Batteries* **2019**, *5*, 48. [CrossRef]

83. Kurland, S.D. Energy use for GWh-scale lithium-ion battery production. *Environ. Res. Commun.* **2019**, *2*, 12001. [CrossRef]

84. EC. *Study on the Review of the List of Critical Raw Materials. Non-Critical Raw Materials Factsheets*; Publications Office of the European Union: Luxembourg, 2017; ISBN 978-92-79-47937-3.

85. Mayer, A.; Haas, W.; Wiedenhofer, D.; Krausmann, F.; Nuss, P.; Blengini, G.A. Measuring Progress towards a Circular Economy: A Monitoring Framework for Economy-wide Material Loop Closing in the EU28. *J. Ind. Ecol.* **2018**, *3*, 62–76. [CrossRef]

86. Kuczenski, B.; Davis, C.B.; Rivela, B.; Janowicz, K. Semantic catalogs for life cycle assessment data. *J. Clean. Prod.* **2016**, *137*, 1109–1117. [CrossRef]

87. Nassar, N.T.; Brainard, J.; Gulley, A.; Manley, R.; Matos, G.; Lederer, G.; Bird, L.R.; Pineault, D.; Alonso, E.; Gambogi, J.; et al. Evaluating the mineral commodity supply risk of the U.S. Manufacturing sector. *Sci. Adv.* **2020**, *6*, 8. [CrossRef] [PubMed]

88. Li, J.; Liu, W.W.; Zhou, H.M.; Liu, Z.Z.; Chen, B.R.; Sun, W.J. Anode material NbO for Li-ion battery and its electrochemical properties. *Rare Met.* **2018**, *37*, 118–122. [CrossRef]

89. *Deloitte Sustainability, British Geological Survey (BGS), Bureau de Recherches Géologiques et Minières (BRGM), Netherlands Organisation for Applied Scientific Research (TNO) Study on the review of the list of Critical Raw Materials*; Critical Raw Materials Factsheets; Publications Office of the European Union: Luxembourg, 2017.

90. CBMM Niobium application in Lithium-ion batteries: Anodes, cathodes and solid-state electrolytes. In Proceedings of the International Conference on Niobium Based Batteries, Beijing, China, 11 August 2019.

91. NiobiumTech. Niobium in Lithium-ion Batteries. Available online: https://niobium.tech/en/Pages/Gateway-Pages/PDF/Briefings/Niobium_in_Li-Ion_Batteries (accessed on 15 February 2018).
92. Nuss, P.; Eckelman, M.J. Life cycle assessment of metals: A scientific synthesis. *PLoS ONE* **2014**, *9*, e101298. [CrossRef] [PubMed]
93. Dolganova, I.; Bosch, F.; Bach, V.; Baitz, M.; Finkbeiner, M. Life cycle assessment of ferro niobium. *Int. J. Life Cycle Assess.* **2020**, *25*, 611–619. [CrossRef]

Article

Chronological Transition of Relationship between Intracity Lifecycle Transport Energy Efficiency and Population Density

Shoki Kosai [1,2]**, Muku Yuasa** [2] **and Eiji Yamasue** [2,*]

[1] Graduate School of Energy Science, Kyoto University, Kyoto 606-8501, Japan; kosai0203@gmail.com
[2] Department of Mechanical Engineering, College of Science and Engineering, Ritsumeikan University, Shiga 525-8577, Japan; admin@yamasue-lab.com
* Correspondence: yamasue@fc.ritsumei.ac.jp; Tel.: +81-77-561-4693

Received: 27 March 2020; Accepted: 18 April 2020; Published: 22 April 2020

Abstract: Interests in evaluating lifecycle energy use in urban transport have been growing as a research topic. Various studies have evaluated the relationship between the intracity transport energy use and population density and commonly identified its negative correlation. However, a diachronic transition in an individual city has yet to be fully analyzed. As such, this study employed transport energy intensity widely used for evaluating transport energy efficiency and obtained the transport energy intensity for each transportation means including walk, bicycle, automobile (conventional vehicles, electric vehicles, hybrid vehicles, and fuel cell vehicles), bus and electric train by considering the lifecycle energy consumption. Then, the intracity lifecycle transport energy intensity of 38 cities in Japan in 1987–2015 was computed, assuming that the cause of diachronic transition of intracity transport energy efficiency is the modal shifting and electricity mix change. As a result, the greater level of population density was associated with the lower intracity transport energy intensity in Japanese cities. The negative slope of its regression line increased over time since the intracity lifecycle transport energy intensity in cities with low population density continuously increased without any significant change of population density. Finally, this study discussed the strategic implications particularly in regional areas to improve the intracity lifecycle transport energy efficiency.

Keywords: metropolitan area; in-city; transport energy intensity; well to wheel; material structure

---

1. Introduction

In the last few decades, the global energy landscape in cities has dramatically changed due to the expansion of the global population and heavy industrialization [1]. Sustainable urban development is difficult to be achieved in view of current climate change and damages to the ecological environment [2]. Given the utmost importance of developing sustainable cities with high energy efficiency for policy makers [3], interests in evaluating energy consumption in cities have increasingly been growing as a research topic [4,5].

In particular, due to the rapid motorization and urbanization, the energy consumption in the city transport sector has been growing in the fastest manner in all energy-intensive sectors [6,7]. The transport sector is considered the most complex sector in which to reduce emissions [8] and the fuel use in the road transportation in 2050 is expected to increase by two times compared to that in 2010 [9]. Therefore, the improvement of intracity transport energy efficiency through the analysis of energy consumption in urban transport is required [10,11].

Various studies evaluated the relationship in the urban transport sector between the energy consumption and the city-related variables [12–26]. Some studies employed the top-down approach at the macro level by using cross-section data in cities (e.g., [16]), while other studies conducted the bottom-up approach by collecting the individual trip data (e.g., [19]). Energy-related indicators include transport energy consumption (e.g., gasoline consumption) and carbon dioxide emissions. Meanwhile, the city-related variables cover various perspectives including the urban form such as population, density, area as well as the social form such as gender, age, income, and job.

Although these relationships have been reviewed and summarized in detail in many studies (see [12,22,24]), in particular, a negative correlation between intracity transport energy consumption and population density was commonly identified in earlier studies.

However, a diachronic transition of energy-related value and urban form in an individual city has yet to be fully analyzed. Some papers assessed the transport energy consumption in a certain time range and identified the growth rate of transport energy consumption and land form in the provincial areas [15,20,26], whereas the transition of an individual city was not considered.

In addition, lifecycle thinking of energy consumption in the city transportation is of significant importance. One of the major lifecycle approaches accounts for the entire fuel consumption covering from well to wheel (WTW) [27]. The WTW fuel consumption has been analyzed for various transportation modes including automobiles [28] (conventional vehicles [29], hybrid vehicles [30], electric vehicles [31] and fuel cell vehicles [32]), trains [33] and bicycles [34]. The other lifecycle approach accounts for the energy consumption during all lifecycle phases including manufacture, operation, maintenance, and end-of-life [35]. These lifecycle approaches need to be integrated in the transport energy consumption in cities for an encompassing understanding of city transport.

One of the approaches for measuring the intracity transport energy efficiency is the consideration of modal split in a given area [36]. By using modal split, Schipper and his colleagues have analyzed the national transport energy intensity as a form of transport energy efficiency in Spain, Australia, France, United Kingdom, Japan and United States [37,38]. While various studies attempted to classify the determinants of urban modal choice from the perspectives of socio demography, spatial characteristics and socio psychology to further explore modal split [39,40], this study uses the modal split in the calculation to evaluate the intracity transport energy efficiency.

This study focuses on the passenger transport energy efficiency in Japanese cities. Limited studies on the urban transport energy consumption in Japan have been reported so far. Waygood et al., evaluated CO_2 emissions in the transport sector in Osaka city considering the modal share [41]. Since Yang et al. pointed out the difficulty of applying the empirical studies in one country to another in this topic [26], the focus on Japanese cities would assist in the exploration of new spatial scope.

As such, the objective of this study is to analyze the chronological transition of relationship between intracity transportation energy efficiency and population density in Japanese cities by considering the lifecycle energy consumption of various transportation means. This study assumes that the cause of diachronic transition of intracity transport energy efficiency is the modal shifting and electricity mix change.

This study is structure as follows: Section 2 illustrates the methodology consisting of boundary of transport energy consumption, introduction of intracity transportation mode, the calculation of intracity transport energy efficiency, and data collection. The transport energy efficiency of each transportation mode is presented in Section 3. Section 4 presents the chronological relationship between intracity transport energy efficiency and population density. Section 5 discusses the potential implications of the current city transportation in Japan based on the obtained results and highlights the research limitations and future perspectives. Section 6 concludes the paper.

2. Materials and Methods

2.1. Boundary of Energy Consumption

A lifecycle phase of transportation means includes the material structure, assembly, operation, maintenance, and disposal. Most existing studies on the transport energy consumption in cities have focused on the tank to wheel (TTW) energy consumption during the operational phase. Meanwhile, the significant impact of energy consumption during the manufacturing phase, especially material structure, has been widely highlighted [28,42–44]. As such, this study considers energy consumption during the phases of material structure and operation among all lifecycle stages. The other phases are out of scope.

In addition, this study extends its boundary from TTW to WTW during the operational phase. In particular, energy consumption from well to tank (WTT) represents the energy input for the fuel production, which is the same range of energy consumption for material structure during the phase of manufacturing.

Although energy input for the development of transport infrastructure is significant and varies depending on the transportation modes, this study does not consider the energy input for the infrastructure.

2.2. Transporation Mode

Transportation modes in Japanese cities are mainly comprised of road, and urban rail. The transportation means and their corresponding fuel types in each of the transportation modes are summarized in Table 1.

Table 1. Transportation modes, means and fuel types.

Transportation Mode	Transportation Means	Fuel Type
	Walk	-
	Bicycle	-
	Automobile	
Road	Conventional vehicle (CV)	Gasoline
	Electric vehicle (EV)	Electricity
	Hybrid vehicle (HV)	Gasoline
	Fuel cell vehicle (FCV)	Hydrogen
	Bus	Light diesel
Urban rail	Electric train	Electricity

2.3. Intracity Transport Energy Efficiency

This study employs the transport energy intensity which has been widely utilized for evaluating the transport energy efficiency [45,46]. The transport energy intensity of transportation means, referring to TEI (J/kg/m), is represented by energy consumption per distance per weights of both passengers and means.

The detailed steps for obtaining transport energy intensity are presented as follows. Subscripts and abbreviations used in this paper are summarized in Table 2.

First, energy consumption for the material structure was calculated based on the weight and rate of the composition of the transportation means and the energy consumption rate of composition, using the following equation:

$$Q_{material} = \sum (Mc_i e_i) \tag{1}$$

Energy consumption for the material structure accounts for the energy input of material production required for a vehicle body. Given that the inclusion of the manufacturing phase causes a change in the transport energy efficiency with operational duration, energy consumption during the operational phase was calculated on the basis of time series. TTW energy consumption during the operational phase per year was obtained by using the following equation:

$$Q_{operation}^{TTW} = \frac{Ld}{k} \tag{2}$$

TTW energy consumption during the operational phase was multiplied with the fuel production rate to obtain the well-to-tank (WTT) energy consumption during the operational phase using the following equation:

$$Q_{operation}^{WTT} = p_f Q_{operation}^{TTW} \tag{3}$$

WTW energy consumption during the operational phase was calculated using the following equation:

$$Q_{operation} = Q_{operation}^{WTT} + Q_{operation}^{TTW} \tag{4}$$

Table 2. Summary of subscripts and abbreviations.

Subscript/Abbreviation	Definition	Unit
TEI	Transport energy intensity of each transportation means	J/kg/m
CTEI	Intracity transport energy intensity	J/kg/m
$Q_{material}$	Energy consumption during the manufacture phase	MJ
$Q_{operation}$	WTW Energy consumption during the operational phase per year	MJ/year
$Q_{operation}^{WTT}$	WTT energy consumption during the operational phase per year	MJ/year
$Q_{operation}^{TTW}$	TTW energy consumption during the operational phase pear year	MJ/year
k	Fuel economy	km/L, km/kWh
t	Operation duration	years
m	Weight of both transportation mean and passenger	kg
M	Weight of transportation mean	kg
P	Maximum capacity	Person
$r_{ave.}$	Average vehicle occupancy	%
L	Average traveled distance per year	km/year
c	Composition rate	%
e	Energy consumption rate	MJ/kg
d	Calorific value	MJ/L, MJ/kWh
p	Fuel production rate	MJ/MJ
s	Modal split in a given city	%
i	Composition of transportation means	-
f	Fuel type	-
x	Transportation means	-
y	Assessed year	

This study considered the entire weight of transportation by summing the weights of both the passengers and means. The average weight of a passenger was assumed to be 60 kg and the average occupancy rate was used for the calculation. The entire weight was obtained using the following equation:

$$m = M + 60Pr_{ave.} \tag{5}$$

Then, the transport energy intensity for each of transportation means was calculated using the following equation:

$$TEI_{t,x} = \frac{Q_{material,x} + tQ_{operation,x}}{m_x t L_x} \tag{6}$$

For the calculation of transport energy intensity in this study, the use duration of transportation means is a major parameter.

Finally, the intracity transport energy intensity was obtained by using the following equation.

$$CTEI_y = \sum_x TEI_{y,x} s_{y,x} \tag{7}$$

38 cities in Japan are selected, and the chronological range assessed in this study is 1987–2015. In addition, for the calculation of intracity transport energy intensity, this study uses the TEI in the case of use duration of lifetime. In addition, since the assessed year range is 1987–2015, this study uses the TEI of conventional vehicles (CV) as a representative value of an automobile.

Two assumptions must be noted. Technical features of each transportation means are assumed to be identical through the assessed years. The WTT energy consumption of gasoline, light diesel and each power generation type is also assumed to be identical during the assessed years. Meanwhile, the transition of WTT energy consumption of electricity with time is calculated on a basis of electricity mix in Japan in each of assessed years. In other words, this study assumes that the cause of diachronic transition of intracity transport energy efficiency is the modal shifting and electricity mix change.

2.4. Data Collection

The energy consumption of material structure was obtained from MiLCA (version 2, Japan Environmental Management Association for Industry, Tokyo, Japan) [47]. The WTT energy consumption including gasoline, light diesel and each power generation type were obtained from Japan Automobiles Research Institute [48].

In the case of roadways, Corolla Fielder (Toyota), LEAF (Nissan), PRIUS (Toyota) and MIRAI (Toyota) represent CV, electric vehicles (EV), hybrid vehicles (HV) and fuel cell vehicles (FCV), respectively. The data for each type was taken from the authors' previous research [28]. ERGA (Isuzu) represents buses and data was obtained from Kudo's work [49]. In the case of railways, Yamanote line 205 system (JR-EAST) data represents electric trains and data was taken from the report of Institute of Energy Economics, Japan [50].

Modal split and population density in 38 Japanese cities in 1987, 1992, 1999, 2005, 2010, and 2015 were taken from Ministry of Land, Infrastructure and Transport [51], and Ministry of Internal Affairs and Communications [52]. Thirty-eight Japanese cities include Sapporo, Hirosaki, Morioka, Sendai, Shiogama, Yuzawa, Koriyama, Utsunomiya, Tokorozawa, Chiba, Matsudo, Tokyo, Yokohama, Kawasaki, Yamanashi, Kanazawa, Gifu, Nagoya, Kasugai, Kyoto, Uji, Osaka, Sakai, Kobe, Nara, Kainan, Matsue, Yasugi, Hiroshima, Kure, Tokushima, Imabari, Kochi, Nankoku, Kitakyusyu, Fukuoka, Kumamoto, and Hitoyoshi.

3. Transport Energy Intensity for Each of the Transportation Means

The energy utilization at the assessed phases is computed and summed to present the transport energy intensity for each of the transportation means. Parametric analysis is executed by changing the use duration.

First, the result in 2015 in the case of use duration of lifetime is shown in Figure 1. Transport energy intensity decreases in the order of automobiles, buses, electric trains, bicycles and walks. For small-scale transportation means including bicycles and automobiles, energy consumption for the material structure has a great contribution to determining the TEI. Meanwhile, the material structure hardly affects the transport energy intensity for large-scale transportation means including buses and electric trains.

In the various types of automobile, its TEI increases in the order of HV, FCV, EV and CV. The CV presents the greatest TTW fuel consumption in the TEI, whereas the EV and FCV have a higher impact of energy consumption for material structure. The installation of additional equipment such as motor and battery and the substituted material for the body such as aluminum replaced with iron in the EV and FCV [53] requires the greater volume of energy input for the material structure. Consequently, the difference of the TEI between CV and EV appears to be slight. Notably, whether to include the weight of vehicle body in the calculation of intensity brings upon different trends of outcomes. The previous author's study [28], in which the transport energy efficiency of these automobile types was computed without considering the weight of vehicle body, showed the EV is the least efficient vehicle type.

In particular, due to the current electricity configuration in Japan, the WTT fuel consumption for the electricity production is significant. Given the lesser amount of energy input required for electricity production by renewables compared to fossil fuels [48], the increase in the share of renewables in the energy mix would contribute to the mitigation of the TEI of EV and electric trains.

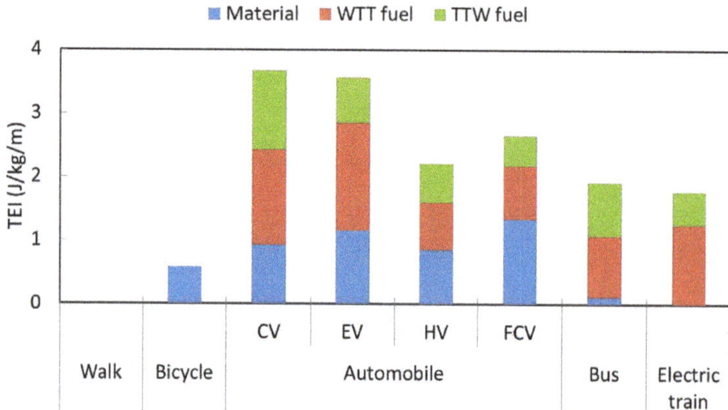

Figure 1. Transport energy intensity for each means of transportation in 2015 in the case of lifetime use duration.

Notably, the virgin materials are only taken into account and the recycled materials are not considered in the calculation of energy consumption under the manufacturing stage. The use of recycled materials would highly contribute to the reduction of energy consumption. For example, the recycling process could mitigate 97% and 65% of energy consumption for producing the virgin Al [54] and virgin Fe [55], respectively. Meanwhile, the recycling rate of Al in Japan has already reached 92.5% in 2017 [54]. The appropriate distribution of virgin and recycled Al is important to meet its incremental demand in the vehicle industry.

Then, the transport energy intensity in 2015, with the change of use duration, is monitored. The result is presented in Figure 2. The transport energy intensity of all of the transportation means decreases in inverse proportion to use duration. Particularly, for automobiles and bicycles, there remains room to improve the transport energy intensity by increasing in the use duration. If the lifetime is extended for another 5 years, the transport energy intensity of bicycle and automobile potentially increases by 38% and 6%, respectively. For electric trains, due to the negligibly small impact of material consumption on the transport energy intensity, its TEI almost remains the same after the three months of operation. This is because the dominator in Equation (6) is significantly greater than the energy consumption for material structure of electric train. The much heavier weight and longer movement distance of electric train compared with other transportation means would highly affect this difference. Given the longer lifetime of railway rolling stock compared with automobiles and buses, electric train would be an appropriate mean for urban transport.

Given that the lifetime extension of both bicycles and automobiles improve the transport energy intensity, promotions of domestic reuse and improvements of material durability are of utmost importance. While the various business models are involved in shipping used bicycles towards Asia and Middle East, its domestic reuse has yet to be fully operated [56]. On the other hand, improvement of material durability needs to be carefully taken into account. Particularly for automobiles, the material structure has been altered to the lightweight design [57]. While the reduction of vehicle weight leads to the improvement of fuel economy, there is a possibility that the energy requirement for a material structure is not mitigated due to the greater energy input for producing the replaced materials. Furthermore, the transition of material composition provokes the issue of raw material criticality. For example, replacement of steel with aluminum mitigates the environmental impacts [58] and relieves the concern of reserve [59] but deteriorates net import reliance [59]. Given that the reliance on raw materials in other countries is highly associated with security of supply and criticality issue [60], the transition of material structure affects not only the transport energy intensity of transportation means but also the criticality of product manufacturing.

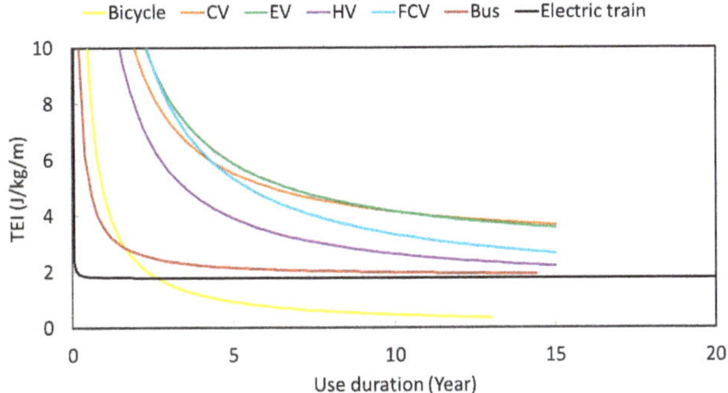

Figure 2. Transport energy intensity in 2015 with the change of use duration.

4. Relationship between Intracity Transport Energy Intensity and Population Density by Year

This Section presents the relationship between intracity transport energy intensity and population density. The obtained intracity transport energy intensities in 38 Japanese cities are presented in Appendix A.

First, the relationship between intracity transport energy intensity and population density in 1987, 1992, 1999, 2005, 2010, and 2015 is illustrated in Figure 3 and its regression model is presented in Table 3. In 1987–2015, the greater level of population density is associated with the lower intracity transport energy intensity in Japanese cities. This negative trend matches the global trend which has been widely reported (e.g., see the most well-known research work relevant to its trend conducted by Newman and Kenworthy [14]). Given the high score of R^2, the high level of negative correlation in the range of 1987–2015 could be seen in Japan's case. Meanwhile, it is seen that the regression coefficient becomes lower with time, which means the negative slope of regression line increases in the course of year.

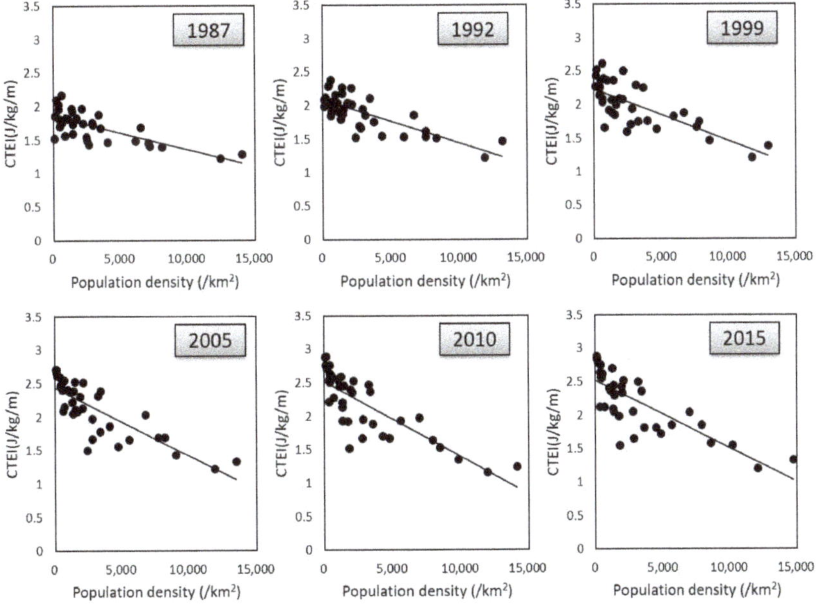

Figure 3. Intracity transport energy intensity and population density.

Table 3. Regression model of CTEI vs. population density.

Year	R^2	Regression Coefficient
1987	0.520	-4.93×10^5
1992	0.583	-6.49×10^5
1999	0.540	-7.71×10^5
2005	0.666	-1.02×10^4
2010	0.668	-1.12×10^4
2015	0.617	-1.01×10^4

Why did the regression coefficient increasingly grow with time? A more in-depth analysis is required to shed light on the mechanism of its relationship in Japan.

After the descriptive characteristics of the relationship between the intracity transport energy intensity and population density by year, a visual inspection with a focus on the transition of each of the assessed cities is executed to reveal the mechanism of its relationship.

The intracity transport energy intensity in the course of change in population density by city is presented in Figure 4. To show the trend of each city, the regression line is presented as well.

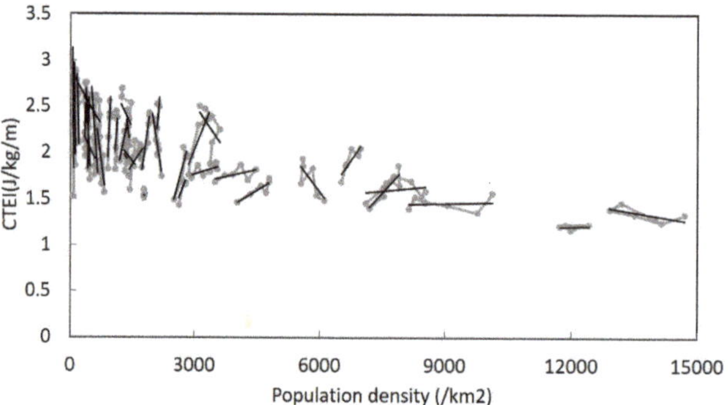

Figure 4. Intracity transport energy intensity and population density.

As for the group of low population density (~2500 persons/km^2), in most of cities intracity transport energy intensity has been increasingly growing with almost no change of population density in 1987–2015. Some cities (e.g., Imabari, Kure, and Yamanashi) drastically decrease in population density while continuously increase in intracity transport energy intensity. The slope of regression line of most cities in this group indicates nearly 80°–90°.

As for the group of medium population density (2500 persons/km^2 ~ 7000 persons/km^2), the degree of increase in intracity transport energy intensity is moderated while the population density has slightly grown in some cities (Uji, Kobe, Kasugai, Nagoya, Matsudo) after 2000. The slope of regression line of these cities indicates 45°–65°. In addition, in some cities (Chiba, Fukuoka, Tokorozawa), the slight increase in intracity transport energy intensity is presented, while the population density has continuously grown due to population influx from regional areas. The slope of regression line of these cities indicates 10°–30°.

As for the group of high population density (7000 persons/km^2~), the slope of regression line in all of cities (Yokohama, Kawasaki, Osaka, and Tokyo) indicates less than 1°. In Kawasaki, Osaka, Tokyo, the intracity transport energy intensity has almost remained the same or even has decreased, while the population density has continuously grown. Meanwhile, the U-shaped figure is observed in Yokohama. It must be noted that the intracity transport energy intensity in Kawasaki and Tokyo has increased from 2010 to 2015 although the modal split has not changed much. This might be because the fossil fuel contributes to the greater share in electricity mix after the Fukushima nuclear accident [61] and this causes the increase in the transport energy intensity of electric train.

Therefore, the main cause of the decrease in regression coefficient with time in 1987–2015 presented in Table 3 would be because the intracity transport energy intensity of each city continuously increased under the case of low population density while it slightly increased under the case of medium population density and remained almost the same in the case of high population density.

5. Discussion

The negative correlation between intracity transport energy intensity and population density was observed in the case of Japan. It was indicated that the negative slope of its regression line increases in the course of time in 1987–2015. In particular, the intracity transport energy intensity in cities with low population density has continuously increased without any significant change of population density. This would be because the reliance on automobiles has increased due to the difficulties in the development and extension of public transport systems from a financial perspective. For example, it is not considerably worthwhile from the financial perspective to invest money on extending the existing urban rail for a few settlements in less densely populated areas. In fact, transportation by rail in regional areas has declined and the operation of the railway in some regional areas has eventually been abandoned due to fiscal deficit in Japan. Meanwhile, the intracity transport energy intensity has slightly increased in cities with medium population density and almost remained the same in cities with high population density. This would be because the public transport system has already been well developed and the modal share of public transport (bus and electric train) has been consistently high.

Japanese population in 2050 is anticipated to decrease by 20% compared to 2012 [62]. Especially, population in cities of regional areas is projected to decrease by more than 20% due to the population influx as previously mentioned [63]. Considering the negative correlation presented in this study, there is a possibility that the intracity transport energy intensity will increase particularly in regional cities.

To address this issue in regional areas, the improvement of transport energy intensity is of significant importance for not only the technical aspects such as fuel economy and service life but also the social and regulatory aspects.

As for technical aspects, the transition of automobile type is important. According to Figure 1, the replacement of CVs with EVs may not significantly change the trend of intracity transport energy efficiency among 38 Japanese cities, whilst its replacement with HVs and FCVs would improve the intracity transport energy efficiency, particularly in regional areas. Considering various challenges in the development of hydrogen infrastructure for FCVs, improving the diffusion of HVs in less densely populated areas would be the first choice.

One of the promising social and regulatory approaches is to introduce the system of ridesharing. It must be noted that the transport energy intensity of automobiles in Japan have a significant room for improvement by increasing the occupancy rate. The number of vehicles privately owned in regional areas has increased in recent decades [64], which may potentially cause a decrease in a number of carried passengers for each of transportation means by trip and a drop in the transport energy intensity of automobiles. The promotion of ridesharing systems might contribute to an increase in occupancy rate. Still, ridesharing is an emerging concept in Japan. Although ridesharing was recently introduced in two underpopulated areas in Japan empirically, the taxi industry staunchly opposes the legalization of ridesharing. In addition, 70% of Japanese people are not willing to make use of ridesharing because of uncomfortableness of sharing vehicles with strangers [65]. To the contrary of negative impressions, ridesharing could deliver various advantages for passengers including reduction of fee and waiting time and free from commute stress [66]. Control of vehicle ownership by taxation [67], establishment of legal system and briefing of ridesharing concept for public hearing would be necessary to increase automobile occupancy rate.

Other conceivable social and regulatory approach would be to obligate elderly drivers to return the driver license. Following a series of frequent fatal traffic accidents caused by elderly drivers, its suitability for making compulsory has been heavily debated. The National Police Agency in Japan has only advised drivers 65 and older to consider surrendering their driver license of their own accord, although it has yet to be legalized [68]. The findings in this research may imply its necessity from the perspectives of transport energy efficiency. The population aging rate in 2050 will increase by 40% compared with 2012 [69]. The automobile modal split for people aged 65 or older has increased from 15% in 1987 to 37.3% in 2010 in three major metropolitan areas as well as from 18.5% in 1987 to 57.6% in 2010 in regional urban areas [70]. Particularly in cities with low population density, this trend would highly affect the increase in the CTEI. Considering that the super-aging society in Japan will potentially exacerbate problems with intracity transport energy efficiency on the basis of the current trend, to obligate elderly drivers to return the driver license would contribute to its improvement.

Notably, this obligation would raise a concern about deteriorating the quality of life for the elderly in regional cities where public transportation is not well developed. People tend to stay at the attached city where they have lived even in an inconvenient area [63]. The establishment of transportation systems such as ridesharing as previously mentioned and small-scale public transportation (e.g., minibus) with financial support as well as city compactification without a forced relocation, would be important.

Future perspectives of this study need to be mentioned.

The chronological transition in technical features in each transportation means and the energy consumption for the production of gasoline, light diesel and each power generation type will be further investigated to aid in more accurate calculation of intracity transport energy intensity.

Other lifecycle stages including maintenance and end-of-life need to be included in the future research. The next generation vehicles such as EV, in particular, need to replace the battery due to the degradation of its quality, and its replacement requires energy to some extent during the maintenance stage. The lifespan of battery used in the next generation vehicles is around 8–10 years, and the battery needs to be replaced even before exhausting the lifetime of vehicle. Besides that, the energy consumption for the infrastructure development for each of the transportation means will be addressed. In spite of the argument in favor of electric trains in urban railways in this study, the inclusion of energy consumption for the development of rail infrastructure might change the result. The extension of the boundary for energy consumption in the transport sector would provide more comprehensive information on the transport energy intensity for each transportation means.

The energy consumption under the operation stage used in this study was on the basis of average fuel economy given in the data lists of representative transport means in a top-down approach, whereas the energy efficiency in the transport sector is somewhat dependent on the local urban environment and driving patterns [71]. The empirical examination in a bottom-up approach would assist in a detailed investigation of the relationships between energy use and city form.

6. Conclusions

This study has first obtained the transport energy intensity for each of the transportation means including walk, bicycle, automobile (CV, EV, HV, FCV), bus and electric train by considering the lifecycle energy consumption. Based on the modal split, the intracity transport energy intensity of 38 cities in Japan in 1987–2015 has been computed. Then, the chronological relationship between intracity transport energy intensity and population density by year has been presented. Finally, to reveal the mechanism of its relationship, the transition of each of assessed cities has been visually monitored.

Findings are as follows:

- Transport energy intensity decreases in the order of automobiles, buses, electric trains, bicycles and walks.
- For small-scale transportation means including bicycles and automobiles, energy consumption for the material structure has a great contribution to determining the transport energy intensity.
- Material structure hardly affects the transport energy intensity for large-scale transportation means including buses and electric trains.
- The greater level of population density is associated with the lower intracity transport energy intensity in Japanese cities, as seen in earlier studies with a focus on other spatial scopes.

- The negative slope of its regression line between intracity transport energy intensity and population density increases over time in 1987–2015.

The main cause of the decrease in the regression coefficient with time in Japan in 1987–2015 between intracity transport energy intensity and population density could be that, in cities with low population density, the intracity transport energy intensity continuously increases without any significant change of population density while, with increasing population density slightly increases in cities with a medium population density and almost remains the same in cities with a high population density.

Author Contributions: Conceptualization, S.K., M.Y. and E.Y.; methodology, S.K., M.Y. and E.Y.; validation, S.K., M.Y. and E.Y.; formal analysis, S.K. and M.Y.; investigation, M.Y.; resources, S.K. and M.Y.; data curation, M.Y.; writing—original draft preparation, S.K.; writing—review and editing, E.Y.; visualization, S.K.; supervision, E.Y.; project administration, E.Y.; funding acquisition, E.Y. All authors have read and agreed to the published version of the manuscript.

Funding: This study was partly supported by research funds from KAKENHI Grants (26281056, and 19H04329) and from the Environment Research and Technology Development Fund (S-16).

Conflicts of Interest: The authors declare no conflict of interest. The funders had no role in the design of the study; in the collection, analyses, or interpretation of data; in the writing of the manuscript, or in the decision to publish the results.

Appendix A

Table A1. Intracity transport energy intensity for each of 38 Japanese cities#.

City	Year					
	1987	1992	1999	2005	2010	2015
Sapporo	1.74	1.95	1.99	2.08	1.92	1.99
Hirosaki	1.83	2.05	2.39	2.55	2.54	2.64
Morioka	1.76	1.84	2.03	2.10	2.22	2.12
Sendai	1.77	1.90	2.07	2.22	2.21	2.30
Shiogama	1.88	2.10	2.25	2.39	2.47	2.50
Yuzawa	1.53	1.98	2.27	2.71	2.88	2.88
Koriyama	1.82	2.30	2.36	2.49	2.61	2.56
Utsunomiya	1.96	2.27	2.36	2.53	2.59	2.69
Tokorozawa	1.46	1.55	1.64	1.56	1.68	1.72
Chiba	1.72	1.86	1.74	1.79	1.89	1.82
Matsudo	1.40	1.53	1.67	1.69	1.65	1.86
Tokyo	1.29	1.46	1.39	1.33	1.24	1.33
Yokohama	1.46	1.62	1.75	1.69	1.53	1.58
Kawasaki	1.39	1.51	1.47	1.43	1.35	1.56
Yamanashi	2.17	2.38	2.61	2.66	2.89	2.85
Kanazawa	1.82	2.15	2.36	2.41	2.55	2.40
Gifu	1.97	2.25	2.50	2.51	2.53	2.52
Nagoya	1.68	1.86	1.87	2.04	1.97	2.05
Kasugai	1.76	1.95	2.29	2.31	2.37	2.36
Kyoto	1.54	1.52	1.60	1.50	1.52	1.55
Uji	1.49	1.70	1.94	1.97	1.95	2.05
Osaka	1.22	1.23	1.21	1.22	1.16	1.20
Sakai	1.48	1.54	1.83	1.66	1.93	1.85
Kobe	1.43	1.66	1.71	1.67	1.67	1.65
Nara	1.60	1.80	1.88	2.05	1.94	2.10
Kainan	1.78	1.91	2.09	2.40	2.59	2.61
Matsue	1.96	2.04	2.14	2.60	2.63	2.75
Yasugi	1.86	2.09	2.43	2.71	2.76	2.84
Hiroshima	1.60	1.88	1.85	2.12	2.13	2.06
Kure	1.57	1.96	1.66	2.16	2.28	2.13
Tokushima	1.93	2.18	2.07	2.38	2.46	2.45
Imabari	1.70	2.02	2.26	2.47	2.51	2.57
Kochi	1.81	2.04	1.92	2.38	2.44	2.37
Nankoku	2.03	2.29	2.16	2.47	2.76	2.75
Kitakyusyu	1.74	2.02	2.08	2.13	2.35	2.34
Fukuoka	1.68	1.75	1.76	1.87	1.70	1.82
Kumamoto	1.83	2.04	2.09	2.30	2.40	2.42
Hitoyoshi	2.10	2.12	2.52	2.61	2.71	2.78

References

1. Parshall, L.; Gurney, K.; Hammer, S.A.; Mendoza, D.; Zhou, Y.; Geethakumar, S. Modeling energy consumption and CO_2 emissions at the urban scale, methodological challenges and insights from the United States. *Energy Policy* **2010**, *38*, 4765–4782. [CrossRef]
2. Dulal, H.B.; Brodnig, G.; Onoriose, C.G. Climate change mitigation in the transport sector through urban planning: A review. *Habitat Int.* **2011**, *35*, 494–500. [CrossRef]
3. Kennedy, C.; Baker, L.; Dhakal, S.; Ramaswami, A. Sustainable urban systems. *J. Ind. Ecol.* **2012**, *16*, 775–779. [CrossRef]
4. Bulkeley, H.; Broto, V.C.; Maassen, A. Low carbon transitions and the reconfiguration of urban infrastructure. *Urban Stud.* **2014**, *51*, 1471–1486. [CrossRef]
5. Rutherford, J.; Jaglin, S. Introduction to the special issue-Urban energy governance: Local actions, capacities and politics. *Energy Policy* **2015**, *78*, 173–178. [CrossRef]
6. Ma, J.; Liu, Z.; Chai, Y. The impact of urban form on CO_2 emission from work and non-work trips: The case of Beijing, China. *Habitat Int.* **2015**, *47*, 1–10. [CrossRef]
7. Wang, Y.; Yang, L.; Han, S.; Li, C.; Ramachandra, T.V. Urban CO_2 emissions in Xi'an and Bangalore by commuters: Implications for controlling urban transportation carbon dioxide emissions in developing countries. *Mitig. Adapt. Strateg. Glob. Chang.* **2017**, *22*, 993–1019. [CrossRef]
8. Brand, C.; Tran, M.; Anable, J. The UK transport carbon model: An integrated life cycle approach to explore low carbon futures. *Energy Policy* **2012**, *41*, 107–124. [CrossRef]
9. IEA. *Technology Roadmap. Fuel Economy of Road Vehicles*; Technical Report; International Energy Agency: Paris, France, 2012.
10. Osorio, C.; Nanduri, K. Energy-Efficient Urban Traffic Management: A Microscopic Simulation-Based Approach. *Transp. Sci.* **2015**, *49*, 637–651. [CrossRef]
11. Zhao, P.J.; Diao, J.J.; Li, S.X. The influence of urban structure on individual transport energy consumption in China's growing cities. *Habitat Int.* **2017**, *66*, 95–105. [CrossRef]
12. Cao, X.; Yang, W. Examining the effects of the built environment and residential self-selection on commuting trips and the related CO_2 emissions: An empirical study in Guangzhou, China. *Transp. Res. Part D Transp. Environ.* **2017**, *52*, 480–494. [CrossRef]
13. Hughes, B.; Chambers, L.; Lansdell, H.; White, R. Cities, area and transport energy. *Road Transp. Res.* **2004**, *13*, 72.
14. Newman, P.W.G.; Kenworthy, J.R. Gasoline Consumption and Cities: A comparison of U.S. Cities with a Global Survey. *J. Am. Plan. Assoc.* **1989**, *55*, 24–37. [CrossRef]
15. Muniz, I.; Galindo, A. Urban form and the ecological footprint of commuting: The case of Barcelona. *Ecol. Econ.* **2005**, *55*, 499–514. [CrossRef]
16. Newman, P. The environmental impact of cities. *Environ. Urban.* **2006**, *18*, 275–295. [CrossRef]
17. Shim, G.E.; Rhee, S.M.; Ahn, K.H.; Chung, S.B. The relationship between the characteristics of transportation energy consumption and urban form. *Ann. Reg. Sci.* **2006**, *40*, 351–367. [CrossRef]
18. Alford, G.; Whiteman, J. Macro-urban form and transport energy outcomes: Investigations for Melbourne. *Road Transp. Res.* **2009**, *18*, 53–67.
19. Brownstone, D.; Golob, T.F. The impact of residential density on vehicle usage and energy consumption. *J. Urban Econ.* **2009**, *65*, 91–98. [CrossRef]
20. Hankey, S.; Marshall, J.D. Impacts of urban form on future US passenger-vehicle greenhouse gas emissions. *Energy Policy* **2010**, *38*, 4880–4887. [CrossRef]
21. Karathodorou, N.; Graham, D.; Noland, R. Estimating the effect of urban density on fuel demand. *Energy Econ.* **2010**, *32*, 86–92. [CrossRef]
22. Su, Q. The effect of population density, road network density, and congestion on household gasoline consumption in U.S. urban areas. *Energy Econ.* **2011**, *33*, 445–452. [CrossRef]
23. Clark, T.A. Metropolitan density, energy efficiency and carbon emissions: Multi-attribute tradeoffs and their policy implications. *Energy Policy* **2013**, *53*, 413–428. [CrossRef]

24. Modarres, A. Commuting and energy consumption: Toward an equitable transportation policy. *J. Transp. Geogr.* **2013**, *33*, 240–249. [CrossRef]
25. Aguiléra, A.; Voisin, M. Urban form, commuting patterns and CO_2 emissions: What differences between the municipality's residents and its jobs? *Transp. Res. Part A Policy Pract.* **2014**, *69*, 243–251. [CrossRef]
26. Yang, W.; Li, T.; Cao, X. Examining the impacts of socio-economic factors, urban form and transportation development on CO_2 emissions from transportation in China: A panel data analysis of China's provinces. *Habitat Int.* **2015**, *49*, 212–220. [CrossRef]
27. Fiori, C.; Ahn, K.; Rakha, H.A. Power-based electric vehicle energy consumption model: Model development and validation. *Appl. Energy* **2016**, *168*, 257–268. [CrossRef]
28. Kosai, S.; Nakanishi, M.; Yamasue, E. Vehicle Energy Efficiency Evaluation from Well-to Wheel Lifecycle Perspective. *Transp. Res. Part D Transp. Environ.* **2018**, *65*, 355–367. [CrossRef]
29. Zhang, B.; Sarathy, S.M. Lifecycle optimized ethanol-gasoline blends for turbocharged engines. *Appl. Energy* **2016**, *181*, 38–53. [CrossRef]
30. Hawkins, T.R.; Gausen, O.M.; Strømman, A.H. Environmental impacts of hybrid and electric vehicles—A review. *Int. J. Life Cycle Assess.* **2012**, *17*, 997–1014. [CrossRef]
31. Casals, L.C.; Martinez-Laserna, E.; Garcia, B.A.; Nieto, N. Sustainability analysis of the electric vehicle use in Europe for CO_2 emissions reduction. *J. Clear. Prod.* **2016**, *127*, 425–437. [CrossRef]
32. Larsson, M.; Mohseni, F.; Wallmark, C.; Gronkvist, S.; Alvfors, P. Energy system analysis of the implications of hydrogen fuel cell vehicles in the Swedish road transport system. *Int. J. Hydrog. Energy* **2015**, *40*, 11722–11729. [CrossRef]
33. Washing, E.M.; Pulugurtha, S.S. Well-to-Wheel Analysis of Electric and Hydrogen Light Rail. *J. Public Transp.* **2015**, *18*, 74–88. [CrossRef]
34. Baptista, P.; Pina, A.; Duarte, G.; Rolim, C.; Pereira, G.; Silva, C.; Farias, T. From on-road trial evaluation of electric and conventional bicycles to comparison with other urban transport modes: Case study in the city of Lisbon, Portugal. *Energy Convers. Manag.* **2015**, *92*, 10–18. [CrossRef]
35. Facanha, C.; Horvath, A. Evaluation of life-cycle air emission factors for freight transportation. *Environ. Sci. Technol.* **2007**, *41*, 7138–7144. [CrossRef]
36. Schipper, L.; Saenger, C.; Sudarshan, A. Transport and Carbon Emission in the United States: The Long View. *Energies* **2011**, *2011*, 563–581. [CrossRef]
37. Kamakaté, F.; Schipper, L. Trends in truck freight energy use and carbon emissions in selected OECD countries from 1973 to 2005. *Energy Policy* **2009**, *37*, 3743–3751. [CrossRef]
38. Mendiluce, M.; Schipper, L. Trends in passenger transport and freight energy use in Spain. *Energy Policy* **2011**, *39*, 6466–6475. [CrossRef]
39. de Witte, A.; Hollevoet, J.; Dobruszkes, F.; Hubert, M.; Macharis, C. Linking modal choice to motility: A comprehensive review. *Transp. Res. Part A Policy Pract.* **2013**, *49*, 329–341. [CrossRef]
40. Limtanakool, N.; Dijst, M.; Schwanen, T. The influence of socioeconomic characteristics, land use and travel time considerations on mode choice for medium- and longer-distance trips. *J. Transp. Geogr.* **2006**, *14*, 327–341. [CrossRef]
41. Waygood, E.O.D.; Sun, Y.L.; Susilo, Y.O. Transportation carbon dioxide emissions by built environment and family lifecycle: Case study of the Osaka metropolitan area. *Transp. Res. Part D Transp. Environ.* **2014**, *31*, 176–188. [CrossRef]
42. Burnham, A.; Wang, M.; Wu, Y. *Development and Applications of GREET 2.7—The Transportation Vehicle-cycle Model*; ANL/ESD/06-5; Center for Transportation Research, Argonne National Laboratory: Argonne, IL, USA, 2006.
43. Chester, M.V.; Horvath, A.; Madanat, S. Comparison of life-cycle energy and emissions footprints of passenger transportation in metropolitan regions. *Atmos. Environ.* **2010**, *44*, 1071–1079. [CrossRef]
44. Sullivan, J.L.; Burnham, A.; Wang, M.Q. Model for the Part Manufacturing and Vehicle Assembly Component of the Vehicle Life Cycle Inventory. *J. Ind. Ecol.* **2013**, *17*, 143–153. [CrossRef]
45. Chung, W.; Zhou, G.; Yeung, I.M.H. A study of energy efficiency of transport sector in China from 2003 to 2009. *Appl. Energy* **2013**, *112*, 1066–1077. [CrossRef]

46. Lipscy, P.Y.; Schipper, L. Energy efficiency in the Japanese transport sector. *Energy Policy* **2013**, *56*, 248–258. [CrossRef]

47. Japan Environmental Management Association for Industry. JEMAI-LCA Pro. Available online: http://www.jemai.or.jp/ (accessed on 12 May 2017).

48. Japan Automobiles Research Institute. Overall Efficiency and GHG Emissions. Available online: http://www.jari.or.jp/Portals/0/jhfc/data/report/2010/pdf/result.pdf (accessed on 14 February 2018). (In Japanese).

49. Kudo, Y.; Ishitani, H.; Matsuhashi, R. Kokyo yuso kikan no raifu saikuru CO_2 haishutsu tokusei no kensho. *KeiO Associated Repository of Academia Resources*. AA12113622-00000116-0001. 2007. Available online: https://core.ac.uk/download/pdf/56669183.pdf (accessed on 21 February 2020). (In Japanese).

50. Oki, Y. Survey on Energy Saving in the Steel Products. Available online: http://eneken.ieej.or.jp/data/pdf/468.pdf (accessed on 9 August 2017). (In Japanese).

51. Ministry of Land, Infrastructure and Transport. Transport and City Planning Report. Available online: http://www.mlit.go.jp/toshi/tosiko/toshi_tosiko_fr_000024.html (accessed on 18 February 2018). (In Japanese)

52. Statistics Japan. *Population Statistics*; Ministry of Internal Affairs and Communications: Tokyo, Japan. (In Japanese)

53. Liedl, G.; Bielak, R.; Ivanova, J.; Enzinger, N.; Figner, G.; Bruckner, J.; Pasic, H.; Pudar, M.; Hampel, S. Joining of Aluminum and Steel in Car Body Manufacturing. *Phys. Procedia* **2011**, *12*, 150–156. [CrossRef]

54. Japan Auminum Can Recycling Association. Recyling. Available online: http://www.alumi-can.or.jp/publics/index/24/ (accessed on 8 September 2018).

55. COSMO RECYCLE CO., Ltd. Iron Scap Recycling. Available online: http://www.cosmorecycle.com/english/ (accessed on 6 September 2018).

56. Ministry of Environmnet. Annual Report on the Environment, the Sound Material-Cycle Society and the Biodiversity in Japan. Available online: http://www.env.go.jp/en/wpaper/2017/pdf/2017_all.pdf (accessed on 10 December 2017).

57. Bai, J.; Li, Y.; Zuo, W. Cross-sectional shape optimisation for thin-walled beam crashworthiness with stamping constraints using genetic algorithm. *Int. J. Veh. Des.* **2017**, *73*, 76–95. [CrossRef]

58. Nuss, P.; Eckelman, M.J. Life Cycle Assessment of Metals: A Scientific Synthesis. *PLoS ONE* **2014**, *9*, e101298. [CrossRef]

59. Graedel, T.E.; Harper, E.M.; Nassar, N.T.; Nuss, P.; Reck, B.K. Criticality of metals and metalloids. *PNAS* **2015**, *112*, 4257–4262. [CrossRef]

60. Kosai, S.; Hashimoto, S.; Matsubae, K.; McLellan, B.; Yamasue, E. Comprehensive Analysis of External Dependency in terms of Material Criticality by Employing Total Material Requirement: Sulfuric Acid Production in Japan as a case study. *Minerals* **2018**, *8*, 114. [CrossRef]

61. Kosai, S.; Unesaki, H. Quantitative Analysis on the Impact of Nuclear Energy Supply Disruption on Electricity Supply Security. *Appl. Energy* **2017**, *208*, 1198–1207. [CrossRef]

62. The Institute of Energy Economics, Japan. Asia/World Energy Outlook. Available online: https://eneken.ieej.or.jp/data/6332.pdf (accessed on 26 November 2017).

63. Ministry of Land, Infrastructure and Transport. Grandline 2050: Appendix. Available online: http://www.mlit.go.jp/common/001050896.pdf (accessed on 17 February 2018). (In Japanese)

64. Ministry of Land, Infrastructure and Transport. Analysis of Current Transport Trend in Japan. Available online: http://www.mlit.go.jp/road/ir/kihon/26/1-1_s1.pdf (accessed on 11 January 2018). (In Japanese)

65. Ministry of Internal Affairs and Communications. *WHITE PAPER Information and Communications in Japan*; Ministry of Internal Affairs and Communications: Tokyo, Japan, 2016.

66. Chan, N.D.; Shaheen, S. Ridesharing in North America: Past, present, and future. *Transp. Rev.* **2012**, *32*, 93–112. [CrossRef]

67. Cervero, R.; Tsai, Y.S. City CarShare in San Francisco, California—Second-year travel demand and car ownership impacts. *Transp. Res. Rec.* **2004**, *1887*, 117–127. [CrossRef]

68. National Police Agency. White Paper on Police. Available online: https://www.npa.go.jp/hakusyo/h29/english/full_text_WHITE_PAPER_2017_E.pdf (accessed on 17 October 2019).

69. Nationl Institute Population and Social Security Research. Population Projections for Japan. Available online: http://www.ipss.go.jp/pp-zenkoku/j/zenkoku2017/pp29_gaiyou.pdf (accessed on 5 March 2018).

70. Ministry of Land, Infrastructure, Transport and Tourism. Human Mobility in City. Available online: https://www.mlit.go.jp/common/001032141.pdf (accessed on 9 July 2019). (In Japanese)

71. Ruzzenenti, F.; Basosi, R. Evaluation of the energy efficiency evolution in the European road freight transport sector. *Energy Policy* **2009**, *37*, 4079–4085. [CrossRef]

Review

Exergetic Life Cycle Assessment: A Review

Martin N. Nwodo * and Chimay J. Anumba

College of Design, Construction and Planning, University of Florida, Gainesville, FL 32611, USA;
anumba@ufl.edu
* Correspondence: nwodomartin@ufl.edu

Received: 8 April 2020; Accepted: 22 May 2020; Published: 26 May 2020

Abstract: Exergy is important and relevant in many areas of study such as Life Cycle Assessment (LCA), sustainability, energy systems, and the built environment. With the growing interest in the study of LCA due to the awareness of global environmental impacts, studies have been conducted on exergetic life cycle assessment for resource accounting. The aim of this paper is to review existing studies on exergetic life cycle assessment to investigate the state-of-the-art and identify the benefits and opportunity for improvement. The methodology used entailed an in-depth literature review, which involved an analysis of journal articles collected through a search of databases such as Web of Science Core Collection, Scopus, and Google Scholar. The selected articles were reviewed and analyzed, and the findings are presented in this paper. The following key conclusions were reached: (a) exergy-based methods provide an improved measure of sustainability, (b) there is an opportunity for a more comprehensive approach to exergetic life cycle assessment that includes life cycle emission, (c) a new terminology is required to describe the combination of exergy of life cycle resource use and exergy of life cycle emissions, and (d) improved exergetic life cycle assessment has the potential to solve characterization and valuation problems in the LCA methodology.

Keywords: exergy; life cycle assessment; sustainability; LCA; review

1. Introduction

Exergy analysis is useful to rationally and meaningfully assess and compare processes and systems [1]. These capabilities are reflected in the two key features of exergy analysis: (a) efficiency to provide a real evaluation of how actual performance tends to or deviates from the ideal, and (b) exergy loss to identify more clearly than energy analysis the types, causes, and locations of thermodynamic losses [1]. Exergy efficiency was included in the 2001 Swiss canton of Geneva as a new parameter for characterizing the energy performance of buildings [2]. As exergy describes the work potential of energy, exergy-based analysis is used in system design or process optimization [3]. Exergy is used for assessment, design, analysis, and improvement of systems; an example is an application of exergy analysis to integrated energy systems such as biomass, geothermal, and steam power systems [1]. Table 1 summarizes the importance of exergy, classified by topic. In life cycle analysis, exergy-based approach is relevant to quantify energy and material resources, determine consumption and depletion of natural resources, and as an indicator of resource utilization efficiency [4–9]. In production processes, exergy method is used to keep inventory of exergy losses and efficiencies on a single scale [10–12]. Exergy has been used in technological processes to achieve sustainability to reflect extent of use of renewable resources, to account for technological efficiency and conversion of waste products into neutral or harmless products [13]. Exergy is used to deepen the understanding of the built environment to develop low-exergy systems for the future [14,15] and for a scientifically based sustainable building assessment tool [16,17]. Exergy analysis optimizes the efficiencies of energy systems [18–21] and in this way, reduces global impacts [22,23].

Table 1. Summary of the importance of exergy classified by topic area.

Topic	Importance	References
Life cycle analysis	Exergy enables the analysis of cumulative consumption of resources	[5,6]
	Exergetic LCA is a more appropriate approach to quantify the environmental problem of the depletion of natural resources	[4,6]
	Exergy provides additional indicator for LCA, energy efficiency, and resource quality need	[7–9]
Production processes	Exergy analysis accounts for exergy losses and efficiencies and thus provides a more accurate inventory	[10]
	Exergy quantifies various results of manufacture, use and disposal of goods, and services on a single scale—exergy loss	[11,12]
Technological sustainability	Exergy enables both qualitative and quantitative evaluations of resource consumptions	[5,13]
Built environment	Exergy concept deepens the understanding of space heating and cooling to develop low-exergy systems for future buildings	[14,15]
Sustainability index	Exergy-based index overcomes the limitations of the subjectively defined weights used in other sustainability assessment tools	[16,17]
Energy systems	Exergy analysis evaluates the performance of energy systems to optimize their efficiencies	[18–21]
Global impacts	By improving the efficiency of a process, exergy analysis reduces global impacts related to the process	[22,23]

Although, most uses of exergy are in the area of metallurgical and chemical process analysis, thermal system design, and energy conversion system design [24], the use of exergy in life cycle analysis is currently emerging. The emergence of exergetic life cycle assessment is probably because of the importance of wholistic analysis of a system, a product or a process for resource accounting over a life span. Szargut et al. [25] first analyzed life cycle of a system based on exergy by developing the concept of cumulative exergy demand. Cornelissen and Hirs [4] further proposed the inclusion of the concept of exergy into Life Cycle Assessment (LCA) as exergetic life cycle assessment. They described exergetic life cycle assessment as a method to measure the depletion of natural resources in LCA and as a tool to evaluate the efficiency of resource use. There are also other studies on exergetic life cycle assessment [5,26–30]. The aim of this paper is to review existing studies on exergetic life cycle assessment to investigate the state-of-the-art and identify opportunities for improvement of the approach. Following the introduction, this paper describes exergy and exergy-based methods, introduces the use of exergy in LCA, analyzes studies on exergetic life cycle assessment, and provides discussion and conclusions from the findings.

2. Description of Exergy

2.1. Definition of Exergy

The capacity of doing work has been accepted as a measure of the quality of energy [31]. Energy quality in a system can be grouped into either available energy or unavailable energy. Exergy is a measure of the possible maximum useful work before a system reaches equilibrium with its environment [32]. Szargut [31] defined exergy as the work obtainable for a matter to be brought to a state of thermodynamic equilibrium with the common elements of the natural surroundings by mechanism of reversible processes, which involves interchange only with the elements of nature. According to Bejan et al. [33], exergy is available when an idealized system (called an environment) interacts to equilibrium with another system of interest, while heat transfer occurs only with the environment. The following can be deduced from these definitions:

- To calculate exergy, the idealized state is specified;
- Only common components such as atmosphere, hydrosphere, and lithosphere can be used as idealized systems because of thermodynamic disequilibrium in the surrounding nature;

- Being a measure of energy quality, exergy is used to investigate technological processes, in addition to analyze the processes of power plants and of other mechanical machines;
- Exergy losses occur from irreversible process, which either cause reduction of the useful results of the process or increase use of energy from whatever source of derivation.

The choice and definition of the reference environment or state is necessary for exergy analysis. This is because the sensitivity of the results to different choices of the reference state might vary with the operative conditions of the system analyzed [20]. Correspondingly, when the state is significantly different from that of the chosen base conditions, its flows are not overly sensitive to the definition of the reference environment. This is the case, for instance, in the analysis of power plants. In turn, when the properties are close to those of the base conditions, results from the analysis have great variations depending on the definition of the chosen base state. This is the case of the analysis of space heating and cooling in buildings.

2.2. Brief Historical Background on Exergy

The historical background on exergy can be traced to the first and second laws of thermodynamics. According to Szargut et al. [25], in the 1840s, James Joule proved that there was an exact numerical equivalence between work and heat (also known as the conservation of energy), in accordance with the first law of thermodynamics. While the second law was based on Carnot's experiments, which demonstrated that the limitations of heat-to-work conversion depend on the temperature at which the heat is available or the 'quality' of the heat. Consequently, the internal energy and entropy functions were defined, followed by the enthalpy, the Helmholtz function, and Gibbs Free Energy [25]. These functions increased the capacity to understand the effects of the laws and their use to effectively solve practical problems. According to Szargut et al. [25], exergy function is introduced to improve our comprehension of thermal and chemical processes by enabling the investigation of processes, whether complex or not, to determine the theoretically most efficient way by which that process could be performed within the environment.

2.3. Relationship between Exergy and Other Energy Qualities

The terms entropy, exergy, and emergy help to articulate the important qualities of energy [32]. While exergy depicts the amount of work a system can exert on its environment, unavailable energy, also known as entropy, cannot be converted into work. According to Shukuya and Hammache [34], exergy measures the potential of dispersion of energy and matter in their environment while entropy quantifies the state of dispersion or the extent of dispersion of the energy and matter (Figure 1). In other words, exergy is used within the process or system to produce entropy. The terms exergy and entropy generation minimization [35] are common in the second law of thermodynamics.

Exergy is related to emergy in that the latter reflects all the exergy retrieved and used from the original state to the present state of a process. Odum [36] defined emergy as: "the available energy of single kind previously used directly and indirectly to make a product. Its unit is the emjoule (ej)" Solar emjoules (sej) are the solar energy equivalents required directly or indirectly to make a product with energy content. Bastianoni et al. [37] stated that emergy can be expressed as a function of exergy, although the goals and boundaries of the reference state differ. While the former retraces the embodied solar energy in a product, the latter assesses the quantity of primary resources that goes into that product. In addition, in the former, the system comprises the whole biosphere, with solar energy as the basic input. In the latter, however, the analyst defines the control volume, according to the goal of the study [37]. Among the major energy qualities, exergy is preferred for this study because of its flexibility in the choice of primary resources for analysis, and its capability to quantify the potential of dispersion of energy and matter into the environment of study.

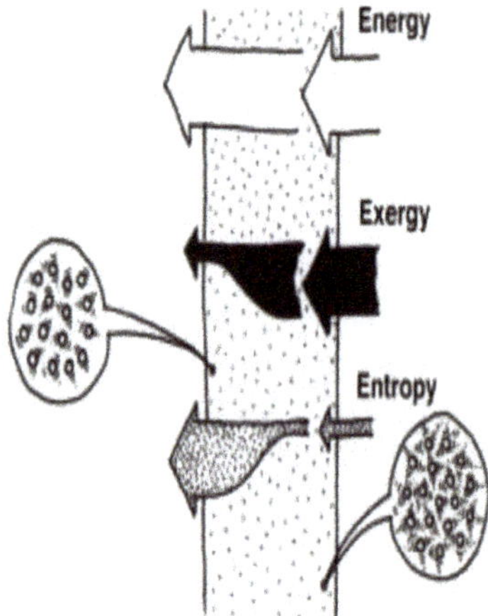

Figure 1. Illustration of energy, exergy, and entropy flow in and out of a building envelope system [34].

3. Exergy-Based Methods and Life Cycle Assessment

3.1. Exergy-Based Methods

Morosuk et al. [8] consider "exergy-based methods" as a general term that includes the conventional and advanced exergetic, exergoeconomic, and exergoenvironmental analyses and evaluations. Exergoeconomic analysis is a technique that combines both exergy and economic analyses to evaluate the costs of inefficiencies in a process, and it is referred to by other names such as thermoeconomics, second-law costing, and cost accounting [38]. Exergoenvironmental analysis combines exergy analysis and life cycle assessment, which is conducted at the component level, to identify the location, magnitude, and the causes of environmental impact [39,40]. Tsatsaronis and Morosuk [41] stated that exergy-based methods describe the situation, the quantity and the sources of inefficiencies, costs, and environmental impacts; and enable analysts to study the interconnections between them. In generic terms, the major exergy-based methods include the following:

3.1.1. Cumulative Exergy Demand (CExD)

The Cumulative Exergy Demand (CExD) was proposed by Szargut et al. [25]. It is defined as the sum of exergy of all supplies required to produce a product or provide a service [7]. CExD is related to Cumulative Energy Demand (CED), but unlike CED, it can account for materials and quality of energy inputs. In addition to this advantage, CExD analysis can provide insight into potential improvements and for comparing alternative products, by accounting for exergy use throughout the life cycle.

3.1.2. Thermo-Ecological Cost (TEC)

The Thermo-Ecological Cost (TEC) method accounts only for the cumulative consumption of non-renewable primary exergy resources. It is expressed in exergy units and not in monetary units. This method was developed with the premise that it is essential to determine and reduce the depletion of non-renewable natural materials in the field of ecological applications of exergy [42].

3.1.3. Cumulative Exergy Extraction from Natural Environment (CExENE)

The Cumulative Exergy Extraction from Natural Environment (CExENE) method is an extension of the boundaries of CExD to include land use. CExD accounts for energetic supplies and materials traditionally considered non-energetic such as mineral, water, and metal, but ignores land use. CED only accounts for materials, which may be used as energy carriers [43]. Therefore, CExENE is quantitatively the most comprehensive resource indicator of the three, because it evaluates energy carriers, non-energetic supplies, and land occupation. Conceptually and qualitatively, CExENE differs from CExD and therefore leads to a different evaluation. CExD measures the exergy that is transferred into the technological system from nature, while CExENE accounts for total exergy that is deprived of the natural system [43] which may or may not be transferred into the technological system.

3.1.4. Industrial/Ecological Cumulative Exergy Demand (ICExD/ECExD)

The Industrial/Ecological Cumulative Exergy Demand (ICExD/ECExD) method is an extension of the CExD method to emphasize on industrial and ecological processes, respectively [44,45]. ICExD reports the exergy of natural wealth consumed by each industrial sector both directly and indirectly, while ECExD reports the exergy used up in ecological systems to produce each natural wealth [46]. For a production chain, ECExD analysis improves on ICExD analysis by including exergy losses in the industrial as well as ecological stages [46]. This method defines the mathematical form of economic and ecological systems through fiscal and physical input-output tables. Like the economic input-output model, the main advantage of this method is probably the availability of the necessary macroeconomic data for each sector. However, there is lack of details of individual processes in these sectors, which can lead to aggregation error [46].

3.1.5. Extended Exergy Accounting (EExA)

As proposed by Sciubba [47], the Extended Exergy Accounting (EExA) method is used to compute a commodity value based on its resource equivalent value instead of its fiscal cost. This method is based on two essential assumptions: (a) that the cumulative exergy content of a product or service is the sum of the exergies of the product's constituents, in addition to a weighted sum of the exergies of the production process of the product, and (b) that non-energetic costs such as labor, capital, and environmental emissions can be reformulated in terms of exergy from global system balances. While (a) is a paraphrase of Szargut's CExD, (b) is the original contribution of EExA. A theoretical and practical advance in EExA can be found in Dai et al. [48].

3.1.6. Comparison of the Exergy-Based Methods

This section presents a comparison of the exergy-based methods. The comparison is based on deductions from the description of the exergy-based methods. Table 2 compares the identified exergy-based methods in terms of scope and limitations. The comparison was based on a desk study.

Table 2. Comparison of the exergy-based methods.

Exergy-Based Method	Scope	Limitations
Cumulative exergy demand	Measures energy quality, exergy losses of materials, and emissions	Limited to exergy losses of natural resource; excludes that of the ecological system
Thermo-ecological cost	Focus on cumulative consumption of non-renewable primary exergy resources	It does not include renewable primary exergy resources
Cumulative exergy extraction from natural environment	Measures quality of energetic and non-energetic resources, and land occupation	It does not track exergy transferred into the technological system
Industrial/ecological cumulative exergy demand	Focus on exergy losses in the industrial and ecological stages of a production chain	It is limited to production processes
Extended exergy accounting	Resource equivalent value of a commodity including labor, capital, and environmental emissions	It is intrinsically limited to time and region

While the scope column emphasizes the focus of each of the methods, the limitations column emphasizes their shortcomings.

According to Dewulf et al. [49], the CExD is by far the most applied method as a measure for environmental impacts. The limitation of CExD method, the exclusion of exergy losses in ecological system, is accommodated in ECExD method but the latter is limited to only production processes. On the other hand, although CExENE method appears to extend boundaries of CExD to include land use, the CExENE method is not designed to track exergy transferred into the technological system. An ideal exergy-method could be an extension of the CExD method to cover ecological systems.

3.2. Life Cycle Assessment

Life Cycle Assessment (LCA) is defined as the aggregation and approximation of inputted and outputted resources, and the potential environmental impacts of a product system, in addition to their processes and designs, over its life cycle [50,51]. This section will focus on the description of LCA methodology. A recent review of articles on "LCA of buildings" can be found in Nwodo and Anumba [52]. The methodology includes the goal and scope definitions, inventory analysis, impact assessment, and interpretation of results [50].

3.2.1. Goal and Scope Definition

The goal and scope definitions entail the aim of conducting the LCA study and expected application, the target audience, the results use, and the specification of system boundary and functional units [53]. During the scope definition process, the system boundary and functional units are determined, indicating the included and excluded processes and quantification of the product's function [53,54]. For example, the system boundary for LCA of a building may consist of either cradle-to-grave process (i.e., raw material production phase to end-of-life phase), cradle-to-gate process (i.e., raw material production phase to construction phase), or gate-to-gate process (i.e., within construction phase).

3.2.2. Life Cycle Inventory Analysis

The Life Cycle Inventory (LCI) is the calculation of the inputted and outputted resources such as energy, materials, carbon emissions, and wastes for each of the stages in the service life of a product [55]. An LCI analysis requires extensive non-duplicated data collection. Finnveden et al. [56] observed that setting up inventory data could be one of the most difficult stages of an LCA. According to ISO [50], the goal and scope definition of a study sets the plan for implementing the LCI phase. For example, with the specified system boundary, the data for each unit process are collected to be included in the inventory. The collected data are utilized to quantify the inputted and outputted resources of a unit process. The operational steps outlined in Figure 2 are performed when executing the plan for an LCI analysis, although some iterative steps are not shown [50]. Over the decades, several national and international databases have evolved, including Swedish SPINE@CPM database, the German ProBas database, the Japanese JEMAI database, the US NREL database, the Australian LCI database, the Swiss ecoinvent database, and the European Life Cycle Database [56].

3.2.3. Life Cycle Impact Assessment

Life Cycle Impact Assessment (LCIA) is the classification and characterization of the LCI results based on environmental impacts or human effects [53–55]. The main impact categories from these classification and characterization include human health impacts such as respiratory organics, respiratory inorganics, global warming potential, and climate change; ecosystem quality impacts such as eutrophication potential, acidification potential, and land use; and resources impacts such as energy and material or mineral use [57–60]. A simplified LCI model for a specific type LCA can be adapted from existing databases such as ecoinvent, Athena, Gabi, and OpenLCA, for calculating LCIA. Figure 3 illustrates the elements of an LCIA while Figure 4 illustrates a simple example of LCA calculation process [61].

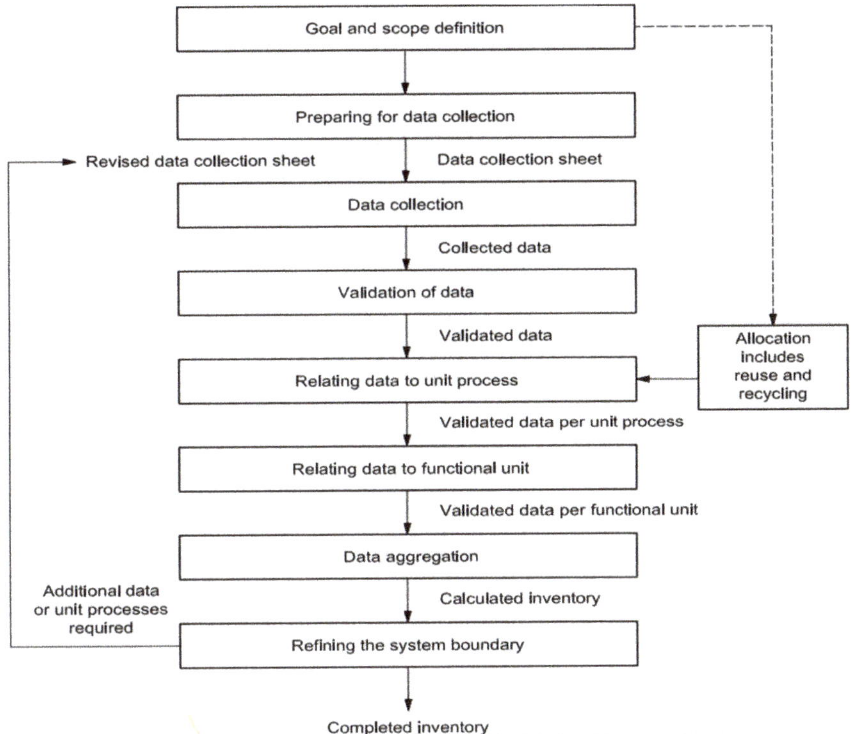

Figure 2. Simplified procedures for Life Cycle Assessment (LCA) inventory analysis [50].

3.2.4. Life Cycle Interpretation

The last process in standard LCA methodology is the interpretation of results. During the interpretation, LCA results are reported in such a way as to evaluate the need and opportunities to reduce the impact of a product on the environment [55]. The life cycle interpretation process comprises elements such as a highlight of essential issues based on LCI and LCIA results; an evaluation of comprehensiveness, sensitivity analysis, and consistency check; and conclusions, limitations, and recommendations. The findings of the interpretation can take the form of conclusions and recommendations to decision-makers, which is consistent with the goal and scope of the study [50]. The interpretation reflects the fact that the LCIA results are based on a relative approach, meaning they indicate likely environmental effects and do not claim to predict actual impacts. The interpretation framework may involve a review and revision of the scope of the LCA in an iterative way, as well as data quality in a manner that is consistent with the goal and scope definition [50]. Validation of results and sensitivity analysis may also be conducted during this process [54]. However, differences in the units of conventional LCA results make life cycle interpretation a difficult process. To avert this challenge, weighting, although optional, is usually employed, which introduces subjectivity to LCA results interpretation.

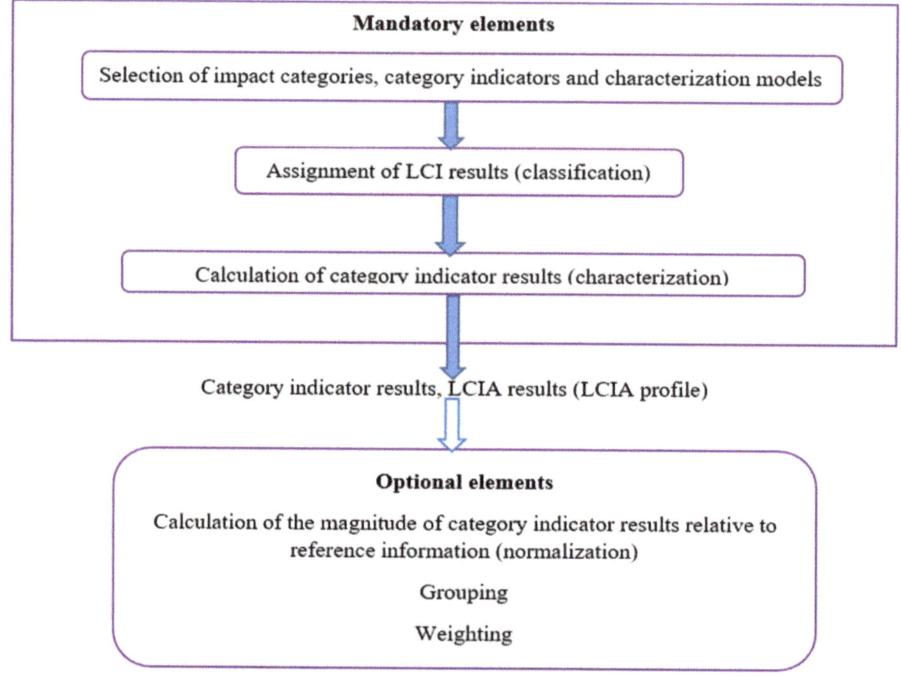

Figure 3. Illustration of the elements of a Life Cycle Impact Assessment (LCIA) [50].

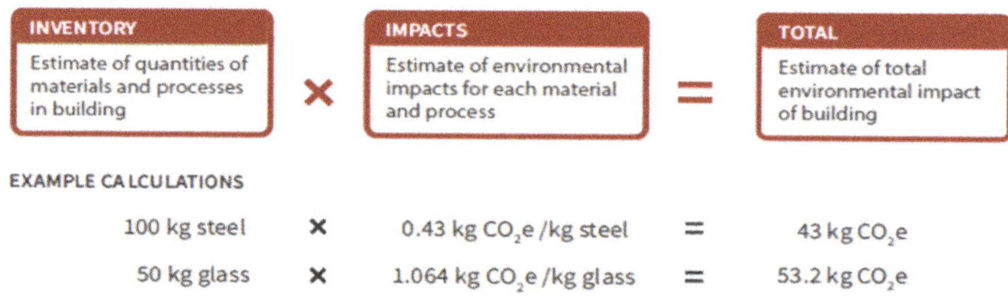

Figure 4. Simple example of LCA calculation process [61].

3.3. Use of Exergy-Based Method in Life Cycle Assessment

An exergy-based method is relevant in LCA since the latter also measures life cycle resource use and corresponding life cycle environmental impacts. De Meester et al. [5] opined that exergy enables the natural energy and material resources to be simultaneously quantified. According to Finnveden et al. [6], a thermodynamic approach based on exergy can be used to measure the use of resources in LCA and in other sustainability assessment methods because the approach can account for energy resources, metal ores, and other materials using their chemical exergies, which are expressed in the same unit. Exergy can complement LCA as an additional impact category indicator [14]. Cornelissen and Hirs [4] demonstrated that, besides LCA, life cycle exergy analysis has value in quantification of the environmental issue of natural resources being depleted.

As mentioned previously, the method of cumulative exergy demand can be used to express the summation of exergies of natural resources expended in all the stages of a technological system or process. Unlike cumulative energy demand, it also measures the chemical exergy of the non-energetic raw materials, which are extracted from the environment. As a result, the CExD method has the capacity to be used in LCA. Currently, CExD method is used in exergetic life cycle assessment for resource accounting by applying any of the following three techniques [25]:

- Process analysis which traces and evaluates exergy for the processes in the manufacturing of a product;
- Balance equations of cumulative exergy demand which utilizes a system of equations to express the CExD of products outcome as a summation of the cumulative exergy of the intermediate products and that of the natural resources extracted directly from the environment;
- Extension from cumulative energy consumption which calculates the CExD based on CED, which, in turn, can be obtained conveniently from commercial LCA tools.

The following section provides a review of previous studies on exergetic life cycle assessment.

4. A Review of Exergetic Life Cycle Assessment

4.1. Methodology for Articles Selection

The procedure involved an analysis of systematically selected articles to investigate the extent of exergy use in LCA and in other sustainability assessments. The reviewed articles were selected from a set of journal articles, which were published for a specified duration between 1990 and December 2018, in addition to recent articles (published since 2019 till date) collected from Google Scholar and/or Scopus. The Web of Science Core Collection is maintained by Clarivate Analytics, and it is a multiple-database platform that includes SCI-Expanded, SSCI, A&HCI, and ESCI. This platform holds more than 12,000 high-impact international journals and it is frequently used by researchers throughout the world [62]. A design was created to search for the articles. To retrieve the articles, three title "TI" record fields and one topic (TS) record field were created as follows:

- TI = "exergy life cycle assessment";
- TI = "exergetic life cycle assessment";
- TI = exerg* life cycle assessment;
- TS = "exerg* life cycle assessment".

The first record field found articles, which have in their titles, the precise phrase "exergy life cycle assessment". Similarly, the second query found articles in which the precise phrase "exergetic life cycle assessment" shows in the title. The third query has a broader scope and found articles that contain at least one of the inputted keyword terms in the title. The "exerg*" term found different forms of that term such as exergy and exergetic. Finally, the fourth record field has the broadest scope and found articles that contain at least one of the inputted keyword terms in the topic field. Table 3 presents an overview of the input data used to search and collect the articles from Web of Science Core Collection database. The combination of the search result sets using the "OR" Boolean operator amounted to 43 articles set for further sorting and manual check. During sorting, some of the articles were found to be irrelevant to the subject and removed, in addition to manually seek for and add relevant cross-referenced articles. A total of 25 articles were finally selected for the tabular analysis.

Table 3. Overview of input data for search of articles on exergy LCA.

Parameter	Setting
Keywords	Exergy, exergetic, exergies, life cycle assessment
Type	Article or review
Time Span	1990–2018 (December)
Citation Index	SCI-EXPANDED, SSCI, A&HCI, ESCI
Language	All languages

4.2. Analysis of the Selected Articles

The 25 articles that discussed the use of exergy or its method in LCA and in other sustainability assessments were analyzed. Table 4 shows the summary of the analysis in terms of aim, method, result/discussion, and relevant conclusion. The result/discussion and conclusion from the collected articles are not limited to those shown in Table 4. However, those presented are deemed the most pertinent to identify the state-of-the-art and relevance of exergy in life cycle analysis. The presented articles essentially investigated the application of exergy methods to sustainability assessments for which LCA is a state-of-the-art technique.

Table 4. Analysis of articles on use of exergy methods in sustainability assessments.

Article	Aim	Method	Result/Discussion	Conclusion
[24]	To develop a multi-objective optimization model for green building structure design	Case study; life cycle analysis methodology; expanded cumulative exergy consumption	The expanded cumulative exergy consumption method enabled LCA to be classified into one objective function	The multi-objective optimization model can be used to locate optimum or near optimum green building designs
[43]	To initiate an extensive resource-based life cycle impact assessment method based on exergy concept	Cumulative Exergy Extraction from the Natural Environment (CExENE)	Fossil resources and land use had high CExENE scores when applied to materials in ecoinvent database	Although they differ in concept, CExENE is like CExD but further includes land use
[7]	To apply CExD indicators to ecoinvent database	CExD; use of resources in the ecoinvent database	In comparison to CED, non-energy resources are likely weighted more strongly by the CExD method	CExD is a more in-depth indicator than CED
[42]	To optimize thermo-ecological cost of a solar collector	Thermo-ecological cost; case study	The depletion of non-renewable natural exergy resources is the objective function for thermo-ecological optimization	The formulated objective function could also be used in economic optimization by introducing purchase prices
[46]	To develop a thermodynamic input-output model of the 1997 U.S. economy	Thermodynamic Input-Output Analysis; Industrial (I)CExD; Ecological (E)CExD	ICExD/money and ECExD/money ratios are useful to perform thermodynamic LCA at economy and ecosystem scales	The model and data encourage the development of sustainable engineering
[5]	To quantify the embodied and operational energy and materials for a family dwelling in Belgium	Case study; CExD; use of resources in the ecoinvent database	Findings show that annual CExD is around 65 GJ_{exergy}/year, with a minimal reliance on the construction type	Reduction of heating requirements is necessary to make family dwellings less fossil resource dependent
[63]	Use of exergy LCA model to assess a 2 × 300 MW coal-fired power plant	Cumulative Exergy Demand (CExD)	Direct exergy (i.e., operational fuel consumption) input accounted for about 93% while indirect exergy input accounted for about 7%	Using exergy as the basic physical parameter made the assessment more objective and reasonable

Table 4. *Cont.*

Article	Aim	Method	Result/Discussion	Conclusion
[23]	Exergy analysis and LCA of solar heating and cooling systems in the building environment in Greece	Exergy model; LCA framework; case study	Solar cooling system has high environmental impacts because of the fan coil units and the cooling tower	The environmental impacts of the systems are significant only at the production phase of their life cycle
[64]	Proposes and implements a framework for exergy-based accounting for land as a natural resource in LCA	Framework; case studies	Site-dependent characterization factors allow for spatial differentiation in exergetic LCA	Using exergy, the framework was able to account for more comprehensive land resources
[48]	To present the sustainability perspective of ecological accounting based on extended exergy	Extended Exergy Accounting (EExA); case study	An extended exergy-based sustainability index system was established to assess the performance of flows in the system	EExA can be used to rate the sustainability level of a place or process
[65]	To determine the environmental effects of a heating system at various stages in building using exergy and LCA	Advanced exergo-environmental analysis	Environmental effects of the exogenous and preventable exergy destruction rates are low	Advanced exergo-environmental analysis provides information not included in the conventional exergy analysis
[66]	To determine the efficiencies of recovering resources from household waste based on exergy analysis and Exergetic LCA	Exergy flow analysis; CExENE for Exergetic LCA	The exergy flow analysis showed scenario efficiencies of between 17% and 27% while Exergetic LCA had about 14%	Cumulative exergy consumption measures in waste LCA should be complimented by other impact categories
[67]	To conduct an optimization of thickness of insulation in a building wall based on Exergetic LCA	Case study	Sensitivity analysis shows that temperature affects total exergetic environmental impact	Walls with lower optimized insulation thickness show increased net savings and fewer payback periods
[6]	Use of case studies to illustrate and compare exergy-based thermodynamic approach with other approaches	Case studies	Different methods produced strikingly different results; this shows the need to be clear about the scope and limitations of the methods	There is a solid scientific base for thermodynamic approach based on exergy; results can be relevant for decision-making
[68]	Exergetic LCA of electricity production from waste-to-energy technology	Hybrid Input-Output method; case study	Primary non-renewable exergy embodied in electricity is non-negligible for both the construction and the operation phases	Joint application of exergy analysis and Exergetic LCA improved the overall thermodynamic performances of the system

Table 4. *Cont.*

Article	Aim	Method	Result/Discussion	Conclusion
[69]	Comparative exergy-based LCA of conventional and hybrid base transmitter stations	Cumulative Exergy Demand	The results elaborated the means of development and sources of environmental impacts during the systems' life cycles	Such details provide the basis for the evolvement and production of sustainable products and processes
[70]	In production phase, to evaluate energy, exergy use, and CO_2 emission of building materials	Thermodynamic method; case study	Although life cycle energy use and life cycle CO_2 emissions are correlated, the latter was higher than the former in production phase	Thermodynamic method practically and significantly improves sustainable building evaluation tools and in making energy policies in building sector
[71]	Environmental sustainability evaluation of an ethylene oxide manufacturing process using CExD and ReCiPe	CExD and ReCiPe	Reduction in environmental impacts expressed in MJ (CExD) and dimensionless (ReCiPe)	CExD is useful in sustainability evaluation of process technology
[72]	Exergy-based study of coal-fired power generation	Case studies	Exergy-based method was successfully used to evaluate the thermal, economic, and environmental benefits	Exergy-based method can improve efficiency of systems
[73]	Exergy-based quantification of resource use during cement clinker production	CExD method	Chemical exergy provided an improved understanding of the resource use	Theoretical gap is needed to be filled in CExD characterization models
[74]	Environmental and economic optimization of insulation thickness	Exergy-based life cycle integrated economic analysis	The approach enabled comparative analysis between environment, economy, and both effects	Exergy-based method enables optimization of environmental and economic effects
[75]	Exergy-based LCA of water injection into hydrocarbon reservoirs	Exergy-based method	Exergy quantified efficiency and CO_2 emission during the process	Exergy method is important even in water driven recovery of oil
[30]	Exergetic LCA of hydrogen production	CExD method	Exergy measures deviation levels of emissions from the reference environment	Exergetic life cycle environmental model is based on LCA and exergy theory
[76]	A thermodynamic assessment with a cradle-to-cradle view	Second law of thermodynamics (Exergy law)	Energy and materials dissipate and deteriorate, and quality is lost irreversibly	Exergy explains the need for global management of the earth's resources
[77]	Assessment of manufacturing sustainability	Hybrid analysis (LCA and Exergy)	While LCA can estimate resource use, exergy includes quality of resource use	Exergy predicts ideal solution, which informs process improvement

4.3. Findings from the Reviewed Studies

The following is a summary of the findings from the reviewed articles:

- There is a solid scientific basis for thermodynamic approach based on exergy [6,70];

- The use of exergy method enables a sustainability assessment to be more objective and reasonable [63];
- The exergy indicator enables LCA to be expressed as a single objective function [24];
- The various exergy-based indicators have unique applications in various disciplines depending on the aim and objective of study [6];
- The CExD provides a more in-depth assessment than the conventionally used CEC [7];
- Use of exergy methods in LCA provides a more comprehensive measure of sustainability by accounting for non-energetic costs such as labor and capital [46];
- Existing studies use exergetic life cycle assessment as a supplement to conventional LCA [4,66].

In addition, the following were observed while studying the bibliometrics of the reviewed articles in Web of Science:

- The use of the term "Exergetic LCA" was more popular than the use of the term "Exergy LCA" to describe the depletion of natural resources in terms of exergy loss over a life cycle;
- Exergetic LCA, as a field of study, is multi-disciplinary;
- The existing number of articles about exergetic life cycle assessment is few relative to the searched timeline, which implies that the subject is still emerging;
- Exergy analysis is most popular in disciplines such as environmental engineering, environmental sciences, energy fuels, thermodynamics, and mechanics;
- In terms of location, most of the publications are based in European countries.

5. Discussion

5.1. Exergetic Life Cycle Assessment and Benefit

Currently, exergetic life cycle assessment, as a thermodynamic approach based on exergy, is utilized to measure resource use in LCA and in other sustainability assessment methods [6]. In other words, exergetic life cycle assessment supplements LCA with a deeper life cycle environmental impact assessment by including the impact of non-energy resources such as mineral ores. One of the categories of environmental impacts that needs consideration is resource use [50], especially, non-energy resources. Exergetic life cycle assessment solves the problem of characterization of non-energy resources in conventional LCA. The exergy-based methods, which were described in Section 3.1, are employed to characterize and quantify the impact of resource use. A review of the current methods to characterize resource use category in LCA can be found in Finnveden et al. [6]. The following two paragraphs state the resource use characterization problem in conventional LCA and the benefit of exergetic life cycle assessment to overcome this methodological problem in LCA.

The resource use characterization problem is encountered in conventional LCA at the LCIA stage in the estimation of environmental impacts for each material and process. Essentially, factors are assigned to each material unit depending on the environmental impact category in consideration. For example, as illustrated in Figure 4, the characterization factor of steel for the global warming potential category is 0.43 per unit mass of steel. Similarly, the characterization factor of glass for the global warming potential category is 1.064 per unit mass of glass. However, these characterization factors mainly depend, amongst others, on the following conditions:

- Fate—the amount of emissions released and the duration of the emitted substances in the environment;
- Exposure—determination of the species in the ecosystem exposed to the emissions;
- Effect—the resulting impact on the species in the ecosystem.

The calculation of characterization factors used in databases for LCIA requires that systematic modeling procedure be followed, which includes multimedia fate and exposure models, in addition to derivation of resulting impacts from experimental toxicity data. Calculation of these factors are localized and complex, even for a controlled study. Exergetic life cycle assessment can be an alternative solution to bypass the complex modeling needed to calculate these characterization factors. As already stated, resource use can be energy or non-energy resources such as mineral ores and materials. For a given energy source, the exergetic life cycle assessment, in form of cumulative exergy demand, can be calculated from the cumulative energy demand and gross calorific exergy-to-energy ratio [24]. For each material (e.g., aluminum, brick, concrete, steel, etc.), exergetic life cycle assessment, in the form of material exergy demand, utilizes the unit exergy of the material to estimate the resource depletion (such as chemical elements, compounds, and ores) due to the material use. Unit exergy of a material is a cumulative of the standard unit chemical exergies of the substances that make up the material. Standard unit chemical exergy of a substance is the unit chemical exergy of the substance when the reference environment is composed of air at 298.15 K of temperature and 101.325 kPa of pressure. Table of values for standard chemical exergies of substances can be found in literature [25].

5.2. Opportunity for a More Comprehensive Exergetic Life Cycle Assessment

Exergetic life cycle assessment describes and measures resource use in LCA and in other sustainability assessment methods. However, in LCA, environmental impacts of both life cycle resource use and life cycle emissions are estimated or predicted. There is, therefore, the need to advance the exergetic life cycle assessment to include the impact of life cycle emissions. The reasoning is that since exergy is a measure of the degree of disequilibrium between a substance (in this case, emission) and its environment [78], then exergy of emissions can also be a measure of the environmental impact potential [24].

In this paper, the term "Exergy-based life cycle assessment (Exe-LCA)" is proffered to describe a measure of both life cycle resource use and life cycle emissions. Full development of the method for Exe-LCA is beyond the scope of this review paper and is discussed elsewhere [79]. Like material exergy demand, which was discussed in Section 5.1, the exergy of life cycle emissions can be estimated using emitted substance, instead. This is the exergy that is lost to the environment due to the emission of substances that cause environmental impacts during the life span of a product or process. Exergy of life cycle emission is a function of emission mass, standard chemical exergy of emission, and molar mass. For example, global warming potential is a measure of the potential of greenhouse gas emissions, such as carbon dioxide and methane, to cause global warming environmental impact. Therefore, the exergy of life cycle emissions that cause global warming environmental impact is a cumulative of the exergies of life cycle carbon dioxide emission, life cycle methane emission, and those of the other greenhouse gas emissions. The same procedure can be followed for other environmental impact categories.

In conventional LCA, the environmental impacts are quantified relative to one selected emission per category. For instance, global warming potential is in carbon dioxide equivalence, and acidification potential is in sulfur dioxide equivalence, etc. Such representations do not portray the contributions of the other underlying emissions e.g., contributions of methane and nitrous oxide to global warming, and the contributions of nitric and hydrochloric acids to acidification. One main advantage of using exergy of life cycle emissions is that the contributions of all the identified chemical emissions during the inventory analysis can fully be included as a cumulative. In addition, exergy of life cycle emissions, as well as exergy of life cycle resource use, express their measurements in the same unit of exergy. This unification capacity results in ease of comparison and other benefits that can make the exergy-based method suitable to achieving life cycle sustainability assessment [80], in addition to being a viable solution to a major methodological problem in LCA.

6. Summary and Conclusions

A review of exergetic life cycle assessment was conducted, using a systematic approach, to investigate the state-of-the-art and relevance of exergy in LCA, and to identify opportunity for improvement of the method. The paper was structured to describe exergy, introduce exergy-based methods and LCA, review and analyze studies on exergetic life cycle assessment, and to discuss the benefits of exergetic life cycle assessment and highlight opportunities for improvement. The methodology involved a literature review, which entailed a systematic selection of journal articles through search of major databases such as Web of Science Core Collection, Scopus, and Google Scholar. The databases were assumed to be comprehensive enough for the collection of the most relevant articles in exergetic life cycle assessment.

The review has shown that exergy-based method can improve on the conventional LCA method. Both exergy and LCA can be used to assess resource consumption and environmental impacts. However, exergy-based LCA goes deeper to assess the quality of resource consumption and environmental impacts. For instance, exergy assesses efficiency of resource use, resource recovery factor, and/or emission rate. Additionally, characterization factors, which are developed using exergy-based method, will be more accurate and robust than that from the conventional LCA method. This is because the former is based on standard thermodynamic properties (e.g., of temperature, and pressure) while the latter is dependent on subjective factors such as fate, exposure, and effects.

In addition, both exergy-based method and LCA can estimate potential environmental impacts from life cycle emissions but report or interpret them in different ways. In conventional LCA, each impact category is reported relative to one reference emission e.g., carbon dioxide equivalence for global warming potential, while in exergy-based method, each emission can uniquely be quantified and summed up. The advantage is that both absolute and relative values can be obtained using exergy-based method for a more robust comparative analysis and decision-making opportunity. It is recommended that exergy-based LCA method, instead of the conventional LCA, be utilized especially in cases where quality of evaluation, single objective values, combined environmental and economic assessment, robust inventory of characterization factors, and benchmarking are required. The critical issues in performing exergy-based LCA include the assumptions in determination of the exergies of the resources and emissions such as standard thermodynamic conditions, pure state of resources and emissions, and difficulty in determination of the individual emission mass. These issues should not be confused with those unique issues in performing exergy analysis of energy systems such as sensitivity of reference environments, choice of exergy efficiency type, unavoidable nature of irreversibility, boundary definition, and choice between steady state and dynamic state conditions, as reported in [20,81].

The following conclusions are deduced from the study:

- Among others, exergy has importance and relevance in life cycle analysis, sustainability, energy systems, and built environment;
- Exergetic life cycle assessment is used for resource accounting in life cycle assessment;
- Exergy-based methods provide a more comprehensive measure of sustainability by accounting for both energetic and non-energetic resources such as labor, and capital;
- The existing studies use exergetic life cycle assessment as a supplement to conventional LCA in resource accounting;
- There is an opportunity for a more comprehensive exergetic life cycle assessment that includes exergy of life cycle emissions;
- A new terminology is required to describe a combination of exergy of life cycle resource use and exergy of life cycle emissions; "Exergy-based Life Cycle Assessment (Exe-LCA)" is proffered;
- Improved exergetic life cycle assessment has the potential to solve characterization and valuation problems in LCA methodology;
- The unification capacity of exergy-based method is a promising technique to achieving life cycle sustainability assessment.

Author Contributions: Conceptualization, M.N.N. and C.J.A.; methodology, M.N.N.; resources, C.J.A.; writing—original draft preparation, M.N.N.; writing—review and editing, C.J.A.; funding acquisition, M.N.N. and C.J.A. All authors have read and agreed to the published version of the manuscript.

Energies **2020**, *13*, 2684

Funding: The authors acknowledge the research sponsorship from The Petroleum Technology Development Fund (PTDF) (grant number: PTDF/ED/PHD/NMA/88/16), Nigeria.

Conflicts of Interest: The authors declare no conflict of interest.

References

1. Dincer, I.; Ratlamwala, T.A.H. Importance of exergy for analysis, improvement, design, and assessment. *Energy Environ.* **2013**, *2*, 335–349. [CrossRef]
2. Favrat, D.; Maréchal, F.; Epelly, O. The challenge of introducing an exergy indicator in a local law on energy. *Energy* **2008**, *33*, 130–136. [CrossRef]
3. Som, S.K.; Datta, A. Thermodynamic irreversibilities and exergy balance in combustion processes. *Prog. Energy Combust. Sci.* **2008**, *34*, 351–376. [CrossRef]
4. Cornelissen, R.L.; Hirs, G.G. The value of the exergetic life cycle assessment besides the LCA. *Energy Convers. Manag.* **2002**, *43*, 1417–1424. [CrossRef]
5. De Meester, B.; Dewulf, J.; Verbeke, S.; Janssens, A.; Van Langenhove, H. Exergetic life-cycle assessment (ELCA) for resource consumption evaluation in the built environment. *Build. Environ.* **2009**, *44*, 11–17. [CrossRef]
6. Finnveden, G.; Arushanyan, Y.; Brandão, M. Exergy as a measure of resource use in life cycle assessment and other sustainability assessment tools. *Resources* **2016**, *5*, 23. [CrossRef]
7. Bösch, M.E.; Hellweg, S.; Huijbregts, M.A.; Frischknecht, R. Applying cumulative exergy demand (CExD) indicators to the ecoinvent database. *Int. J. Life Cycle Assess.* **2007**, *12*, 181. [CrossRef]
8. Morosuk, T.; Tsatsaronis, G.; Koroneos, C. Environmental impact reduction using exergy-based methods. *J. Clean. Prod.* **2016**, *118*, 118–123. [CrossRef]
9. Wang, Q.; Ma, Y.; Li, S.; Hou, J.; Shi, J. Exergetic life cycle assessment of Fushun-type shale oil production process. *Energy Convers. Manag.* **2018**, *164*, 508–517. [CrossRef]
10. Costa, M.M.; Schaeffer, R.; Worrell, E. Exergy accounting of energy and materials flows in steel production systems. *Energy* **2001**, *26*, 363–384. [CrossRef]
11. Dewulf, J.; Van Langenhove, H. Assessment of the sustainability of technology by means of a thermodynamically based life cycle analysis. *Environ. Sci. Pollut. Res.* **2002**, *9*, 267–273. [CrossRef] [PubMed]
12. Aleksic, S.; Mujan, V. Exergy-based life cycle assessment of smart meters. *ELEKTRO* **2016**, *2016*, 248–253.
13. Dewulf, J.; Van Langenhove, H.; Mulder, J.; Van den Berg, M.M.D.; Van der Kooi, H.J.; De Swaan Arons, J. Illustrations towards quantifying the sustainability of technology. *Green Chem.* **2000**, *2*, 108–114. [CrossRef]
14. Shukuya, M. Exergy concept and its application to the built environment. *Build. Environ.* **2009**, *44*, 1545–1550. [CrossRef]
15. Sakulpipatsin, P.; Itard, L.C.M.; Van der Kooi, H.J.; Boelman, E.C.; Luscuere, P.G. An exergy application for analysis of buildings and HVAC systems. *Energy Build.* **2010**, *42*, 90–99. [CrossRef]
16. Zmeureanu, R. Exergy-based index for assessing the building sustainability. *Build. Environ.* **2013**, *60*, 202–210.
17. Romero, J.C.; Linares, P. Exergy as a global energy sustainability indicator. A review of the state of the art. *Renew. Sustain. Energy Rev.* **2014**, *33*, 427–442. [CrossRef]
18. Sangi, R.; Müller, D. Exergy-based approaches for performance evaluation of building energy systems. *Sustain. Cities Soc.* **2016**, *25*, 25–32. [CrossRef]
19. Meggers, F.; Ritter, V.; Goffin, P.; Baetschmann, M.; Leibundgut, H. Low exergy building systems implementation. *Energy* **2012**, *41*, 48–55. [CrossRef]
20. Torio, H.; Angelotti, A.; Schmidt, D. Exergy analysis of renewable energy-based climatisation systems for buildings: A critical view. *Energy Build.* **2009**, *41*, 248–271. [CrossRef]
21. Schlueter, A.; Thesseling, F. Building information model-based energy/exergy performance assessment in early design stages. *Autom. Constr.* **2009**, *18*, 153–163. [CrossRef]
22. Cassetti, G.; Colombo, E. Minimization of local impact of energy systems through exergy analysis. *Energy Convers. Manag.* **2013**, *76*, 874–882. [CrossRef]
23. Koroneos, C.; Tsarouhis, M. Exergy analysis and life cycle assessment of solar heating and cooling systems in the building environment. *J. Clean. Prod.* **2012**, *32*, 52–60. [CrossRef]

24. Wang, W.; Zmeureanu, R.; Rivard, H. Applying multi-objective genetic algorithms in green building design optimization. *Build. Environ.* **2005**, *40*, 1512–1525. [CrossRef]

25. Szargut, J.; Morris, D.R.; Steward, F.R. *Exergy Analysis of Thermal, Chemical, and Metallurgical Processes*, 1st ed.; Hemisphere Publishing Corporation: New York, NY, USA, 1988.

26. Lombardi, L. Life cycle assessment (LCA) and exergetic life cycle assessment (ELCA) of a semi-closed gas turbine cycle with CO_2 chemical absorption. *Energy Convers. Manag.* **2001**, *42*, 101–114. [CrossRef]

27. Peiró, L.T.; Lombardi, L.; Méndez, G.V.; i Durany, X.G. Life cycle assessment (LCA) and exergetic life cycle assessment (ELCA) of the production of biodiesel from used cooking oil (UCO). *Energy* **2010**, *35*, 889–893. [CrossRef]

28. Sui, X.; Zhang, Y.; Shao, S.; Zhang, S. Exergetic life cycle assessment of cement production process with waste heat power generation. *Energy Convers. Manag.* **2014**, *88*, 684–692. [CrossRef]

29. Guven, S. Calculation of optimum insulation thickness of external walls in residential buildings by using exergetic life cycle cost assessment method: Case study for Turkey. *Environ. Prog. Sustain. Energy* **2019**, *38*, e13232. [CrossRef]

30. Li, Q.; Song, G.; Xiao, J.; Hao, J.; Li, H.; Yuan, Y. Exergetic life cycle assessment of hydrogen production from biomass staged-gasification. *Energy* **2020**, *190*, 116416. [CrossRef]

31. Szargut, J. International progress in second law analysis. *Energy* **1980**, *5*, 709–718. [CrossRef]

32. Srinivasan, R.; Moe, K. *The Hierachy of Energy in Architecture: Emergy Analysis*, 1st ed.; Routledge: New York, NY, USA, 2015.

33. Bejan, A.; Tsatsaronis, G.; Moran, M.J. *Thermal Design and Optimization*, 1st ed.; John Wiley & Sons: New York, NY, USA, 1995.

34. Shukuya, M.; Hammache, A. *Introduction to the Concept of Exergy for a Better Understanding of Low-Temperature Heating and High-Temperature Cooling Systems*; VTT Technical Research Centre of Finland: Espoo, Finland, 2002.

35. Bejan, A. *Entropy Generation Minimization: The Method of Thermodynamic Optimization of Finite-Size Systems and Finite-Time Processes*, 1st ed.; CRC Press: New York, NY, USA, 2013.

36. Odum, H.T. *Environmental Accounting: Emergy and Environmental Decision Making*; John Wiley & Sons: New York, NY, USA, 1996.

37. Bastianoni, S.; Facchini, A.; Susani, L.; Tiezzi, E. Emergy as a function of exergy. *Energy* **2007**, *32*, 1158–1162. [CrossRef]

38. Rosen, M.A.; Dincer, I. Exergoeconomic analysis of power plants operating on various fuels. *Appl. Therm. Eng.* **2003**, *23*, 643–658. [CrossRef]

39. Meyer, L.; Tsatsaronis, G.; Buchgeister, J.; Schebek, L. Exergoenvironmental analysis for evaluation of the environmental impact of energy conversion systems. *Energy* **2009**, *34*, 75–89. [CrossRef]

40. Morosuk, T. Exergoenvironmental analysis is a new tool for evaluation of an energy conversion system. Енергетика та автоматика **2013**, *4*, 3–14.

41. Tsatsaronis, G.; Morosuk, T. Understanding and improving energy conversion systems with the aid of exergy–based methods. *Int. J. Exergy* **2012**, *11*, 518–542. [CrossRef]

42. Szargut, J.; Stanek, W. Thermo-ecological optimization of a solar collector. *Energy* **2007**, *32*, 584–590. [CrossRef]

43. Dewulf, J.; Bösch, M.E.; Meester, B.D.; Vorst, G.V.D.; Langenhove, H.V.; Hellweg, S.; Huijbregts, M.A. Cumulative exergy extraction from the natural environment (CEENE): A comprehensive life cycle impact assessment method for resource accounting. *Environ. Sci. Technol.* **2007**, *41*, 8477–8483. [CrossRef]

44. Hau, J.L.; Bakshi, B.R. Expanding exergy analysis to account for ecosystem products and services. *Environ. Sci. Technol.* **2004**, *38*, 3768–3777. [CrossRef]

45. Yang, S.; Yang, S.; Qian, Y. The inclusion of economic and environmental factors in the ecological cumulative exergy consumption analysis of industrial processes. *J. Clean. Prod.* **2015**, *108*, 1019–1027. [CrossRef]

46. Ukidwe, N.U.; Bakshi, B.R. Industrial and ecological cumulative exergy consumption of the United States via the 1997 input–output benchmark model. *Energy* **2007**, *32*, 1560–1592. [CrossRef]

47. Sciubba, E. From engineering economics to extended exergy accounting: A possible path from monetary to resource-based costing. *J. Ind. Ecol.* **2004**, *8*, 19–40. [CrossRef]

48. Dai, J.; Chen, B.; Sciubba, E. Ecological accounting based on extended exergy: A sustainability perspective. *Environ. Sci. Technol.* **2014**, *48*, 9826–9833. [CrossRef] [PubMed]

49. Dewulf, J.; Van Langenhove, H.; Muys, B.; Bruers, S.; Bakshi, B.R.; Grubb, G.F.; Paulus, D.M.; Sciubba, E. Exergy: Its potential and limitations in environmental science and technology. *Environ. Sci. Technol.* **2008**, *42*, 2221–2232. [CrossRef] [PubMed]

50. ISO 14040. *Environmental Management–Life Cycle Assessment–Principles and Framework*; British Standards Institution: London, UK, 2006.

51. Grant, A.; Ries, R.; Kibert, C. Life cycle assessment and service life prediction. *J. Ind. Ecol.* **2014**, *18*, 187–200. [CrossRef]

52. Nwodo, M.N.; Anumba, C.J. A review of life cycle assessment of buildings using a systematic approach. *Build. Environ.* **2019**, 106290. [CrossRef]

53. Bovea, M.D.; Powell, J.C. Developments in life cycle assessment applied to evaluate the environmental performance of construction and demolition wastes. *Waste Manag.* **2016**, *50*, 151–172. [CrossRef]

54. Rashid, A.F.A.; Yusoff, S. A review of life cycle assessment method for building industry. *Renew. Sustain. Energy Rev.* **2015**, *45*, 244–248. [CrossRef]

55. Bayer, C.; Gamble, M.; Gentry, R.; Joshi, S. *AIA Guide to Building Life Cycle Assessment in Practice*; The American Institute of Architects: Washington, DC, USA, 2010.

56. Finnveden, G.; Hauschild, M.Z.; Ekvall, T.; Guinée, J.; Heijungs, R.; Hellweg, S.; Koehler, A.; Pennington, D.; Suh, S. Recent developments in life cycle assessment. *J. Environ. Manag.* **2009**, *91*, 1–21. [CrossRef]

57. Seo, S.; Tucker, S.; Newton, P. Automated material selection and environmental assessment in the context of 3D building modelling. *J. Green Build.* **2007**, *2*, 51–61. [CrossRef]

58. Bare, J. TRACI 2.0: The tool for the reduction and assessment of chemical and other environmental impacts 2.0. *Clean Technol. Environ. Policy* **2011**, *13*, 687–696. [CrossRef]

59. Margni, M.; Curran, M.A. Life cycle impact assessment. In *Life Cycle Assessment Handbook: A Guide for Environmentally Sustainable Products*; Curran, M.A., Ed.; John Wiley & Sons: New York, NY, USA, 2012.

60. Hollberg, A.; Ruth, J. LCA in architectural design—A parametric approach. *Int. J. Life Cycle Assess.* **2016**, *21*, 943–960. [CrossRef]

61. Huang, M. *Life Cycle Assessment of Buildings: A Practice Guide*, 1st ed.; The Carbon Leadership Forum, Department of Architecture, University of Washington: Seattle, WA, USA, 2018.

62. Zeng, R.; Chini, A. A review of research on embodied energy of buildings using bibliometric analysis. *Energy Build.* **2017**, *155*, 172–184. [CrossRef]

63. Wang, Y.; Zhang, J.; Zhao, Y.; Li, Z.; Zheng, C. Exergy life cycle assessment model of "CO_2 zero-emission" energy system and application. *Sci. China Technol. Sci.* **2011**, *54*, 3296–3303. [CrossRef]

64. Alvarenga, R.A.; Dewulf, J.; Van Langenhove, H.; Huijbregts, M.A. Exergy-based accounting for land as a natural resource in life cycle assessment. *Int. J. Life Cycle Assess.* **2013**, *18*, 939–947. [CrossRef]

65. Acıkkalp, E.; Hepbasli, A.; Yucer, C.T.; Karakoc, T.H. Advanced exergoenvironmental assessment of a building from the primary energy transformation to the environment. *Energy Build.* **2015**, *89*, 1–8. [CrossRef]

66. Laner, D.; Rechberger, H.; De Soete, W.; De Meester, S.; Astrup, T.F. Resource recovery from residual household waste: An application of exergy flow analysis and exergetic life cycle assessment. *Waste Manag.* **2015**, *46*, 653–667. [CrossRef]

67. Ashouri, M.; Astaraei, F.R.; Ghasempour, R.; Ahmadi, M.H.; Feidt, M. Optimum insulation thickness determination of a building wall using exergetic life cycle assessment. *Appl. Therm. Eng.* **2016**, *106*, 307–315. [CrossRef]

68. Rocco, M.V.; Di Lucchio, A.; Colombo, E. Exergy life cycle assessment of electricity production from waste-to-energy technology: A hybrid input-output approach. *Appl. Energy* **2017**, *194*, 832–844. [CrossRef]

69. Milanovic, B.; Agarski, B.; Vukelic, D.; Budak, I.; Kiss, F. Comparative exergy-based life cycle assessment of conventional and hybrid base transmitter stations. *J. Clean. Prod.* **2017**, *167*, 610–618. [CrossRef]

70. Xie, H.; Gong, G.; Fu, M.; Wang, P.; Li, L. A thermodynamic method to calculate energy & exergy consumption and CO_2 emission of building materials based on economic indicator. *Build. Simul.* **2018**, *11*, 235–244.

71. Ghannadzadeh, A.; Meymivand, A. Environmental sustainability assessment of an ethylene oxide production process through Cumulative Exergy Demand and ReCiPe. *Clean Technol. Environ. Policy* **2019**, *21*, 1765–1777. [CrossRef]

72. Zhang, Q.; Gao, J.; Wang, Y.; Wang, L.; Yu, Z.; Song, D. Exergy-based analysis combined with LCA for waste heat recovery in coal-fired CHP plants. *Energy* **2019**, *169*, 247–262. [CrossRef]

73. Sun, B.; Liu, Y.; Nie, Z.; Gao, F.; Wang, Z.; Cui, S. Exergy-based resource consumption analysis of cement clinker production using natural mineral and using calcium carbide sludge (CCS) as raw material in China. *Int. J. Life Cycle Assess.* **2020**, *25*, 667–677. [CrossRef]

74. Açıkkalp, E.; Kandemir, S.Y.; Altuntaş, Ö.; Karakoç, T.H. Exergetic approach to determine optimum insulation thickness for cooling applications with life cycle integrated economic analysis. *Int. J. Exergy* **2020**, *30*, 307–321. [CrossRef]

75. Farajzadeh, R.; Zaal, C.; van Den Hoek, P.; Bruining, J. Life-cycle assessment of water injection into hydrocarbon reservoirs using exergy concept. *J. Clean. Prod.* **2019**, *235*, 812–821. [CrossRef]

76. Antonio, V.C.; Alicia, V.C. Thanatia: Thermodynamics of mineral resources. In *The Destiny of the Earth's Mineral Resources: A Thermodynamic Cradle-to-Cradle Assessment*, 1st ed.; Yun, A., Ed.; World Scientific Publishing: Hackensack, NJ, USA, 2014; pp. 253–290. [CrossRef]

77. Dassisti, M.; Semeraro, C.; Chimenti, M. Hybrid Exergetic Analysis-LCA approach and the Industry 4.0 paradigm: Assessing Manufacturing Sustainability in an Italian SME. *Procedia Manuf.* **2019**, *33*, 655–662. [CrossRef]

78. Rosen, M.A.; Dincer, I. Exergy analysis of waste emissions. *Int. J. Energy Res.* **1999**, *23*, 1153–1163. [CrossRef]

79. Nwodo, M.N. Investigation of Exergy-Based Life Cycle Assessment of Buildings. Ph.D. Thesis, University of Florida, Gainesville, FL, USA, 2020.

80. Kloepffer, W. Life cycle sustainability assessment of products. *Int. J. Life Cycle Assess.* **2008**, *13*, 89–95. [CrossRef]

81. Simpson, A.P.; Edwards, C.F. An exergy-based framework for evaluating environmental impact. *Energy* **2011**, *36*, 1442–1459. [CrossRef]

MDPI
St. Alban-Anlage 66
4052 Basel
Switzerland
Tel. +41 61 683 77 34
Fax +41 61 302 89 18
www.mdpi.com

Energies Editorial Office
E-mail: energies@mdpi.com
www.mdpi.com/journal/energies